Russia

East Asia

pe

r Middle East

South
Asia

Southeast
Asia

aharan Africa

Australia &
Pacific Islands

Cities of the World

Cities of the World

World Regional Urban Development

FIFTH EDITION

EDITED BY
STANLEY D. BRUNN, MAUREEN HAYS-MITCHELL,
AND **DONALD J. ZIEGLER**

ROWMAN & LITTLEFIELD PUBLISHERS, INC.
Lanham • Boulder • New York • Toronto • Plymouth, UK

Published by Rowman & Littlefield Education
A wholly owned subsidary of The Rowman & Littlefield Publishing Group, Inc.
4501 Forbes Boulevard, Suite 200, Lanham, Maryland 20706
http://www.rowmanlittlefield.com

Estover Road, Plymouth PL6 7PY, United Kingdom

British Library Cataloguing in Publication Information Available

Library of Congress Cataloging-in-Publication Data
Cities of the world : world regional urban development / edited by Stanley D. Brunn, Maureen Hays-Mitchell, and Donald J. Zeigler. — 5th ed.
     p. cm.
   Includes bibliographical references and index.
   ISBN 978-1-4422-1284-8 (cloth : alk. paper) — ISBN 978-1-4422-1286-2 (electronic)
   1. Cities and towns.   2. City planning.   3. Urbanization.   4. Urban policy.   I. Brunn, Stanley D.
II. Hays-Mitchell, Maureen.   III. Zeigler, Donald J., 1951–
   HT151.C569 2011
   307.76—dc23
                                        2011033234

∞ ™ The paper used in this publication meets the minimum requirements of American National Standard for Information Sciences—Permanence of Paper for Printed Library Materials, ANSI/NISO Z39.48-1992.

Printed in the United States of America

To the citizens of our planet: may they create and enjoy caring geographies, cherished environments, and humane urban lives.

Contents

List of Illustrations

Boxes

Tables

Preface

There are three major reasons why it is important to study the world's cities. First, over half of the world's population (7 billion in mid-2011) now lives in them; that threshold was attained in the middle of 2007. Second, most of the world's problems, whether economic, social, political, or environmental, are associated with cities, and especially the largest cities and those that are national capitals. Third, the impacts of globalization are greatest in cities; this generalization applies not only to the largest or megacities on the planet, but also medium-sized and smaller cities.

When one considers major headline items in the past five years, many of these are associated with cities. Three examples illustrate this point. One major global problem of megaproportions was (and still is) the financial crises or economic slowdown; some scholars would even use the word depression. The impacts of these recession economies were global in extent, not limited to the major international financial centers of New York, Tokyo, and London, but in many major regional centers as well: Cairo, Singapore, Cape Town, Sydney, Mexico City, Moscow, Dubai, Mumbai, Buenos Aires, Shanghai, Seoul, and countless others. An in-depth inquiry into the geographies of the world's recessions reveals the intricate and continental networks of financial institutions and governing bodies around the world. A second major headline item is associated with environmental crises. The environmental theme is a major thread throughout the regional chapters in this book. Again, it is cities that spotlight these crises; examples include the devastating earthquake in Port au Prince, anomalous winter storms in central Europe, a megavolcanic eruption east of Reykjavik, brush fires in east Australian cities, massive flooding on the Mississippi River in 2011, and the BP megaoil spill in the Gulf of Mexico a year earlier. Technological crises, such as the nuclear disaster in Japan in early 2011, had ripple economic effects through much of urban Japan. A third global headline item concerns cities that have been the center of political and social conflicts. Most protests against despotic governments or globalization/westernization or conflicts that are sectarian in nature occur in major cities. Kabul and Baghdad remain much in the news, but other cities have also emerged on the world conflict map in the past few years, including many in the Greater Middle East associated with the 2011 "Arab Spring." Border cities, for example, in the southern United States and southern Europe represent another group of cities whose social and political mix is changing with refugees and asylum seekers.

As editors, we are cognizant of and sensitive to the new economic, social, and political currents that are evident in all major regions and especially in major cities. We are also aware of the increased importance of environmental, security, and conflict issues. Because of their growing importance, we decided to devote additional discussions to these three issues In our fifth edition. We have retained the previous chapter sections on the history of urbanization in each region, discussions of selected representative or distinctive cities, and also coverage of major urban problems. What the regional authors have added are materials on urban sustainability and environmental planning, security and privacy issues, and urban conflict. The conflict discussions will, depending on the author team, address environmental conflicts (land use, uses of space, etc.) or social inequalities and justice, or violence or armed conflict. Woven into these discussions are new maps and photos, many not appearing in previous editions, and also new boxes about urban life and urban living. We consider the visual additions, especially photographs, important in providing insights into a region's or city's current economy and society or its future.

The organization of this edition is similar to the previous four. The "book end" chapters look at contemporary world urbanization (Chapter 1) and the future of cities (Chapter 13). The remaining 11 chapters are devoted to urbanization and cities in major world regions, the same regions that are covered in many world regional geography texts. Each chapter begins with two facing pages; on the left side, a regional map that shows the major cities and, on the right, a table of basic statistical information about cities and urbanization in each region and a list of 10 salient points about that region's urban experience are provided. The regional chapters conclude with a list of references that can be used by the student and instructor for additional information about cities in that region or specific cities.

This edition includes some new faces to the author teams. There are new members for these nine chapters: United States and Canada, Middle America and the Caribbean, Europe, Russia, Greater Middle East, Sub-Saharan Africa, South Asia, Southeast Asia, and East Asia. All chapters are written by regional specialists, many who have done extensive field work in their region and also traveled extensively in both rural and urban areas. We are also very pleased that of the 29 authors contributing to this edition, 14 are women.

With the increased interest in global and international themes and programs in many universities, we believe this book can be used in a variety of different courses. These would Include World Urbanization, Cities of the World, Urban Planning, Global Economics, Globalization, and even World Regional Geography, should the instructor wish to focus on urbanization and global environmental issues. Individual chapters might also be used in specific classes, for example, on Urban History, Globalization, The Urban South, Regional Planning, Environmental Sustainability, Public Health, Human Security, or Urban Futures. Regional anthropology, history, economics, and geography classes could also use one or more chapters.

Finally, we wish to thank all chapter authors for providing timely and well-written chapters. We have worked with some in previous editions, and welcome those new authors who agreed to become important parts of this book. We also want to thank other specific individuals who played major roles in having this edition see the light of day. We especially want to thank Susan McEachern of Rowman & Littlefield for her longstanding support for this volume and previous ones. She has been a bulwark of support for introducing new ideas and repacking the contents to make them more appealing to students and teachers. Susan's team at Rowman & Littlefield worked with us to ensure that this edition be of high quality and a useful final product of which we are all proud. We are indebted to this fine team for their devoted commitment throughout the process. We also wish to thank Don Emminger of Old Dominion University who provided valuable cartographic support, and Lesley Parrish of Colgate University who prepared the statistical information that opens each chapter and the appendix. We are grateful for their timely support and attention to detail. Finally, we thank our families whose enthusiastic and selfless support made this project enjoyable and possible.

As always, we welcome feedback from students and teachers on ways to ensure that subsequent editions will make learning about the world's cities and urbanization more useful, appealing, challenging and rewarding.

Stanley D. Brunn
Maureen Hays-Mitchell
Donald J. Zeigler

Cities of the World

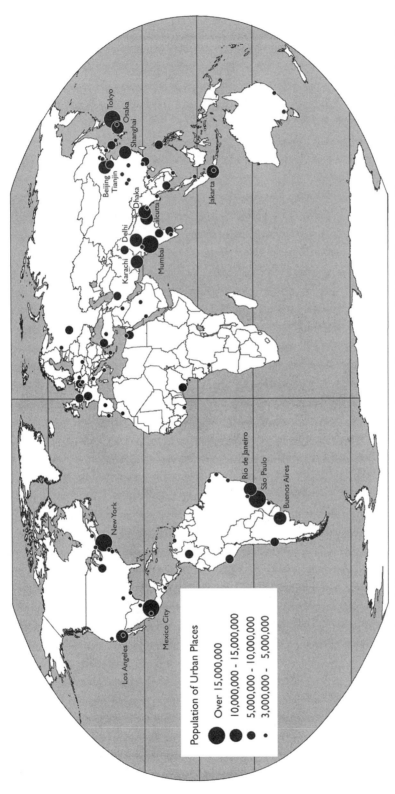

Figure 1.1 Major Cities of the World, 2005. *Source*: UN, *World Urbanization Prospects: 2005 Revision*, http://www.un.org/esa/population/
publications/WUP2005/2005wup.htm

1

World Urban Development

DONALD J. ZEIGLER, MAUREEN HAYS-MITCHELL, AND STANLEY D. BRUNN

KEY URBAN FACTS

Total World Population	7.1 billion
Percent Urban Population	51%
Total Urban Population	3.62 billion
Most Urbanized Countries (100 percent)	Anguilla
	Bermuda
	Cayman Islands
	Gibraltar
	Holy See
	Monaco
	Nauru
	Singapore
Least Urbanized Countries	Burundi (11.0%)
	Uganda (13.3%)
	Liechtenstein (14.3%)
	Sri Lanka (14.3%)
	Trinidad & Tobago (14.3%)
	Ethiopia (16.7%)
	Niger (17.1%)
Annual Urban Growth Rate:	
2000–2005	1.18%
2005–2010	1.11%
Number of Megacities	21 cities
Number of Cities of More than 1 Million	442 cities
Largest Urban Agglomerations	Tokyo (36.7 million)
	Delhi (22.2 million)
	São Paulo (20.3 million)
	Mumbai (20.0 million)
	Mexico City (19.5 million)
World Cities	~50
Global Cities	3 (New York, London, Tokyo)

KEY CHAPTER THEMES

1. In 2007, Earth became a majority-urban planet, yet the proportion of people living in cities varies widely from 10 percent in Burundi to 100 percent in Singapore.
2. The world's population is growing rapidly, but the world's urban population is growing four times as fast.
3. Megacities, cities over 10 million in population, are increasing in number but only in the developing world which now accounts for 13 of the 20 largest urban agglomerations.
4. Geography—and its subfields of environmental, economic, political and cultural geography—provides a holistic framework for understanding cities.
5. The scale of urbanization is increasing as evidenced by the emergence of conurbations and megalopolises around the world.
6. Some countries' patterns of urbanization are following the rank-size rule, while other countries are characterized by urban primacy or dual primacy.
7. The evolution of cities is best understood as a three-stage process: preindustial cities, industrial cities, and postindustrial cities.
8. Cities are usually classified by function as market centers, transportation centers, or specialized service centers.
9. Four classic models have been proposed to explain the spatial organization of land uses within cities: concentric zone model, sector model, multiple nuclei model, and inverse concentric zone model.
10. Urban management issues revolve around population size, growth rate, and geographic distribution; governance and the provision of services; accommodating globalization; and the natural environment, among others.

If one were to compare a map of the world as it was around the year 1900 with one of the world in 2012, two changes would be strikingly apparent: (1) the proliferation of independent nations and (2) the mushrooming numbers and the sizes of cities. A century ago, the world was divided up by a dozen or so major empires; today, the world is home to 194 independent states, most carved out of those empires. The continued disintegration of imperial realms gave birth to the world's most recent newly independent state, South Sudan, and its capital city of Juba. Likewise,

a century ago, the world's major cities (few in number) were concentrated in the industrialized countries of Europe, North America, and Japan. Today, the greatest numbers of cities, and the largest cities, are found in former colonial regions of the developing world (fig. 1.1). Around the year 1800, perhaps 3 percent of the world's population lived in urban places of 5,000 people or more. The proportion rose to more than 13 percent by 1900 and skyrocketed to more than 47 percent by 2000. Then, in 2007, the "blue marble" on which we live became an "urban marble." For the first time in human history, more than half of Earth's people made their homes in urban areas. With urbanization has come globalization, the outcome of technological advances in transportation and communication. Our urban marble is increasingly tied together by the flows of people, goods, services, and capital that interconnect its cities. Globalization, the hallmark of 21st century economic geography, is driven by urban institutions. Yet, our urban marble also pulsates with problems at the human-nature interface. They range from local problems of air quality, to regional problems of water shortages, to global problems materializing as the result of climate change.

Yet, increasingly, we have come to realize that problems need to be approached on a multitude of scales. Climate change may be taking place worldwide (albeit differentially); but the consequences of climate change are going to be felt at the local level, in coastal cities confronted by rising sea level and in cities that can expect average temperatures to climb over the century ahead. Chicago, for instance, has already funded the planting of more shade trees and has chosen species that do well in warmer climates. The city is also considering air conditioning for its schools.

The present phenomenon of worldwide urbanization is as dramatic in its revolutionary implications for the history of civilization as were the earlier agricultural and industrial revolutions. In the industrial countries of Europe, North America, some of Asia, and Australia—the more developed countries (MDCs) as the U.N. calls them—urbanization accompanied and was the consequence of industrialization. Although far from being utopias, cities in those regions brought previously undreamed-of prosperity and longevity to millions. Industrial and economic growth combined with rapid urbanization to produce a demographic transformation that brought declining population growth and enabled cities to expand apace with economic development. In the developing countries of Latin America, Africa, and most of Asia—the less developed countries (LDCs)—urbanization has occurred only partly as the result of industrial and economic growth and in many countries it has occurred primarily as the result of rising expectations of rural people who have flocked to the cities seeking escape from misery (and often not finding it). This march to the cities, unaccompanied until very recently by significant declines in natural population growth, has resulted in the explosion of urban places in the LDCs.

Urban nations tend to be more economically developed and more politically powerful. The relationship lags in a few regions, notably South America where levels of urbanization approach those of North America and Europe but where indicators of social well-being do not indicate commensurate levels of well-being. Yet, most nations that are highly urbanized experience high standards of living, whether measured in per capita income or other standard-of-living measurements. For those still climbing the urbanization ladder, the challenge will be to bring rates of urbanization in line with rates of economic growth.

However, even in the MDCs, where life for the vast majority of urban residents is incomparably better than for those living in the cities of the LDCs, there are serious concerns about the future of the city. What, for example, is the optimum size of a city? Are cities in general getting too large to permit effective administration and to ensure an humane and uplifting urban environment in which to live?

Is the megalopolis or superconurbation to be the norm for the 21st century? What will be the impact of ever-expanding urban agglomerations on human society, on life-sustaining environmental systems, on resource development, and on governments facing increased social disparities, cultural pluralism, and diversity of political expression? Can urbanization and sustainable development work hand-in-hand to improve life for all of Earth's inhabitants? Are cities compatible with nature?

THE WORLD URBAN SYSTEM

As the 21st century entered its second decade, world population stood at almost seven billion. In was only in 1800, however, that it stood on the brink of reaching its first billion. It took 130 years to reach its second billion, but only 11 years to add its last billion. Between 1950 and 2008, the world's population increased more than 2.5 times, but the world's urban population increased almost 4.5 times (fig. 1.2). Increases in urban populations have been felt worldwide, but the pace of urban change has been most dramatic in the world's developing regions, the "Global South" (fig. 1.3).

The United Nations' *World Urbanization Prospects: 2009 Revision* provides the most complete synopsis of urbanization trends. Among its major findings are the following:

1. The world's population is more urban than rural, but it will still take another decade for half of the population of the less developed regions to live in urban areas.
2. The world urban population is expected to increase by 84 percent by 2050, from 3.4 billion in 2009 to 6.3 billion in 2050. Virtually all of the expected growth in the world

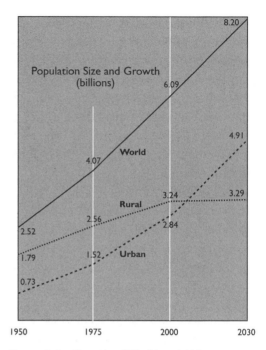

Figure 1.2   Growth of World and Urban Population, 1950–2030. *Source:* UN, *World Urbanization Prospects: 2005 Revision*, http://www.un.org/esa/population/publications/WUP2005/2005wup.htm

population will be concentrated in the urban areas of the less developed regions.
3. The world rural population is expected to reach a maximum of 3.5 billion in 2020 and to decline slowly thereafter, to reach 2.9 billion in 2050. These global trends are driven mostly by the dynamics of rural population growth in the less developed regions.
4. The rate of growth of the world urban population is slowing down. Between 1950 and 2009, the world urban population grew at an average rate of 2.6 percent per year and increased nearly fivefold over the period. During 2025–2050 the urban growth rate is expected to decline further to 1.3 percent per year. The sustained

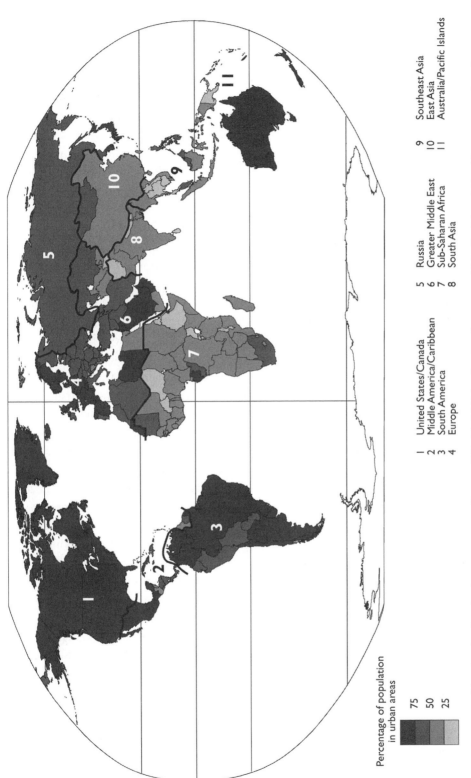

Percentage of population
in urban areas

75
50
25

1 United States/Canada
2 Middle America/Caribbean
3 South America
4 Europe

5 Russia
6 Greater Middle East
7 Sub-Saharan Africa
8 South Asia

9 Southeast Asia
10 East Asia
11 Australia/Pacific Islands

Figure 1.3 World Urbanization by Country, 2005. *Source:* UN, *World Urbanization Prospects: 2005 Revision,* http://www.un.org/esa/
population/publications/WUP2005/2005wup.htm

increase of the urban population combined with the pronounced deceleration of rural population growth will result in increasing proportions of the population living in urban areas. Globally, the level of urbanization is expected to rise from 50 percent in 2009 to 69 percent in 2050.

6. The world urban population is not distributed evenly among cities of different sizes. Over half of the world's 3.4 billion urban dwellers live in cities or towns with fewer than half a million inhabitants.

7. In 2009, cities with fewer than 100,000 inhabitants accounted for one-third of the world urban population, amounting to 1.15 billion. Cities with fewer than 500,000 inhabitants account for 51.9 percent of the urban population.

8. In contrast, the 21 megacities in the world, each with at least 10 million inhabitants, accounted for 9.4 percent of the world urban population. The number of megacities is projected to increase to 29 in 2025 and account for 10.3 percent of the world urban population.

9. Until 1975 there were just three megacities in the world: New York, Tokyo and Mexico City. Today, Asia has 11 megacities, Latin America has 4, and Africa, Europe, and Northern America have 2 each. By 2025, the number of megacities is expected to reach 29.

10. Tokyo, the capital of Japan, is today the most populous urban agglomeration. If it were a country, it would rank 35th in population size. Tokyo, the megacity, is actually an urban agglomeration that comprises not only Tokyo but also 87 surrounding cities and towns.

11. Following Tokyo, the next largest urban agglomerations are Delhi in India with 22 million inhabitants, São Paulo in Brazil and Bombay in India, each with 20 million inhabitants, and Mexico City in Mexico and New York-Newark in the United States, each with about 19 million inhabitants.

12. In 2025, Tokyo is projected to remain the world's most populous urban agglomeration, with 37 million inhabitants, although its population will scarcely increase. It will be followed by the two major megacities in India: Delhi with 29 million inhabitants and Mumbai with 26 million.

13. Megacities are experiencing very different rates of population change. The populations of 9 of the 21 megacities of 2009 are expected to grow at rates ranging from a very low 0.02 percent per year to at most 0.51 percent per year during 2009–2025. The megacities exhibiting such relatively slow rates of population growth include all those located in developed countries plus the four megacities in Latin America.

14. Megacities represent the extreme of the distribution of cities by population size. They are followed by large cities with populations ranging from 5 million to just under 10 million, which in 2009 numbered 32 and are expected to number 46 in 2025. Three quarters of these "megacities in waiting" are located in developing countries.

15. Cities in the next size class, with more than a million inhabitants but fewer than 5 million, are numerous (374 in 2009 increasing to 506 in 2025); and they account for 22 percent of the urban population. Smaller cities, with populations ranging from 500,000 to one million inhabitants, are even more numerous (509 in 2009 rising to 667 in 2025); but they account for just 10 percent of the overall urban population.

16. The distribution of the urban population by city size class varies among the major regions. Europe is exceptional in that 67 percent of its urban dwellers live in urban

centers with fewer than 500,000 inhabitants and only 8 percent live in cities with 5 million inhabitants or more. Africa has a distribution of the urban population by size of urban settlement resembling that of Europe. In Asia, Latin America and the Caribbean, and Northern America, the concentration of the urban population in large cities is marked: about one in every five urban dwellers in those major areas lives in a large urban agglomeration.

17. Historically, the process of rapid urbanization started first in today's more developed regions. In 1920, just under 30 percent of their population was urban and by 1950, more than half of their population was living in urban areas. In 2009, high levels of urbanization, surpassing 80 percent, characterized Australia, New Zealand, and Northern America. Europe, with 73 percent of its population living in urban areas, was the least urbanized major area in the developed world.

18. Among the less developed regions, Latin America and the Caribbean has an exceptionally high level of urbanization (79 percent), higher than that of Europe. Africa and Asia, in contrast, remain mostly rural, with 40 percent and 42 percent, respectively, of their populations living in urban areas. By mid-century, Africa and Asia are expected still to have lower levels of urbanization than the more developed regions or Latin America and the Caribbean.

The more developed regions of the world had a higher percentage of their populations living in urban areas in both 1950 and 2010. In absolute numbers, however, there were more urban dwellers in the more developed regions in 1950; by the end of the century the tables had turned (table 1.1). Unfortunately,

Table 1.1　Urban Patterns in More Developed Regions (MDRs) and Less Developed Regions (LDRs): 1950, 2000, and 2005

	1950		
	Total Population (billions)	*Urban Population*	*% Urban*
World	2.5	0.73	29
MDRs	0.8	0.42	52
LDRs	1.7	0.31	18
	2000		
	Total Population (billions)	*Urban Population*	*% Urban*
World	6.1	2.84	47
MDRs	1.2	0.87	73
LDRs	4.9	1.97	40
	2005		
	Total Population (billions)	*Urban Population*	*% Urban*
World	6.5	3.15	49
MDRs	1.2	0.90	74
LDRs	5.3	2.25	43

Source: UN, *World Urbanization Prospects: 2005 Revision*, http://www.un.org/esa/population/publications/WUP2005/2005wup.htm

throughout Middle and South America, Africa and the Middle East, and much of Asia, urban development has not kept up with urban growth. Latin America and the Caribbean, for instance, have caught up to the world's MDCs in level of urbanization (now 79 percent urban), but economic development, health care, and education continue to lag. Sub-Saharan Africa remains the least urbanized region in the world—and the least developed. Latin America's urban explosion may be over, but in Africa, India, and China, the urban population explosion continues. Only in 2010 did China reach urban majority status.

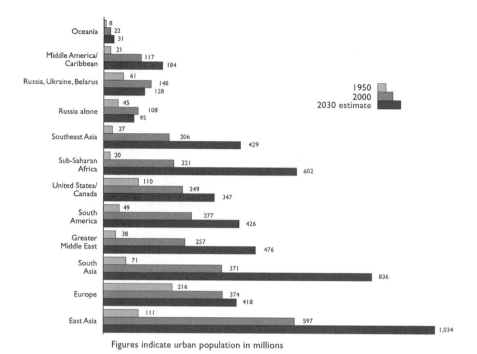

Figures indicate urban population in millions

Figure 1.4   Urban Population of World Regions, 1950, 2000, 2030. *Source:* UN, *World Urbanization Prospects: 2005 Revision*, http://www.un.org/esa/population/publications/WUP2005/2005wup.htm

The greatest number of cities and the greatest number of large cities (3 million and above) are now found in LDCs (fig. 1.4). This can also be seen in a list of the world's largest urban areas, where those in the LDCs now outnumber those in the MDCs (table 1.2). Of the 20 largest urban agglomerations in 1950, 13 were in the MDCs and 7 were in the LDCs. The 20 largest urban agglomerations in 2000 included only five in the MDCs, and those five were located in only three countries: Japan (Tokyo and Osaka), the United States (New York and Los Angeles), and France (Paris). Mexico City is larger than three-quarters of the world's independent states. Less developed regions are urbanizing at much more rapid rates since 1975, compared to more developed regions, in terms of population class of urban settlements for the two periods, 1975–2000 and 2000–2015 (fig. 1.5). Urbanization is a process involving two phases: (1) the movement of people from rural to urban places, where they engage in primarily nonrural occupations, and (2) the change in lifestyle that results from leaving the countryside.

CONCEPTS AND DEFINITIONS

Geographers approach the study of cities through the discipline's major subfields: environmental geography, economic geography, political geography, and cultural geography (fig. 1.6). They all also require an understanding of concepts and definitions used in urban geography.

Table 1.2 The 30 Largest Urban Agglomerations, Ranked by Population Size, 1950–2015

	1950			1975	
Rank	*Agglomeration and Country*	*Population (millions)*	*Rank*	*Agglomeration and Country*	*Population (millions)*
1	New York–Newark, USA	12.338	1	Tokyo, Japan	26.615
2	Tokyo, Japan	11.275	2	New York–Newark, USA	15.880
3	London, United Kingdom	8.361	3	Ciudad de México (Mexico City), Mexico	10.690
4	Shanghai, China	6.066	4	Osaka–Kobe, Japan	9.844
5	Paris, France	5.424	5	São Paulo, Brazil	9.614
6	Moskva (Moscow), Russian Fed.	5.356	6	Los Angeles–Long Beach–Santa Ana, USA	9.926
7	Buenos Aires, Argentina	5.098	7	Buenos Aires, Argentina	8.745
8	Chicago, USA	4.999	8	Paris, France	8.630
9	Kolkata (Calcutta), India	4.513	9	Kolkata (Calcutta), India	7.888
10	Beijing, China	4.331	10	Moskva (Moscow), Russian Fed.	7.622
11	Osaka–Kobe, Japan	4.247	11	Rio de Janeiro, Brazil	7.557
12	Los Angeles–Long Beach–Santa Ana, USA	4.046	12	London, United Kingdom	7.546
13	Berlin, Germany	3.338	13	Shanghai, China	7.326
14	Philadelphia, USA	3.128	14	Chicago, USA	7.160
15	Rio de Janeiro, Brazil	2.950	15	Mumbai (Bombay), India	7.082
16	Saint Petersburg, Russian Fed	2.903	16	Soul (Seoul), Republic of Korea	6.808
17	Ciudad de México (Mexico City), Mexico	2.883	17	Al-Qahirah (Cairo), Egypt	6.450
18	Mumbai (Bombay), India	2.857	18	Beijing, China	6.034
19	Detroit, USA	2.769	19	Manila, Philippines	4.999
20	Boston, USA	2.551	20	Tianjin, China	4.870
21	Al-Qahirah (Cairo), Egypt	2.494	21	Jakarta, Indonesia	4.813
22	Manchester, United Kingdom	2.422	22	Philadelphia, USA	4.467
23	Tianjin, China	2.374	23	Delhi, India	4.426
24	São Paulo, Brazil	2.334	24	Saint Petersburg, Russian Fed.	4.325
25	Birmingham, United Kingdom	2.229	25	Tehran, Iran (Islamic Republic of)	4.273
26	Shenyang, China	2.091	26	Karachi, Pakistan	3.989
27	Roma (Rome), Italy	1.884	27	Hong Kong, China, Hong Kong SAR	3.943
28	Milano (Milan), Italy	1.883	28	Madrid, Spain	3.890
29	San Francisco–Oakland, USA	1.855	29	Detroit, USA	3.885
30	Barcelona, Spain	1.809	30	Krung Thep (Bangkok), Thailand	3.842

(continued on next page)

Source: UN, *World Urbanization Prospects: 2005 Revision,* http://www.un.org/esa/population/publications/WUP2005/2005wup.htm

Urbanism

Urbanism is a broad concept that generally refers to all aspects—political, economic, social—of the urban way of life. Urbanism is not a process of urban growth, but rather the end result of urbanization. It suggests that the urban way of life is dramatically different

Table 1.2 (*continued*)

	2000			2015	
Rank	*Agglomeration and Country*	*Population (millions)*	*Rank*	*Agglomeration and Country*	*Population (millions)*
1	Tokyo, Japan	34.450	1	Tokyo, Japan	35.494
2	Ciudad de México (Mexico City), Mexico	18.066	2	Mumbai (Bombay), India	21.869
3	New York–Newark, USA	17.846	3	Ciudad de México (Mexico City), Mexico	21.568
4	São Paulo, Brazil	17.099	4	São Paulo, Brazil	20.535
5	Mumbai (Bombay), India	16.086	5	New York–Newark, USA	19.876
6	Shanghai, China	13.243	6	Delhi, India	18.604
7	Kolkata (Calcutta), India	13.058	7	Shanghai, China	17.225
8	Delhi, India	12.441	8	Kolkata (Calcutta), India	16.980
9	Buenos Aires, Argentina	11.847	9	Dhaka, Bangladesh	16.842
10	Los Angeles–Long Beach–Santa Ana, USA	11.814	10	Jakarta, Indonesia	16.822
11	Osaka–Kobe, Japan	11.165	11	Lagos, Nigeria	16.141
12	Jakarta, Indonesia	11.065	12	Karachi, Pakistan	15.155
13	Rio de Janeiro, Brazil	10.803	13	Buenos Aires, Argentina	13.396
14	Al-Qahirah (Cairo), Egypt	10.391	14	Al-Qahirah (Cairo), Egypt	13.138
15	Dhaka, Bangladesh	10.159	15	Los Angeles–Long Beach–Santa Ana, USA	13.095
16	Moskva (Moscow), Russian Fed.	10.103	16	Manila, Philippines	12.917
17	Karachi, Pakistan	10.020	17	Beijing, China	12.850
18	Manila, Philippines	9.950	18	Rio de Janeiro, Brazil	12.770
19	Soul (Seoul), Republic of Korea	9.917	19	Osaka–Kobe, Japan	11.309
20	Beijing, China	9.782	20	Istanbul, Turkey	11.211
21	Paris, France	9.692	21	Moskva (Moscow), Russian Fed.	11.022
22	Istanbul, Turkey	8.744	22	Guangzhou, China	10.420
23	Lagos, Nigeria	8.422	23	Paris, France	9.858
24	Chicago, USA	8.333	24	Soul (Seoul), Republic of Korea	9.545
25	London, United Kingdom	8.225	25	Chicago, USA	9.469
26	Guangzhou, China	7.388	26	Kinshasa, Dem. Republic of the Congo	9.304
27	Tehran, Iran (Islamic Republic of)	6.979	27	Shenzhen, China	8.958
28	Santa Fé de Bogotá, Colombia	6.964	28	Santa Fé de Bogotá, Colombia	8.932
29	Lima, Peru	6.811	29	London, United Kingdom	8.618
30	Tianjin, China	6.722	30	Tehran, Iran (Islamic Republic of)	8.432

in all respects from the rural way of life; as people leave the country and move to the city, their lifestyles and livelihoods change.

Urbanization

The important variables in the first phase are population density and economic functions. A place does not become urban until its workforce is divorced from the soil; trade, manufacturing, and service provision dominate the economies of urban places. The important variables in the second phase are social, psychological, and behavioral. As a population becomes increasingly urban, for instance, family size becomes smaller because the value placed upon children changes.

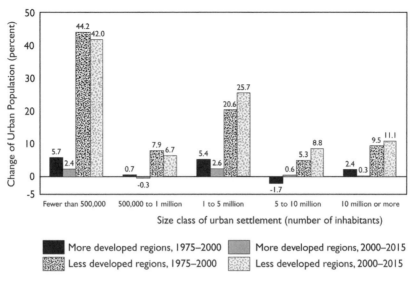

Figure 1.5   Urban Population in MDCs vs. LDCs by Size Class of Urban Settlement, 1975–2015. *Source:* UN, *World Urbanization Prospects: 2001 Revision*, http://www.un.org/esa/population/publications/wup2001/wup2001dh.pdf

Urban Place

As a place increases in population, it eventually becomes large enough to assume that its economy is no longer tied strictly to agriculture or to other primary activities. At that point, a rural place becomes an urban place. Translating the dividing line between rural and urban into a minimum population size varies significantly from country to country. In Denmark and Sweden only 200 people are required for a place to be classified as urban; in New Zealand the figure is 1,000; in Argentina 2,000; in Ghana 5,000; and in Greece 10,000. In the United States, places are defined as urban by the Bureau of the Census if they have at least 2,500 people. Smaller places are defined as rural.

City

The term *city* is essentially a political designation referring to a large, densely-populated place that is legally incorporated as a municipality. However, a settlement of any size may call itself a city, whether it is large or small. Towns are generally smaller than cities.

Megacity

"Megacity" is a used colloquially rather than formally to designate the very largest urban places, usually conceptualized as an urban core and its peripheral expansion zone. A city with more than 10 million inhabitants may be called a megacity. In 1950, only New York City exceeded 10 million. Today, 21 cities worldwide fit that category, and one out of every 20 people worldwide lives in a megacity.

World cities function as the command-and-control centers of the world economy (box 1.1). They offer advanced, knowledge-based producer services (businesses serving businesses), particularly in the fields of accounting, insurance, advertising, law, technical expertise, and the creative arts. The top-tier cities, defined by their financial centrality, are called *global cities*, of which there are three:

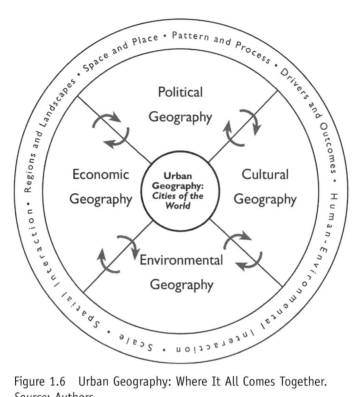

Figure 1.6   Urban Geography: Where It All Comes Together.
Source: Authors

New York, London, and Tokyo. One rung lower are the second-tier *world cities*: Paris, Frankfurt, Los Angeles, Chicago, Hong Kong, and Singapore, among others (figs. 1.7 and 1.8). Beyond that are several dozen more cities—Amsterdam, Moscow, Sydney, Toronto, San Francisco, and others—which draw their strength from particular megaregions or from particular cultural and economic niches. Even cities such as Mecca and Jerusalem may be termed world cities because their influence is felt worldwide within particular religious niches.

Urban Area

As cities have expanded, the boundary between urban and rural has become increasingly blurred, especially in industrialized countries where automobile transportation has fostered urban sprawl. Thus, the urban (or urbanized) area is defined as the built-up area where buildings, roads, and essentially urban land uses predominate, even beyond the political boundaries of cities and towns. The urban area is basically a city and its suburbs.

Conurbation

The word is of 20th-century, European origin. Hence it is common to speak of the Randstad conurbation in the Netherlands or the Rhine-Ruhr conurbation in Germany. In the United States, the Dallas-Fort Worth Urban Area is a good example of a conurbation. As urban areas expand, they engulf smaller cities in the urban expansion zone, turn nearby towns

Box 1.1 Globalization and World Cities

Peter Taylor

Although there is a large literature on "world" or "global" cities, little evidence has been gathered on what actually makes such cities so important: their connections with other cities across the world. Thus, if world cities are indeed the crossroads of globalization, then we need to consider seriously how we measure intercity relations. It was just such thinking that led to the setting up of the Globalization and World Cities (GaWC) Study Group and Network at Loughborough University, United Kingdom, as a virtual center for world cities research. GaWC currently carries out three strands of research:

- Comparative City Connectivity Studies: These focus on relations between chosen cities as they respond to particular events. In one study, Singapore, New York, and London were compared in the way in which their service sectors responded to the 1997 Asian financial crisis. In another study, relations between London and Frankfurt were studied in the wake of the launch of the euro currency. The generic finding of this work is that city competitive processes are generally much less important than cooperative processes carried out through office networks within the private sector.
- Elite Labor Migrations between Cities: Moving skilled labor around to different world cities is found to be a key globalization strategy for financial firms wanting to embed their businesses into the world-city network. For instance, London firms regularly send staff to Paris, Amsterdam, and Frankfurt to provide "seamless" service across European cities. The prime finding of this research is that a transnational space of flows is produced as a necessary prerequisite for firms accumulating financial knowledge.
- Global Network Connectivity of Cities: A world-city network is an amalgam of the office networks of financial and business service firms. This network has three levels: a network level of cities in the world economy, a nodal level of cities as global service centers, and a subnodal level of global service firms that are the prime creators of the world-city network. This specification directs a data collection that enables the global network connectivities of world cities to be calculated. These results are based on a uniquely large set of data for the year 2000 covering one hundred firms in 316 cities across the world.

GaWC research goes beyond world city formation to study world city network formation. The focus has been on this complex process within economic globalization. For a full global urban analysis, further research is required within other important strands of globalization. The Study Group's website is: www.lboro.ac.uk/gawc.

into full-fledged cities, sometimes stimulate the development of new cities, and bump into other expanding urban areas.

Megalopolis

Megalopolis is a 20th-century word of North American origin. The term was first applied in 1961 by geographer Jean Gottmann to the urbanized northeastern seaboard of the United States from Boston to Washington. Its coinage focused attention on a new scale of urbanization. Today, megalopolis is used as a generic term referring to urban coalescence of metropolitan areas at the regional scale. That coalescence is channeled along transportation corridors connecting one city with another. It is evident in the magnitude of vehicle traffic, telephone calls, e-mail exchanges, and air transport among cities strung along a megalopolitan corridor. Metropolis and megalopolis are both derived from the ancient Greek word for city, *polis*.

Metropolis and Metropolitan Area

The term *metropolis* originally meant the "mother city" of a country, state, or empire. Today, it is used loosely to refer to any large city. A metropolitan area includes a central city (or cities) plus all surrounding territory—urban or rural—that is integrated with the urban core (usually measured by commuting patterns). In the United States, the term Metropolitan Area (MA) is officially used, of which there are three types: Metropolitan Statistical Areas (MSAs), Primary Metropolitan Statistical Areas (PMSAs), and Consolidated Metropolitan Statistical Areas (CMSAs). The U.S. Bureau of the Census has been designating metropolitan areas since 1950. While terminol-

ogy and criteria have changed since then, the core definition has remained the same in that metropolitan areas (1) have an urban core of at least 50,000 people, (2) include surrounding urban and rural territory that is socially and economically integrated with the core, and (3) are built from county (or county-equivalent) units. In Canada, their counterparts are officially termed Census Metropolitan Areas (CMAs), which must have an urban core of at least 100,000 people.

Site and Situation

Why are cities located where they are? How and why do they grow? Concepts used by urban geographers to answer these and related questions are *site* and *situation*. Site refers to the physical characteristics of the place where the city originated and evolved. Surface landforms, underlying geology, elevation, water features, coastline configuration, and other aspects of physical geography are considered site characteristics. Montreal's site, for instance, is defined by the Lachine Rapids, historically the upstream limit of ocean-going commerce. Paris's site is defined by an island in the Seine River, known as the *Île de la Cité*, which gave the city defensive advantages and offered an easy bridging point across the Seine. New York City's site is defined by a deepwater harbor. A city's origin is often wrapped up in site characteristics.

Situation, by contrast, refers to the relative location of a city. It connotes a city's connectedness with other places and the surrounding region. Some cities are centrally located at the junction of trade routes, while others are isolated. A city's growth and decline is more dependent on situation than it is on site characteristics. In fact, a good relative location, or situation, can compensate for a poor site. Venice, for instance, triumphed as a center of the Renaissance not because of its site (so water saturated it is sinking), but because of its relative location at the head of the Adriatic Sea with access to good passes through the Alps. New York City emerged as the United States' most populous city in the early 19th century, not because of a superior site, but because of a superior situation. After the opening of the Erie Canal in 1825, New York had easy access to the resource-rich interior of the United States via the Great Lakes.

Urban Landscapes

Urban landscapes, visible and invisible, are the manifestations of the thoughts, deeds, and actions of human beings. They are charged with clues to the economic, cultural, and political values of the people who built them (fig. 1.7). At the macroscale, geographers may look at the vertical and horizontal dimensions of the landscape—at city skylines and urban sprawl. At the microscale, they may look at architectural styles, signage, activity patterns near busy intersections, or urban foodways (box 1.2). Interpreting, analyzing, and critiquing the landscape is one of the traditional themes of urban geography.

Capital City

"Capital" comes from the Latin word for "head," *caput.* Capital cities are literally "head cities," the headquarters of government functions. Every country has one and a few (*e.g.,* South Africa, Netherlands, Bolivia) have more than one. Capital cites are seats of political power, centers of decision making, and loci of national sovereignty. Their landscapes are

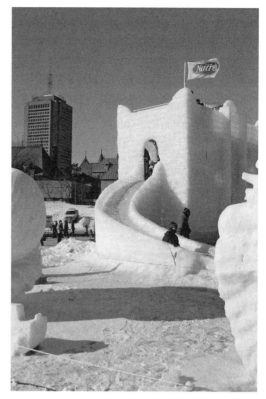

Figure 1.7   The annual Winter Carnival brings the world to Quebec City and bolsters the winter tourism industry. Here on the Plains of Abraham, the *built environment* is built of ice. (Photo by Donald Zeigler)

charged with the symbols of solidarity, real or imagined; their museums are the attics of the nation; their locations are symbolic of the central role they play in the national urban system. In some countries, national capitals share power with provincial capitals. As a class of cities, they are among the best known in the world.

Preindustrial City

The preindustrial city—sometimes referred to as the traditional city—identifies a city that was founded and grew before the arrival

Box 1.2   Urban Gastronomy

Timothy C. Kidd

The gastronomical scene of a city is often a significant cultural marker showcasing changes in ethnic composition, shifting taste preferences, and availability of ingredients. Cities around the world often have their own unique dishes, restaurants, drinks, and street food. Street food ranges from the doner kebab in Istanbul, to hot dogs in New York, and balut in Manila. They will typically be sold ready to eat from stalls or vans in parts of the city with high foot traffic. The Food and Agriculture Administration estimated in 2007 that 2.5 billion people eat such street food everyday because it is faster and cheaper than a restaurant meal. As the world continues to become more urban, the number of street food vendors continues to grow and their offerings become more varied. With diffusion of corporate fast food franchises, the nature of the fast take-away meal has certainly been altered. While there are often local variations on foreign-based fast food menus, there is still the homogenized feel to such establishments from the food itself, to the uniforms worn by the employees, to the furnishings, and signage of the restaurant. In some cases, like London, such restaurants coupled with changing demographics can be deleterious for traditional street food purveyors.

In London's historically poor East End, the remnants of uniquely English fast food still exist. Perhaps most famous of all traditional dishes are eels, particularly jellied eels. The Thames River estuary provided an excellent habitat for eels, and fish traps known as eel bucks lined the waterway to reap the bounty. The plentiful eels became something of a delicacy among the poor East Enders, many of whom were immigrants. By the late 1600s, Dutch eel boats became a major supplier of eels. In great markets like Billingsgate and Spitalfields, fishmongers sold them to restaurants and street hawkers known as "piemen." Stalls of vendors selling eel pies or jellied eels became ubiquitous. By the first half of the 1800s, however, London's industrial expansion brought severe pollution that had a decidedly negative effect on the Thames fishery.

The decline in fish stocks was to herald the beginning of a new East End dining tradition: the eel, pie, and mash shop. These eateries served meat pies made from beef or mutton with mashed potatoes ("mash") doused with parsley-infused, green eel gravy called liquor. Often the restaurants were similar in their interior décor, sporting tiled floors and walls, marble counter tops, and wooden benches. From the Victorian era until the end of World War II, eel, pie, and mash shops did a booming business. Today, there are still many pie and mash shops in the East End, though they number probably fewer than 100; and many menus feature less than the traditional pie versions (such as vegetarian pies).

The culinary landscape in the East End changed markedly in the 20th century. Considerable numbers of Bangladeshi moved into the area (especially near Brick Lane) and immediately impacted the dining options. Soon, curries and kedgeree became more popular than the traditional jellied eels. However, in the 1990s and 2000s the traditional English fare had a bit of a renaissance when celebrity chefs like Gordon Ramsey began serving upscale versions in their restaurants. Meanwhile, the Thames had become much cleaner and eel harvesting improved. Unfortunately, the good news was short-lived as the Zoological Society of London revealed in January 2010 that the European eel was endangered, with Thames estuary stocks having declined by 98 percent from 2005–2010. The cause of this steep decline is not certain; but it cannot be welcomed news for consumers and producers of eel pies and jellied eels.

Jellied eels can still be found at some seaside towns in the United Kingdom, as well as a few street stalls in London's East End. Chopped into bite size pieces with bones left in and boiled with herbs, the eels are left to cool in their own gelatinous stock and then served with chili vinegar. It is said to be an acquired taste akin to pickled herring. Perhaps preferences for pizza, hamburgers, or chicken tikka are proving more popular for younger generations than this most traditional English dish. Nevertheless, the legacy of the East End's street food lives on with vendor's like Tubby Isaac's near Aldgate, and in toponyms like the culturally famous Eel Pie Island at Twickenham, and even in music with the Eel Pie Recording Studios founded by guitarist Pete Townsend of The Who.

Figure 1.8   Jellied eels remain a part of the East End's cultural landscape in London. (Photo by Donald Ziegler)

of industrialization in the 19th and 20th centuries and thus typically had quite different characteristics from industrial cities. Elements of the traditional city are still part of urban landscapes, particularly in the developing world, even though there are no longer any pure preindustrial cities in existence. Remnants of the traditional city include central markets (hearty survivors in Europe and making a come-back in the United States), pedestrian quarters where the streets are too narrow for cars, walls and gates now serving as visual reminders of the past, and intimidating architecture (palaces and cathedrals) that preceded industrialization.

Industrial City

An industrial city has an economy based on the production of manufactured goods, sometimes light industrial products (*e.g.,* food, textiles, footwear) and sometimes heavy industrial items (*e.g.,* motor vehicles, appliances, ships, machinery). Factories and foundries anchor their urban landscapes. Although small-scale manufacturing characterized even pre-industrial cities, the invention of the steam engine begat ever larger factories and the cities which provided them with workers and services.

Postindustrial City

A relatively new type of city has emerged, particularly in the world's wealthiest countries. It is the postindustrial city. Its economy is not tied to a manufacturing base but instead to high employment in the services sector. Cities that are mainly the headquarters for corporations or for governmental and intergovernmental organizations are examples, as are those specializing in research and devel-

opment (R&D), health and medicine, and tourism/recreation. With an increase in the number of people employed in tertiary and quaternary occupations, especially in fields such as finance, health, leisure, R&D, education, and telecommunications and in various levels of government, the cities with concentrations of these activities have an economic base in sharp contrast to those cities that originated in industrial economies.

Primate City

A type of city defined solely by size and function is the *primate city* (fig. 1.9). The term was coined by geographer Mark Jefferson in the late 1930s to refer to the tendency of countries to have one city that is exceptionally large, economically dominant, and culturally expressive of national identity. A true primate city is at least twice as large as the second-largest city, but the gap is often much larger. Paris, for instance, is seven times larger than France's second-largest city, Lyon. In general, however, primacy is more typical of the developing world. In a few instances, countries may be characterized by *dual primacy*, where two large cities share the dominant role, such as Rio de Janeiro and São Paulo in Brazil. The presence of a primate city in a country usually suggests an imbalance in development: a progressive core, defined by the primate city and its environs, and a lagging periphery on which the primate city may depend for resources and migrant labor. Some see the relationship between core and periphery as a parasitic one.

Rank-Size Rule

The rank-size rule represents an alternative to primacy. This concept evolved out of empiri-

Figure 1.9   Mexico City, overwhelmingly the primate city of Mexico and one of the world's largest cities, seems to sprawl forever. With giant size come giant problems, too. (Photo by Robert Smith)

cal research on the relationships among cities of different population sizes in a country. Simply put, the rule states that the population of a particular city should be equal to the population of the country's largest city divided by its rank. In other words, the fifth-largest city in a country should be one-fifth the size of the largest city. A deviation from this ranking may mean that the urban system is unbalanced.

Colonial City

Although virtually gone from the face of the earth now, the colonial city has had a profound impact on urban patterns throughout much of the world, starting around 1500 A.D.

and culminating with the global dominance of European imperial powers in the 19th and early 20th centuries. The colonial city was unique because of its special focus on commercial functions, its peculiar situation requirements, and the odd blend of Western urban forms with traditional indigenous values and practices. There were two distinct types of colonial city, depending on the age of the European or colonial enclave in relation to the age of the native or indigenous settlement. In one type, the European city was created virtually from scratch on a site where no other significant urban place had existed. This would then lead to in-migration of local peoples drawn by the economic opportunity created under colonial rule. Examples include Mumbai (Bombay), Hong Kong, and Nairobi. In the other type, the European city was grafted onto an existing indigenous urban place and then became the dominant growth pole for that city, typically swallowing up or overwhelming in size and importance the original indigenous center. Examples include Shanghai, Delhi, and Tunis. Either type of colonial city would eventually give rise to a *dual city* consisting of one modern, Western, part and another more traditional, indigenous part.

Socialist City

Cities that evolved under Communist regimes in the former Soviet Union, Eastern Europe, China, North Korea, Southeast Asia, and Cuba have given us the concept of the *socialist city* (box 1.3). Communism was characterized by massive government involvement in the economy, coupled with the absence of private land ownership and free-markets. Communism produced cities that were distinct in form, function, and internal spatial structure (fig. 1.10). Although most Communist regimes collapsed in the late 20th century, central planning and the command economy have left a lasting, visible impression on urban landscapes. Now, however, most socialist cities of the world are experiencing rapid change; and a post-socialist city is emerging. Although China is still the province of Communist Party rule, competitive enterprise is transforming its urban landscapes. Only North Korea, and to some extent Cuba, continue to maintain cities under the principles of communism.

Post-Socialist City

Cities evolving under *post-socialist* regimes are breaking away from the urban plans so

Figure 1.10   These heroic statues in front of the opera house in Novosibirsk, Russia, are typical of former socialist cities. Statues, paintings, posters—all were designed to inspire the populace to sacrifice lives of personal comfort for the sake of national welfare. (Photo by Donald Zeigler)

strictly enforced by communist/socialist governments. Socialism's largely compact, comprehensively planned cities were structured internally and regionally to be self-sufficient, but this is changing today as individuals and businesses make their own decisions about where to locate residences and businesses in freer market economies. Three growth trends alter the urban form, function, and spatial structures of post-socialist cities; these trends are reforming socioeconomic and political processes in addition to the built environment.

Box 1.3   Where and When Does Soviet Urban Influence End?

Jessica Graybill

When people think about the former Soviet Union, they often imagine the territory occupied today by the Russian Federation. Although Russia remains the largest country in the world, stretching from Europe in the west to the Pacific in the east, and from semi-arid desert and steppe regions in the south to the Arctic in the north, its territorial extent and influence has been greater in the past several centuries, first as the Russian and then the Soviet empire expanded. A powerful Soviet political bloc encompassed Eastern Europe, the Caucasus, Central Asia, and Cuba. Other Asian, African, and Latin American countries were also supported by the Soviets in the effort to spread communism worldwide. Since the USSR collapsed in 1991, the former Soviet states are now sometimes called the Commonwealth of Independent States (CIS) or the Newly Independent States (NIS). The former Soviet republics today, from west to east, are located in three main regions: Eastern Europe (Estonia, Latvia, Lithuania, Belorussia, Ukraine, Moldavia), the Caucasus (Georgia, Armenia, Azerbaijan) and Central Asia (Kazakhstan, Turkmenistan, Uzbekistan, Tajikistan, Kirghizia). They should not be confused with the wider political arena, the Soviet bloc.

Under communist rule, a new style of urban settlement arose across Soviet territories, as they built a new world order based on socialist principles that would, ideally, lead to the development of true communism. In addition to industrialization of largely rural territories, the Soviets developed socialized goods and services (e.g., housing, medicine, education) available to all citizens, especially in urban areas, as the Soviet regions progressively fulfilled 5- and 10-year development plans. Rapid industrialization required rapid urban growth, and plans for entire cities were developed centrally (in Moscow) and distributed for adoption around the Soviet Union. In many cases, they were also used in the Soviet bloc countries, as not only their ideology but also their economies were tightly tied to that of the Soviet Union. A network of industrial factories producing universal construction materials supplied the demand for region-wide urban growth. Prefabricated, multistory apartment buildings generally constructed from large concrete panels arose everywhere, providing free housing for people moving to cities and inexpensive construction for the Soviet government. Centralized planning of entire neighborhoods (*microrayons*; see fig 6.7) created a cookie-cutter built urban environment that remains in every urban setting, small and large, across the former Soviet bloc today.

Many of the nations under former Soviet influence have developed new political, economic, or cultural allegiances since 1991, but their urban landscapes remain similar, even intact in some places. Residents stay despite the often-crumbling, ghetto-like appearances of many apartments (due to shoddy construction and materials) because of the economic hardships faced by individuals and governments during post-socialist transformation. It is, quite simply, better household economics to privatize a formerly owned state apartment than to finance a new apartment or suburban house. As people remain *in situ*, so do basic government services and businesses catering to everyday life needs (e.g., post and doctor's offices, grocery stores). For these reasons, Soviet urban influence on the built environment remains outside the boundaries of today's Russia and looks to stay for quite a while.

First, emerging land markets and commercial real estate spaces transform the urban fabric as new housing, shopping and industrial developments are created within city limits and in suburban or exurban locations. Second, increased automobile and cargo truck ownership causes new kinds of movements in and around cities, especially with congestion resulting from the movement into cities. Third, as suburban growth develops, there is a tendency for previously compact post-socialist cities, often radial or quadrangular in form, to become linear in form, as economic activities occur along arterial routes out of cities into the surrounding countryside. Redevelopment and growth from city centers to peripheries often occurs chronologically, where the inner city redevelops first, followed by the periphery and suburbia. Thus, post-socialist city growth is both vertical (upward) and horizontal (outward), as it integrates industrial, commercial, and residential development in new ways in city centers and peripheries.

New Town

The *new town*, narrowly interpreted, is a phenomenon of the 20th century and refers to a comprehensively planned urban community built from scratch with the intent of becoming as self-contained as possible by encouraging the development of an economic base and a full range of urban services and facilities. New towns have come into existence for any number of reasons: relieving overcrowding in large cities; helping to control urban sprawl; providing an optimum living environment for residents; serving as growth poles for the development of peripheral regions; creating or relocating a national or provincial capital. The modern new-town movement began in Britain and later diffused to other European countries, the United States, the Soviet Union, and in the post-World War II era, to many other newly independent countries. The idealized form of the new town, in the West at least, tends to follow the Garden City concept of the British, with its emphasis on manageable population size, pod-like housing tracts, neighborhood service centers, mixed land uses, much green space, pedestrian walkways, and a self-contained employment base (akin to premodern villages). After a century of experimentation, however, most countries have found new towns to be extremely difficult to establish and sustain. Three types of new towns have been developed with some degree of success: (1) suburban-ring cities such as Reston, Virginia, in the United States; (2) new capitals such as Brasilia, Brazil; and (3) economic growth poles such as Ciudad Guyana, Venezuela.

Green Cities

Green cities are those that are on the way to environmental sustainability. No city is yet a green city, but thousands of cities around the world are trying to lessen their environmental impacts by moving away from fossil fuels, lowering emissions of greenhouse gases, reducing consumption, recycling urban wastes, utilizing water conservation technologies, erecting energy-efficient buildings, reinvigorating mass transportation systems, encouraging walking and cycling, increasing population densities to reduce sprawl, expanding open space—planting trees and flowers, and cultivating reliance on local food delivery chains. Curitiba, Brazil, is often regarded as the premier environmentally sustainable city. In the United States, Portland, Oregon, and Seattle, Washington, were early leaders.

WORLD URBANIZATION: PAST TRENDS

*Early Urbanization
(Antiquity to 5th Century A.D.)*

The first cities in human history were located in Mesopotamia, along the Tigris and Euphrates Rivers, probably about 4000 B.C. Cities were founded in the Nile Valley about 3000 B.C., in the Indus Valley (present-day Pakistan) by 2500 B.C., in the Yellow River Valley of China by 2000 B.C., and in Mexico and Peru by 500 A.D. Estimates of the size of these early cities suggest they were relatively small. Ur in lower Mesopotamia, for instance, was the largest city in the world 6,000 years ago but probably had a population of only 200,000. In fact, most cities of antiquity remained in the range of 2,000 to 20,000 inhabitants and the number of cities did not increase significantly. The largest ancient city was Rome, which Peter Hall has called "the first great city in world history." In the 2nd century A.D., Rome may have had 1 million inhabitants, making it the world's first great city of that size. Between the 2d and 9th centuries, however, Rome's population declined to less than 200,000. In fact, the world's largest cities in 100 A.D. were completely different from the largest cities in 1000 A.D. (fig. 1.11 and table 1.3).

Ancient cities appeared where nature and the state of technology enabled cultivators to produce more food and other essential goods than they needed for themselves and their families. That surplus set the stage for a division of labor among specialized occupations and the beginning of commercial exchanges. Cities were the settlement form adopted by those members of society whose direct presence at the places of agricultural production was not necessary. They were religious,

administrative, and political centers. These ancient cities represented a new social order, but one that was dynamically linked to rural society.In these ancient cities were specialists working full-time, such as priests and service workers, as well as a population that appreciated the arts and the use of symbols for counting and writing. Other attributes of these early cities included taxation, external trade, social classes, and gender differences in the assignment of work. Each city was surrounded by a countryside of farms, villages, and even towns. Ancient cities demonstrated the emergence of specialization. Rural life limited the exchange of goods, ideas, and people, as well as the complexity of technology and the division of labor. Thus, trade was a basic function of ancient cities, which were linked to the surrounding rural areas and to other cities by a relatively complex system of production and distribution, as well as by religious, military, and economic institutions (fig. 1.12).

The Middle Period (5th–17th Century A.D.)

From the fall of the Roman Empire to the 17th century, cities in Europe grew only slowly or not at all. The few large cities declined in size and function. Thus, the Western Roman Empire's fall in the 5th century A.D. marked the effective end of urbanization in western Europe for over six hundred years.

The major reason for the decline of cities was a decrease in spatial interaction. After the collapse of Rome and the dissolution of the empire it commanded, urban localities became isolated from one another and had to become self-sufficient in order to survive. From their very beginnings, cities have survived and increased in size because of trade with their rural hinterlands and with other

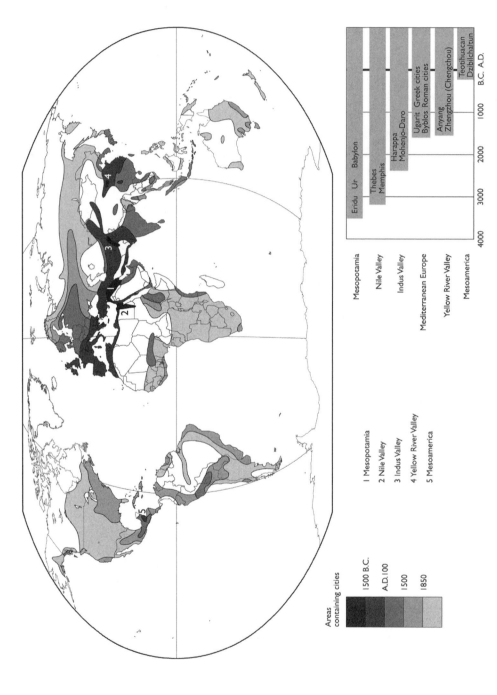

Figure 1.11 Spread of Urbanization, Antiquity to Modern Times. *Source:* Adapted from A. J. Rose, *Patterns of Cities* (Sydney: Thomas Nelson, 1967), 21. Used by permission.

Table 1.3 The Largest Cities in History

	Largest Cities in the Year 100				*Largest Cities in the Year 1000*	
1	Rome	450,000		1	Cordova, Spain	450,000
2	Luoyang, China	420,000		2	Kaifeng, China	400,000
3	Seleucia (on the Tigris), Iraq	250,000		3	Constantinople (Istanbul), Turkey	300,000
4	Alexandria, Egypt	250,000		4	Angkor, Cambodia	200,000
5	Antioch, Turkey	150,000		5	Kyoto, Japan	175,000
6	Anuradhapura, Sri Lanka	130,000		6	Cairo, Egypt	125,000
7	Peshawar, Pakistan	120,000		7	Baghdad, Iraq	125,000
8	Carthage, Tunisia	100,000		8	Nishapur (Neyshabur), Iran	125,000
9	Suzhou, China	n/a		9	Al-Hasa, Saudi Arabia	110,000
10	Smyrna, Turkey	90,000		10	Pata (Anhilwara), India	100,000

	Largest cities in the year 1500				*Largest cities in the year 2000*	
1	Beijing, China	672,000		1	Tokyo, Japan	34,450,000
2	Vijayanagar, India	500,000		2	Ciudad de México (Mexico City), Mexico	18,066,000
3	Cairo, Egypt	400,000		3	New York–Newark, New York	17,846,000
4	Hangzhou, China	250,000		4	São Paulo, Brazil	17,099,000
5	Tabriz, Iran	250,000		5	Mumbai (Bombay), India	16,086,000
6	Constantinople (Istanbul), Turkey	200,000		6	Shanghai, China	13,243,000
7	Guar, India	200,000		7	Kolkata (Calcutta), India	13,058,000
8	Paris, France	185,000		8	Delhi, India	12,441,000
9	Guangzhou, China	150,000		9	Buenos Aires, Argentina	11,847,000
10	Nanjing, China	147,000		10	Los Angeles–Long Beach–Santa Ana, USA	11,814,000

Sources: Historical cities: Tertius Chandler, *Four Thousand Years of Urban Growth: An Historical Census* (St. David's University Press, 1987; http://geography.about.com/library/weekly/aa01201a.htm; Year 2000 data: UN, *World Urbanization Prospects: 2005 Revision*, http://www.un.org/esa/population/publications/WUP2005/2005wup.htm

cities, near and far. The disruption of the Roman transportation system, the spread of Islam in the 7th and 8th centuries, and the pillaging raids of the Norse in the 9th century almost completely eliminated trade between cities. These events, plus periodic attacks by Germanic and other groups from the north, resulted in an almost complete disruption of urban and rural interaction. Both rural and urban populations declined, transportation networks deteriorated, entire regions became isolated, and people became preoccupied with defense and survival.

Although urban revival did occur 600 years after the fall of the Roman Empire via fortified settlements and ecclesiastical centers, growth in population and production remained quite small. The reason is simply that exchange was limited—conducted largely with people of the immediate surrounding region. Most urban residents spent their lives within the walls of their cities. Thus, urban communities developed very close-knit social structures. Power was shared between feudal lords and religious leaders. The economically active population was organized into guild—for crafts persons,

Figure 1.12   Countless ancient cities have ended up like these ruins of a former city in China's Xinjiang province. The city was once a thriving commercial center along the famous Silk Road, but once that economic base disappeared, the city withered and died also. (Photo by Stanley Brunn)

artisans, merchants, and others. One's social status was determined by one's position in guild, family, church, and feudal administration. Gender roles were also well defined. Merchants and the guilds saw innovative possibilities in "free cities," where a person could reach his or her full potential within a community setting.

Over time, commerce expanded and linked the city to expanding state power, resulting in a system called *mercantilism*. The purpose of mercantilism was the use of the power of the state to help the nation develop its economic potential and population. Mercantile policies protected merchant interests through the control of trade subsidies, the creation of trade monopolies, and the maintenance of a strong armed force to defend commercial interests. Cities were mercantilism's growth centers, and specialization and trade kept the system alive.

Mercantilism, though based on new economic practices, had one important element in common with the system of the previous period. It restrained and controlled individual merchants in favor of the needs of society. However, the rising new middle class of merchants and traders were against any restrictions on their profits. They opposed economic regulation and used their growing power to demand freedom from state control. They desired an end to mercantilism. As the power of the capitalists increased, the goal of the economy became expansion, with economic profit the function of city growth. While the new market economy provided the means to social recognition, the social costs were high. The greatest hardship fell on those receiving the fewest benefits—women, poor farmers, and members of the rising industrial working class. The new force of capitalism pushed aside the last vestiges of feudal life and created a new central function for the city—industrialization. It was capitalism that ushered in the Industrial Revolution and led to the emergence of the industrial city. While Europe was going through this process of decline and then rebirth in its cities, areas of the non-Western world experienced quite different patterns. In East Asia, for example, the city did not suffer the decline it went through in medieval Europe. In China, numerous cities founded before the Christian era remained continuously occupied and economically viable down through the centuries. Moreover, long before any city in Europe again grew to a size to rival ancient Rome, very large cities were thriving in East Asia. Changan (present-day Xi'an), for example, reputedly had more a population of more than 1 million people when it was the capital of Tang China in the 7th century. Kyoto, the capital of Japan for over a thousand years (and modeled after ancient Changan), had a population exceeding 1 million by the middle of the 18th century. Although most of the ancient cities of Asia had populations of less than 1 million, they were still far larger than cities in Europe until the commercial/industrial revolutions there. The principal explanation for this historical pattern of urban growth lies in the very different cultures and geographical environments of the great Asian civilizations. Although empires waxed and waned in Asia, just as in Europe, premodern cities in Asia continued to serve as vital centers of political administration, cultural and religious authority, and markets for agricultural surplus. It was not until the arrival of Western colonialism that those societies and their cities began to be threatened. The several centuries of Western colonialism in Asia added a new kind of

city to the region, a Western commercial city sometimes grafted onto a traditional city, sometimes created anew from virgin land. In either case, these new cities came eventually to dominate eastern Asia's urban landscape. That dominance has continued right into the contemporary period.

In the Greater Middle East, the traditional city also existed and thrived down through the centuries, long before Europeans began to claim pieces of the region as colonial territory. But once colonialism was fully asserted in the region, the same process of grafting and creating new Western commercial cities occurred, with consequences similar to those in eastern Asia.

In Sub-Saharan Africa and Latin America, the urban experience varied somewhat from that of much of Asia and the Middle East. In the case of Latin America, the traditional city—and the societies that created that city, such as the Mayan, Incan, and Aztec—was obliterated by Spanish conquest and colonization. The Spanish, as well as the Portuguese, thus created new cities in the vast realm of Latin America, cities that reflected the cultures of Europe. In Sub-Saharan Africa, the indigenous cities of various African kingdoms, such as Mali, Songhay, Axum, and Zimbabwe, had existed for centuries, but they also felt the impact of European colonialism. By the 19th century, they were largely destroyed, and Europeans created new commercial cities, usually coastal, that quickly grew to dominate the region.

Hence, as it materialized in Europe and was exported with the creation of colonial empires after 1500 A.D., the European-created city became the model for urban growth and development worldwide. In some regions, it was imposed on indigenous societies that were exterminated or shoved aside (as in North and South America, Australia, and the Pacific). In regions with long histories of indigenous cultures and urban life, it existed alongside of and transformed indigenous cities (as in most of Asia, the Middle East, and Africa).

Industrial and Postindustrial Urbanization (18th Century to the Present)

Only after the Industrial Revolution, which began around 1750, did significant urbanization occur. It was not until the 19th century, however, that cities emerged as important places of population concentration. By 1900, only one nation, Great Britain, could be regarded as an urbanized society in the sense that more than half of its inhabitants resided in urban places. During the 20th century, however, the number of urbanized nations increased dramatically. In the United States, the census of 1920 first revealed that a majority of Americans lived in cities. Only Africa and parts of Asia continue to lag behind the rest of the world in urbanization.

The city is not a static entity, but a system in flux. Within the city, some sectors may decline and die as investment is withheld, while others may grow and prosper as investments increase. Every change in urban function has both positive and negative multiplier effects. Adjustment to the changed situation may occur slowly. If the change is a reduction in functions, poverty levels will usually rise as unemployment spreads through the city in a cumulative manner, which produces greater effects in some neighborhoods than in others. Whether the city is growing or declining, spatial change within it will occur as a result of the decline of older neighborhoods, an influx of poor rural families or of immigrants from foreign areas, labor-shedding as a result of automation, the development of new suburbs or satellite towns, and the migration of businesses and industries to the suburbs. Those without access to opportunities, skills, or transportation will be left out by the operation of the market system. Whenever and wherever the market system operates, it serves the affluent rather than the poor. Thus, the problems of urbanization are in a sense the problems of marginalized peoples and unheard voices.

CITY FUNCTIONS AND URBAN ECONOMIES

City Functions

Some cites come into being because a strategic location needs to be defended; others serve the demands of trade and commerce; others serve the needs of governmental administration or religious pilgrimage; and still others serve the need to turn primary commodities into manufactured goods. Geographers have traditionally classified cities into three categories based on their dominant functions: (1) market centers (trade and commerce); (2) transportation centers (transport services); and (3) specialized service centers (such as government, recreation, or religious pilgrimage). Some cities may serve a single function—the "textile cities" of the southeastern United States, for instance—but functional diversity is more often the case.

Cities categorized as market centers are also known as central places, because they perform a variety of retail functions for the surrounding area. Central places offer a variety of goods and services (from grocery stores and gas stations to schools and corporate headquarters). Small central places, or market centers, depend less on the characteristics of a particular site and more on being centrally located with respect to their market areas. These centers tend to be located within the trade areas of larger cities: people living in small cities must go to larger cities to make certain purchases for which there is not a sufficient market locally. There is thus a spatial order to the settlements and their functional organization. Central place theory, as revealed by geographer Walter Christaller in the 1930s, is concerned with explaining the regular size, spacing, and functions of urban settlements as they might be distributed across a fertile agricultural region, for instance. In central place theory, the largest cities, or highest-order centers, are surrounded by medium-sized cities that are in turn surrounded by small cities, all forming an integrated part of a spatially organized, nested hierarchy. The locational orientation of market centers is quite different from the locational orientation of transportation and specialized function cities.

Transportation cities perform break-of-bulk or break-in-transport functions along waterways, railroads, or highways. Where raw materials or semi-finished products are transferred from one mode of transport to another—for example, from water to rail or rail to highway transport—cities emerge either as processing centers or as transshipment centers. Unlike central places, whose regularity in location is accounted for by marketing principles, transportation cities are located in linear patterns along rail lines, coastlines, or major rivers. Frequently, major transport cities are the focus of two or more modes of transportation, for example, the coastal city that is the hub of railways, highways, and shipping networks. Today, of course, almost all cities have multiple

transportation linkages. Exceptions tend to be isolated towns, such as mining centers in Siberia that may have only air or rail connections or primitive seasonal roads to the outside.

Cities that perform a single function, such as recreation, mining, administration, or manufacturing, are labeled specialized-function cities. A very high percentage of the population participating in one or two related activities is evidence of specialization. Oxford, England, is a university town; Rochester, Minnesota, is a health-care town; Norfolk, Virginia, a military town; Canberra, Australia, a government town; and Cancún, Mexico, a tourist town. Specialization is also evident in cities where the extraction or processing of a resource is the major activity. Cities labeled as mining and manufacturing cities have much more specialization than those with diversified economic bases.

Sectors of the Economy

The economic functions of a city are reflected in the composition of its labor force. Preindustrial societies are associated with rural economies; these economies have the largest percentages of their labor force engaged in the primary sector. Primary economic activities are agriculture, fishing, forestry, and mining. Preindustrial cities have historically been commercial islands in seas of rural-oriented populations. The Industrial Revolution triggered the emergence of cities oriented to manufacturing. The secondary sector of the economy, another name for manufacturing, expanded and so did the demand for labor to work in the factories. As larger percentages of the population began living in cities, retail trade, transportation, and all kinds of services began to flourish. The service sector,

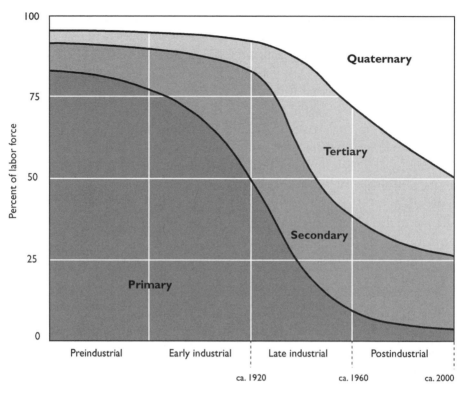

Figure 1.13   Labor Force Composition at Various Stages in Human History. *Source:* Adapted from Ronald Abler et al., *Human Geography in a Shrinking World* (Belmont, Calif.: Wadsworth, 1975), 49.

or tertiary economic activities, grew at the expense of the primary and secondary sectors, which declined in their proportions of the total labor force. The quaternary sector, a more advanced stage of the service sector, consists of information- and intellect-intensive services, which are playing an increasingly important role in the world economy. The mix of primary, secondary, tertiary, and quaternary activities within urban regions, as it has changed over time, also helps us to identify specific stages in humanity's economic evolution (fig. 1.13).

The association between urbanization and industrialization has been characteristic of Europe, North America, Japan, Australia, and New Zealand. That is, cities and industries grew in synchronization with each other. In many parts of Africa, Asia, and Latin America, many countries that recently have been increasingly urban have not experienced a corresponding increase in the manufacturing sector of the economy. Rather, their service sectors have provided jobs for growing urban populations. Included in service-sector employment are small retailers, government servants, teachers, professionals, and bankers. Also included are many service workers in the informal economy, including those members willing to perform odd jobs (watching parked cars or cleaning houses) and those working in unskilled service occupations, such as street

Figure 1.14 This textile retailer in Kashgar, China, caters primarily to local residents, but tourist business helps, too. The cell phone, a visible sign of globalization, helps the owner stay competitive. (Photo by Stanley Brunn)

vending, scavenging, and laboring at construction sites. In the informal sector, barter and the exchange of services often take the place of monetary exchanges, thus bypassing government accounting and taxation.

Basic and Non-Basic Economic Activities

Economic functions are keys to the growth of cities. The economic base concept states that two types of activities or functions exist: those that are necessary for urban growth and those that exist primarily to supplement those necessary functions. The former are called basic functions or city-forming activities. They involve the manufacturing, processing, or trading of goods or the providing of services for markets located outside the city's boundaries. Economic functions of a city-servicing nature are called nonbasic functions. Grocery stores, restaurants, beauty salons, and so forth are nonbasic economic activities because they cater primarily to residents within the city itself (fig. 1.14).

Of the two, the basic functions are the key to economic growth and prosperity. A city with a high percentage of its labor force in the production of such items as automobiles, furniture, and electronic equipment depends on sales beyond the city's boundaries to bring money into the community. Income generated by the sales of those industrial goods is channeled back into the city's nonbasic sector, where employees in those industries purchase groceries, gasoline, insurance, entertainment, and other everyday needs and wants.

The economic base of some cities is grounded in manufacturing industries, the secondary sector of the economy. Manchester, England, and Pittsburgh, Pennsylvania, are prime examples of older industrial cities whose growth and prosperity depended upon world markets, for cotton textiles and steel, respectively. Since World War II, both cities have lost their manufacturing base and have been challenged to find service industries for which there is a larger market. The economic base of postindustrial cities, in fact, is to be found in the tertiary and quaternary sectors of the economy. Silicon Valleys developed around the world in the late 20th century to service the needs of the computer industry. In the early 21st century, biotech valleys are becoming the economic base of choice. Cities such as Geneva, Singapore, San Francisco, and Boston are competing to have biotechnology firms move into their regions. Money from biomedical and pharmaceutical research provides an economic base tied to high-level applications of technology and brainpower. Banking, accounting, architecture, and advertising are other service industries that some cities depend upon for their economic bases.

As a city's economic base increases, it has a multiplier effect throughout the community. Growth (and conversely, decline) becomes a cumulative process in which growth begets growth. This is known as the *principle of circular and cumulative causation*. For instance, one of the major ways cities grew in the past was by attracting more manufacturing enterprises. Each new factory in an urban area stimulated general economic development and population growth. Business output increased due to a greater demand for products. Rising profits increased savings, causing investments to rise. Increased productivity resulted in greater wealth. The growing population then reached a new level, or threshold, resulting in a new round of demands. Larger cities are able to offer a greater number and variety of services than smaller cities. Conversely, cities stagnate because they lose industries and population, conditions that create a negative circular and cumulative causation—a downward spiral. Thus, it is easy to understand why city mayors and chambers of commerce work so hard to promote their respective cities as favorable sites for investment and new business locations.

THEORIES ON THE INTERNAL SPATIAL STRUCTURE OF CITIES

In addition to the origin and growth of cities, geographers have long been intrigued by their internal spatial structure. Components of that structure include industrial zones, commercial districts, warehouse rows, residential areas, parks and open space, and transportation routes, among others. A variety of theories has been developed to describe and explain the pattern of land use and the distribution of population groups within cities. The four most widely accepted theories, or models, of city structure are the concentric zone model, the sector model, the multiple nuclei model, and the inverse concentric zone model (fig. 1.15). All evolved from observations of urban landscapes that suggested that different land uses were predictably, not randomly, distributed across the city.

The Concentric Zone Model

The concentric zone theory was first conceptualized by Friedrich Engels (coauthor of the Communist Manifesto) in the mid-19th century. Engels observed that the population of Manchester, England, in 1844 was residen-

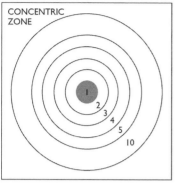

I   Central business district
2   Wholesale light manufacturing
3   Low-class residential
4   Medium-class residential

5   High-class residential
6   Heavy manufacturing
7   Outlying business district

8   Residential suburb
9   Industrial suburb
10 Commuter zone

Figure 1.15   Generalized Patterns of Internal Urban Structure. *Source:* Adapted from various sources.

tially segregated on the basis of class. He noted that the commercial district (offices plus retail and wholesale trade) was located in the center of Manchester and extended about half a mile in all directions. Besides the commercial district, Manchester consisted of unmixed working people's quarters, which extended a mile and a half (2.3 km) around the commercial district. Next, extending outward from the city, were the comfortable country homes of the upper bourgeoisie. Engels believed this general pattern to be more or less common to all industrial cities.

Engels may have described the pattern first, but most social scientists consider E. W. Burgess, a University of Chicago sociologist, to be the father of the concentric zone model. According to Burgess, the growth of any city occurs through a radial expansion from the center so as to form a series of concentric rings, in essence, a set of nested circles that represents successive zones of specialized urban land use. The five zones Burgess described during the 1920s, before the automobile transformed

Chicago, were: (1) the central business district (CBD), with its retail and wholesale areas; (2) the zone of transition, characterized by stagnation and social deterioration; (3) the zone of factory workers' homes; (4) the zone of better residential units, including single-family dwellings and apartments; and (5) the commuter zone, extending beyond the city limits and consisting of suburbs and satellite communities. The process Burgess used to explain these concentric rings was called invasion and succession. Each type of land use and each socioeconomic group in the inner zone tends to extend its zone by the invasion of the next outer zone. As the city grows or expands, there is a spatial redistribution of population groups by residence and occupation. Burgess further demonstrated that many social characteristics are spatially distributed in a series of gradients away from the central business district. Such characteristics include the percentage of foreign-born groups, poverty, and delinquency rates. Each tends to decrease outward from the city center.

The Sector Model

The sector model was developed in the 1930s by Homer Hoyt, an economist. Hoyt examined spatial variations in household rent in 142 American cities. He concluded that general patterns of housing values applied to all cities and that those patterns tended to appear as sectors, not concentric rings. According to Hoyt, residential land use seems to arrange itself along selected highways leading into the CBD, thus giving land-use patterns a directional bias. High-rent residences were the most important group in explaining city growth, because they tended to pull the entire city in the same direction. New residential areas did not encircle the city at its outer limits, but extended farther and farther outward along a few select transportation axes, giving the land-use map the appearance of a pie cut into many pieces. The sectoral pattern of city growth can be explained in part by a filtering process. When new housing is constructed, it is located primarily on the outer edges of the high-rent sector. The homes of community leaders, new offices, and stores are attracted to the same areas. As inner, middle-class areas are abandoned, lower-income groups filter into them. By this process, the city grows over time in the direction of the expanding high-rent residential sector.

The Multiple Nuclei Model

In 1945, two geographers, Chauncy Harris and Edward Ullman, developed a third model to explain urban land-use patterns, the multiple nuclei model. According to their theory, cities tend to grow around not one but several distinct nodes, thus forming a polynuclear (many-centered) pattern. The multiple nuclei pattern is explained by the following factors:

1. Certain activities are limited to particular sites because they have highly specialized needs. For example, the retail district needs accessibility, which can best be found in a central location, while the manufacturing district needs transportation facilities.
2. Certain related activities or economic functions tend to cluster in the same district because they can carry on their activities more efficiently as a cohesive unit. Automobile dealers, auto repair shops, tire shops, and auto glass shops are examples.
3. Certain related activities, by their very nature, repel each other. A high-class residential district will normally locate in a separate area from the heavy manufacturing district.
4. Certain activities, unable to generate enough income to pay the high rents of certain sites, may be relegated to more inaccessible locations. Examples may include some specialty shops.

The number of distinct nuclei occurring within a city is likely to be a function of city size and recentness of development. Auto-oriented cities, which often have a distinct horizontal as opposed to vertical appearance, include industrial parks, regional shopping centers, and suburbs layered by age of residents, income, and housing value. Rampant urban sprawl is likely to be reflected in a mixed pattern of industrial, commercial, and residential areas in peripheral locations. Geographer Peirce Lewis describes this sprawling urban landscape as the *galactic metropolis* because the nucleations resemble a galaxy of stars and planets. Some of those nucleations become cities in the suburbs, what some have called *edge cities*. These edge cities are, in effect, the CBDs of newly emerging urban centers scattered through the suburban ring surrounding older central cities. This pattern reinforces what has been described as the typical urban spatial model for most U.S. urban areas, the doughnut model. In this model, the hole in the doughnut is the central city (with poor, mostly nonwhite, blue-collar, working-class residents, large numbers of whom are on welfare, and a declining tax base and economy) and the ring of the doughnut is the suburbs (rich, mostly white, middle- and upper-class, white-collar employment, and an expanding tax base and economy). Old, often declining manufacturing tends to be found in the hole; new, often high-tech manufacturing tends to be located in the suburbs, in the edge cities. In the 21st century, however, center cities are reinventing themselves by building more residential quarters, upgrading infrastructure, invigorating amenities (including shopping), and emphasizing mixed land-use development.

The Inverse Concentric Zone Theory

The preceding three theories of urban spatial structure apply primarily to cities of the MDCs and to American cities in particular. Many cities in the LDCs follow somewhat different patterns. A frequent one is the inverse concentric zone pattern, which is a reversal of the concentric zone model. Cities where this pattern exists have been called preindustrial; *i.e.*, they are primarily administrative and/or religious centers (or were at the time of their founding). In such cities, the central area is the place of residence of the elite class. The poor live on the periphery. Unlike most cities in the MDCs, social class in these places is inversely related to distance from the center of the city.

The reasons for this pattern are twofold: (1) the lack of an adequate and dependable transportation system, which thus restricts the elites to the center of the city so they can be close to their places of work, and (2) the functions of the city, which are primarily administrative and religious/cultural, are controlled by the elite and concentrated in the center of the city (with its government buildings, cultural institutions, places of worship, etc.).

As many developing countries have begun to industrialize, newer growth industries have tended to locate not in city centers but on the periphery, often in industrial parks or enterprise zones established by the government for the purpose of attracting domestic and foreign investors. The city centers tend to be far too congested for industrial plants of any considerable size. Moreover, the elites in the city centers often do not want large industrial plants near their place of work and residence. Hence, emerging gradually in many of the larger cities of the LDCs is the pattern of the multiple nuclei model, with new industrial parks serving as the nuclei. In other words, the inverse concentric zone pattern, while still valid in many LDCs, is merging with the multiple nuclei pattern.

As useful as these four models are, they must be viewed with caution as generalizations of the extremely complex mix of factors that influence the use of land within cities. One can commonly find elements of more than one model present in a given city. Moreover, each of the models must be viewed as dynamic. There are changes going on all the time in economic functions, social and administrative services, transportation, and population groups that will alter the size and shape of specific sectors or zones (fig. 1.16). Furthermore, the complexities of applying these theories multiply severalfold when

Figure 1.16   The Frontenac Hotel was built by the Canadian Pacific Railway. As the most well-known signature landscape in Quebec City, it still functions as a tourist magnet even though most visitors no longer come by train. (Photo by Donald Zeigler)

working with non-Western cultures and economic systems. Nowhere is this more apparent than in China and the former Communist countries of Eastern Europe, where various forms of the so-called socialist city were being created and where internal spatial structures were quite unlike those described by any of the four theories mentioned previously. The legacy of those socialist patterns lingers on, as free-market forces transform those cities.

URBAN CHALLENGES

Managing Population Size and Growth

Excessive size of urban regions, both in population and in geographical area, might more properly be described as a cause of problems than a problem in itself. It presents a particularly severe challenge in less developed regions, where the economic base of cities is inadequate to cope with the stresses created by more and more people. A concomitant of excessive size is overcrowding, meaning too many people occupying too little space, but it does not always equate to population density. Some cultures are more adapted to high densities than others. For Americans or Europeans, though, it is sometimes difficult to fully comprehend the magnitude and effects of really severe urban overcrowding. Seeing or being caught up in the tidal wave of humanity that one can find in the larger cities such as Manila, Shanghai, or Cairo is a vivid lesson in the consequences of excessive size that one never forgets.

The rate of population growth—or decline—may present cities with challenges as well (box 1.4). Some cities, especially in the developing world, are growing so fast that economic development cannot keep up. *Hyperurbanization* is sometimes used to describe what is happening in the world's most rapidly growing cities. Conversely, some cities in the developed world may be stagnating or declining in population. Whether in Russia, Germany, or the United States, the problems of no-growth cities are very similar. They are often home to companies that have outdated technology, high production costs, expensive labor, an aging workforce, and products with declining demand. In the developing world, these cities are generally victims of deindustrialization. They have not kept up with the transition from manufacturing to a service-based economy.

Managing Urban Services

With so many people in urban settings, city governments are hard-pressed to provide all of the human services that residents need—education, health care, pharmacies, clean water, sewage disposal, garbage pick-up, police and fire protection, disaster relief, public parks, mass transit, and numerous other essential services. How can a city in the developing world that is doubling in population every ten years or so maintain the economic growth needed to provide for so many new arrivals, particularly when those new residents are poor? While these problems exist around the world, cities in the developed world are more likely to have the resources to deal with them. Yet, even in the world's most developed countries, providing services to sprawling, energy-inefficient suburban and exurban regions can strain municipal budgets.

Managing Slums and Squatter Settlements

Most cities of the world have slums or squatter settlements, poorer communities that are not fully integrated, socially or economically, into the development process (fig. 1.18). Slums tend to be found in old, run-down areas of inner cities (sometimes, paradoxically, on very valuable land) throughout the world. Squatter settlements are typically new, but made up of makeshift dwellings erected without official permission on land that the squatters do not own. These settlements are usually located on the outskirts of cities in the developing world. They may be constructed of cardboard, tin, adobe bricks, mats, sacks, or any other available materials. They tend to lack essential services, sometimes even electricity. Squatter settlements go by various names in different countries: *barriadas* in Peru, *favelas* in Brazil, *geçekondu* in Turkey, *bustees* in India, and *bidonvilles* in former French colonies.

Managing Social Problems

Perhaps one of the most insidious effects of hyperurbanization throughout the world is a reduction in people's sense of social responsibility. As more and more people compete for space and services, the competition tends to breed antisocial, even sociopathic, attitudes. City life can bring out the worst in human behavior. People exhibit social pathologies when they resist waiting in line for services; think nothing of despoiling public property; disregard traffic regulations; or show a disregard for the rights of fellow citizens. To the extent that large cities provide neither a sense of community not a respected police presence, crime soars. Social norms that hold people in check in rural areas, may be absent in cities.

Box 1.4   The Subtracted City

Deborah E. Popper and Frank J. Popper

Detroit is the ultimate symbol of urban depopulation. A third of its 139 square miles (360 square kilometers) lies vacant. In the 2010 Census, Michigan's powerhouse city lost a national-record-setting quarter of the people it had in 2000: a huge dip to the city and to potential private- and public-sector investors.

Is Detroit an epic outlier or just a dramatic example of Industrial Age population decline? Cleveland lost 17 percent of its population, Birmingham 13 percent, and Buffalo 11 percent. Their losses and the losses of smaller cities like Braddock, Pennsylvania, and Cairo, Illinois, transcend simple population declines. In recent decades, houses, businesses, jobs, schools, entire neighborhoods—and hope—keep getting removed.

The subtractions occur without plan, intention, or control. They are haphazard, volatile, and unexpected; and so they pose big risks. Population growth or stability, by contrast, seems manageable and politically palatable. But no American city plan, zoning law, or environmental regulation anticipates subtraction. A city can buy deserted houses and factories and return them to use. Yet which use? If the city cannot decide, how long should the properties stay idle before the city razes them? How prevalent must abandonment become before it needs systematic neighborhood or citywide solutions instead of lot-scale ones? Subtracted cities have struggled for two or three generations, since the consumer boom that peaked after World War II. Thousands of neighborhoods in hundreds of cities lost their American Dream.

The nation has little idea how to respond. Subtracted cities rarely begin concerted action until half the population has left. Generations separate the first big losses and substantial action. Usually, new, local leadership must emerge to work with or around loss instead of against it. By then, the tax base, public services, budget troubles, and morale are dismal. To reverse the long downward spiral requires extraordinary effort.

Yet, subtracted cities must try to reclaim control of their destinies. They could start by training their residents to value, salvage, restore, and market unused sites and the materials found there. They might supplement school drug-free zones with "subtraction-action zones," reacting fast when nearby empty properties show neglect that could harm children. Just planting a few trees improves a deserted lot.

Subtracted cities should encourage gardens on the abundant vacant lots. Community gardens rouse people for positive local efforts rather than against city hall. Plants can sometimes go directly into the ground or into raised beds. Community gardens upgrade the food supply, teach business and social skills, provide equipment that can be lent for home repair, and may create new enterprises.

The downtowns, main streets, and public works of subtracted cities still scream walkability, public transit, and efficient energy use. They support entertainment, retail sales, and services. Cleared lots on the downtown's edge can create new parks, outdoor amphitheaters, and sports facilities.

Subtracted cities should not obliterate their past or the contributions of their people. They should reuse old structures, with factories becoming restaurants, apartments, and business incubators, and rail lines becoming hiking trails. The answer to desertion begins with embracing the on-the-ground subtraction without pretending that it will go away.

Figure 1.17   Urban farming can often take advantage of land resources that become available in "subtracted cities," as in this Earth Works Urban Farm, which is being promoted as a Garden of Unity. (Photo by Deborah Popper)

Managing Unemployment

Virtually everything else connected with the city is related in one way or another to the economic health of its population, and economic well-being is dependent on people having jobs. In capitalist economies, however, employment is not guaranteed. The result is often unemployment and underemployment. In the developing world, too many people may be competing for too few jobs, driving down the cost of labor. In the developed world, large segments of city populations may lack the skills necessary to find jobs in the high-end service sector of the economy. Sometimes, the result is underemployment: people take jobs that are not commensurate with their skills. These jobs pay less than a living wage meaning that some have to take more than one job to survive and others have to supplement their income with employment in the informal sector of the economy with long hours and no fringe benefits. In cities of the developing world, unemployment rates of 30 percent to 40 percent or more are not

Figure 1.18   Scavengers push their carts past squatter shacks in Smokey Mountain, a once-notorious slum of Manila that got its name from the continuously burning garbage. It has been the target of redevelopment in recent years (see chapter 10). (Photo by James Tyner)

uncommon. Women and children, recent migrants, and the elderly are often the most victimized by problems of employment and underemployment.

Managing Ethnic Issues

Unemployment, underemployment, and other factors breed a variety of subsidiary problems related to ethnicity and class status. For example, relative economic prosperity in the United States has produced a tidal wave of illegal immigrants, primarily from Mexico and other Latin American countries, who come seeking a better life. These people, along with large numbers of legal immigrants and refugees, commonly settle in the cities, as did Cuban refugees in Miami, Florida. Refugees and new migrants may come into conflict with (1) community elites who find their power diluted, (2) groups, often other minorities, with whom they compete for jobs, and (3) majorities who have completely different languages, religions, and worldviews. Throughout the world, many cities must manage severe centrifugal forces generated by cultural diversity (fig. 1.19).

Managing Modernization and Globalization

One phenomenon that is sweeping the world's cities, especially the larger ones, is the dilemma of Westernization versus modernization. The problem facing the LDCs is how to raise standards of living without completely abandoning traditional cultural values and ways of life. Some might argue that tradition and modernization are incompatible, that modernization automatically entails change, and that change is likely to take the form of Westernization. To be sure, there are ample signs of this Westernization (some might wish to call it homogeni-

Figure 1.19 The square in Rotorua, a city famous for its geothermal pools, has been brought to life with a modern rendition of Maori art. To the Western world, Rotorua is in New Zealand; to the indigenous Maoris, it is in Aotearoa. (Photo by Donald Zeigler)

zation or globalization) of the world's major cities in the forms of skyscrapers, modern architecture, the automobile society, advertising, the focus on high mass consumption, and so forth. Nonetheless, as anyone who has lived in cities of the LDCs for any length of time can attest, traditional cultural values and lifestyles do somehow manage to persist even in the most modern metropolis.

Almost the entire world—rural and urban—is adjusting to changes in global economies. Globalization means the movement of products, money, information, and human talent around the world in ever-larger quantities at ever-lower costs and in ever less time.

Mayors and governing councils must now think globally as well as locally. Companies produce items for world, not local or regional, markets. Money is transferred electronically from major financial centers in Europe to Asia and from North to South America. Trade barriers are being reduced between countries. Transnational corporations and nongovernmental organizations (NGOs) promote a "world without borders." The net result is an easier flow of people, money (including credit), and products across boundaries that once separated those with different ideologies and economies. In all of this, there are winners, losers, and adjustments demanded of society. And, there are also potential problems because change does not come easily.

Managing Privacy

The tentacles of wireless communication penetrate deeply into our private lives, especially as more and more information about us is stored in what is coming to be known as "the cloud," those reservoirs of information that exist on computer servers in cyberspace. We voluntarily give up so much information today to both the public and private sectors and assume that it is appropriately used and secure. Yet, once "out there" it remains beyond our control. Likewise, there are also involuntary means of collecting information about us. We may be forced to yield biometric data, making it possible, for instance, for cities to keep track of their individual inhabitants. With e-tracking devices such as cell phones, information is created about our whereabouts almost 24 hours a day, and Geographic Information Systems (GIS) can be used (not by us but by unknown "others") to map it all out and store it permanently. Moreover,

Figure 1.20   Banksy is a well-known graffiti artist whose works materialize on urban landscapes while no one is watching. In London, his unauthorized critique of CCTV appeared overnight on Royal Mail Service property. (Photo by Donald Zeigler)

surveillance in cities is becoming increasingly common and commonly accepted. We have learned to like Closed-Circuit TV cameras on our streets because we feel they make us safe. But, they also deprive us of privacy (fig. 1.20)

Managing the Environment

Pollution of air and water, excessive noise levels, visual blight, and hillside clearance for urban expansion are among the many serious environmental problems in cities around the world (box 1.5). Moreover, global climate change adds a new dimension to the environmental problems experienced by cities as water becomes scarcer; heat waves more frequent, prolonged, and severs; and sea levels rise higher. Cities in the richer parts of the world at least have the means to do something about these problems, but cities in the poorer countries often regard such concerns as less important in the face of more immediate life-and-death issues. For instance, the cities with the greatest air pollution are no longer London (formerly nicknamed "the Big Smoke") and Los Angeles

Box 1.5   Tackling Urban Environmental Problems with GIS

Joseph J. Kerski

People have always been fascinated with investigating their home, the Earth. For centuries, maps have stirred imaginations and inspired explorations of the unknown. Navigating one's way through expanding cities and expanding commerce made city maps a natural application of cartography in the centuries since the Babylonians first mapped the city of Nippur about 3,500 years ago. However, by the latter part of the 20th century, the complex and interconnected nature of urban areas brought an end to the old "city street map" as an effective planning tool. Geographic Information Systems (GIS) and associated geotechnologies, especially Global Positioning Systems (GPS) and Remote Sensing, transformed the ways in which cities managed environmental, transportation, zoning, and other planning issues. The organizational transformation brought about by the use of GIS also meant that all city departments could work from a common mapping framework, eliminating duplication and increasing efficiency within city government departments and across multi-government metropolitan areas. As GIS became embedded in the internal information technology infrastructure of these organizations, it was increasingly relied upon for daily decision making.

While the advent of GIS has certainly not eliminated urban environmental problems, technicians and managers can now visualize relationships between population growth, demographic characteristics, climate, vegetation, landforms, river systems, underground cables, land use, soils, natural hazards, crime, security, and other components of the urban infrastructure. Plus, they can model how that infrastructure is changing. With the advent of web-based GIS tools at the dawn of the 21st century, coupled with the beginnings of a worldwide set of high-resolution vector and raster data sets, geotechnologies began to be applied to identify and address urban environmental problems. People began to realize the efficiencies that could be achieved if data and models of spatial analytical procedures were shared to address problems of mutual concern. These problems operated at scales of regions, hemispheres, and even the planet. For example, air pollution from one megalopolis could have impacts on human health thousands of kilometers away, while urban sprawl in one country could affect deforestation rates in a faraway country if trees are cut to support construction of the new homes in sprawling subdivisions.

Researchers use GIS in many ways: (1) collecting data, (2) analyzing data, and (3) communicating data. Data collection can range from air quality to vehicle counts, but all data must be geocoded, either by street address, latitude-longitude locations, mile or kilometer markers, or by other means so that they can be mapped and understood. Data analysis can range from querying the number of houses that would be submerged in a 100-year flood to the amount of additional energy use required to keep homes cool if global temperatures rose by 1°C over the next 40 years. Analysis depends upon accurate spatial data sets and the ability to apply different models to study the changes that are continually occurring within cities and across the global network of cities. For example, consider a map produced within a GIS environment showing which cities in the southwest part of the North Atlantic Ocean could be submerged if sea levels rose by 50 meters (164 feet) (fig. 1.21). It was produced by one specific model of climate change, and based on digital elevation information and coastline information collected under spatial data standards. If and when the model and the data change, the GIS must be able to reflect those changes.

Data communication is still conducted via paper maps plotted from a GIS and via public presentations. Increasingly, it is also conducted over the Internet—beginning with posting static maps for viewing only, and later, serving spatial data that others could use and modify. These developments may herald the beginnings of a global urban spatial information infrastructure.

Thus, GIS is encouraging an information infrastructure both within cities and between cities. Because urban environmental issues are interdisciplinary by their very nature, and because GIS was created as an interdisciplinary, problem-solving tool, GIS found a natural home in urban environmental analysis. GIS data in urban analysis is increasingly tied to ground monitoring stations, and collectively form a kind of "nervous system" for the city and, increasingly, the planet. The goal in using a GIS is better and more coordinated decision making to build a sustainable urban future.

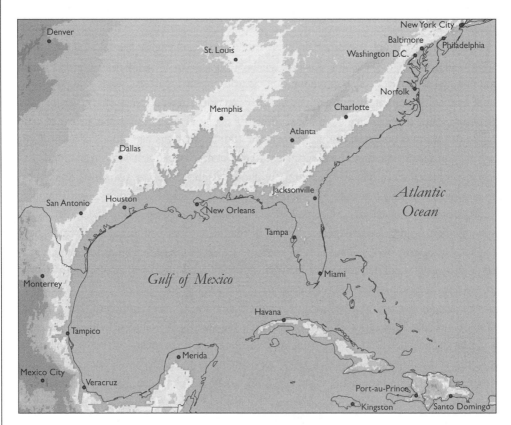

Figure 1.21   The Impact of Hypothetical Sea Level Rise of 50 meters on the Caribbean and Atlantic Coasts of North America. *Source:* Joseph Kerski

Figure 1.22 Portland, Oregon, offers a lesson in many aspects of successful urban development, including its trolley system to reduce reliance on private automobiles. (Photo by Judy Walton)

(once dubbed "Smog Central"), but cities such as Shanghai, Mexico City, and São Paulo. An additional environmental problem arises from the expansion of urban areas into agriculturally productive land close to the city. China has been estimated to be losing 2 million acres (809,600 hectares) of farmland to urban expansion each year, and the United States almost 1 million acres (404,600 hectares).

Managing Traffic

Another obvious effect of urbanization, produced in large part by the growing number of motor vehicles, is traffic congestion. Superficially, this might be viewed as a mere nuisance, an aggravation of much less consequence than survival-level problems such as employment, housing, and social services. Nonetheless, traffic congestion is a serious dilemma that is choking many cities to a standstill in terms of the movement of people and goods. Consequences include economic inefficiency (loss of time and waste of resources), social stress, and pollution, all of which diminish a city's development potential (fig. 1.22).

Managing Urban Governance

Urban governments around the world face challenges related to the balance between revenues and expenditures. Few governments have the funds to meet every need, so priorities must be established. Those priorities may be set by higher levels of government (sometimes authoritarian) or by the democratic process locally. In any case, the task of governing or administering services to a mushrooming city is daunting, whether the city is New York or Mumbai (Bombay). In some countries, like the United States, problems arise because urban areas are fragmented among so many jurisdictions, some overlapping. In other countries, many in the developing world, government bureaucracies are bloated with excess employees, are suspected of serving elites alone, and generally have a hard time combating pressing urban problems.

Problems or Management Challenges?

Some countries in the developing world, and a great many in the developed, are successfully attacking the problems of urban life and urban growth. The proposed solutions tried or planned are far too numerous and complex to even summarize here. There are pessimists who undoubtedly contend that it is too late to solve the urban ills of humankind. Optimists hope this is not so. In the following chapters, we will observe both the problems and the promises of urbanism in major world regions.

SUGGESTED READINGS

Amin, Ash, and Nigel Thrift. 2002. *Cities: Reimagining the Urban.* Cambridge, England: Polity, 2002. *Challenges the notion that the contemporary city is separate from the country and offers a model of the New Urbanism.*

Caves, Roger W. 2005. *Encyclopedia of the City.* London, New York: Routledge. *Experts from many disciplines define and interpret the city.*

Davis, Mike. 2007. *Planet of Slums.* New York: Verso. *A documentation of poverty in cities of the developing world.*

Hall, Peter. 1998. *Cities in Civilization.* New York: Pantheon. *Looks at the world's great cities during their golden ages, with an emphasis on culture, innovation, and the arts.*

Knox, Paul L., and Linda M. McCarthy. 2005. *An Introduction to Urban Geography.* Upper Saddle River, NJ: Prentice Hall. *A basic textbook on the city that captures the dynamism of both urbanization and urban geography.*

LeGates, Richard T., and Frederic Stout. 2011. eds. *The City Reader.* New York: Routledge. *An anthology of classic and contemporary titles about urban history, design planning, social, and environmental problems.*

Lynch, Kevin. 1960. *The Image of the City.* Cambridge, MA: M.I.T Press, 1960. *Seminal work on the "visual quality" of cities and how to read the urban landscape.*

Sassen, Saskia. 2001. *The Global City: New York, London, Tokyo,* 2nd ed. Princeton, NJ: Princeton University Press. *An exploration of the world's three leading centers for international transactions and their impact on the global urban hierarchy.*

Soderstrom, Mary. 2006. *Green City: People, Nature and Urban Places.* Montreal: Véhicule. *An examination of 11 cities and their interactions with the natural environment.*

Vance, James E., Jr. 1990. *The Continuing City: Urban Morphology in Western Civilization.* Baltimore, MD: Johns Hopkins University Press. *Explores the role of the city in Western society and its changing form through time.*

Whitfield, Peter. 2005. *Cities of the World: A History in Maps.* Berkeley: University of California Press, 2005. *Through their maps, we see how cities have perceived themselves overtime.*

SUGGESTED WEBSITES

Center for International Earth Science Information Network (CIESIN)
www.ciesin.columbia.edu
Data and research on population and the environment, especially climate change.

Cities.com
www.cities.com
The latest news from cities all over the world.

City Population
www.citypopulation.de
Presents population statistics and maps for cities around the world.

Cyburbia
www.cyburbia.org

A large directory of Internet resources relevant to planning, architecture, urbanism, growth, sprawl, and other aspects of the built environment.

Demographia
www.demographia.com/
Data and full-text reports on cities and urban processes.

ESRI Community Showcase
resources.esri.com/showcase/
Showcases resources from Environmental Systems Research Institute (ESRI) and web-based GIS portals built and hosted by city and regional governments.

GaWC—Globalization and World Cities
www.lboro.ac.uk/gawc/
Includes an inventory of world cities, data presentations, commentaries on specific cities, and articles of scholarly interest.

Geographically Yours
www.geographicallyyours.blogspot.com
Landscape photos from around the world, including many from cities featured in this book.

NASA "Search the Cities from Space" Collection
city.jsc.nasa.gov/cities/
Color photographs of the world's cities taken by National Aeronautics and Space Administration (NASA) astronauts during spaceflights. Voluminous information on population, including the annual report on world urbanization prospects.

U.S. Bureau of the Census
www.census.gov/
Copious data on towns and cities in the United States.

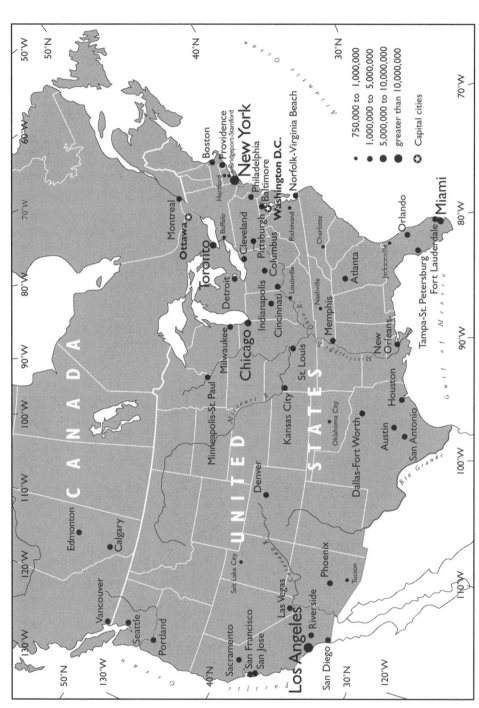

Figure 2.1 Major Cities of the United States and Canada. *Source:* UN, *World Urbanization Prospects: 2009 Revision,*
http://esa.un.org/unpd/wup/index.htm

2

Cities of the United States and Canada
LISA BENTON-SHORT AND NATHANIEL LEWIS

KEY URBAN FACTS

Total Population	352 million
Percent Urban Population	82%
Total Urban Population	289 million
Most Urbanized Country	United States (82.3%)
Least Urbanized Country	Canada (80.6%)
Annual Urban Growth Rate	1.20%
Number of Megacities	2 cities
Number of Cities of More Than 1 Million	48 cities
Three Largest Cities	New York, Los Angeles, Chicago
World Cities	New York, Los Angeles, Chicago, Washington, San Francisco, Atlanta, Miami, Toronto, Montreal, Vancouver
Global Cities	New York

KEY CHAPTER THEMES

1. After 1950 the United States and Canada became metropolitan societies, as the majority of their populations came to live in metropolitan areas.
2. The United States and Canada comprise one of the most urbanized regions of the world, with a developed urban hierarchy composed of a multitude of small, medium, large, and multi-million cities and even several megacities.
3. Although there are variations in urban land-use patterns, the most pronounced pattern is of a declining core and an expanding suburban area; however, in some cities the core has seen some resurgence.
4. Cities in the United States and Canada have been recently shaped by the intensification of economic globalization and heightened competition in the global urban hierarchy.

5. The reliance on the automobile and weak investment in public transportation has resulted in low population densities and increased sprawl.

6. Both U.S. and Canadian cities have responded to economic changes since the 1980s by employing a variety of redevelopment schemes designed to increase investment through tourism, sports, historical districts, and cultural events.

7. A long history of industrialization has left cities dealing with numerous environmental issues, including air, land, and water pollution, all of which threaten to erode the quality of life for many urban residents.

8. Immigration, now from a more diverse set of countries than ever before, is transforming numerous cities in North America as immigrants gravitate toward long established immigrant magnets as well as to newly emerging gateway cites.

9. In a post-9/11 world, heightened concerns about security have begun to transform urban space through the installation of security cameras and the fortressing of selected spaces with security features.

10. The housing foreclosure crisis that began in 2008 has resulted in a wider economic recession, a bust in the construction business, and a plunge of housing prices for many cites.

Many urban scholars suggest the world is in the midst of the Third Urban Revolution, a complex phenomenon that began in the middle of the 20th century and is marked by a massive increase—in both absolute and relative terms—in urban populations, the development of megacities and giant metropolitan regions, and the global redistribution of economic activities. As former manufacturing cities decline, new industrial cities, service-sector hubs, and tech-poles emerge elsewhere. The cities of the United States and Canada embody the dynamic and challenging trends of this Third Urban Revolution (fig. 2.1). In both countries, a rising percentage of the population resides in cities. In 2010, 82 percent of the U.S. population lived in urban areas. As of 2006, 81 percent of the Canadian population lived in urban areas with half of this population in the three largest urban areas: Toronto, Montreal and Vancouver. Nearly three out of four of North Americans live in cities, and experts predict that by 2030 this percentage could be as high as 87 percent. Without a doubt, urbanism is the norm in the United States and Canada (hereafter, "North America" in this chapter). North American cities are being transformed by this third revolution. Central cities have characteristically become sites of new urban spectacle: inner cities are peppered with sites of gentrified renaissance as well as rampant poverty and criminality; inner suburbs are showing signs of decline; and exurban development continues apace as gated communities and mixed-use developments sprawl into the former countryside. The new lexicon that has emerged to describe many North American cities—"post-modern," "global," "networked," "hybrid," "splintered"—offers some hint as to the rich complexity and deep contradictions of the Third Urban Revolution. Yet, much remains to be said and done before we can make sense of the new forms of urbanism that characterize the 21st-century North American city.

Table 2.1   Megalopolitan Areas of the United States and Canada

Area	Anchor Cities
Cascadia	Vancouver, Seattle, Portland, Eugene
NorCal	San Francisco, San Jose, Oakland, Sacramento
Southland	Los Angeles, San Diego, Las Vegas
Valley of the Sun	Phoenix, Tucson
I-35 Corridor	Kansas City, Oklahoma City, Dallas, San Antonio
Gulf Coast	Houston, New Orleans, Mobile
Piedmont	Birmingham, Atlanta, Charlotte, Raleigh
Peninsula	Tampa, Orlando, Fort Lauderdale, Miami
Midwest	Chicago, Madison, Detroit, Indianapolis, Cincinnati
Northeast	Richmond, Washington, Philadelphia, New York, Boston
Golden Horseshoe	Toronto, Hamilton, Buffalo

Source: Adapted from the Metropolitan Institute at Virginia Tech.

Large city regions are emerging as the new building blocks of national and global economies. The largest one in the United States is the urbanized Northeastern seaboard, a region named Megalopolis by Jean Gottmann. Megalopolis stretches from south of Washington, D.C., north through Baltimore, Philadelphia, and New York to north of Boston. It is responsible for 20 percent of the nation's gross domestic product. In 1950 Megalopolis had a population of almost 32 million people. By 2010 the population had increased to well over 50 million.

North American megalopolitan regions, conceptually modeled after Gottmann's Megalopolis, are defined as clustered networks of metropolitan regions that have populations of more than 10 million. They comprise at least two contiguous metropolitan areas. A megalopolitan area is created when big cities merge into expansive urban networks. In North America there are 11 megalopolitan regions (table 2.1). Collectively they constitute only 20 percent of the nation's land surface yet comprise 67 percent of the population, and approximately three-quarters of all predicted growth in population and construction from 2010 to 2040.

Yet another term to describe the ways that global investment, sophisticated communications, and widespread corporate and personal mobility are transforming cities in North America into a new type of urban form is the *global city-region*. In an era of globalization, cities increasingly function as the nodes of the global economy; and it is more appropriate to see cities or networks of cities in their regional (rather than local) contexts. As a result of globalization, city-regions such as Los Angeles, San Diego, Seattle, and New York are making a transition from national or regional economic capitals to more integrated cities of the world. Toronto, which has traditionally been the focus of an economic and cultural East-West Canadian economy, is increasingly turning its attention to the North-South economic opportunities created by the North American Free Trade Agreement. Distinguishing features of global city-regions are that they tend to be socioeconomically elite, globally oriented corridors that contrast sharply with disadvantaged, insular residential cities.

Figure 2.2   Skyscrapers, such as the Wrigley Building in Chicago, became the cathedrals of urban commerce as steel-frame construction and the elevator allowed architects to design ever taller buildings. (Photo by Donald Zeigler)

There are numerous transformations occurring in contemporary North American cities that contrast with long held myths about them. One myth is that cities in the United States and Canada lack historic character. Yet, many cities, especially Savannah, Charleston, Boston, Montreal, and Quebec City, have active preservation and restoration programs. Another myth generalizes about the form and function of North American cities—the ubiquitous skyscrapers, freeways, shopping malls, office parks, and bland suburban "boxes"—are often cited as "distinct" to North American cities (fig. 2.2). In the age of globalization, however, these features can be found in cities throughout the world.

URBAN PATTERNS AT THE REGIONAL SCALE

U.S. and Canadian cities vary tremendously in size, form and fortune. In the U.S., recent urban growth has been robust in western and southwestern cities, while some cities in the East and Midwest such as Detroit and Buffalo have seen economic and demographic decline. In Canada, the fastest growing metropolitan areas are in Calgary and Edmonton. At the local level (within urban regions), there has been a deconcentration of population in central cities, and growth and development in suburbs and exurban zones (*i.e.,* former rural areas).

One of the distinguishing features of urban North America is the tremendous range of sizes. The New York metropolitan region has 22 million people, Los Angeles has 18 million, and Chicago has 10 million. Metropolitan regions of 5 million or more include Miami, Atlanta, Toronto, Houston, Washington, D.C., and Philadelphia. Regions around two million inhabitants include Denver, Portland, Baltimore, Vancouver, and Ottawa; and those around 1 million include Tulsa, Nashville, Calgary, Winnipeg, and Edmonton. Historical Geography of Urban Development Cities of North America are rather recent developments in terms of world urban history. Almost all cities are less than 300 years old; many developed only in the last one hundred years. The contemporary North American city is primarily the product of industrial developments of the 19th and 20th centuries.

Colonial Mercantilism (1700-1840)

Beginning in the late 16th century, the British, French, Spanish, and Dutch established colonies in eastern North America. Colonial mercantilism was the result of attempts to keep the European states prosperous. Each European power in North America exercised various controls over commerce and industry and intervened in regional markets. These regulations resulted in an export-based market. For example, the production of American commodities such as sugar, timber, and other staples was designed to satisfy Europe's changing patterns of consumption. In the colonial era, cities were very small in both population and physical size. They served as trading centers that were essentially export centers for raw materials including fish, furs, timber, and agricultural products destined for Europe. The largest cities during this era were found along the Atlantic coast (*e.g.,* Boston and

Philadelphia) and along rivers (e.g., Quebec City and Montreal). The St. Lawrence River's gateway at Montreal controlled the northern route into the center of North America, and the Mississippi River's gateway at New Orleans controlled the southern route into the continent.

The growth of cities during the colonial era was greatly influenced by the types of exports from the region. Quebec City was founded by Samuel de Champlain in 1608. The French settlement was at first sparsely inhabited, serving mainly fur traders and missionaries. The French settlers established relations with the Algonquins, who traded beaver pelts in return for metal knives, axes, cloth, and other goods. Beaver pelts were highly prized and expensive in European markets, thus fueling the continued hunting of beavers. The physical layout of Quebec is typified by narrow, winding streets with a city wall, complete with watchtowers, built by the conquering British.

The city of New Orleans owes its origins and economic rationale to the Mississippi River. French traders, Jesuit priests and functionaries traveled along the Mississippi in search of pelts, converts and allies. The city of New Orleans was founded by a French merchant company in 1718. The attempt to build a trading city close to the mouth of the river encountered the watery geography of a giant delta, half marsh, half mud, a floating, spongy raft of shifting vegetation. The city was located 120 mi (73 km) from where the river flows into the Gulf of Mexico, at a bend in the river close to Lake Pontchartrain, a site that enabled the portage of goods from the lake to the city. It was easier to ship goods to the lake and transport them to the city than to sail up the ever-shifting Mississippi River. The city was an outpost of the French empire, part of a global network of colonial possessions that stretched from the Americas to Africa and Asia. And yet the city grew slowly. A 1764 map shows that one-third of the blocks were empty. One distinguishing feature, however, was the imposition of the French tradition of "long lots" fronting on the river, rather than irregular or square parcels of land.

In contrast to the French influence in New Orleans and Quebec City, Philadelphia was initially designed on a grid system, with a series of four large market squares. Philadelphia was founded in 1681 as an English Quaker settlement at the confluence of the Delaware and Schuylkill rivers. The city's founder, William Penn, designed the plan in reaction to the disorder of his hometown, London. His symmetrical, orderly plan became the template for Philadelphia's later growth. The city became a busy shipping port, for both external goods (feed, food, and tobacco destined for England) and internal products such as rifles and Conestoga wagons. It was also a major banking center and home to the first U.S. stock exchange (1790).

For much of the colonial era, cities were "walking cities." They rarely covered more than a few square miles or had more than 100,000 people. The outward expansion of many North American cities was limited by elements of their geographic sites, including water features and topography. In some cities, the high ground was avoided because it was difficult to pump water or to get horse-drawn fire services uphill. Early forms of public transport, primarily carts and carriages, also found steep hills hard going. Economic growth would later bring new forms of transportation, and cities would begin to grow outwards and upwards.

Industrial Capitalism (1840–1970)

The era of industrial capitalism marks the transformation of the U.S. and Canadian economy from ones based on trade in natural resources to ones that processed raw materials and manufactured products. An industrial economy is one dominated by mechanized factory production. In 1800, approximately 7 percent of the U.S. population was urban. By 1900, more than 40 percent of the population was urban. Urban growth went hand-in-hand with the industrialization of the economy. The economic foundations of the industrial city were the coal mine, the vastly increased production of iron, and the use of reliable mechanical power—the steam engine. All of this industrialization was accompanied by an unprecedented population increase in the numbers of cities and the subsequent enlargement of urban areas. By 1830, New York, Philadelphia, and Baltimore were the main industrial cities in the United States, while Toronto and Montreal dominated industry in what would become Canada. Interior urban development was also fueled by industrial expansion. Cities located along rivers and lakes took advantage of advances in transportation technology, such as canals and railroads, to become important hubs in the distribution of goods (fig. 2.3). By the 1860s, Buffalo, Pittsburgh, St. Louis, Chicago, and Cincinnati emerged as key gateways. In Canada, Winnipeg became the hub of rail service for the west, and Edmonton and Calgary emerged as major regional service centers by the 1870s.

From 1885 to 1935, the U.S. economy completed its transformation from an agricultural and mercantile base to an industrial-capitalist one. In the early 20th century, the emergence of powerful national corporations and large-scale assembly-line manufacturing prompted robust economic growth. While many of the biggest cities were still located in the Northeast, Midwestern cities such as Chicago, Detroit, and Cleveland grew into vital industrial centers by the 1920s. Also during this time Canada's economy experienced

Figure 2.3   The Erie Canal, passing through downtown Syracuse, New York, was critical in helping to establish New York City as the leading port and city of the United States.

major transformation and urban growth in the West, spurred by manufacturing growth in Canada and the rapid growth of the petroleum and natural gas industry in the urban centers of Calgary and Edmonton.

Industrialization was more than just the proliferation of factories. Many key inventions changed the look of cities and transformed spatial patterns. The use of iron, then steel, in construction launched the era of skyscrapers. In the 1880s, the electric street-trolley helped to make mass transit possible and laid the foundations for 20th-century suburbanization by allowing people to live farther away from city centers. Consequently, most industrial cities grew outwards at the edges as well as upwards in the center.

The end of World War II marked a significant turning point for cities in North America. Many U.S. and Canadian national corporations merged or expanded into large multinational

corporations and achieved dominance in the North American market. The U.S. emerged as the world's largest and richest economy. The late 1930s also began the era of widespread automobile ownership and suburbanization. Reliance on private vehicles and weak investment in public transportation resulted in low population densities and increased sprawl, particularly for cities in the West. Cities such as Los Angeles, San Diego, Houston, Phoenix, Dallas, Denver and Vancouver grew horizontally as much as vertically. From 1950 onward, the United States simultaneously became more urban and more suburban. While urban regions continued to grow, there was an exodus of people moving out from the core to the surrounding regions. To lure suburbanites back to the city, municipalities undertook immense urban renewal and infrastructure projects that included highways, bridges, and civic centers. By the late 1960s, however, three significant changes materialized that would have profound changes on cities throughout North America: globalization, deindustrialization, and decentralization

Postindustrial Capitalism (1975-present)

We have entered into an era shaped by global capitalism. By the 1970s, many corporations were moving out of North America to developing countries where lower labor costs and tax breaks promised higher profit margins and larger market shares. In cities such as Pittsburgh, Syracuse, Buffalo, Akron, Cleveland, and Detroit, companies fired or relocated workers, closed factories, and moved out of the region or country. These cities were transformed from "industrial" to "Rustbelt," from vibrant manufacturing centers to ghost towns of despair. Even growth cities such as Los Angeles and San Francisco struggled to cope with the social and economic consequences of a decline in manufacturing-based employment. This decline marked a critical shift in the North American economy. Michael Moore's 1989 documentary *Roger and Me* chronicled the massive job losses and factory closings in Flint, Michigan, home to General Motors. General Motors laid off 40,000 people in Flint between 1980 and 1989, a figure comprising 50 percent of Flint's GM workforce and one of the largest layoffs in American history. Flint continues to face hard times. In 2010, Flint's unemployment rate was about 13 percent, much higher than the national average. For many industrial cities, high unemployment rates continue to impact local economies. All the while, political and economic leaders generate strategies to bolster employment and investment, especially in sectors such as services and tourism.

At the same time that many cities in the United States saw economic decline, others experienced rapid growth. Cities such as Seattle, Orlando, Miami, Phoenix, and San Diego successfully blended an existing industrial base with an expanding service sector. Newer city-regions, such as Atlanta, Charlotte, Dallas-Fort Worth, and Silicon Valley (an urban techno-pole between San Francisco and San Jose), came into their own at this time. Silicon Valley is home to Apple and Hewlett-Packard, leaders in the high-technology sector that emerged after the 1980s.

Many cities also became economically and demographically decentralized during this era. Decentralization occurs when city centers lose either population or jobs. A decline in the tax base, which limits city funding of social services, results in disinvestment in education and infrastructure. City residents and jobs leave the center of the city for the suburbs or other metropolitan and even non-metropolitan areas. 1970 marks the first time there was an actual decline in central-city populations, and by 1980 the trend was intensifying. Most people were relocating to the suburbs.

The rise of the service sector has also played a critical role in urban development. Wholesale and retail trade, finance, insurance, real estate, information and communication technologies, education, and medical services, among others, have replaced manufacturing employment as key components of urban economies. The service sector is highly diverse. Some jobs are low-level and pay at or slightly above minimum wage; these jobs include data entry, cleaning services, and retail (salesclerks at The Gap, or waitstaff at a restaurant). Other jobs—often referred to as quaternary activities—generate far higher wages. These include research and development, brokerage services, banking, medicine, law, advertising, computer engineering and software development. Richard Florida has referred to those employed in these types of service sector jobs as the "creative class." Cities have expended tremendous efforts to attract the "creative class" to their cities (box 2.1).

Box 2.1 "Creative" Cities in North America

As the Fordist economies of mass production and heavy industry declined in most North American cities by the 1980s, municipal governments began looking for ways to compete in a global economy that demanded proficiency in high-technology industry, knowledge-based services, and the creative arts. In a theory popularized by Richard Florida, many geographers and economists argued that the key to growing these economic sectors was attracting a mobile pool of talented, highly educated individuals. To do this, cities would need to invest not only in knowledge-based sectors, but also in the type of urban environment that would ostensibly attract the "creative" individuals to work in those sectors. The "creative city" theory has since acted as a powerful force in the urban development schemes of U.S. and Canadian cities. In order to align with the "3 Ts" of talent, technology, and tolerance, many cities have attempted to fashion themselves as diverse, bohemian cities with a wide array of entertainment, dining, museums, galleries, and other amenities.

In many cities, municipal governments and economic development corporations have employed plans to make their cities "cool" and "creative." Toronto, for example, has developed both a "Culture Plan for the Creative City" and a "Creative City Planning Framework" to capitalize on its momentum in the arts and sciences and move from a "second-tier city" a to "world city." The city has invested heavily in the Art Gallery of Ontario, the Royal Ontario Museum, the *Nuit Blanche* arts festival, waterfront redevelopment, and other projects intended to position culture as a "fourth pillar" of development. In Michigan, the state's housing development authority has launched the "Cool Cities Initiative" to promote downtown redevelopment projects, loft condominium construction, film festivals, and other projects in post-industrial cities ranging from Detroit to Ann Arbor.

The creative city theory has recently been critiqued by geographers as a fuzzy concept that promotes elitism and misguided development as much as economic growth. Although many cities have a renewed interest in downtown living and the cultural components of urban life, critics worry that the creative city theory promotes development for one highly educated, wealthy, mobile class of people while leaving less privileged groups behind. Others have claimed that smaller cities, which often lack the resources to invest in arenas, galleries, technology parks, and other staples of the "creative economy," can never truly compete as creative cities. Indeed, the premise of the creative city may need some redefinition. In the context of a volatile global economy, rising unemployment, and growing environmental pressures in urban areas, cities may need to focus on measures to improve livability and sustainability. Improving housing stock and public transportation while reducing the ecological footprint may ultimately prove to be better investments that constructing the next arena or condominium development.

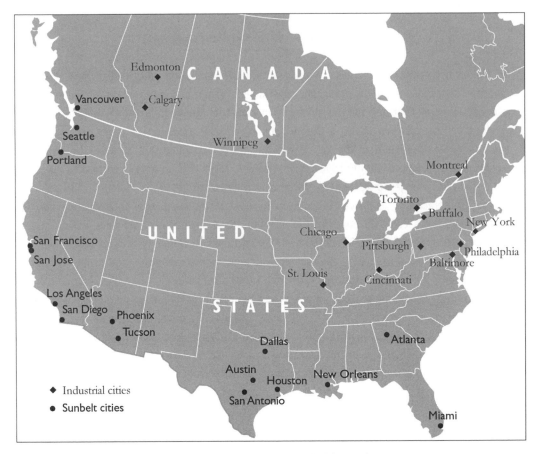

Figure 2.4   Industrial and Sunbelt Cities. *Source:* Compiled by authors

The Sunbelt, a region that stretches from Florida to California, has seen the most dramatic growth in the past thirty years. The term *Sunbelt* refers less to an absolute geographic area and is used as shorthand to describe cities in the South, West, and Pacific Northwest that have seen increased economic growth, population, and symbolic importance in the urban hierarchy (fig. 2.4). Sunbelt cities have been successful primarily because their economies were more diverse and did not rely on traditional manufacturing (fig. 2.5). The Sunbelt region grew primarily as a result of the expansion of the service sector and benefited from significant government investment in aerospace (Houston, Seattle, Los Angeles), petrochemicals (Houston, Dallas) and information technologies (Silicon Valley, Research Triangle Park, North Carolina).

MODELS OF URBAN STRUCTURE

Although no single model completely explains why things are located where they are within cities, we can generalize about the region's patterns of urban land use (box 2.2). There are two general features that characterize most North American cities. The first is the grid system. The second is sprawl or the horizontal spread

Figure 2.5   In the Sunbelt suburbs, golf carts are "second cars" for many. The Villages, an hour north of Orlando, Florida, is an affluent, planned suburban retirement community surrounded by golf. (Photo by Kathy Schauff)

of development from the central city. This is different from many other world regions where cities have remained densely developed.

The average city in the United States and Canada has a land-use pattern that breaks down as follows: residential (30 percent); industrial/ manufacturing (10 percent); commercial (4 percent); roads and highways (20 percent); public land, government buildings, and parks (15 percent); and vacant or undeveloped land (20 percent). The layout of these various land-use categories differs according to the age of the city. For example, cities established prior to 1840 tend to have very dense or compact cores. In cities that developed later during the 20th century, industrial activities might be located outside the core area to take advantage of advances in transportation facilities such as railroads. Still other cities that saw develop-ment occur after 1950 have more deconcen-trated cores with expansive residential zones due to the automobile and the emergence of

suburbs. The majority of North American cit-ies, however, have both high-rent and low-rent residential areas in the inner core, and moving outward into the suburbs the price (and size) of single-family homes tends to increase. Pat-terns of expansion and land-use often follow the concentric-zones model.

The grid plan, which imposes a rectangular street grid on urban space, dates from antiq-uity. Some of the earliest planned cities were built using grids. The grid provides a simple, rational format for allocating land by setting streets at right angles to one another. Despite a multitude of geographies, topographies, altitudes, and latitudes, many cities laid out on a grid share common design features: a lack of sensitivity to the physical environ-ment, the imposition of the grid regardless of topography, a focus on straight lines (geom-etry over geography), and an underlying sense of the ability to control urban space. The 1734 map of Savannah shows the rigid adoption

Box 2.2 Using GIS to Engage with the Community

Urban geography has been revolutionized by Geographic Information Systems. Computerized spatial analysis or GIS techniques allow geographers to interpret spatial patterns and trends within and across urban areas. GIS is used to analyze segregation (including housing, immigration, gentrification), flows and linkages (commuter flows, migration), politics (gerrymandering), locations for economic activities (such as retail, manufacturing and services), and to define the urban spatial structure such as rural-urban boundaries, suburbanization, and land-use change. GIS can also be used to explore the changing structure of urban areas and particularly the urban periphery. Land-use–change maps, which show the transition of open areas and farmland to housing developments, are useful in determining patterns of suburbanization.

Using GIS, geographers can also explore the internal structure of cities. For example, the location of hazardous waste facilities in relation to low-income or minority neighborhoods can take advantage of GIS to study the patterns and processes of the natural environment and how contaminants influence the quality of life. Another example is using census data and census tracks to show various demographic changes over time. These could include the change in income, ethnicity, home ownership, and rates of poverty.

For students of urban geography, using GIS can reveal unexpected spatial relationships. The map below was made by undergraduate students in an urban geography course at George Washington University as part of a service-learning project. This service-learning project challenged students in the course to map critical social services such as soup kitchens, shelters, and other services and to deliver a set of maps for a non-profit community partner, So Others Might Eat. Students divided into teams, gathered data in the field using GPS devices, and collectively made a range of thematic maps for the clients of So Others Might Eat. The composite map in figure 2-6 shows that social services tend to be concentrated in the downtown area; however, most of the poorest areas of D.C. are to the east of the downtown. The techniques used to map D.C. neighborhoods are fairly simple; but this type of project shows that even basic GIS skills can be used to enhance student learning about cities and, at the same time, create a tangible benefit to the community. Students commented that this type of GIS project helped them to see the city in a new and better informed way.

of the grid (fig. 2.7). Similarly, San Francisco imposed a grid on what must be considered among the most dramatic topographies of any North American city.

Due to the influence of European culture, the grid plan was nearly universal in the construction of new towns and cities in the United States and Canada. Adopting the grid plan, however, also allowed for the rapid subdivision of large parcels of land. As U.S. cities have grown outward, particularly after the mid-20th century, the grid has become less prevalent. The suburbs present a more organic pattern of growth, often designed with cul-de-sacs and winding lanes. As a result, many suburbs provide a purposeful contrast to the grid-plan

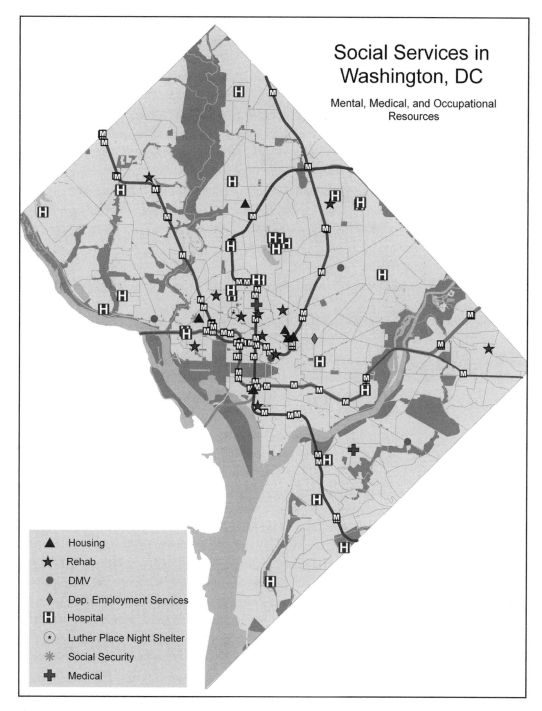

Figure 2.6   Map of Social Services in Washington, D.C. *Source:* Lisa Benton-Short.

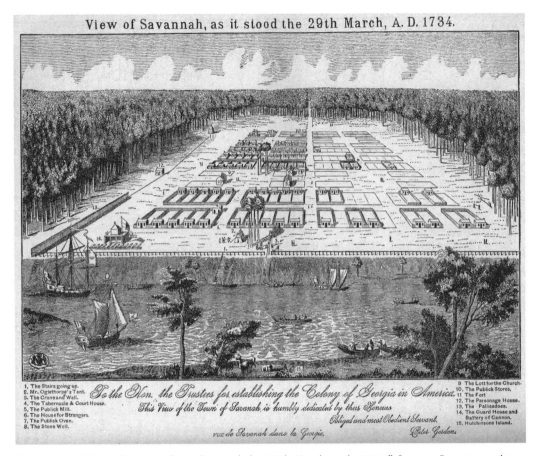

Figure 2.7   "View of Savannah, as it stood the 29th March, a.d. 1734." Source: Report on the Social Statistics of Cities, Compiled by George E. Waring, Jr., U.S. Census Office, Part II, 1886. Courtesy of the University of Texas Libraries, University of Texas at Austin (http://www.lib. utexas.edu/maps/historical/savannah_1734.jpg)

city. When cities and suburbs merge, the grid merges into a series of loops, curves, and open spaces; and its rigidity begins to disappear.

In contrast, some recent suburbs have returned to the grid system, under the influence of New Urbanism. New Urbanism nostalgically longs for lost community and the older high-density cities of the past. New Urbanist design principles emphasize walkability, mixed-use "town centers" and higher-density residential areas. Ironically, New Urbanism, which is a response to suburban sprawl, has had less impact on redesigning cities than it has on the redesign of suburbs. The most cited example

of New Urbanism is the town of Celebration, Florida, a community initially planned and built by the Disney Corporation. Celebration's design elements include low-rise, high-density residential areas where garages are at the back of the residences, walkways and porches that allow pedestrian movement, and a vibrant, car-free, mixed-use downtown.

The emergence of *edge cities* provides yet another model of urban structure. Edge cities consist of predominately large-square-footage office space located beyond the central city. The journalist Joel Garreau coined the term "edge cities" and defined them as having:

- more than five million square feet of office space, enough to house up to 50,000 office workers (as many as some traditional downtowns).
- more than 600,000 square feet of retail space, the size of a medium shopping mall.
- more jobs than bedrooms.
- been nothing like a city before 1960.

These edge cities attract large numbers of service-sector workers during the day, but empty each night as residential areas are scarce. Garreau identified 123 places as being true edge cities, including two dozen such areas in greater Los Angeles, 23 in metro Washington, D.C., and 21 in greater New York City. Tyson's Corner, Virginia, west of Washington, is an example of an edge city. Other examples are found in Dallas-Forth Worth, Orlando, and Atlanta.

Edge cities, however, may become a phenomenon relegated to the 20th century. Since edge cities are built in and around major highway intersections, traffic congestion has become a problem. Pedestrian access is poor and public transportation is nearly absent. Ironically, edge cities may stimulate redevelopment of the downtown core as people seek to leave their cars and commute by public transportation. Additionally, as development continues in and around edge cities, they may "merge" into megalopolitan areas with nodes of residential areas and nodes of commercial/business activity.

THE CHANGING CHARACTER OF U.S. AND CANADIAN CITIES

Globalization and the Urban Hierarchy

A major factor underlying urban change today is the tie of cities to global trends. The past two decades have seen intensified competition among cities at the global scale. Globalization has restructured cities spatially. In the most competitive and successful world cities, new financial districts, luxurious residential areas, and unprecedented property booms are indicators of the benefits of globalization and a competitive position in the urban hierarchy. Almost all cities are impacted by globalization, but not all cities become world cities. For example, many urban boosters no longer focus solely on recruiting or attracting domestic firms, but attempt to secure investment internationally. The presence of international banks, department stores, and other retail establishments provides a visual cue that most large cities now bear the imprint of globalization.

Consider the changing urban location of *Fortune 500* company headquarters. In the 1950s and 1960s the largest cities in the Northeast and Midwest, including Chicago, Boston, Philadelphia, Pittsburgh, and Toronto, were home to the world's largest industrial companies. In 1960 New York was home to six of the top ten Fortune 500 headquarters including Standard Oil, General Electric, U.S. Steel, Mobil Oil, Texaco, and Western Electric. Today the number of *Fortune 500* corporate headquarters in New York has fallen by half. Firms such as General Electric and Xerox have moved to the suburbs or the Sunbelt. Recently, Seattle-based Boeing decided to move its headquarters to Chicago, while Volkswagen announced it would leave Detroit and move to Washington, D.C. The biggest growth in corporate headquarters is occurring in Orlando, West Palm Beach, Greensboro, Atlanta, Dallas, and Houston. This changing geographic distribution is the result of many factors: the relocation of a company, the rise and fall of local firms, or the merging of companies. The lower costs for office space and housing in medium-sized cities such as West Palm Beach and Greensboro provide another reason for a company to relocate. Most firms, however, have chosen metropolitan areas of at least one million.

Because the key globalization arenas have been concentrated in Europe, North America, and East Asia, cities in these regions tend to dominate the global urban hierarchy. Cities such as Miami, Phoenix, San Diego, Los Angeles, San Francisco, Washington, D.C., Toronto, and Vancouver have become key urban centers in the global economy and the North American urban hierarchy. Some cities have benefited from globalization and have eclipsed their rivals. For example, many corporate headquarters have moved from Montreal to Toronto, which has become the major conduit between Canada and the international capital markets, while Calgary and Edmonton serve as crucial links to the more specialized international petroleum industry. Quebec City has also transitioned to a more diverse, post-industrial economy. The North American Free Trade Agreement of 1992 has resulted in an increase of exports from Quebec City to the United States. In addition to a fairly robust aerospace industry, tourism, information technology, and biotechnology compose some of the economic growth sectors for Quebec City.

Other cities have fallen down the hierarchy. Detroit, Cleveland, Buffalo, and Pittsburgh have experienced a decline, as factories have closed. These cities struggle to compete for coveted global linkages and networks that promise to reinvigorate their economies. In Canada, Thunder Bay, St. John's, and Halifax contend with the challenges of an urban economy based on natural resources.

There are various articulations of the urban hierarchy. Some cities compete for financial command functions—stock markets, banks, multinational corporate headquarters, and other forms of capital exchange. New York is the most important city in this regard, followed by Chicago and Toronto. Other articulations include multi-cultural command centers such as Vancouver, Miami, and Los Angeles, which make important linkages to other regions through their immigrant populations. Some cities have found a niche in the hierarchy by establishing important air route connections through the global airline network. These cities include Toronto, Los Angeles, Seattle, Memphis (home to Federal Express), Atlanta (home to UPS), and Anchorage, a stop along military and cross-Arctic air routes. Some cities are resurrecting their place in the urban hierarchy. Cleveland and Pittsburgh, which experienced economic decline and massive job losses associated with deindustrialization, have successfully realigned their economies with global markets. While some of the old industrial companies remain, revamped economies in these cities now capitalize on health care facilities that have gained national reputations. To some extent, Cleveland and Pittsburgh have moved back up the hierarchy—albeit in a more specialized role.

Cities and International Spectacles

Emerging as a world-class city is one of the main goals of hosting an international event such as a World Cup Soccer tournament, a World's Fair, or the Olympic Games. In some cases, cities purposefully use international spectacles as a way to showcase themselves to a global audience. The city becomes more physically connected to the rest of the world while positive images circulate through the mass media. Hosting an international spectacle allows the city to achieve global recognition and the possibility of increased tourism and investment.

New York City, for example, hosts "Fashion Week" biannually to provide an opportunity for the world's best fashion designers to showcase both their fall and spring collections. Beginning in 1933, Fashion Week was housed in tents at Bryant Park. As attendance grew, however, the city relocated Fashion Week to Damrosch Park at the Lincoln Center complex—also home to the Metropolitan Opera and the American Ballet Theatre. In 2010, its first year at Lincoln Center, more than 100,000 attendees entered on a red carpet, while behind the scenes a lucky handful of interns from city fashion schools such as Parsons School of Design, toiled as volunteers. Local fashion fans, unable to attend in person, tuned in to watch full coverage on local channel TV 25, which devoted more than 150 hours to covering the event.

The Olympic Games in particular have a wide global television audience, and hosting the Games is highly coveted. Recently, New York City's bid for the 2012 games was unsuccessful (they were awarded to London) and in 2010 Chicago lost out to Rio de Janeiro to host the 2016 Summer Games. Still, North American cities have served as Olympic host cities almost as often as those in Europe, reflecting their reputation and influence in the global urban hierarchy.

Hosting a major event such as a World Cup, a World's Fair, or the Olympic Games involves not only a creative vision but also a physical restructuring of the city and an opportunity for urban renewal. Such changes typically require a major investment in infrastructure. The construction of an Olympic Village, projects such as new roads and sewer systems, and the creation or improvement of historic parks and plazas link the Games with urban renewal.

While considered a general success, Vancouver's 2010 Winter Olympics were in many ways plagued by their geography. Whereas most Winter Olympics have been held in cold, alpine or high-latitude climates, Vancouver and its surroundings are in a milder and moister climate zone. Many found it unsurprising, then, when the ski slopes at Whistler had not received any snowfall during the month leading up to the games. Although snowmakers were able to produce the snow needed for daily events, rain delays and sludgy slopes prompted many to rename Vancouver's so-called "Green Olympics" the "Brown Olympics." Vancouver's sea-level elevation (several thousand feet lower than the last Winter Olympics in Torino, Italy) also played havoc with some events. Speed skaters, used to training and competing in high-elevation, low-oxygen environments, recorded record times well below those of the previous Olympics.

The unique topography of the area also created some transportation challenges. The most difficult of these was getting competitors and staff for the ski events from Vancouver to the town of Whistler—an elevation change of over 2,000 feet (609 meters) in a distance of 76 miles (123 kilometers). The route between the two places, known as the Sea to Sky highway, required extensive improvements to accommodate the traffic. The road, known for its unstable weather conditions and frequent accidents, received wider lines, electronic weather monitoring systems, stronger bridges, and new sets of passing lanes. The improvements were expected to raise British Columbia's GDP by over $300 million CDN between 2010 and 2015.

Much of the development associated with the 2010 Olympics, however, was more contentious. In a city known for both extremely high housing costs and endemic homelessness, drug use, and poverty on its lower East Side, many of the residents felt that the construction of an athlete village and numerous venue renovations were a poor use of government funds. Recurring protests in the years leading up to the

Figure 2.8   Vancouverites protest the use of public funds for the development of the Olympic Village—which they did not see as social housing. (Photo by Elvin Wyly)

games ultimately forced the Vancouver Organizing Committee to refrain from holding the large public ceremonies intended to drum up excitement and support for the games (fig. 2.8). To prevent large-scale interference during the event, police were heavily armed, and nearly 1,000 closed-circuit surveillance cameras were installed throughout the city. The branding of the Olympics was also criticized. In a country that struggles with its past treatment of Aboriginal peoples, many felt that adoption of the inukshuk (an Inuit symbol) as the Olympic logo and the integration of totem poles, ceremonial dress, and a "mystical" interpretation of nature was an appropriation of native culture. Many argued that the Olympics had ripped Inuit and First Nations cultures from their true contexts and re-packaged them for consumers and competitors.

The large and expensive nature of facility provision raises the issue of post-Games usage of such facilities. In the case of the 1996 Atlanta games, the specially built 85,000-seat Olympic Stadium was converted into the 52,000-seat home of the Atlanta Braves baseball team. The 15,000-resident Olympic Village became student housing for Georgia State University. There were also some public-space legacies. The most notable was Centennial Olympic Park in the middle of downtown Atlanta. The park was handed over to the state of Georgia after the Games and has become a popular public park, a centerpiece of downtown activities that hosts 160 events and receives 1 million visitors a year.

However, not all cities have been successful at converting Olympic venues into vibrant urban centers. In Montreal, the main Olympic Stadium is deteriorating and is widely seen as a monument to a costly experience (fig. 2.9). Sadly this deterioration is not unique; the city was unable to retain its baseball team, the Expos, which relocated to Washington, D.C., in 2005, with the promise of a new state-of-the-art stadium.

Figure 2.9   The Olympic venue in Montreal, built for the 1976 Summer Games, now stands neglected. (Photo by John Short)

International spectacles can be a catalyst for urban renewal, environmental remediation, and improvements to a city's infrastructure that can make a city more competitive on the world stage. They can also reposition the city in the global imagination. A successful event can promote a positive global image of the city and stimulate tourism and investment. Hosting an international event has become an important goal for many urban political leaders around the world because these events provide huge development opportunities with the possibility to change how the city is perceived.

Greening the City

Before the 1970s, ecology had little role in urban planning and landscape design. By the 1980s however, urban planners noted the importance of nature in the city and recognized that ecology has powerful implications for how a city should be designed, built, and maintained. There are numerous examples of

ways that cities have begun to reconnect to the natural world within the urban landscape. Greening the city can involve tree-planting programs, heritage preservation, smart buildings, urban farms and urban forests, ecosystem restoration, bicycle-friendly cities, improved recycling programs, restricting the use of cars, and expanding open spaces. Whether known under the moniker of green cities, sustainable cities, or sustainable urban development, planning in U.S. and Canadian cities today more fully embraces urban ecology.

For many years, cutting-edge architecture and sustainable design have, to a large extent, existed in separate camps. More recently, however, green builders, architects and interior designers have created avant-garde designs for green buildings and, at a smaller scale, green houses. Since buildings consume enormous quantities of the Earth's resources in both their construction and daily operation, they are tremendous opportunities to showcase innovative, eco-friendly design. Green buildings incorporate features that support the conservation of the environment, including roofs and windows designed and oriented to minimize summer afternoon solar heat gain and optimize winter solar heat gain. Some may use solar energy as an alternative to fossil fuels, and green construction materials (produced without harming the environment) can be recycled. Many green builders also select materials made without formaldehyde or toxic chemicals in order to improve indoor air quality. Another example is installing landscaping rather than paved surfaces, which impede storm-water infiltration. Even interiors incorporate materials and products that have high levels of renewability or reusability, such as bamboo flooring or cork tiles. Seattle, Portland, and Vancouver have been lauded as leaders in green buildings; there are over 30 green buildings in Seattle alone. Chicago and Toronto have won numerous awards for advancing green roofs. Toronto initiated a "Green Roofs for Healthy Cities" program and has proposed a series of "green walls" where vegetation will grow on the sides of buildings. In addition to the cities mentioned above, the other cities included among the "greenest" by National Geographic's *Green Guide* are Austin, Boulder, Madison, Minneapolis, Oakland, and San Francisco. Forty U.S. cities today have comprehensive sustainability plans that address a wide diversity of environmental issues such as air pollution, water pollution, parks and open spaces, climate change, urban farming, and community gardens.

Beginning in the 1960s and accelerating in the 1980s, many U.S. and Canadian cities have sought to "reclaim" the waterfront for their urban populations by pursuing waterfront redevelopment. De-industrialization resulted in abandoned warehouses and unused port facilities on city waterfronts. Containerization meant many older port facilities were inad-equate for the new technology and became obsolete. Cities were forced to adjust to new economic impacts and to create new spaces out of old industrial sites. Vacant lands became opportunities. Today, waterfront redevelopment is widespread (fig. 2.10). Many cities have transformed their waterfronts into vibrant, public spaces that attract locals and tourists. Baltimore's Inner Harbor is often cited as a model U.S. waterfront redevelopment project. It has become the city's gathering place: home to the national aquarium, two sports stadiums, hotels, restaurants, museums, high-rise condominiums and hotels. In Boston, Pittsburgh, Toronto, and Vancouver waterfronts have become the new festival spaces filled with sports stadiums, restaurants, and hotels. Even smaller cities such as Ottawa, Syracuse, Buffalo, Savannah, Victoria, Charleston, Austin, and Cleveland have transformed their harbors, lakes, or riverfronts

Redevelopment is not without contestation: how and for what purposes these waterfronts are redeveloped can generate fierce debate and are not without high price tags. The reconstruction of Baltimore's Inner Harbor cost $2.9 billion. Some criticize the social costs as well. The diversion of funds to Baltimore's Inner Harbor contrasts with the city's poor public school system and the perceived decline of many public services. In some cases, waterfront development valorizes the parts of the urban landscape that will boost real-estate interests at the expense of social welfare programs. While Baltimore's Inner Harbor flourishes, many inner-city neighborhoods continue to experience high crime rates, population loss, and housing abandonment.

Despite the substantial costs, waterfront transformations represent a dramatic story of urban rebirth—economically and environmentally. In addition to new hotels and an aquari-

Figure 2.10   The locks at the northern terminus of the Rideau Canal sit on the Ottawa side of the river while both a paper factory (left) and the Museum of Civilization (right) can be seen on the Gatineau side. Confederation Boulevard provides pedestrian access to attractions and historical sites on both sides of the Ottawa River. (Photo by Nathaniel Lewis)

um, Boston's waterfront has been transformed by the eliminating the old elevated portion of Interstate 93. The freeway was demolished and moved underground as part of the well-known "Big Dig." The resulting open space has become the Rose Kennedy Greenway, a linear park. Waterfront redevelopment has helped to restore the centers of cities to economic, social, and ecological health.

DISTINCTIVE CITIES

New York City: A Global Metropolis

New York City is the largest city in the United States and the one most frequently identified as the cultural and financial capital of the country. At 8.4 million people, and over 19 million in the metropolitan area, New York is the only metropolis in North America to rival the megacities of Asia and Latin America in

terms of population. A true "global city," New York is a point from which global and economic trends diffuse outward. Recent events, however, have also shown that New York—like any other city—is far from invincible. The World Trade Center attacks of September 2001, the blackout of August 2003, and the real estate crisis of 2008 and 2009 have shown that life in New York City, regardless of its global connectedness and prominence, can also be brought to a halt by external actors.

Despite these setbacks, New York continues to lead the world in commerce and industry. The port of New York/New Jersey, once the main connection between the Atlantic Ocean and the St. Lawrence Seaway (via the Erie Canal), now handles the third-highest tonnage in the United States and is counted among the world's top-20 busiest ports. New York—along with Los Angeles—continues to dominate the United States manufacturing-based industries with over 20,000 establishments, but New York has especially thrived in the so-called "creative industries" that have gained popularity in the past two decades. These sectors tend to be spatially concentrated in different portions of the city, with music and theater in Times Square, fashion in the Garment District, interior design and architecture in Chelsea and SoHo, and advertising on Madison Avenue. These industries often draw from the talent pools of nearby educational institutions, such as Julliard (music), the Pratt Institute (design), and the Fashion Institute of Technology. The city government has also contributed by boosting tax incentives for the television and film sectors.

The city sits in a strategically important location where mainland New York State meets with the Atlantic Ocean, Long Island, southwestern Connecticut, and northeastern New Jersey. This region is commonly called the "tri-state area." The hard metamorphic rock making up Manhattan Island has allowed for the centralized vertical development (*i.e.*, skyscrapers) that characterizes New York. Central Park, a product of the urban parks movement of the Progressive Era, is the only extensive open space on the island of Manhattan. First inhabited by the Algonquin Indians, Manhattan Island was purchased and settled by the Dutch in 1624. By 1800, New York City, with its 60,000 residents, had become the largest city in the country. It has maintained this position since then, forming the center of the Boston-to-Washington conurbation that geographer Jean Gottmann later called the Megalopolis. Along with Tokyo and London, New York is one of the three traditional "global cities" that anchor world trade, commerce, banking, and stock transactions.

Even as New York becomes ever more globally interconnected, the city is in many ways socially, economically, and spatially fragmented. Despite its world city status, New York has always struggled with its reputation for poor sanitation and high crime. Many geographers have argued that these problems, rather than being truly corrected, have merely been suppressed or relocated and confined to particular areas of the city. New York has frequently located environmentally hazardous projects, such as expressways and incinerators in poor and racialized neighborhoods (*e.g.*, South Bronx, Sunset Park) where public visibility is reduced and resistance is less likely. The city has also tried to reconfigure itself as a safe, livable city and tourist destination. Under the administration of Mayor Rudolph Giuliani (1994-2001), New York implemented "quality of life" laws that criminalized panhandling and homelessness while policing and securitizing (*e.g.*, through closed-circuit cameras) highly trafficked and visible public spaces. Even places that previously served as important sites of public protest, such as the steps of City Hall, are now frequently fenced off or guarded.

Gentrification has undoubtedly driven the fragmentation of New York City. Although gentrification is commonly understood as the product of gradual, localized, ground-up improvements to neighborhood properties, much of the gentrification in New York since 2000 has been driven by corporate developers and municipal interests. The New York Urban Development Corporation, for example, has forcibly purchased many Times Square properties for redevelopment. Privately managed business improvement districts (BIDs) are now responsible for many of the public space improvements (*e.g.*, beautification, signage) in New York. As many economic geographers have observed, the competitive advantage of the creative industries located in New York is highly contingent on the "branding" of the neighborhoods in which they are located. The result has been an intra-city competition of neighborhoods striving to be the city's next arts destination or nightlife center and the rapid gentrification of previously decaying neighborhoods such as the Meatpacking District (Manhattan) and Williamsburg (Brooklyn). Although these processes have expanded the living space available for the upper class and the upwardly mobile, they have priced out the middle-class residents of neighborhoods such as Park Slope (Brooklyn), Harlem, and the Lower East side of Manhattan. Critics of gentrification have claimed that rapid development, securitization, and branding of urban space serve to create "Disneyfied" fantasy cities and "entertainment machines" that put tourists and business tenants ahead of residents.

Immigration has been central to both the identity of New York and the individual identities of its neighborhoods. Over 12 million immigrants from Ireland, Germany, Italy, Poland, Greece, and elsewhere arrived in New York City during the late 1800s and early 1900s. During the past three decades, however, immigrants from Latin American and Asia comprised the bulk of newcomers. Despite being one of the United States's most ethnically diverse areas, New York City is still highly segregated. Real estate agents, who serve as gatekeepers to the city's properties, often sort new immigrants toward neighborhoods dominated by their particular ethno-racial group. Such practices, however, reinforce self-segregating tendencies, ethno-racial ghettos (*e.g.*, Puerto Ricans in the Bronx, Dominicans in Washington Heights), and mutual antipathies between groups. New York thus remains a city of extremes. In a relatively small area, extreme wealth meets with extreme poverty, global integration encounters local fragmentation, and individualistic economic opportunity sits side-by-side with the increasing management and regulation of public space. These dichotomies, however, are likely to guarantee New York's place as a fascinating site of geographic study for years to come.

Los Angeles: Fifty Suburbs in Search of a City

Los Angeles is a city that contrasts sharply with New York City. The large, sprawling city covers 498 square miles (1290 sq. km.) has multiple business districts (*e.g.*, Hollywood, Beverly Hills), and depends on complex, often congested, networks of roads to connect the various "nuclei" of the city. The sprawl of the city, typified by concrete structures, superhighways, and low-lying residential and commercial developments (*e.g.*, strip malls) extends north through the San Fernando Valley ("The Valley) and eastward into the "Inland Empire" of San Bernardino and Riverside Counties. With a metropolitan region population of 12.9 million people as of 2009, the city of Los Angeles itself is the nation's second largest, a position that it took from Chicago during the 1980s. Yet like New York, Los Angeles is also a city of extreme contrasts. Both the immense wealth of Beverly Hills and the endemic poverty and disorder of South Central Los Angeles have been fixtures in the U.S. media and the American cultural imagination. A young, politically liberal population in the arts and entertainment sector stands in stark contrast to the Republican families (and the conspicuous consumption featured in the *Real Housewives*) of Orange County. And in a city perhaps less beholden to dominant notions of social class and pedigree than New York City, Boston, or Chicago, gated communities—intended to protect and contain wealth—abound in many areas. Finally, the extent of and the rapidity of human-made development is intermittently interrupted by the natural forces of the region: earthquakes, wildfires, and mudslides.

Originally inhabited by the Shoshone Indians, the area around Los Angeles was settled by the Spanish in the late 1700s and renamed El Pueblo Nuestra Señora la Reina de Los Angeles de Porciùncula (The Town of Our Lady the Queen of the Angels of Porciùnula). The area was Spanish until 1848, when the United States won the Mexican-American War and promptly annexed the town and its environs through the Treaty of Guadalupe Hidalgo. Anglo settlers from the Eastern United States followed by the mid-1800s, but the town did not exceed 2,000 inhabitants until later in the century. Much like the Erie Canal in New York, it was the arrival of the Southern Pacific and Santa Fe railroad lines in 1876 and 1885 that ultimately led to a population explosion. Los Angeles had 100,000 residents by 1900, but played a relatively minor role in the U.S. economy compared with commercial centers like New York and Philadelphia, and industrial complexes such as Chicago, Detroit, Pittsburgh, and Cleveland.

It took yet another development in transportation, the construction of an artificial harbor at San Pedro in 1914, to create a West Coast metropolis that would truly rival New York and Philadelphia. The port, now referred to as Los Angeles-Long Beach, provided an export hub for the West's growing oil industry as well as an entrepôt for cruise ships and imported goods from Asia. As depicted in literary works such as Steinbeck's *Of Mice and Men*, southern California became a destination for thousands of unemployed workers during the Great Depression of the 1930s. These internal migrants, many from Oklahoma, Kansas, and other "Dust Bowl" areas afflicted by drought and famine, came to work in the oil fields, chemical factories, and automobile plants. After serving as the center of the United State's Pacific arena campaign during World War II, Los Angeles had become not only the primate city of the western United States, but the center of a growing West Coast economy with industrial-commercial centers in San Diego, San Francisco, Portland, and Seattle. It was then, after 1945, that Los Angeles would gain equal status as a lifestyle center, characterized by warm weather, suburban housing development, conspicuous consumption, and the influence of the entertainment industry.

Today, Los Angeles is known for being a national leader in manufacturing—especially clothing and luxury goods—and a global leader in aerospace, film, and television. The city's port handles more trade with Japan, Southeast Asia, Oceania, and Latin America than any other U.S. port and Los Angeles International Airport is the third busiest nationwide (after O'Hare in Chicago and Hartsfield in Atlanta). The strong industrial base and transportation networks situated in Los Angeles have also attracted several ancillary sectors, including tourism and conventions, federal government contract work, and medical services including plastic surgery.

Although Los Angeles has a shorter history of international immigration than New York—most of the city's migrants were from within the United States before the 1960s—the city is one of the most diverse in the United States. More than a third of the city's population is foreign-born, with the most recent waves of immigrants coming from Latin America—particularly Mexico—and Asia. The newness of immigration also poses challenges for Los Angeles. Most immigrants are not English-speaking, and both Los Angeles and California have debated whether to continue integrating foreign languages (especially Spanish) into school curricula and signage, or to make English the sole official language. New immigrants in the Los Angeles area also tend to be poor and experience discrimination in housing, employment, and education. While these immigrants help support the quintessentially "southern California lifestyle" by working in service industries ranging from gardening to dry cleaning, they often live in spatially and economically marginalized neighborhoods such as East Los Angeles and Compton.

Los Angeles is also marked by a number of environmental problems, such as traffic congestion and pollution. In this sprawling, automobile-dependent city, most residents commute and more than 7 out of 10 workers drive to work alone. All of the streetcar lines in Los Angeles were closed in 1963 in favor of freeway development. Although the city reinstituted a commuter rail system in 1990, its five lines—designed mostly to transport riders from downtown Los Angeles to the surrounding suburbs—are insufficient to provide an adequate alternative to car transport. The traffic congestion, coupled with an unfavorable basin location and dry climate, has made Los Angeles the smoggiest city in the country. Although smog alerts in Los Angeles have decreased since the 1970s—when there were almost 100 alerts per year—the National Lung Association has consistently ranked Los Angeles as the first or second most polluted city in the United States. Los Angeles also has insufficient water resources to support its population. Transfers of water from northern California and the Colorado River have reached their limits, and when combined with the frequent droughts, the only alternatives seem to be conservation and a turn toward the sea. In 1990, the first desaliniza-

tion plant opened along the California coast. Finally, the ever-present threat of earthquakes is announced by several low-grade tremors each year. The potential for devastation is real, and emergency planning is a priority item for schools, businesses, and police.

Detroit and Cleveland: Shrinking Cities

Overall the U.S. national economy has seen growth and prosperity; however, some cities confront a declining or stagnant economy and a shrinking population. In 1950, the population of municipal Cleveland was 900,000, but by 2010 it had fallen to 397,000. Detroit was once home to both GM and Ford and was dubbed "Motor City" or "Motown." It was home to 1.8 million residents in 1950. By 2010, its population had fallen to 714,000. These two former urban industrial giants have experienced a reversal of fortune.

In Detroit, as high-paying manufacturing jobs became scarce and unemployment high, many residents lost their homes or apartments to foreclosure or eviction. At the same time, unemployment plagued the city, the crack cocaine epidemic of the 1980s and 1990s led to drug-related violence and property crimes, all of which gave Detroit unwelcome notoriety as one of the most crime-ridden cities in North America. Detroit's woes have continued. In 2010 it was estimated that about one-third of the city, some 40 square miles (104 sq km), is vacant. The current joke is that the only expanding business in Detroit is demolition.

Redevelopment has been a buzzword since the 1990s, but redevelopment strategies have garnered mixed results. In the mid-1990s, three casinos opened up in Detroit's downtown. In 2000, Comerica Park replaced historic Tiger Stadium as the home of the Detroit Tigers, and in 2002 the NFL Detroit Lions returned to a new downtown stadium, Ford Field. The 2004 opening of "The Compuware" gave downtown Detroit its first significant new office building in a decade. The city hosted the 2005 Major League Baseball All-Star Game and Super Bowl XL in 2006, both of which prompted more improvements to the downtown area. Currently Detroit is constructing a riverfront promenade park similar to the one directly across the Detroit River in Windsor, Ontario, replacing acres of train tracks and some abandoned buildings with several miles of uninterrupted parkland.

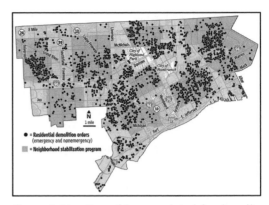

Figure 2.11   Map of houses slated for demolition in Detroit, Michigan. *Source:* Lisa Benton-Short.

Yet, new infrastructure has not necessarily resulted in improved economic growth. Detroit remains one of the nation's poorest cities. In 2009 more than one-third of residents were living below the poverty line, and the multiracial population (76 percent African Americans, 17 percent White, and 7 percent Hispanic) was highly segregated. Abandoned housing ranks as one of the city's most persistent problems. In 2010 a total of 78,000 housing units were vacant or abandoned; of that number 55,000 were in foreclosure (fig. 2.11). Detroit, which was already hit hard by abandoned housing in the 1980s and 1990s, is one of the U.S. cities hit hardest by the 2008 foreclosure crisis (box 2.3). With so many vacant properties, Detroit's mayor set a goal of demolishing 10,000 vacant houses by 2013.

Cleveland has also had to struggle with the legacy of deindustrialization to reinvent itself in the more competitive global economy. Initiatives to rebuild Cleveland have replicated the formula that many cities have employed: new museums, sports stadiums, convention centers, the renovation of old industrial warehouse districts for housing and retail, and waterfront development. Pundits dubbed these efforts "the Cleveland Comeback." One of the most successful projects has been the

Rock and Roll Hall of Fame and Museum, which opened to the public in 1995. The building, located on the shore of Lake Erie, was crucial in the redevelopment of Cleveland's waterfront area. New downtown sports stadiums for the professional teams have also aided revitalization. The Gateway sports complex cost $360 million and included an open-air stadium for baseball, and an indoor arena for basketball. Currently, the city is redeveloping the waterfront along both Lake Erie and the Cuyahoga River as a destination for tourists and locals alike. The city has also become a regional and national player in health services by capitalizing on the wealth of educational and medical facilities in the region to produce economic growth. Both the Cleveland Clinic and University Hospitals have announced billions of dollars in investment in new facilities, such as a new heart center for the Clinic, as well as a cancer center and a new pediatric hospital. Despite these efforts, some experts claim that the Cleveland comeback has stalled. Between 2000 and 2007, Cleveland suffered one of the largest proportional population losses in the country: the city shrank by 8 percent. In addition, many of Cleveland's inner

Box 2.3 Blogging about the Housing Crisis

Like many people I am appalled at the prospect of bailing out banks and feckless homeowners, and the pundits are having a field day with the notion of a rule-governed, tax-paying people—that's us—subsidizing scoundrels and incompetents—that's them. But much of the anger is based on the mistaken notion that our economic self-interest is undermined by such intervention. This individualistic ideology is particularly strong in the United States. There is a very real issue of the moral hazard of the public subsidization of reckless private behavior; but we also have to be aware that our financial security and economic health is crucially dependent on other people. Let's consider the housing crisis as an example. As an individual owner-occupier, I am concerned with the value of my home. But this value is based not only on the characteristics of the individual dwelling but also on the going price of my neighbors' homes. If they go down in value, so does mine. The price of any home is a function of the homes around it. We can refer to this as the neighborhood effect. Foreclosures increase the number of vacant and abandoned properties and so surrounding home values, including mine, decrease. It is in our economic self-interest to have the mortgage crisis solved. And even if you live in a neighborhood untouched by foreclosures—less of a possibility as the crisis worsens—the housing market is based on long chains of purchases. House sales form chains from the top to the bottom of the market. When someone buys an entry level property, that enables the existing owner to sell and use the proceeds to buy a more expensive house, and that in turn allows the owner of a more expensive home to buy another place. A broken link in this chain of purchases has effects further up the chain.

Neighborhood effects and housing chains are just some of the ways that we as individual homeowners and purchasers are enmeshed in wider connections and ties. We need to remember this so that the debate can more effectively be about the details of stemming foreclosures and minimizing their neighborhood effects rather than being about the principle of government intervention as an inherently bad thing. Underlying much of the criticism of economic policy is a mistaken assumption that we are economic isolates. We need to be aware of the social nature of even a capitalist economy, and more especially, of a functioning capitalist economy.

Source: http://Johnrennieshort.blogstpot.com/2009_03_01_Archive.html. Originally published March 3, 2009.

suburbs continue to decline and overall urban growth remains negligible (box 2.4). The case studies of Detroit and Cleveland show that both cities continue to experience mixed results in efforts to realign and reinvigorate their economies.

Halifax, Nova Scotia: Small City in Transition

In contrast to shrinking cities, North America is also home to cities experiencing growth. A mid-sized city on Canada's east coast, Halifax occupies a strategic position that has given

Box 2.4 Suburbs in Crisis

In October 2008, the suburb of Garfield Heights outside Cleveland was declared in fiscal emergency by the state of Ohio. The suburb was $3.4 million in the red and the local government had overspent its tax revenue. The suburb was bankrupt. The mayor blamed the fiscal crisis on the loss of income taxes due to rising unemployment and delinquent property taxes because of the spike in housing foreclosures.

Garfield Heights had been in a downward spiral since the 1970s. Poverty increased gradually and incomes levels declined over the past four decades. The population of Garfield Heights peaked at 41,417 in 1970, and it has steadily dropped ever since, as Garfield Heights lost residents to suburbs much further out on Cleveland's metropolitan fringe.

The story of Garfield Heights is not unique. It is part of a larger story of population loss and economic stagnation affecting many suburbs, especially those immediately adjacent to U.S. cities. During the postwar period, inner suburbs such as Garfield Heights experienced a residential boom. This was a period of mass suburbanization. More than 60 years has passed, and the housing stock in postwar suburbs has become outdated, it now requires large-scale capital for revitalization. Typically, declining suburbs are located closest to the metropolitan core. There is suburban decline in Philadelphia, Baltimore, Atlanta, Detroit, Chicago, Cleveland, Camden, Los Angeles, and Miami.

The traditional model of metropolitan American posits poor cities juxtaposed with wealthier suburbs. This model is becoming more dubious with geographic shifts in poverty and decline. The new metropolitan reality challenges the conventional image of suburbia as comprised solely of upper- and middle-income families. Some suburbs are affluent suburban successes; other are at-risk, poor, and struggling. Over the past decade, the geographical distribution of high-poverty neighborhoods has shifted from cities to the suburbs. A recent study of 2005 Census data by the Brookings Institution found that, for the first time in U.S. history, the number of poor suburbanites was larger than the number of poor city dwellers by one million people. Once focused in central cities, poverty is now very much a part of suburban reality.

Source: Hanlon, Bernadette, John Rennie Short and Thomas Vicino. *Cities and Suburbs: New Metropolitan Realities in the US* (New York: Routledge, 2010), 175-177.

it a rich military and commercial history. In a country subject to the "MTV" (Montreal, Toronto, and Vancouver) syndrome, however, the role of smaller cities in Canadian culture, industry, immigration, and heritage is often overlooked in favor of further research on the three largest cities. Halifax provides a counterpoint to the MTV cities with its historically resource-based economies and novel approach to urban development and preservation. With about 400,000 residents in the Halifax regional municipality (an amalgamation of Halifax and surrounding areas) and 282,000 in the principal city, Halifax is the largest Canadian coastal community after Vancouver.

The area around Halifax was first inhabited by the Mik'maq (Micmac) First Nations people and settled by the British in 1749. The British felt that they needed to establish a strong counterpoint to the increasing French settlement in the area, and particularly the French fort at Louisburg on Cape Breton Island in northern Nova Scotia. Following the onset of the Seven Years War (1756–1763) and a British victory, French (Acadian) settlers from the area were deported to French colonies such as Louisiana. Given that Halifax was founded primarily for military purposes rather than economic ones, the city has grown more slowly than the industrial centers of Quebec and Ontario. Although Halifax never experienced the volume of immigration seen in New York or Montreal, a steady flow of immigrants arrived throughout the 1700s and 1800s. Most came from Great Britain, Ireland, Germany, and the United States—many of whom were U.S. settlers who wanted to stay loyal to the British crown or slaves seeking freedom via the Underground Railroad. With the advent of a railroad connection to the rest of Canada in the 1870s, Halifax eventually became a center for cotton processing, sugar refining, and other activities associated with the Atlantic economy's "triangular trade" between Europe, North America, and the Caribbean.

The influence of the sea has always been central to Halifax's development. Initially the domain of the Atlantic cod fisheries, Canada's east coast eventually became a hub of shipping, shipbuilding, and even sea-based military operations. During World War I, Halifax acted not only as the point of export for the war goods manufactured in Quebec, Ontario, and elsewhere, but also an important lifeline to the British when many of their trade connections with European countries were blockaded by German forces. Serving as a center of the Allied war effort, however, had an unfortunate and unexpected impact on Halifax. In 1917, a French ship carrying explosives through Halifax harbor crashed into a Belgian ship. The resulting explosion killed 2,000 people, catapulted debris several miles into the city, and destroyed Halifax's North end. Although the reconstruction of the city led to some innovations, such as the development of Hydrostone (one of Canada's first social housing projects), the next three decades were a period of relatively slow growth in Halifax.

Since the 1950s, Halifax has refashioned itself as a center for wholesale distribution, transport and shipping, government, education, and maritime-related industries such as oceanographic research. Most firms in Halifax, however, are branch locations governed by headquarters located elsewhere in Canada—usually Toronto. Changes in the urban form of Halifax have transformed it from a relatively peripheral outpost to an Atlantic metropolis. The construction of the MacDonald (1955) and MacKay bridges (1970) across Halifax harbor joined the city with its smaller mid-sized neighbor, Dartmouth. An unfortunate outcome of the MacKay bridge construction was the forcible eviction and demolition of Africville—a mostly black community just outside of Halifax. The increased access to cheap, unused land on the urban periphery led to considerable industrial park development in the 1970s and 1980s, while better connectivity in the core led to construction of several highrise office buildings. Perhaps the biggest recent change in Halifax has been its political geography. Like many Canadian cities, Halifax was amalgamated with over 50 towns and rural areas into the Halifax Regional Municipality in 1996. The municipality occupies an area of 2,353 square miles (6,094 sq km) and is governed by a mayor and a regional body of 23 councilors.

In spite of its recent growth and past mistakes, Halifax has maintained a focus on heritage and historic preservation. Unlike many Canadian cities (*e.g.*, Toronto), Halifax resisted the proposed development of a freeway along the waterfront, choosing recreational and tourist development instead. The city has also consistently developed downtown shopping areas and improved its retail mix to compete with out-of-center venues that provide options and amenities such as free parking. The city has more recently positioned itself as a leader in the knowledge economy by capitalizing on the innovations in health, technology, engineering, and other fields at universities such as Dalhousie, Mount Saint Vincent's, and St. Mary's. The thriving music scene, which acts as an alternative to the more competitive, commodified music sectors in Toronto and Montreal, often references the region's Celtic heritage. Though often overlooked in Canadian urban geography, Halifax has followed its own development path and now offers a livable mid-sized alternative to Canada's largest cities.

Ottawa: A Capital of Compromise

More a tool of political compromise than an economic boomtown, Canada's capital city of Ottawa has long acted as a nexus of compromise between the many facets of Canadian identity: French and English, urban and rural, Upper Canada (Ontario) and Lower Canada (Quebec and the Atlantic provinces), and the interests of federal bureaucrats and a burgeoning technology sector. Ottawa and its Quebec neighbor, Gatineau, comprise the fourth-largest census metropolitan area (CMA) in Canada (1.13 million residents).

Ottawa was designated the capital of Canada by Britain's Queen Victoria in 1857. (Canada would remain part of the British Empire until confederation in 1867.) Already an important lumber town at the northern terminus of the Rideau Canal linking the Ottawa River and Lake Ontario, Ottawa was chosen for its defendable strategic position (*i.e.*, away from the U.S. border) and its location on the Ontario-Quebec border, midway between Toronto and Quebec City. The location of compromise is evident in the linguistic diversity of Ottawa-Gatineau today. About 40 percent of the population declares itself French-English bilingual, while another 45 percent claim to speak English only and 15 percent claim to speak French only. But living in an urban area straddling two provinces also creates challenges. Many residents live in one province and work in another, and frequently negotiate the traffic laws, rental practices, and tax structures of two different jurisdictions. In addition to legislative and linguistic differences in the area, there are political differences between downtown Ottawa residents and the rural and suburban residents that joined the city following amalgamation in 2001. Ottawa politicians frequently avoid projects or developments that may irritate more politically conservative residents in West Carleton, Osgoode, Greely, and other towns that pay Ottawa taxes but do not always receive Ottawa municipal services.

An additional challenge is the dominance of the federal government in the development of the area. The Ottawa-Gatineau metropolitan area is also part of the National Capital Region, a federal jurisdiction managed by a corporation called the National Capital Commission (NCC). Some see the NCC, which manages the federal government's vast properties in the area, as a hindrance to developing Ottawa into a world-class city. Buildings cannot exceed the elevation of the National Parliament and most development has been geared toward creating a "green capital" dotted with parks and fringed with a belt of woodlands and farmlands. Ottawa, however, has refashioned its historic and national emblems as focal points of development. In 2006, Confederation Boulevard was built to help pedestrian access attractions on both sides of the Ottawa River, including the Parliament, the Rideau Canal locks, and the Museum of Civilization in Gatineau. The historic Byward market in the center of downtown Ottawa—originally a supply stop for the lumber camps around Ottawa—is now the center of a revitalized downtown nightlife and shopping district.

Perhaps the most notable collaboration in Ottawa has been the cooperation of government and industry to refashion the city-region as a tech-pole now known as "Silicon Valley North." The Ottawa-Gatineau hi-tech sector, which is focusing on information technology, telecommunications, and nanotechnology, has employed anywhere between 50,000 and 85,000 workers per year during the past decade and includes well-known firms such as Corel, Nortel, and Adobe. Although the "government town" is usually seen as the antithesis of an open-market environment fostering technology and innovation, the Canadian federal government has actually been central in the development of the region's tech-pole. The government is not only the largest user of information technology in the region, it has actively funded hi-tech research and

innovation in the laboratories of the National Research Council. Other public-private bodies, such as the Ottawa-Carleton Research Institute and the Ottawa Capital Network, provide the funding and programming to ensure that local innovations can be leveraged into the creation of local startup firms. By capitalizing on these networks and Ottawa's highly educated talent pool (30 percent have university degrees), Ottawa has developed its own version of Silicon Valley— one committed to local research, development, and re-investment.

Washington, D.C.: A New Immigrant Gateway

No longer are the world's immigrants going to the older, established destinations of the 19th and early 20th centuries. In the 21st century, many immigrants choose to settle in cities where immigration is a relatively new phenomenon, but where economic growth demands both high-skilled and low-skilled workers. Washington, D.C., has become home to hundreds of thousands of new immigrants in the past 15 years. Despite the economic recession of 2008, Washington, D.C., saw modest economic growth as a result of federal jobs and contract work (government externalities) as well as the information technology sector. Military-funded aerospace firms such as Northrop-Grumman and Lockheed Martin, have established East Coast headquarters in D.C. to be close to the Department of Defense. More recently, the Dulles "High Tech Corridor," reaching from northern Virginia westward toward Dulles International Airport, has attracted high-skilled software engineers and other high-technology workers. The firm AOL employs 5,000 in its headquarters there. A shortage of high-skilled workers in the 1990s resulted in the creation of an H1 Visa specifically to target foreigners with college degrees in computer-related fields. As a result, many skilled immigrants from India, South Korea, Hong Kong, and mainland China moved to the Washington, D.C. region during this time. Prosperity has led simultaneously to an increased demand for domestic labor. Many immigrants from El Salvador, Bolivia, Peru, Brazil, Mexico, and Guatemala have also arrived in D.C. to work as nannies, landscapers, construction workers, and in the hotel/hospitality sector.

Washington has become a magnet to a range of immigrants from many diverse countries. In 1970, only 4.5 percent of the population of Washington, D.C., was foreign-born. By 2009, Washington was home to 1 million foreign-born residents, which accounted for 20 percent of the urban population. Other cities in the United States and Canada are emerging as new gateways. Las Vegas, for example, continued to see population growth throughout the first part of the 21st century. In 2009, it was a metropolitan area of 1.9 million, and home to more than 400,000 foreign-born residents. Like D.C., most of the newcomers in Las Vegas have arrived since 1990. In 1970 Las Vegas was only 3.9 percent foreign-born residents; by 2009 nearly 22 percent of its residents were foreign-born despite the recent economic crisis. Because immigration is a relatively new phenomenon in these cities, Washington and Las Vegas, together with Charlotte (NC), Orlando, and Atlanta, are considered newly emerging gateways for immigrants.

These new gateways, however, do not appear to follow the same sort of settlement patterns and processes seen in more established gateway cities such as New York or Chicago. Contrary to the conventional model of spatial assimilation that assumes that immigrants first cluster with their co-ethnics in center-city enclaves such as Chinatowns and Little Italys and—as they gain higher levels of education and income—leave these enclaves to reside in suburban areas with higher social status and larger homes, in Washington and Las Vegas many immigrants have located first in the suburbs, not in the central-city. Immigrants in these new gateways are thus becoming a vital force in shaping the suburban future. In addition, many immigrants to Washington live in moderate- to high-income neighborhoods, not in the poorest ones. In fact, many are settling in places that only thirty years ago were mostly White and had very few foreign-born residents. The implications are interesting. For Washington the historic image of the city polarized into "Black and White" no longer holds true, and city leaders and residents are grappling with how an increase in diversity has impacted their communities (fig. 2.12).

New Orleans: Vulnerable City

Cities that have experienced disasters remind us that many cities are vulnerable to environmental forces. Hurricanes and floods are not new to New Orleans. The city was originally located where the Mississippi River flows into the Gulf of Mexico, at a riverbend close to Lake Pontchartrain. The original city was sited on a relatively high piece of land where French traders had already been encamped close to a Native American trail. This became known as the French Quarter. The low water table created difficulties in construction, and the river threatened the city with regular flooding. Improvements in pumping technology in the early 20th century encouraged more development in the lower lying areas of the city, which tended to be settled by working-class,

Figure 2.12   Migrants make their presence felt in numerous ways. In this case, there are sufficient Brazilian immigrants to form a Brazilian service at this Baptist Church outside Washington, D.C. (Photo by Lisa Benton-Short)

poor, and minority populations. Still, the river that gives the city economic prominence also threatens to destroy it.

In 2005, the City of New Orleans was devastated by Hurricane Katrina. Hurricanes produce downpours of up to 25 inches of rain in 30 hours, high winds that threaten to tear buildings and other structures away from their moorings, and damage to infrastructure ranging from washed out or flooded roads to power lines that are torn apart by wind and waves. Storm surges associated with the high winds can reach over 20 feet (6.96 meters) above normal sea level, causing massive flooding. For most coastal cities flooding is a serious problem. For a city below sea level, it is potentially catastrophic. Much of New Orleans is below sea level.

It was not the ferocious winds of Katrina that damaged the city, but the storm surge that breached the levees in the city. The city was flooded when portions of the levees at 17th Street and Industrial Canal collapsed. Almost 80 percent of the city was flooded, in some cases by water over 20 feet (6.1 m) in depth. An estimated 1,000 people were killed, most of them drowned by the rapidly rising floodwaters. Much of the city was destroyed in the flooding that followed the hurricane.

At first glance Hurricane Katrina seemed like a natural disaster. A hurricane is a force of nature. But Katrina was a force of nature whose impacts and effects were mediated through the prism of socio-economic conditions. The flooding of the city was caused by poorly designed levees that could not withstand a predictable storm surge. The levees were poorly constructed with pilings set in unstable soils. Pilings that should have been 15 feet (5.22 meters) high had settled in many places to only 12 to 13 feet (3.7 m to 4 m) above sea level. It was not Katrina that caused the flooding but shoddy engineering, poor design, and inadequate funding of these vital public works.

As Katrina approached New Orleans, evacuation orders were finally given. Those with cars were able to exit the city, but there was little provision made for the care and safety if the city's most vulnerable residents. Those without access to private transport were abandoned; many disabled, poor, Black, and elderly residents were trapped. Between 50,000 and 100,000 people were left in the city when the hurricane struck and the levees failed. Some made their way to the Superdome and the Convention Center, which for several days afterward housed between 30,000 and 50,000 people. They remained there for days, a stunning indictment of social and racial inequality.

Even the effects of Hurricane Katrina on the city were socially and racially determined. Flooding disproportionately affected the poorest neighborhoods of the city. The more affluent, predominantly White sections, such as the French Quarter, Audubon Park, and the Garden District were built at higher elevations and escaped flood damage. The flooded areas were 80 percent non-White, and most of the high-poverty tracts were flooded. The racial and income disparities in the city were cruelly reflected in the pattern of flood damage. In closer detail, the "natural" disaster appears to have been a social disaster as well. Environmental disasters become social in the way they are handled and how the distribution of their effects reflects social differences.

By 2010, five years after Katrina, the damage of the hurricane lives on in abandoned houses and their scattered owners, in empty storefronts, and in the ongoing debates about how to rebuild the city. While most of the rest of the country has long moved on to other headlines, the rebuilding of New Orleans is far from over. The good news:

Figure 2.13 Hurricane Katrina damaged areas of the Lower Ninth Ward in New Orleans.

almost 150,000 residents have returned. Tourism, an important part of the economy, is close to what it was before the storm. The official blight statistic, which encompasses unoccupied residential addresses and vacant lots, has dropped from 34 percent of New Orleans' total addresses to 27 percent. And in 2010, the city's beloved NFL Saints were crowned Super Bowl champions. The bad news: the population of New Orleans is still down about 100,000 from what it was before Katrina. Block after block in the Lower Ninth Ward remains empty. Overgrown weeds and empty concrete slabs stand in place of the homes that once were there (fig. 2.13). In contrast to the stumbling and delay characterized by government rebuilding, nonprofit organizations have become an integral part of redevelopment efforts. The Make It Right NOLA Foundation, established by actor Brad Pitt, has had tremendous success working directly on the ground. The Foundation focuses on rebuilding affordable homes for working families. By 2010, the Foundation had rebuilt more than 50 eco-friendly homes, allowing more than 200 people to return to this Ward.

CHALLENGES FACING CITIES IN THE UNITED STATES AND CANADA

Nature and the City

Urban studies have ignored the physical nature of cities; the emphasis has been on the social, political, and economic factors rather than the ecological. And yet cities are ecological systems; and these systems impact the social, political, and economic realms of the city. The city itself can be seen as an ecosystem with inputs of energy and water and outputs of noise, greenhouse gases, sewerage, garbage, and air pollutants. Water, for example, is an essential ingredient of life, especially within the city. One of the largest urban differences is between cities with clean, easily accessible water and cities where water is expensive, inaccessible, and polluted. Many North American cities have undertaken immense engineering projects to provide cheap, clean water. As cities have grown, water catchments have extended outwards and the engineering sophistication of the sytems piping water into the cities has grown and deepened. The availability of fresh water is a determinant of the limits of urban growth. In the arid west of the United States, for example, urban growth has been predicated upon massive federal subsidies and expensive engineering projects that have provided fresh water at low cost to urban consumers. Cities that have benefited include Las Vegas, Tucson, Phoenix, and Los Angeles. The ecological limits are always more flexible than the environment seems to suggest, but they are not infinitely extendable. We may be reaching the "water" limits of urban growth in the arid U.S.

Cities also modify the environment. Human activity in the city produces pollutants. Industrial processes and auto engines emit substances that include carbon and sulfur oxides, hydrocarbons, dust, soot, and lead. In the Canadian province of Ontario, for example, which has a population of 11.9 million, air pollution costs citizens at least $1 billion annually in hospital admissions, emergency room visits, and worker absenteeism. In 2005, Toronto had 48 "smog days" of unhealthy air, the highest since 1993. The pollutants of cities are not only harmful to the health of individuals, they also cause more general damage; cities are in part a major cause of global warming and ozone depletion.

Cities also reference nature, and since the 19th century there has been an explicit urban parks movement. Landscape architects such as Frederick Law Olmstead have left a permanent legacy on cities. It is difficult to imagine New York without Central Park, San Francisco without the Golden Gate Bridge, Vancouver without Stanley Park, or Washington, D.C. without the National Mall. Today city parks are developed as much for their recreational opportunities as for their aesthetic appeal. Urban planners realize that successfully referencing nature is an important element in creating the right atmosphere, and it is often linked with the promise of economic redevelopment. Increasingly, urban residents are recognizing the various scales of nature: from vest-pocket parks, urban gardens, greenways, and rooftop gardens, to large expansive megaparks. Whether it is in the beaches of southern California, lakeshore of Chicago, the parks of Vancouver, or the community gardens in Montreal, a commonly accepted attractive feature of urban life is the successful (re) incorporation of nature into the urban lifestyle, and the way in which natures defines and enhances the city's image and the metropolitan experience (fig. 2.14).

Figure 2.14   Reflecting Whistler's image as environmentally oriented, a place that values nature, sports and recreation, this couple opted to bike to the top of the mountain to get married. (Photo by Kathy Schauff)

Migration and Increasing Diversity

Immigration is a window through which to view the reconfiguring of urban and global networks as millions of economic migrants settle in select cities around the world. While we tend to think of immigrants as coming from and going to countries, most immigrants go to cities. Although these localities or "immigrant gateways" take on different forms, many are hyper-diverse and globally linked through transnational networks. Hyper-diverse immigrant cities are those places where the percentage of foreign-born residents exceeds the national level and where immigrants come from many regions of the world with no single dominant country of origin. *Immigrant gateway cities* are growing in number because of

globalization and the acceleration of migration driven by income differentials, social networks, and various state policies to recruit skilled and unskilled laborers as both temporary workers and permanent residents (box 2.5).

As large numbers of foreign-born residents and ethnically distinct people are thrown into the mix, cities become the places where global differences are both celebrated and contested. Immigrants can add to a city's global competitiveness by enhancing a city's diversity and talent, thus making such places better able to compete in a global age. With birth rates declining in both Canada and the United States, migration is now a more important determinant of differential urban growth (or decline).

While it is true that immigration is a global phenomenon, it is also true that some regions

Box 2.5 Anchorage, Alaska

With a 2010 population of 292,000, Anchorage is the largest city in Alaska. It is also home to over 40 percent of the state's population. The city has seen increased international immigration and currently there are some 21,000 foreign-born residents in Anchorage. This represents 8.2 percent of the population, which is lower than the U.S. average but higher than average for the state of Alaska at 5.9 percent. Similar to cities such as Las Vegas and Washington, D.C., most of the foreign-born residents in Anchorage have arrived since 1990. The major countries of origin for Anchorage immigrants include the Philippines, Korea, Mexico, the former Soviet Union, and Canada.

As with many U.S. cities, the history of economic development in Anchorage is a history of boom-bust cycles. Located at the head of Cook Inlet, this relatively new city was founded in 1915 as a construction port for the Alaska Railroad. Anchorage remained a relatively small frontier town until the beginning of World War II. During World War II, Anchorage became a key aviation and defense center. The influx of federal defense spending during the 1950s increased Anchorage's population and its business community. During the 1950s, Anchorage became a boomtown whose growth was led primarily by federal government investment. The 1970s brought yet another "boom" to Anchorage: the development of the Prudhoe Bay oil fields in northern Alaska. In 1972, Congress authorized the trans-Alaska pipeline system. Oil discovery and pipeline construction of the pipeline fueled a modern-day boon when oil and construction companies set up headquarters in Anchorage. Construction began in 1974, with oil flowing from the North Slope to the ice-free port of Valdez in 1977. The petroleum industry also provided skilled employment opportunities for thousands. Population, office space and housing tripled by the 1980s. Oil revenue to the Alaskan state treasury was used to develop the city's infrastructure: between 1980 and 1987 nearly $1 billion worth of capital projects were constructed in the city including a new library, civic center, sports arena, and performing arts center.

Since 2000 the increase in the price of oil has, once again, made Anchorage a boomtown. Seeking to avoid the boom-bust cycles, Anchorage is gradually broadening its economic base with more retail trade and a larger service sector to serve tourists. Although the U.S. government and the oil industry have been integral to the Anchorage economy, new economic opportunities have prospered including construction, light manufacturing, high technology, software development, commercial fishing, and seafood processing.

of the world receive significantly more immigrants than others. In North America, long an established region of immigrant settlement, the volume of immigration is among the highest in the world. Although the three largest immigrant destinations in the region are New York City, Toronto, and Los Angeles, there are sixty other metropolitan regions with more than 100,000 foreign-born residents. In Canada immigrants primarily go to one of three cities: Vancouver, Montreal, and Toronto. But smaller Canadian cities such as Ottawa and Calgary are places where well over 20 percent of the population is foreign-born. Similarly in the United States, immigrants are targeting newer gateways such as Washington, D.C., Phoenix, Charlotte, and Atlanta, where the relative increase in the foreign-born population is much higher than in traditional gateways.

Yet crude measurements of the foreign-born populations in urban North America tell us little about the composition or distribution of the foreign-born residents within cities. Some cities are home to a "hyperdiverse" foreign-born population; others are home to a large number of foreign-born residents, but are not particularly diverse. Two of the most hyperdiverse cities in the world are New York and Toronto. Together, they are home to millions of foreign-born residents who have come from every region of the world.

At the dawn of the 20th century, when New York City was the premier immigrant gateway in the United States, nearly all immigrants were European. The city was linguistically and ethnically diverse but not so racially. In the first years of the 21st century, New York City is one of the most racially and ethnically diverse places on the planet. Of the top ten sending countries to metropolitan New York, only one is European. The top-ten sending countries represent half of the city's foreign-born residents and include immigrants from the Dominican Republic, China, Jamaica, Mexico, Guyana, Ecuador, Haiti, Colombia, and Italy (fig. 2.15).

A similar pattern holds true for Toronto. In 2006, 49.9 percent of the city's population was foreign-born, one of the highest percentages for any major metropolitan area. About 70,000 immigrants from approximately 170 countries arrive in the city annually. No one group dominates Toronto's immigrant stock. Nine countries account for half of the foreign-born population, led by China, then India, the United Kingdom, Italy, the Philippines, Jamaica, Portugal, Poland and Sri Lanka (fig. 2.16). Other hyperdiverse metropolitan areas in North America include Washington, D.C., San Francisco, and Seattle. Due to the globalization of migration, there is a tendency for immigrants to come from a broader range of sending countries and, in the process, to create cities that are more racially and ethnically diverse.

Not all immigrant gateways are hyper-diverse. Mexican immigrants account for about half of the foreign-born population in Los Angeles, Chicago, Houston, and Dallas. These millions of Mexicans often impact their cities in unexpected ways. For example, four of the top ten television shows in the Los Angeles market are Spanish-language broadcasts. Similarly, foreign-born Cubans dominate in Miami. Immigrants from mainland China and Hong Kong account for over 25 percent of the foreign-born population in Vancouver. It is fair to say that North American cities will continue to be home to many of the world's immigrants well into the 21st century.

Security and Urban Fortification

Responses to the potential of terrorism, whether domestic or international, have had a profound impact on many North American cities (fig. 2.17). Intense surveillance and security have changed both the physical and symbolic landscape of many cities. It is not unusual in times of war and conflict for visible forms of fortification to appear on urban landscapes. Fortification of urban space is not merely a recent phenomenon. Since September 11, 2001, however, security has become much more visible in most North American cities. Given that many of our barricaded urban spaces are valued public places with important symbolic and spatial connections to local,

New York's Foreign-Born Population

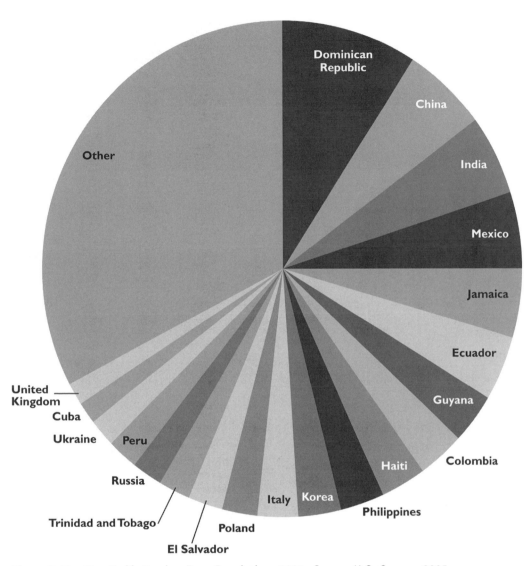

Figure 2.15 New York's Foreign-Born Population, 2005. *Source:* U.S. Census, 2005

regional, or national identity, this trend is of significant concern to many citizens.

The rise of security policing and new forms of surveillance and control of urban spaces are not new subjects. Mike Davis's 1990 book, *City of Quartz*, diagnosed what he calls *fortress cities* as a response to per-

ceived urban disorder and decay, primarily from domestic sources. He noted that the car bomb could well become the ultimate weapon of crime and terror; and he has predicted that the urban authorities might create fortress style rings of steel as a counter response. The fortress metaphor describes

Toronto's Foreign-Born Population

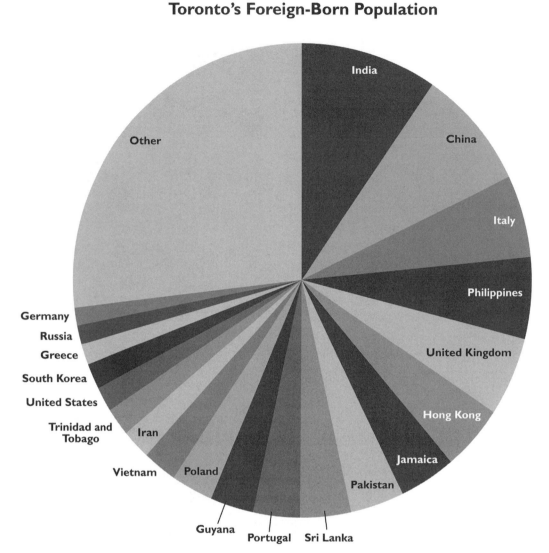

Figure 2.16   Toronto's Foreign-Born Population, 2006. *Source:* Canadian Census, 2001

a landscape that is demarcated by physical barriers such as gates and walls, as well as by often-invisible surveillance devices such as closed-circuit television (CCTV) cameras that watch city streets, parks, and gated communities. This is a vision of a city that can be controlled. For some residents, security and surveillance offer reassurance in an uncertain age; for others these measures are the architecture of paranoia.

Although cities have long had police forces and emergency plans, prior to the early 1990s many did not have a comprehensive security and defensive strategy. Attempts to design out terrorism occurred after an event or a direct threat raised the issue of vulnerability. Today

Figure 2.17   A downtown Vancouver sign reminds residents that unsanctioned shelter or commercial activity is not permitted. (Photo by Elvin Wyly)

terrorists target high profile cities to attract global media publicity. And since 9/11 it is clear that symbolic targets—such as monuments, memorials, landmark buildings and other important urban public spaces—are increasingly at risk. The responses have been highly visible counter-terrorist measures. In New York, Toronto, Los Angeles, and Philadelphia, for example, bollards, bunkers, and other barriers have been placed around selected "high-risk" targets and buildings. Fortifications such as barriers and fences have impacted people's access to public spaces such as museums, monuments, memorials, and parks. Many people see heightened fortification as a threat to public space. When security measures are particularly visible in capital cities, such as Washington, D.C., the symbolic impact of

fortress architecture is elevated to represent a national discourse dominated by the awareness of war, fear, and entrenchment.

Washington, D.C. is one of the most visibly fortified cities in North America. Miles of fences, jersey barriers, and bollards surround federal buildings, monuments, and memorials throughout the city. In addition, CCTV cameras keep an eye on the streets, sidewalks, and public spaces and are mounted on roofs and the sides of libraries, shopping malls and banks. Security cameras are even mounted on some of the most public of spaces: the monuments and memorials on the National Mall (fig. 2.18). The prevalence of security cameras in many cities has become commodified in the popular culture as those "caught in the act" provide hours of video entertain-

Figure 2.18   Here, framed by a black box, a security camera has been positioned atop the Jefferson Memorial in Washington, D.C. What messages do surveillance cameras convey in a public space which memorializes freedom, liberty and independence? (Photo by Lisa Benton-Short)

ment for television shows and video websites like YouTube. Regardless of whether security cameras actually deter crime or not, more and more urban residents are photographed, videoed, and live much of their daily lives under surveillance. This raises troubling questions about what happens to old video footage, whether police use video surveillance to create criminal profiles, and whether law-abiding residents should be concerned that they are often under surveillance, knowingly or not.

Memorials and monuments are not merely ornamental features in the urban landscape, they are also highly symbolic signifiers that confer meaning to urban spaces and thus represent the politics of power. Changes or alterations to these memorials via CCTV, fences, bollards, or other barriers also provide a glimpse into the competing agendas of security and terrorism, as well as into the power relationships that can ultimately determine its realization. Perhaps more contentiously,

security cameras do more than inconvenience public access to memorials and other spaces. They also inhibit many types of radical political protest such as civil rights protests, antiwar marches, and protests from marginalized members of society whose voices emerge in public spaces in cities around North America. Although the stakes have changed, it is as yet unclear how the need for improving security translates into acceptable levels of fortification, the loss of civil liberties, inhibited public access to public space, and the implied messages of a new urban culture dominated by a sense of hyper-security. The ambiguous nature of security remains unresolved.

Preserving and Recreating Urban Historical Landscapes

In North America, there has always been some appreciation of our cultural and architectural heritage, but only in the last few decades have we become increasingly aware of the significance of our urban historic structures and sites. The unprecedented urban changes of the 20th century, particularly under urban renewal, the growth of the interstate highway system in the United States and other massive public works, and the skyscraper building boom provided momentum for the emergence of a preservation movement.

In the United States, the preservation movement evolved along two distinct paths. The private-sector path focused on important historical figures and landmark structures. One of the first major historic preservation undertakings from the private sector, for instance, was that of George Washington's plantation estate, Mount Vernon, along the Potomac River not far from Washington, D.C.

The public-sector path was involved with establishing the national parks, but also included the restoration of historic buildings. The public sector was successful at establishing some historic districts, such as Charleston, South Carolina (1931), the Vieux Carré (French Quarter) section of New Orleans (1936), and Alexandria, Virginia (1946). In 1949, several organizations evolved into the British-inspired National Trust for Historic Preservation. The purpose of the Trust was to link the preservation efforts of the private side with the activities of the federal government, particularly the National Park Service activities. The most important piece of historic preservation legislation was the National Historic Preservation Act of 1966. Preservation no longer focused on saving single landmarks; instead, entire areas within established cities were delineated as historic districts and this designation became an important tool of urban revitalization during the 1970s and 1980s. Historic Districts are defined as areas of cultural and physical

distinction that contribute to a local identity. Cities can have multiple historic districts. For example, there are more than 20 historic districts in Chicago, including the Pullman district (the first planned industrial town) and the Black Metropolis district (a series of 9 buildings that were home to numerous nationally-prominent, African-American-owned-and-operated businesses and cultural institutions). Many historic districts cover a wide geography. The Freedom Trail in Boston is a 2.5 mile (4 km) red-brick walking trail that winds through the city and leads the visitor to 16 historic sites: a collection of museums, churches, meeting houses, burying grounds, parks, a ship, and historic markers that tell the story of the American Revolution. Today there are thousands of local preservation associations and thousands of designated historic sites, buildings, and other structures in cities in the United States.

Canadian preservation history follows a similar trajectory. The primary conservation tool is the Canada's Historic Places Initiative. This program was the result of a major collaborative effort among federal, provincial, territorial, and municipal governments, heritage conservation professionals, heritage developers, and many individual Canadians. As a pan-Canadian collaboration, it is intended to reinforce the development of a culture of conservation in Canada. So far, over 1500 places, persons, and events have been commemorated.

Today, historic preservation is composed of a variety of strategies: preservation, restoration, reconstruction, and rehabilitation. Preservation refers to the maintenance of a property without significant alteration to its current condition. When preservation is the guiding strategy, the only intervention is normal maintenance or special work needed to protect the structure against further damage. An example of innovation in preservation is Pike Place Market in Seattle. The old city market was threatened with demolition to make way for an urban renewal project. In a city-wide vote, however, residents elected to save the market as an important part of their city's life and culture. In order to prevent the loss of its original character as a working everyday market run by local farmers, fisherman, and small entrepreneurs, Seattle developed an ordinance that not only protected the structure, but also the activities within it.

Restoration refers to the process of returning a building to its appearance from a specific time period. It can involve cleaning, minor repair, major repair, and the replacement of parts of the building that may have been damaged or missing. For example, nearly all 21 of the Spanish missions along the California mission trail (located along the urban corridor that begins in the south with San Diego, proceeds north to Los Angeles and along the coast to San Francisco, and ends in Sonoma) have been restored to varying degrees. The missions, a legacy from the Spanish era of the 18th century, needed to be restored after the ravages of time, weather, earthquakes, and neglect took their toll. In some cases, restoration may require removing features from other periods in its history and the reconstruction of missing features from the restoration period. Critics believe this can lead to a more contrived picture of a building's "original condition." This aspect of restoration is more common in historic homes, farms, or churches. This strategy, however, is not without criticism, as it casts doubt on the authenticity of restored structures.

The term reconstruction refers to the building of a historic structure using repli-

cated design and/or materials. This approach is taken when a historic structure no longer exists. In some cases reconstruction goes hand-in-hand with restoration. The earliest and best example is Williamsburg, Virginia. In 1926 John D. Rockefeller was persuaded to fund the restoration of the entire colonial town of Williamsburg. The primary problem was that much of the original town had been lost over the centuries, and while many of the historic buildings remained in the town, a few central buildings from the original town layout were missing. Planners decided to reconstruct the Governor's Palace (destroyed by a fire in 1781). The efforts to reconstruct Williamsburg were not without controversy—as some buildings were removed to make way for reconstructed ones. Yet it remains one of the most visited historic districts in the United States. In addition to both restoration and reconstruction, Colonial Williamsburg presents live recreations of historic events by actors in period costumes. This way of presenting historical places and artifacts, often referred to as "living history museums," has become increasingly popular, as seen in the animated "theater" performed in Independence Square in Philadelphia (fig 2.19).

Lastly, many buildings no longer perform their original function or use but retain their architectural integrity. For these structures, a common strategy is rehabilitation, sometimes referred to as adaptive use. The purpose is to modify or update portions of the structure and adapt the building for a new purpose. Numerous examples abound—from abandoned factories that are adapted into microbrewery pubs, museums, or residential lofts. Increasingly, rehabilitation is a strategy employed by those seeking to revitalize old areas of the city.

Figure 2.19   History is "performed" here in Independence Square in Philadelphia. Around the city are designated spots, called "Once Upon A Nation," where actors animatedly perform short skits about the Revolutionary era. (Photo by Lisa Benton-Short)

Historic preservation and recreation is increasingly at the center of urban redevelopment efforts, as cities search for ways to celebrate their past while looking toward the future. In some cases, the financial investment in preservation and recreation efforts can provide economic returns in increased tourism and the (re)attraction of commercial and residential interests. Cities are learning that preservation and recreation can make good economic sense.

Smog Cities

Residents in many North American cities confront the reality of air pollution. Since the 1970s the U.S. and Canadian governments have taken steps to control pollution emissions from automobiles and factory smokestacks. Catalytic converters capture much of the chemical pollutants emitted in automobile exhaust, and vapor traps on gas pumps help prevent the evaporation of carbon dioxide into the air. Recent efforts to develop hybrid and zero-emission vehicles (such as electric cars) are another way both private and public interests are using technology to alleviate air pollution. However, new sources of pollution, combined with increased use of fossil fuels for a myriad of needs, has meant that air pollution for many U.S. and Canadian cities has continued to increase despite regulatory efforts.

Smog represents the single most challenging air pollution problem in most North America cities. Smog is produced by the combination of pollutants from many sources including smokestacks, cars, paints, and solvents that interact with ground-level ozone. Smog is often worse in the summer months when heat and sunshine are more plentiful. Short-term exposure can cause eye irritation, wheezing, coughing, headaches, chest pain, and shortness of breath. Long term exposure scars the lungs, making them less elastic and efficient, often worsening asthma and increasing respiratory tract infections. Because ozone penetrates deeply into the respiratory system, many urban residents are at risk including the weak and elderly, but pollution also affects those who engage in strenuous activity. In the United States, a quarter of the population lives in urban areas where air quality concentrations for ozone exceed acceptable levels. According to the American Lung Association, six out of ten Americans live in urban areas where air pollution can cause major health problems. Although there has been substantial progress against air pollution in many urban areas, nearly every major city continues to be burdened by air pollution. Despite America's growing "green" movement, the air quality in many cities remains poor. The worst U.S. cities for smog in 2009 were Los Angeles, Bakersfield, Fresno, Houston, Sacramento, Dallas, Charlotte, and Phoenix. In Canada, Windsor, Toronto, Montreal, and Vancouver are cities where acceptable ozone levels are exceeded on an average of ten or more days in the summer.

Smog is one pollutant that is often exacerbated by geography. Cities located in basins and valleys—such as Los Angeles—are particularly susceptible to the production of smog. Denver, the Mile High City, suffers from smog and other air pollutants that are made worse by its elevation in the Rocky Mountains. Because of Denver's high altitude, the city experiences frequent temperature inversions when warm air is trapped under cold air and cannot rise to disperse the pollutants to a wider area. As a result, smog may hover in place for days at a time, generating a "smog soup" that envelopes the city. Denver has worked to turn around its air pollution problem, but despite these efforts, as of 2010 Denver remained out of compliance with EPA standards. Another issue of "smog geography" occurs in both Montreal and Toronto. These cities are downwind of major industrial cities in the American Midwest, and some of the air pollutants that generate smog in Toronto and Montreal originate across the border.

Many of the improvements in air quality in North America's cities have been offset by increases in their populations that demand more energy as well as increased automobile use. Despite decades of regulation and good intention, cities in North America remain far from eliminating the threat of air pollution to both public health and environmental quality.

CONCLUSIONS

Cities in the United States and Canada entered the 21st century facing numerous and complex challenges. One was the challenge of dealing with rapid economic, social, and environmental change. Economic restructuring, which has included deindustrialization and the rise of a diverse service-sector economy, has resulted in uneven development. Those cities unable to tap into the global circuits of capital struggle to realign and expand their economies in the wake of deindustrialization. Cities are in greater competition with each other in the urban hierarchy to retain center-city populations, to attract domestic and international businesses and investment, and to develop a diverse economy. Urban boosters have responded to the underlying economic restructuring with a number of redevelopment and growth strategies that include raising their visibility by hosting Olympic Games, redeveloping waterfronts, and preserving and recreating historic landscapes for both local and tourist usage. Yet economic diversification is no guarantee of success. It remains to be seen what the driving engine of the next boom-bust cycle to impact the urban economy will be.

Social change is occurring as well. Suburban sprawl continues. Yet not all suburbs are wealthy; some are in decline and experiencing increased poverty. Issues of immigration (both legal and undocumented) have become part of a wider public debate around citizenship, race, ethnicity, and gender (box 2.6). Traditional immigrant cities, such as New York and Toronto, continue to see the influx of immigrants; but cities without a long history of immigration have begun to attract large numbers of foreign-born residents. These new immigrant gateways often lack the institutional mechanisms to cope with increased cultural diversity and are challenged to provide a range of social services (such as English language instruction in schools and translation services in hospitals). Diversity is also an issue as lesbian, gay, bisexual, transgendered, and queer communities press for the right to marry legally. They are also creating new urban spaces that reflect their lives and experiences. Finally, the continuing war on terror has resulted in physical changes to the urban landscape as cities attempt to deal with safety, security, and the vulnerability of urban populations. Cities must now grapple with protecting "national security" and building fortifications, yet widespread public debate about the

trade-offs of security versus public access and free speech has yet to occur.

Last, environmental factors are transforming the urban landscape in many ways. The impact of Hurricane Katrina on New Orleans and the widespread economic and ecological impacts from the 2010 Gulf Oil Spill are forceful reminders that many cities are vulnerable to environmental events and disasters. Any geographic location has risk, as the many challenges of preparing for and recovering from disasters such as hurricanes, earthquakes, floods and droughts have revealed. Urban growth patterns illustrate that the environment can be a constraining element; but in the 21st century we often fail to recognize this, and instead continue to develop along coasts, river valleys, deltas, and earthquake fault lines and in areas like Phoenix, where fresh water is disappearing. As much as cities are subject to environmental forces, they are also agents of environmental change. Cities emit tremendous amounts of pollutants into the air, land, and water. In some cities air pollution has significant health impacts on its urban residents, despite efforts to reduce and control pollution since the 1970s. In recognizing the interrelated human-environment nexus, many cities have developed plans to green their city through the development of parks and waterfronts, often reclaiming once-polluted land and waterways.

North American cities are in constant motion; social, economic, political, and environmental changes ensure that, as current challenges continue, new ones will arise. Many cities have taken positive and often creative steps to address these challenges. The cities in the United States and Canada remain some of the most interesting and dynamic cities in the world; and they often lead the way in

Box 2.6 Queer Space in North American Cities

As sites where difference is negotiated and often contested, cities have long been focal points for identity formation, community building, and advancement among lesbian, gay, bisexual, transgendered, and queer (LGBTQ) people. The establishment of queer space, often in the form of so-called gay villages has been central to the project of queer liberation and queer activism in the United States. The Stonewall Inn pub, in the Greenwich Village area of Manhattan (long considered one of New York City's many gay and gay-friendly neighborhoods), is often thought to be the place where the gay liberation movement began during a 1969 police raid and riot. During the 1970s and 1980s, queer people from across the United States and Canada migrated to areas that were earning reputations as safe, open places where queer people could "come out" and become part of a community. Neighborhoods such as the Castro (San Francisco), Christopher Street (New York), and Boys Town (Chicago) became entrenched in the American imagination as promised lands for the inclusion and advancement of LGBTQ people.

Gay neighborhoods gained further importance during the 1980s as centers for the fight against HIV/AIDS. Although the disease significantly impacted many of these communities, they became sites of research, advocacy, and activism for the victims of a disease that had for some time gone unrecognized by medical professionals. But as the AIDS epidemic began to wind down, the 1990s marked a new era of gay neighborhoods marked by redevelopment, consumer activity, and heavy branding. As social acceptance of non-normative sexualities grew in urban areas of North America, municipal governments became interested in capturing the "pink dollar," marketing their cities to queer tourists, and solidifying the gentrification already being led by queer people in many neighborhoods. Often, this involved a clearer delineation of queer space through signage and symbolism. While "Le Village" in Montreal added a rainbow-clad metro station (Beaudry), Chicago's Boys Town added 11 pairs of rainbow-colored Art Deco pylons along Halsted Street. While these processes quite literally concretized queer space in North American cities, many geographers critiqued them as forwarding highly commodified, normalized visions of gay communities that were geared more to tourists than to queer people themselves.

In more recent years, geographers have expanded the idea of what constitutes queer space. Noting that the traditional gay village posits a hegemonic, cosmopolitan, upwardly mobile, and mostly male vision of queer identity, many geographers have argued that lesbians, transgendered people, and others who have felt excluded from the village, have formed alternative communities in particular venues or private spaces, or in locations outside the city center. This trend is evident in many cities as traditional village-based hubs of services and social outlets begin to disperse. In Toronto, the village along Church Street is now competing with a set of queer and queer-friendly establishments on Queen Street West, casually called "Queer West." Still others have argued that even the mundane spaces of the home, the suburb, the office, and the day care center have increasingly been "queered" by out individuals and families who willingly assert their identities outside of delineated urban queer spaces.

recognizing and dealing with issues that other cities in other regions are just beginning to confront.

SUGGESTED READINGS

Anisef, Paul, and Michael Lanphier, eds. 2003. *The World in a City.* Toronto: University of Toronto Press. *Analyzes the challenges of immigrants in Toronto and the value of municipal policies that provide resources to aid in settlement and integration.*

Fogelson, Robert M. 2003. *Downtown: Its Rise and Fall.* New Haven: Yale University Press. *Describes the evolution of urban centers in America from 1880 to 1950 including the influence of construction, business, and transport.*

Florida, Richard. 2005. *The Flight of the Creative Class: The New Global Competition for Talent.* New York: Harper Business. *Details the pool of skilled workers crucial for urban growth and the forces of globalization that pull these workers away from the U.S.*

Greenberg, Miriam. 2008. *Branding New York: How a City in Crisis was sold to the World.* New York: Routledge. *Traces the history of New York City as a "brand" and the resultant transformations in urban politics.*

Hanlon, Bernadette, John Rennie Short, and Thomas Vicino. 2010. *Cities and Suburbs: New Metropolitan Realities in the US.* New York: Routledge. *Explores cities and suburbs, growing urban diversity, environmental consequences of development, and introduces the term "suburban gothic" to describe suburbs in crisis.*

Hartman, Chester, and Gregory D. Squires, eds. 2006. *There is No Such Thing as a Natural Disaster: Race, Class and Hurricane Katrina.* New York: Routledge. *Scholarly essays on the impact and implications of Hurricane Katrina for urban planning and social policy.*

Kahn, Matthew E. 2006. *Green Cities: Urban Growth and the Environment.* Washington, DC: Brookings Institute Press. *Uses macro-level economic analysis to explore the worldwide trend in "greening" urban environments.*

Knox, Paul. 2011. *Cities and Design.* New York: Routledge. *An examination and critical appraisal of the complex relationship between design and urban environments.*

Price, Marie, and Lisa Benton-Short, eds. 2007. *Migrants to the Metropolis: The Rise of Immigrant Gateway Cities.* Syracuse, NY: Syracuse University Press. *Examines immigration to 13 cities around the world and investigates the ways in which immigrants are transforming urban space.*

Vale, Lawrence J., and Thomas J. Campanella, eds. 2005. *The Resilient City: How Modern Cities Recover From Disasters.* New York: Oxford University Press. *A collection of essays analyzing the response and ultimate recovery of cities hit by natural and non-natural disasters throughout history.*

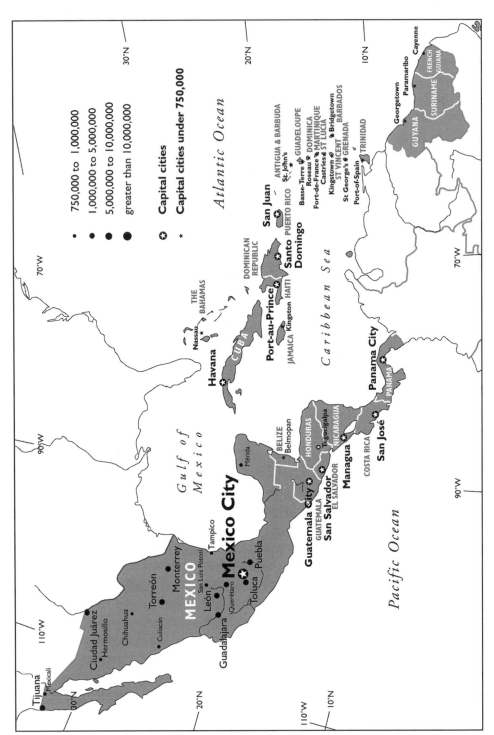

Figure 3.1 Major Cities of Middle America and the Caribbean. *Source:* UN, *World Urbanization Prospects, 2009 Revision,* http://esa.un.org/unpd/wup/index.htm

3

Cities of Middle America and the Caribbean

JOSEPH L. SCARPACI, IRMA ESCAMILLA, AND TIM BROTHERS

KEY URBAN FACTS

Total Population	197 million
Percent Urban Population	71%
Total Urban Population	139 million
Most Urbanized Countries	Cayman Islands (100%)
	Puerto Rico (98.8%)
	Guadeloupe (98.4%)
	Virgin Islands (95.3%)
Least Urbanized Countries	Trinidad and Tobago (13.9%)
	Montserrat (14.25%)
	Saint Lucia (28.0%)
Annual Urban Growth Rate	1.8%
Number of Megacities	1 city
Number of Cities of More Than 1 Million	19 cities
Three Largest Cities	Mexico City, Guadalajara, Monterrey
World Cities	Mexico City

KEY CHAPTER THEMES

1. The Mexican urban system was established in large measure by the Aztec pattern of urbanization. It was militarily subjugated by the Spanish to facilitate the colonizers' dual mission of proselytizing and mining.
2. It was not until the second half of the 19th century that new important regional centers emerged, stimulated by foreign investment and the development of railways and highways.

3. Today, urban growth in Mexico is occurring in intermediate cities located close to large cities, in cities along the U.S. and Mexican border, and in independent cities in remote regions far from the large agglomerations.

4. The urban systems of Central America and the Caribbean developed under various European powers and followed an agricultural-driven model of colonial and post-colonial growth.

5. Today, Central America is nearly 70 percent urban, ranging from approximately 45 percent in Honduras and Guatemala to 59 percent and 66 percent in Costa Rica and Panama respectively. National poverty rates are inversely proportional to urbanization rates in that the poorest countries are the least urbanized countries.

6. Social and geographic segregation has deepened in Central America's cities; crime and violence is a serious problem there and in Mexico.

7. Four patterns highlight contemporary urbanization in the Caribbean: urban primacy characterizes every island, cities with one to five million residents have more than doubled, mid-size cities (500,000 to 1,000,000 residents) have held the same relative proportion of urban residents while smaller cities have declined, and insularity is a key constraint on many islands.

8. Since the mid-19th century, Cuba has taken the most divergent path to urban and national development in the past half century with its variant form of socialist cities.

9. Natural disasters in the Caribbean, Central America, and Mexico compound the challenges of urban poverty.

10. Cross-border urbanization unfolds unevenly across the region. The San Diego-Tijuana model contrasts with the Dominican Republic-Haitian example. However, both processes driving urbanization result from unevenly sized economies, from the demand for unskilled low-cost labor, and different commodity and retail pricing.

The European conquest of the Americas imposed the most dramatic landscape modification in the history of the human race, while unleashing a tragic chapter of inter-continental slavery and the annihilation of millions of Native Americans. The human drama that unfolded over the ensuing five centuries built upon and significantly transformed preexisting patterns and processes of urbanization throughout the region (fig. 3.1). This chapter shows that the Caribbean's urbanization would be shaped by the plantation system and slave trade, whereas the urban development of the mainland of Mexico and Central America unfolded differently. Some Spanish settlements in Central America, Mexico, and the Caribbean replaced indigenous ones (Mexico City) while others served as strategic transshipment points (San José, Costa Rica, and Colón, Panama) or military outposts as part of a network of defensive safeguards (Cuba's original seven *villas*). Urban design came from "military engineers" who aimed to follow guidelines on street width and length, block size, and land use, all of which derived from versions of the colonial Law of the Indies. Some cities developed thus began as uniform grid patterns while others evolved more organically by succumbing to the demands of topography or to the whim of elites. In all instances, a spatially and socially segregated settlement emerged, whose irascible imprint persists half a millennium later.

The Mexican urban system was forged in large measure by the Aztec pattern of urbanization, which was subjugated militarily by the Spanish so the colonizers' dual mission of proselytizing and mining could proceed. Mexico, both in the 15th and 21st centuries, became the economic and urban anchor to mainland production. Its pre-Columbian mining and agricultural system allowed the colonial and independent nation of Mexico to enter the Industrial Revolution before the rest of the region. The urban geographies of Mexico City and Monterrey, explored later in this chapter, highlight the relationship between these resource endowments and industrial-led urbanization.

Meanwhile, the urban corridors of Central America and the Caribbean followed an agricultural model of colonial and post-colonial growth. Primate city functions in two capital cities–San José and Panama City—would be deepened once rail lines and the canal, respectively, opened up to these city's and their hinterlands to world markets. Caribbean urbanization developed slowly and was confined in large measure by limited flat lands in the Caribbean and tied to the fortunes of monocultural exports such as sugar, bananas, and spices. The urban geography of Havana, Cuba, highlights this region's external dependency on trade, sugar, and slavery that, in turn, shaped the process of urbanization.

HISTORICAL GEOGRAPHY OF MIDDLE AMERICAN AND CARIBBEAN URBANIZATION

Mexico

The historical antecedents of the urban system in Mexico date back to the pre-colonial era when the cities were first established. Many pre-Columbian cities remain to this day, and Tenochtitlan (now Mexico City) is the most renowned. At the time of the Spanish conquest in the sixteenth century, Tenochtitlan, located in the Valley of Mexico, had a population of approximately 300,000. It was the most important urban settlement of the Aztec Empire (also known as *Culhua-Mexica)* and the largest pre-Columbian settlement in the western hemisphere. The empire of the Aztecs stretched across a large part of Meso-America. Important settlements co-existed, including the Mayan population in the Yucatan Peninsula, the Tarascos in the present-day states of Michoacán, Jalisco, Colima, and Guanajuato, and the Zapotecas and Mixes in the state of Oaxaca.

Two particular aspects of these patterns of population settlement stand out. First, these large population agglomerations adopted a "city-state" model of organization, whereby a large commercial and religious settlement dominated rural communities and other smaller political-religious localities within their hinterland. Second, the major urban cultures were particularly prominent in the central region of Mexico. Estimates place this dispersed population in 1521, the time of contact, at 2.5 million. This region played a historically significant role in the formation of the subsequent urban agglomerations of the Spanish. Tenochtitlan, re-founded as Mexico City, became the capital of the Spanish Empire.

Mining and agricultural centers constituted the first phase in the colonization of the northern region of the country. Spanish mining towns were founded close to important silver mines, whose indigenous settlements included Taxco, Pachuca, Zacatecas and Guanajuato. These centers and company towns functioned as enclave economies. Bajio, in west-central Mexico, was (re-)constructed during the colonial period as a key base of the agriculture and livestock sector. Abundant natural resources in this region—its fertile plains supported food and fiber for the colonial government—were key factors in its colonization and in the establishment of conditions favorable to future urban growth.

It was not until the second half of the 19th century, following Mexico's independence in 1821 and during the presidencies of Porfirio Díaz (1876–80 and 1884–1911), that new important regional centers emerged. During this period, moderate regional growth was stimulated through foreign investment and through the development of railways and highways. Until the 1910 Mexican Revolution, foreign investment concentrated in railways and mining. Port development linked the railway network to maritime trade. Together, these technological and commercial links led to a proliferation of mining centers in northern Mexico, which, in turn, triggered regional markets and urban growth.

Railroad expansion played a crucial role in stimulating urban growth in various cities in the central and northern regions of the country. Mérida (the hub of commercial sisal plantations) and Guadalajara, Veracruz, Monterrey, and San Luís Potosí (all with direct transport links to Tampico on the Gulf of Mexico) grew rapidly. Old mining towns in the north gave way to new cities. Monterrey, at one time known as the "Pittsburgh of Mexico," became a major center of heavy industry. Veracruz, a principal transport node, handled nearly all import and export shipping cargo.

The economic geographic and political changes that took place during the Porfirio Díaz presidencies had long-term implications for Mexico's urban system. A communication network facilitated interaction between the central and northern regions of the country. High dependency on exports to the United States largely inhibited the formation of a balanced urban system; and those cities that were the largest agglomerations at the start of the 20th century retained their economic and political dominance in subsequent years.

Particular national and international events slowed urban growth in the first decades of the 20th century. The revolutionary movement within Mexico of 1910–1921 and the global economic depression of the 1930s curtailed exports and urban growth. Nevertheless, between 1900 and 1940, the urban population grew at a rate far greater than the total population, increasing from 1.4 to 3.9 million inhabitants. Although the number of cities increased, most urban growth was concentrated in the larger cities of Mexico. In 1900, there were only two cities with populations greater than 100,000. Yet, these made up one-third of urban Mexico and represented 10.5 percent of the national population. By 1940, there were six cities of this size, accounting for 12 percent of the urban population and 20 percent of the total population. The population of Mexico City had reached 1.5 million, and its primacy index had increased; it was nearly seven times larger than the second largest city, Guadalajara.

At the beginning of the 1970s, a shift toward metropolitan expansion emerged as a new form of urban growth in Mexico City and in some secondary cities. There was a massive rural-urban migration flow, with approximate-

Table 3.1   The U.S.-Mexican Border Twin Cities Phenomenon: Population and Employment, 2009*, 2010**

City	Population	Formal Employment
El Paso, Texas*	751,296	313,882
Ciudad Juárez, Chihuahua**	1,062,913	396,911
Laredo, Texas*	241,438	79,008
Nuevo Laredo, Tamaulipas**	373,725	75,210
McAllen, Texas*	741,152	213,458
Reynosa, Tamaulipas**	589,466	191,158
Brownsville, Texas*	396,371	115,855
Matamoros, Tamaulipas**	449,815	126,458

Source: *Table 5. Estimates of Population Change for Metropolitan Statistical Areas and Rankings: July 1, 2008 to July 1, 2009 (CBSA-EST2009–05). U.S. Census Bureau, Population Division. Release Date: March 2010

Source: **INEGI, XIII Censo General de Población y Vivienda, 2010 y Censo Económico 2009

ly 3 million migrants moving to Mexico City in the 1960s. This gave the capital an annual growth rate of 5.7 percent, which was a historic high at that time. Eleven secondary cities experienced notable metropolitan expansion; three of these—Monterrey, Guadalajara, Puebla—had populations over half a million. Three border cities—Tijuana and Mexicali in Baja California, and Ciudad Juárez in Chihuahua—expanded significantly and strengthened their relationships with twin cities across the border (table 3.1).

Mexico's border cities grew in importance when the demand for contractual migrant labor during World War II cast these cities as "staging areas" for border crossings of laborers into the U.S. Between the 1940s and the early 1960s, the *bracero* workers' program (named for the day-laborers who were contracted) brought specified numbers of Mexican laborers into U.S. corporate farm operations. When the program was discontinued in the 1960s, concern grew for the service industries that had developed on the Mexican side of the "twin cities" and for potential unemployment

problems. In response, *maquiladora* factories were established as part of the Border Industrialization Program. This arrangement allowed American companies to import manufacturing parts to Mexican cities, have them assembled in *maquiladora* (*i.e.*, piece-meal assembly) plants, and re-import the finished products into the U.S. while paying only value-added tax. However, with the creation of the North American Free Trade Act in 1992, the relative locational advantage of being close to the U.S. dissipated, as trade barriers excluding the rest of Mexico fell for trade with the United States and Canada. Today, the border cities identified in Table 3.1 retain high levels of manufacturing and service workers and, except for the Mexican twin city of Reynosa-Tamaulipas, even have larger labor markets than their U.S. counterparts. Accordingly, it is more appropriate to think of these twin cities as a single conurbation, working in similar manufacturing and service sectors, rather than as discrete cities divided by an international boundary (box 3.1).

Between 1950 and 1970, Mexico's urban population grew at a rate of almost 5 percent

Box 3.1 Industrial Free Zones and Transnational Urbanization

In the era of globalization, even small cities become international. Goods are imported and exported across international boundaries not just in finished form but often as components of products that are truly international, whatever their apparent country of origin. Perhaps the most obvious example in the Central American and Caribbean context, as in much of the developing world, are the industrial free zones that assemble clothing, electronics, medical supplies, and other goods for shipment to the United States. Although local arrangements vary, the components for assembly are commonly imported from the United States partly processed and duty free, assembled by wage laborers in industrial enclaves subsidized by the Central American and Caribbean host countries, then re-exported—again, duty free—for sale in the United States. These industrial free zones, called *zonas francas*, *maquiladoras*, or *zones franches* in the non-English-speaking countries of the region, are often set apart from the rest of the urban landscape by walls or fences and by acres of single-storied white buildings.

Industrial free zones have attracted special attention along the United States-Mexico border, but they can also be found along the border between the Dominican Republic and Haiti. An industrial free zone has been established on the poorer Haitian side of the border near the sister cities of Dajabón (Dominican Republic) and Quanaminthe (Haiti) (fig. 3.2). To the casual observer, the free zone seems to be on the Dominican side of the border, on the outskirts of Dajabón. In fact, the international border, which has followed the Massacre River down out of the Central Cordillera to the south, here diverges from the river to enclose a small slice of the "Dominican" side of the river in Haiti. The free zone sits on this narrow island between the river and the border, so that the Haitians who work there cross a special bridge across the river each day to arrive at work. Visitors who enter through the main gate on the Dominican side are crossing the international border, though they might not know it. Like Mexican *maquiladoras*, this free zone takes advantage of the large pool of cheap labor on the Haitian side of the border, though here the assembly plants are partly owned by Dominican entrepreneurs, not just American companies.

Dajabón and Quanaminthe are opposite sides of a deep political divide. The Dominican Republic and Haiti have a long history of antagonism; and their border is officially open at only a few points, all of which are well staffed by Dominican soldiers. The Massacre River, named for an early Spanish massacre of Amerindians, was the site of the brutal 1937 massacre of thousands of Haitians at the order of the Dominican dictator, Rafael Trujillo, an event still in the living memory of many Haitians in Quanaminthe. And yet the two cities are ever more closely tied by commerce. The border is opened twice a week for market days, when hundreds of Haitians cross to buy goods for later resale in Haiti. Hundreds more Haitians cross every day to work in the free zone. Haitians have flocked to Quanaminthe from the mountains and coastal plain to seek work; and its population grew from about 7,200 persons in 1982 to

Figure 3.2   Satellite image of the "sister" cities Quanaminthe (left) and Dajabón (right). The border between Haiti and the Dominican Republic follows the Massacre River in the bottom two-thirds of the image but leaves it in the top third to run more directly north. The industrial free zone, visible as the row of large white buildings near the river at the top of the image, lies in a political no man's land between the border and the river. (*Source:* Google-Earth)

about 40,000 in 2003, without any urban planning and with few city services. Lena Poschet, of the École Polytechnique Fédérale of Lausanne, found that by 2004 Quanaminthe had five times the population density of Dajabón, with one-twelfth the urban budget. This growth is likely to continue, particularly as international aid agencies plan more industrial free zones for Haiti in the wake of the January 2010 earthquake.

per annum, while the rural population (in settlements of fewer than 2,500 inhabitants) grew only at an average rate of 1.5 percent per year. From 1950 to 1970, most demographic factors signaled improvements in the quality of life. Despite this progress, there were significant gaps between urban and rural areas. Millions left the countryside in search of work in cities. Almost half of the rural migrants ended up in Mexico City, and one-fifth went to Guadalajara and Monterrey.

By the start of the 1980s, a process of urban growth deconcentration was underway, as intermediate cities in various regions began to experience greater growth rates than larger cities. This process took advantage of the opportunities offered by medium-sized cities located close to large cities. Such amenities included lower costs

Figure 3.3   The demand of (primarily the young) population to "be connected" has meant a number of Internet cafes opening throughout the city,
particularly in areas with schools nearby. (Photo by Irma Escamilla)

of land and housing, newer infrastructure, more parks and open space, and less congestion.

By 2006, 70 percent of Mexico's 103 million people lived in cities. Although a persistent trend nationwide is the concentration of population in an increasing number of large cities, it is important to note that growth rates have dropped for cities of all sizes. Since the 1970s growth rates of cities in excess of one million residents have been consistent with the overall national population growth.

The expansion of various cities and the growth of certain metropolitan areas characterize Mexican urbanization today. These cities generate 75 percent of GDP. The most recent population census of 2005 identified 56 metropolitan areas. Located in 29 of the 32 states in Mexico, these metropolitan areas include almost 58 million inhabitants, or just over half (56 percent) of the Mexican population, and nearly three-quarters (78.6 percent) of the total urban population.

Central America

If the conquest of the Aztec population in the Valley of Mexico was facilitated by a large city, whose leadership was quickly subjugated by Spanish rule, the smaller more dispersed towns of the lands south of Mexico delayed conquest. Although pre-Columbian Guatemala was an important center for the Mayan civilization, the growth of significant cities in Central America dates to the colonial era when Spain was responsible for the politico-administrative division of the region. The Captaincy General of New Spain (Spanish Empire) first used the city of Antigua, Guatemala, as its base; but after a series of earthquakes devastated Antigua, the capital was moved to present-day Guatemala City. Each captaincy (colonial jurisdiction) had a provincial capital: San Salvador in El Salvador, Comayagua in Honduras, Granada in Nicaragua, Cártago in Costa Rica, and Panama La Vieja in Panama. Shortly after 1821, the year in which independence from Spain was achieved, most of these provincial capitals became national capitals.

The climatic conditions in Central America in the early colonial period proved unfavorable to agricultural development. Land on the windward Atlantic side received more precipitation than its Pacific counterpart, was more prone to hurricanes, and consisted of more jungle and swampland. Coastal settlements were especially vulnerable to attacks by pirates, an important factor in locating settlements in higher elevations where the bulk of the indigenous population resided. Central American cities were dependent on agricultural production in their hinterlands, as previously many settlements had been created to group together dispersed agricultural producers in order to charge taxes and also for purposes of religious indoctrination.

The urbanization process in Central America can largely be divided into two main phases. The first period, from 1821 to 1930, includes the first century of independence from Spain and the foundation and subsequent peak of agricultural export economies. The second period dates from the 1930s until the present day and marks a transition in both the economic model in the region and a new phase of accelerated urbanization.

The era of independence shifted hegemonic control of the region from Spain to Great Britain and opened new external markets for the region's agricultural produce, which significantly influenced the nature of urbanization in the region. The early decades of independence marked a transition for some countries from the export of various agricultural products to nearly exclusive export of coffee. In 1835, San José became the capital of Costa Rica, first due to tobacco production and later to coffee production as it became home to the country's coffee oligarchy. The colonial export base of Guatemala shifted from grain production and pig farming to coffee in the second half of the 19th century. By the end of the 19th century, practically all countries in the region depended largely on income from coffee exports. The coffee boom consolidated the Central American capitals. This was especially apparent in Guatemala City (Guatemala), San Salvador (El Salvador), and San José (Costa Rica), where national governments expanded to fill this new political and economic role and city populations and physical expanse grew accordingly.

Meanwhile, North American investment in banana plantations, first in Costa Rica then in Guatemala and finally in Honduras, accelerated urbanization in these regions. Together with the coffee economy, banana production actively produced social differentiation through the need for agricultural, transport, and dock laborers in cities, ports, and hinterlands, and for salaried employees in the emerging urban centers. International capital co-opted and monopolized the communication, transportation, and commercial infrastructures of much of the region and its cities. For example, between 1875 and 1885, the banana enclave appropriated more than 300 miles (483 kilometers) of railroad in Guatemala; this infrastructure had been built with national funds. Multinational companies also controlled the docks and port installations of Puerto Barrios in Guatemala and San José in Costa Rica. Foreign railroad tycoons, such as the American Minor Keith, wielded considerable economic and political influence in

Figure 3.4   Most Guatemalan indigenous women still wear a traditional blouse called the huipil. This ornate garment is woven on traditional looms and then hand embroidered. Villages and towns have their own unique design, which is particularly useful at periodic and regional markets, so both vendors and buyers can readily identify the community of the seller, which in turn means that certain goods and produce might be available there. (Photo by Bobby Bascomb)

expanding the railroads and developing international trade throughout Central America.

Yet, the Central American urbanization process was most prolific around the middle of the 20th century as rural dwellers came to cities in search of work and improved quality of life in terms of education, healthcare, personal security, housing, transportation, and communications. In some cases, their expectations have been met; in other cases, they have encountered disappointment.

In the early 21st century, Central America is more than 70 percent urban (table 3.2), ranging from the high forty-percent ranges in Honduras and Guatemala, to 64 and 74 percent in Costa Rica and Panama, respectively. National poverty rates seem to mirror urbanization rates. Poverty is highest in Honduras (75 percent) and Guatemala (65 percent), and it is lowest in Costa Rica (20 percent) and Panama (35 percent). Coincidentally, Guatemala has the highest indigenous population of any country in the region, standing at approximately five million, which is 80 percent of the total indigenous population of Central America (fig. 3.4). Although levels

Table 3.2   Levels of Urbanization in Central America, 1970–2015

Country	*Level of urbanization (%)*				
	1970	*1980*	*2000*	***2010*	***2015*
Panama	47.6	50.4	65.8	74.8	77.9
Costa Rica	38.8	43.1	59.0	64.3	66.9
Belize	51.0	49.4	47.7	52.7	55.3
El Salvador	39.4	44.1	58.4	61.3	63.1
Nicaragua	47.0	50.3	57.2	57.3	59.0
Guatemala	35.5	37.4	45.1	49.5	52.0
Honduras	28.9	34.9	44.4	48.8	51.4
*Total: Latin America & Caribbean	57.2	65.1	75.4	79.4	80.9
Central America	53.8	60.2	68.8	71.7	73.2
Caribbean	45.4	52.3	62.1	66.9	69.3

Countries are ordered by level of urbanization in 2000

*Urban population as a percentage of total population

**Based on 2009 projections, Source: http://esa.un.org/unup/p2k0data.asp, accessed February 11, 2011

Source: Data from UN, *World Urbanization Prospects, 2009 Revision,* http://esa.un.org/unpd/wup/index.htm

of urbanization in Central American cities have increased over the past 30 years (by 71 percent on average), they still lag behind the Latin American average in almost 80 percent of those countries (table 3.2).

The largest and most important urban centers in Central America essentially correspond to the capital cities of the seven countries, together with other cities important for their economic integration and population levels, such as Antigua in Guatemala and León in Nicaragua. Next are a series of medium-sized and small cities, such as Cártago and Puntarenas in Costa Rica, Acajutla and New San Salvador in El Salvador, Chichicastenango and Esquipulas in Guatemala, San Pedro Sula and Copán in Honduras, Chinandega in Nicaragua, and Portobelo in Panama.

City and population distribution conform in part to the constraints of Central American region's physical geography, which is crisscrossed with extensive mountain ranges and volcanoes, as well as innumerable rivers, waterfalls, and lakes. Central America is also influenced by geological faults where unstable continental plates collide far below the surface. Together with the confluence of weather phenomena, these environmental conditions combine to make most of the human settlements of the region vulnerable to natural disasters, including earthquakes, volcanic eruptions, landslides, floods, and hurricanes. Unfortunately, most countries lack the resources to prevent, prepare for, and/or manage these hazards. Yet, it was precisely the availability of water and land resources that induced the growth of settlements on volcanic soil and flood plains, and which in turn allowed the development of agriculture as the basic economic activity and the consequent urbanization.

Figure 3.5   The Panama Canal is a great engineering feat, which allows deep-draft ships to transport merchandise of every type between the Atlantic and Pacific Oceans. It is also one of the main tourist attractions of the city. (Photo by Jorge González)

Except for El Salvador, which has only a Pacific coastline, and Belize, which borders the Caribbean Sea, the remaining Central American countries straddle the coastlines of both the Pacific and Atlantic Oceans. This allowed seaports to be established which were of vital economic importance to the individual countries and the region as a whole, though export revenues did not always 'trickle down' to national and provincial treasuries. Panama City is a case in point. Following independence, it traded its links with Spain for the geopolitical project of *Gran Colombia* (Great Colombia). Its subsequent growth was based on the development of inter-ocean commu-

nication, initially via railroad and later via the canal projects driven first by France and later the United States (figs. 3.5, 3.6). In 1999, the Canal Zone reverted to Panama after 85 years of foreign administration.

Construction of the Panamerican Highway, dating back to the 1930s, has also promoted urban development throughout Central America. The Panamerican Highway connects all the capital cities but its engineering remains precarious as evidenced by serious landslides. In September 2010, torrential rain at the beginning of the month combined with the prior effects of Tropical Storm Agatha in May to softened mountainsides and created landslides at various

Figure 3.6   Modern Panama City, with its dense cluster of skyscrapers, is seemingly a vision of modernity and prosperity.

points along the main route into Guatemala City. According to local news sources, in a section of 30 miles (48 kilometers), there were more than 30 landslides, cutting off the Panamerican Highway, one of the main roads leading to the Guatemalan capital. Despite these problems, the Panamerican Highway remains the "backbone" of Central America, off which secondary roads connect with other cities and towns.

Urbanization has played a crucial, although not always proactively planned, role in providing the necessary infrastructure to allow ease of movement between a range of traditional tourist and newer eco-tourist sites. The rise of adventure tourism in recent years has generated multipliers in the economies of local communities, especially those of indigenous peoples who are able to organize the control and development of tourist activity. Tourism in all its variations attracts both regional and international tourists, and thus provides an important source of income for the region.

Despite the economic, cultural, and political development that has occurred within and around many cities, the urban panorama in Central America is not promising. Priority is often given to property development serving the high-earning population. Export processing zones took hold in the 1970s throughout the isthmus, and these enclave centers represent a new form of urbanization. Child and adolescent labor is rampant as families press their children into petty commerce, service provision, and begging. Many families face the difficult decision of pushing one or more family members to migrate, either to larger cities or even to the United States. It is common among Central American women, particularly from El Salvador, to make the long trip to Europe, where they characteristically labor in domestic or care-giving work in an effort to support their families economically.

In recent decades, the widespread poverty and exclusion that affects large sections of the population in Central American cities has led to a proliferation of urban gangs. Gang members are typically young and tend to live in peripheral zones of large cities such as Guatemala City (Guatemala), Tegucigalpa (Honduras), and San Salvador (El Salvador). The United Nations Development Program reports that in 2008 gang-caused deaths per 100,000 were highest in Honduras (58), El Salvador (52), and Guatemala (48). The principal gangs are the "Mara Salvatrucha" and the "Mara 18." They define themselves by extensive tattoos on the face and body. They are linked with criminal acts such as the trafficking of drugs and people, assassinations, rape, and assaults. It is estimated that more than 100,000 young people have joined these gangs—40,000 in Honduras, 60,000 in Guatemala, 10,000 in El Salvador (box 3.2). The influence of the *maras* has spread beyond Central America into Mexican, Spanish and North American cities. The social and economic instability in Central America, evidenced in scarce educational and job opportunities and family disintegration, leaves many urban youths to believe they have only two viable options: attempt to migrate to the United States, or join a gang.

The present homicide rate in Guatemala (46 per 100,000) is double the rate in Mexico and ten times higher than that of Honduras and El Salvador. Guatemala, Honduras, and El Salvador are referred to as the "Northern Triangle" (Triángulo Norte) in Central America and these three countries manifest the highest levels of violence. But even the three more tranquil Central American nations (Nicaragua, Costa Rica, and Panama) have experienced increases in violence in recent years. Since 2005, it is estimated that 45 percent of students in secondary education drop out and join gangs, which has created a 50 percent increase in gang membership since then. Evidence now shows that the "maras" have expanded into South America as well. In Argentina, for instance, they have appeared and the high levels of unemployment there are fertile ground for potential gang growth.

The Caribbean

In 1496, Bartolomé Colón established what might well be the first permanent European settlement in the Americas, at Santo Domingo on the southeast coast of Hispaniola. However, it was Governor Nicolás De Ovando, arriving in 1502 with 2,500 new Spanish colonists, who established the enduring outlines of the colonial city. For some two centuries, the Spanish focused their energies on the Greater Antilles, avoiding the fierce Carib Indians to the south. Caribbean ports were scoped for harbor protection, fresh water and provisions, and nexuses between what little mineral and agricultural wealth might be extracted from port hinterlands.

Before colonization of the Caribbean had taken hold, however, the Spanish conquest of the Aztec and Inca empires diverted attention to the mainland. Spanish colonists left the islands in droves to seek their fortunes in gold and silver, and the Caribbean colonies languished. The only exceptions were ports on the routes of the Spanish treasure fleets, such as Havana (Cuba) and San Juan (Puerto Rico), which were

Box 3.2 Gangs: A Violent Urban Social Development

The rise of gangs throughout the cities of Central America is a consequence of many factors. Some observers hold that gangs reflect the struggle of some young people in search of an identity. Others argue that the gangs are the outcome of widespread and persistent poverty and political disenfranchisement. Most observers concur that, in seeking to improve the quality of their lives and acquire what is otherwise unattainable, some youths resort to gang violence. Gangs are associated with such violent/criminal activities as organized crime, arms trafficking, forgery, gangsterism, rape, kidnapping, extortion, and the sale and consumption of drugs. Some gangs demand "taxes" from bus drivers in order to pass through their territory, while others extort protection money from small business owners who operate on their turf.

In Central America, the most notorious and violent type of gangs are known as *maras*, the most infamous of which is the *Mara Salvatrucha* or MS 13. It is made up primarily of young men between the ages of 12 and 25. Although the *Mara Salvatrucha* is dominant in El Salvador, where it represents approximately 70 percent of all youth gangs, it has spread throughout the Americas from Canada to Colombia. It has taken particular hold in impoverished border regions of Mexico and cities of Central America where alternative sources of fulfillment are conspicuously absent. These gangs are particularly distinctive in their highly visible use of tattoos, with many gang members having identifying gang tattoos on their faces, necks, chests and hands.

The word *mara* has become the generic term for youth gangs in Central America. *Mara Salvatrucha* was founded on the streets of Los Angeles by immigrant Salvadoran youths fleeing the Salvadoran civil war. It is alleged that *Mara Salvatrucha* was formed in response to the discrimination and victimization Salvadoran youths experienced at the hands of ethnic gangs proliferating in Los Angeles in the 1970s. Later, other Central American immigrants were integrated into the gang. The word *Salvatrucha* refers to one who is a "shrewd Salvadoran." It is widely thought that the current proliferation of violent gangs in Central and parts of South America is related to large-scale repatriations from the U.S., including many gang members who find fertile conditions in the poverty that is so prevalent in the region's cities. Gangs have come to represent (at least in the popular and political imagination) one of the most serious threats to security and democracy in the region. The formal political power vacuum created by many Central American governments enhances the power of gangs. In many cities, virulent attacks have become an issue of national security. The spread of the *maras* has undermined the authority of the police and weakened the ability of governments to protect communities.

Figure 3.7   In keeping with popular culture, "pirate-endorsed" refreshments are widely promoted throughout the Caribbean. Even Castro's Cuba promotes Bucanero (buccaneer) beer with a striking resemblance to a Morganesque image, adjacent to a Red Bull energy-drink vending machine near Plaza Vieja, Havana, Cuba. (Photo by Joseph L. Scarpaci)

protected by massive stone fortresses. Spanish control of the rest of the Caribbean, an area of more than a million square miles (2.7 million km²), was tenuous at best. Pirates and privateers sallied from undefended islands along the Spanish shipping routes—Antigua, Barbados, Isle de la Tortue, the Bahamas—to attack Spanish ports and shipping. The romanticized exploits of such notorious figures as Bluebeard, Francis Drake, and Henry Morgan are part of the popular image of the Caribbean and its ports (fig. 3.7).

By the 1600s, England, France, and the Netherlands began to stake claims to those parts of the Caribbean not firmly under Spanish control. This coveted real estate included western Hispaniola (now Haiti), Jamaica, and the Lesser Antilles. They were not after gold and silver, which were scarce in the Caribbean,

but forests, farmland, and salt. The pace of colonization and urban settlement quickened dramatically after development of the sugar plantation system on Barbados during the 1640s. As sugar spread rapidly to other British and French colonies, the colonial powers grappled with each other, and often with Carib Indians, to control the best "sugar islands." At the same time, they imported African slaves in immense quantities; of nearly 10 million slaves brought to the Americas before 1870, nearly half came to the Caribbean. Despite horrendous mortality rates, African slaves came to outnumber whites in most of the British and French colonies by ratios of well over 10:1. The Spanish colonies, where slaves and sugar had first been introduced from the Old World, came late to this revolution, but by the late 19th

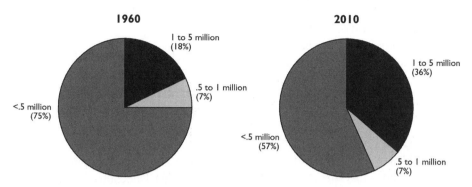

Figure 3.8   Caribbean Urbanization by City Size, 1960 and 2010. *Source:* UN, *World Popula-tion Prospects: The 2004 Revision,* http://www.un.org/esa/population/publications/WPP2004/wpp2004.htm, and *World Urbanization Prospects, 2005 Revision,* http://www.un.org/esa/population/publications/WUP2005/2005wup.htm

century Cuba, the Dominican Republic, and Puerto Rico had also become centers of the sugar industry.

Together, the plantation system and the slave trade established the basis for a distinctive Caribbean geography and set the fundamental settlement pattern of Antillean cities, which were first established on protected leeward harbors as trading centers. Raw sugar, molasses, and rum left these ports for Europe; European foods, machinery, and capital passed through them to the interior plantations. These early sugar ports—Bridgetown, Fort-de-France, Kingston, Port-of-Spain, and Charlotte Amalie—have remained important even as tourism and industrial free zones have supplanted sugar.

Four striking patterns highlight the contemporary urbanization and settlement patterns of the Caribbean. First, no Caribbean island is without its primate city; and, with the exceptions of Havana (Cuba) and San Juan (Puerto Rico), most primate cities are located on the leeward coast, immune from the steady trade winds and often nestled along a deep and protected bay. These historic ports were well suited for anchorage or for loading sugar and unload-

ing slaves, machinery, and provisions. Colonists built gun sites and forts on commanding hilltops and ridges to protect the locals from marauding pirates or rival European colonists.

Second, Caribbean urbanization in the past half-century shows that mid-size cities (500,000 to 1,000,000 residents) have held the same relative proportion of urban residents, while those cities with fewer than half a million residents have declined (fig. 3.8).

Third, places with one to five million residents have more than doubled (fig. 3.8). These trends are particularly striking given the limited amount of low-lying land along the bay fronts, coastal plains, and river valleys etched in the islands' landscapes that can accommodate city growth.

Fourth, beyond the Greater Antilles (Cuba, Hispaniola, Puerto Rico, Jamaica), insularity is a key constraint. With the exception of Barbados and Trinidad, most of the islands are small, and many of them are restricted by mountainous topography, particularly in the volcanic Lesser Antilles. The scale of Caribbean cities therefore pales in comparison to that of other Latin American cities.

Spanish and English settlement patterns provide historical backdrops to contemporary urbanization. Spanish settlements needed to defend the windward approaches into the Caribbean and major ports on the route of the treasure fleets. These locations marked early landfalls for those ships riding the trade winds, and they relied on nearby forests for shipbuilding and repair. Spanish towns in the Caribbean followed the grid-style settlement plan dictated by the Law of the Indies that characterized town design throughout Spanish America. Towns were centered on the main plaza, usually anchored by a government building (*cabildo*) and church at each end. Block size and street width were predetermined; locally unwanted land uses such as garbage dumps, slaughterhouses, and cemeteries were sited at the periphery of the new towns. Early Spanish Caribbean ports facilitated the extraction of mineral wealth from Mexico and other parts of the mainland, and little urban growth took place in Caribbean ports of the sixteenth century.

Non-Spanish settlements were less orthodox and more haphazard in form. In Britain's settlements, for instance, royal favor was doled out to loyalists by the Proprietary System. Caribbean settlers from England had learned from Atlantic seaboard settlements in North America. Accordingly, their priorities entailed clearing land for timber and agriculture, constructing fortresses, and coming to terms with indigenous peoples. Although the British originally planted tobacco and cotton on their settlements, they would gradually turn to the sugar monoculture. In both British and Spanish settlements, little colonial architecture has survived until today other than a few military structures, a few churches, and some fortified (brick and stone) sugar plantations because of fire, tropical storms, and rebuilding. Subsequent discussion of Charlotte Amalie, U.S.

Virgin Islands, and Havana, Cuba will illustrate these historical settlement patterns.

REPRESENTATIVE CITIES

Mexico City: Ancient Aztec Capital, Contemporary Megacity

Mexico's *Distrito Federal,* or capital city (Mexico City), was founded in the 14th century by the Aztecs. Called Tenochtitlan, it soon became the center of the largest empire in pre-Columbian Mesoamerica. As the current capital of the United States of Mexico, it is also the country's largest urban center, and it serves as the nation's economic, social, educational, and political hub. With 20.5 million residents in 2010, Mexico City is the second-largest urban agglomeration in the Western Hemisphere, after São Paulo but larger than New York-Newark. Its population in the 20th century burgeoned from 3.4 million in 1950 to about 9 million in 1970, and was just shy of 15 million by 1990.

In local parlance, "Mexico City" (known as "day efay" for D.F. in Spanish) refers to the entire metropolitan area, which not only covers the Federal District, but also includes parts of the states of Mexico and Hidalgo. At the beginning of the 21st century, its metropolitan area stretched over an area of more than 5,000 kms^2 (1,900 sq. miles or four times the size of the state of Rhode Island), of which the Federal District accounted for nearly 30 percent. Mexico City is located in a closed drainage basin at an altitude of approximately 2,250 meters (7,283 ft.) above sea level. Mexico City is encircled by six mountain ranges and two striking landmarks: Volcano Iztacíhuatl and Volcano Popocatépetl.

The *Zócalo* or main square—now officially called the *Plaza de la Constitución*—is the

Figure 3.9   The Zócalo (Main Square) in Mexico City is surrounded by colonial buildings, most notably the Metropolitan Cathedral and the headquarters of the Federal and Capital Governments. (Photo by Irma Escamilla)

traditional center of the city. On the northern side of the square, close to the ancient site of the main Aztec temple, is the Metropolitan Cathedral. Spanish conquistadores frequently subjugated the Native American population by having them rebuild churches atop the ruins of indigenous temples. To the east is the *Palacio de Gobierno* (main government building); built on the ruins of the ancient Aztec emperor's palace; it is another symbolic replacement of political power (fig. 3.9). The colonial city extended in an orderly fashion for various blocks around this square, as prescribed by guidelines specified in the Law of the Indies. These orders, first issued in Spain in 1494, became the military engineer's template and mandated the location of many colonial and independence-era buildings in this zone. Many of the original structures and buildings of the traditional urban core—known as the *Centro Histórico,* or Historical Center—remain intact.

In general, Mexico City typifies the urbanization patterns and processes for Middle America and the Caribbean, where the historic quarters of most cities have been safeguarded. The more 'modern' aspects of 20th century urbanization developed just beyond the *Centro Histórico.* Unlike the Anglo American and European models of urbanization, the national elite in Spanish America placed more social value on centrality, until the 20th century when congestion, the automobile, and the need for new construction encouraged a slow process of middle- and upper-income suburbanization *a la norteamericana.* As a result, Mexico has dozens of World Heritage Sites that celebrate these colonial quarters. The poor in most Middle American and Caribbean cities tend to concentrate at the city edge where land values are cheaper, and self-help housing develops. In the case of present-day Mexico City, the wealthy districts are concentrated in

Figure 3.10   The colonial architecture of the streets and houses of the colonia Pedregal de San Ángel has been preserved to this day. The area is one of the most exclusive zones in the south of the city, and is also an important tourist attraction. (Photo by Irma Escamilla)

the west and various zones in the south, and in *colonias* such as Lomas de Chapultepec, Polanco and Pedregal de San Ángel (fig. 3.10). These districts contrast sharply with the poverty in the northern zones and the illegal settlements in the eastern edges, beyond the Benito Juárez International Airport, where many communities in zones such as Chalco and Ixtapaluca lack basic services (box 3.3).

As in other corners of the world, suburban retailing challenges the traditional role of the city center as the main retailing district. Examples include the *Plaza Satélite* in the north of the city, *Perisur* in the South (fig. 3.11), and Sante Fé in the west. The market, *La Merced*, which had been the main food market for the city since colonial times, was replaced in the 1980s with a modern market in the east of the city (fig. 3.12).

The import substitution industrialization strategy implemented in the 1940s created conditions of stability and prosperity that made Mexico City the most important industrial center in the country. Today, it is responsible for 30 percent of national industrial production. In the second half of the 20th century, encouraged by the Border Industrialization Program of 1964, heavy industry began moving from the capital to border cities of the north. Just as many U.S. manufacturing towns lost jobs to lower-wage labor in *maquiladoras*, so too did Mexico City. As a result, the under- and unemployed work in informal commerce.

Mexico City is also the economic center, accounting for 45 percent of the country's commercial activity, and remains the hub of the national transport system. Five main highways link the capital to the different regions

Box 3.3 GIS and the Solution of Urban Problems

Mexico City has been expanding since the second half of the 20th century. Since 1995, a range of technological innovations has been employed to map irregular human settlement in former conservation lands in the southern part of the city. GIS and remote sensing systems aid in identifying biophysical, economic, and social conditions that cannot be observed with the naked eye.

One analysis identified 8,017 areas of squatter settlements (shantytowns) and examined a sample of 493 of these areas. It determined that shantytowns expanded into the eastern and western extremes of the conservation area at an annual rate of 13.2 percent between 1995 and 2000, and to the west-central and south-central zones at a rate of 4.6 percent between 2000 and 2005. The study found that expansion was taking place beyond settlement borders. It further detected fragmentation of the properties as well as differences in relief variation among the settlements under study due to road construction. In some traditional settlements there are areas of relatively continuous building, while in others there is notable dispersion. Squatters account for as much as half the total area of all new expansion of traditional settlements (*i.e.,* legally zoned and regulated).

The use of GIS and remote-sensing techniques may contribute to the design of policies appropriate for both these vulnerable populations and ecologically fragile land.

of the country, as well as with Guatemala and the United States through the now 70-year old Pan American Highway. There is an extensive intra-city transport network, including the metro system that is used by over 2 million people daily (fig. 3.13), and a range of different types of buses.

Mexico City has always been one of the most important cultural centers in Latin America. It boasts a number of major cultural sites, and its cinema, film, theatrical, and television industries rival those of Buenos Aires, Argentina. The *Palacio de Bellas Artes* in the center of the city is an important opera and concert venue, and the Cultural Center of the National Autonomous University of Mexico in the south hosts the National Library, a large concert hall, and various theatres. The National Museum of Anthropology is considered one of the most important of its kind,

and some monuments, such as the Chapultepec Castle and the Independence Monument, are considered national symbols (fig. 3.14). Mexico City is a megacity of significant history, impressive scale and striking contrasts.

Monterrey: Mexico's "Second City"?

Monterrey is the capital of the state of Nuevo León, and it is situated approximately 200 km (124 miles) from the Texan border. It is the third-largest city in Mexico, with a 2010 population of 3.7 million. Urbanization in Monterrey derives in large measure from its proximity to the U.S. border. Its relative location is also advantageous because it is located where the plains of the Gulf of Mexico and the eastern Sierra Madre meet. A pass through these mountains gives Monterrey direct access to the Caribbean Sea at the Tampico-Altamira port.

Figure 3.11   Perisur is one of the large shopping malls in the south of Mexico City, where the presence of globally recognized prestigious stores provides stark evidence of globalization. (Photo by Irma Escamilla)

Monterrey was founded in 1579 as *San Luís Rey de Francia*. In 1596, it became known as the Metropolitan City of Monterrey. It was built, on the west bank of the Santa Catarina River, according to the grid system dictated by the Law of the Indies, with streets crossing at right angles around a central plaza, which at the time was the *Plaza de Zaragoza*. Given its distance from the center of the country, Monterrey's population remained stationary until the 18th century, and it was not until the 19th century that the city began to grow significantly. When Mexico ceded the territory of Texas to the United States in 1848, the new border region around the Rio Bravo (called the Rio Grande in the United States) began to prosper. Contraband and the cotton trade during the American Civil War promoted the growth of Monterrey and that of many other border cities. Between 1882 and 1905, railroads linked Monterrey with Laredo in Texas and Tampico, Matamoros, and Mexico City, establishing the base for Monterrey's industrial development.

Monterrey's greatest industrial development took place between 1890 and 1910. About 40 members of Monterrey's original ten powerful families have been linked to more than 260 corporations involved in diverse economic activities. After the Mexican Revolution of 1910, most of these family members formed part of the famous Monterrey Group.

Figure 3.12   The Central de Abasto in Mexico City is the city's main collection and distribution point for retailers and wholesalers, which means thousands of daily transactions in the buying and selling of goods as diverse as fruit and vegetables, cereals, seeds, cleaning products, pharmaceutical goods, sweets, and cigarettes. (Photo by Irma Escamilla)

The economic expansion of the 1940s spurred industrial integration for the Monterrey Group. Industrial investment fortified the banking and financial sectors, and in 1943 the Monterrey Technological Institute of Advanced Study (*Instituto Tecnológico de Estudios Superiores de Monterrey*, ITESM) was founded to provide future generations of executives and administrators of Monterrey's industry with a high-quality university education. By 1960, the electronics industry was established, with pronounced growth in transport and car manufacturing. Factories spread along the edge of the city and its transportation corridors. The automobile industry began supplying the growing Mexican automobile industry as well

as the traditional core in Detroit, Michigan. The oil boom of the late 1970s stimulated development of the petrochemical industry in Monterrey through productions of petroleum-derived materials and fibers, marine exploration rigs, and submarine ducts.

Economic liberalization in the late 1980s ended advantages previously granted to Monterrey by the government. The implementation of the North American Free Trade Act eliminated any comparative advantages that cities located near the border had as maquiladora assembly sites--value-added taxes are now available for products assembled anywhere in Mexico and then sent to the United States. As a result, Monterrey looked to the export market

Figure 3.13   The Metro is the backbone of the city's public transport system, transporting over 2 million passengers a day via 11 lines. (Photo by Irma Escamilla)

which allowed various corporations to sell off holdings, develop strategies of co-investment, and form strategic capital alliances with the United States, Europe, and Asia.

In addition to hosting key educational, cultural, healthcare, and business enterprises, Monterrey boasts entertainment attractions that are significant for both national and international tourism. One of the most famous landmarks of the city is the *Cerro de la Silla*, a hill that is famous for its likeness to a riding saddle (fig. 3.15). The *Macroplaza*, situated in the heart of Monterrey, is one of the largest plazas (though it is more of a park) in the world. It stretches 40 hectares (99 acres) and contains a playful mix of green areas, monuments, and buildings Commercial activity and services cluster near shopping malls, most of which host various "shoper-tainment" services including cinemas and restaurants, and attract shoppers from within and beyond the city limits.

It is noteworthy that Monterrey, like other cities in Mexico, has been battered by a wave of drug-related violence among rival drugs gangs and with the police. Increasingly, innocent civilians have also become victims of this violence. Public authorities are not immune to this violence either; the assassinations of the mayors of the Santiago and Dr. González municipalities, near Monterrey, attest to this. Ongoing attacks against police stations heighten levels of insecurity, fear, and concern about public safety among residents. Narco-violence is having a dampening effect on how citizens think about using Monterrey's prolific public spaces.

Figure 3.14   Among the most distinctive monuments in Mexico City is the Angel of Independence, which located on the Paseo de la Reforma. It is here that the city's inhabitants congregate in collective celebration, for example of sporting victories. (Photo by Irma Escamilla)

San José, Costa Rica: Cultural Capital Beset by Problems

San José is the political and economic capital of Costa Rica. Like most Latin American cities, San José is laid out in a grid pattern anchored by a series of town squares fronted by churches. In 2006, the Union of Spanish American City Capitals declared San José the cultural capital of Spanish America. Costa Rica's relative economic prosperity and political stability has made San José the safest city in the region. In recent years however, crime has risen; and it is a serious concern.

San José is the primate city in Costa Rica. It is more than twice the size of Limón, Costa Rica's second largest city located on the Atlantic/Caribbean coast. In contrast to most primate cities in Latin America, San José is located in the geographic center of Costa Rica. The semi-humid and temperate climate and fertile soils of the region favor intensive agriculture, and high-quality export products such as specialty leaf tobacco do well here. Since the colonial era, settlement and development have been concentrated in this part of Costa Rica. With time, settlement gradually spread outward toward the coastal plains, a pattern that runs counter to the patterns experienced in most Latin American countries where settlements first took hold at navigable ports on the coasts and gradually moved inland.

A series of hills within the Central Valley has not curtailed San Jose's expansion, whose metropolitan zone today encompasses the adjacent communities of Alajuela, Cártago and Heredia. This conurbanization constitutes the "Central Region" and spills into adjoining valleys and mountains regions. Although the Central Region includes just 15 percent of the country's land area, more than half of the national population lives there. Wealth generated from Costa Rica's mining and agricultural sectors has supported business investment in San José and the subsequent expansion of its metropolitan region. Urban sprawl has overtaken small towns and outlying villages to such an extent that some peripheral zones lack basic services such as housing, jobs, and schools.

Metropolitan San José, like most primate cities, contains the most important and largest industries, businesses, and residential areas of the country. This concentration implies changes in land use, private-sector investment, and the distribution of wealth. San José consists of 14 *cantones* (administrative units similar to counties in the U.S.). Most *cantones*

Figure 3.15   A satellite view of Monterrey illustrates the process of metropolitanization and shows the distinctive physical feature of this Northern city, the Cerro de la Silla in the upper right, and the Macroplaza in the the lower center. (Photo by Google-Earth)

are residential areas that function as bedroom communities and are distant from most places of work, retail commerce, and medical and educational facilities. There is a growing demand in the more distant *cantones* for jobs, housing, and infrastructure to accommodate the city's growth.

Agglomeration reinforces San José's primacy and disadvantages other regions of Costa Rica that are less densely populated and suffer from poor public services. Urban sprawl imposes high economic costs, necessitates the consumption of fossil fuels, and

exacts human costs in the form of long and stressful commutes. A road network unable to accommodate present usage exacerbates these problems. Moreover, San José's sprawl threatens rich agricultural and protected lands in the Central Valley. In general, rapid growth and congestion threaten the sustainability of this capital city.

In recent years, Costa Rica has developed its tourism industry. It takes advantage of a variety of natural resources, including an excellent climate and such spectacular physical features as mountains, volcanoes, beaches and rainforests. Tourism has energized the national economy as it has attracted hard currency expended by visitors from North America, Europe, and Asia. Most tourist ventures start in San José and fan out to the interior of the country. Although this creates economic multipliers for San José and its hinterland, it also creates economic and environmental stress. The tourism infrastructure (*e.g.,* expansive networks of hotels, restaurants, and land and air transportation) must be maintained and upgraded continually to meet international expectations. San José experiences the financial, infrastructural, and environmental pressures that accompany Costa Rica's international notoriety as a safe, secure, and high-quality tourist destination.

Havana: The Once and Future Hub of the Caribbean?

Diego de Velázquez de Cuéllar founded San Cristóbal de Habana in 1519 as one of seven military outposts (*villas*) around the island of Cuba. Just two of these original settlements— Camaguey (then Puerto Principe) and Santiago—were founded on good harbors. Havana was first located in 1514 on the Broa Inlet, at the Gulf of Batabanó, on the island's southern (Caribbean) side. The shallow port and the generally swampy (and unhealthy) site forced colonists to relocate directly north, to the Atlantic side of the narrow island. Although Santiago de Cuba reigned as the official island capital until the late 16th century, Havana's relative location was enhanced by the discovery of the Bahamian Channel, which served as a key transshipment route for goods exchanged between the Americas (chiefly Mexico) and Europe. Lacking mineral wealth and a large native population to enslave or evangelize, Havana served as a strategic refurbishing port

and a temporary holding place for precious metals coming from Mexico and Andean South America.

Military engineers enhanced this colonial port by building a network of fortresses over the next two and one-half centuries. Flotillas carrying wealth out of the ports of Cartagena and Santa Marta in Colombia, Nombre de Diós in Panama, and Veracruz in Mexico would dock in the safe waters of Havana before crossing the Atlantic for Seville, Spain. Ranching, timber, shipping, and allied services would define the colonial city's economy. It lacked the wealth of Lima, Peru and Mexico City, but Havana served as a vital link in the Spanish colonial empire. Securing a steady freshwater supply was a prerequisite to Spain's granting the title of 'city' on its American settlements. The completion of a major aqueduct in 1592 brought fresh water from the Almendares River to the west and sealed Havana's fate as an official city. It soon surpassed Santiago de Cuba in political and economic significance in the region, and it has not ceded that position in over four centuries.

Havana's location on a pocket-shaped bay made it an ideal warehouse and transshipment point. It is so narrow at the entrance to the harbor from the Florida Straits that military officers often drew chains across it at night to entrap intruders. Located on a plain with mild marine terrace escarpments that yield to a gently undulated topography, the city is unconstrained by topographic barriers except for the bay (which curtailed growth to the east until a tunnel was completed in 1957). A wall on the western edge was started in 1663 and confined growth. By the time the last stone was laid in 1740, a polygon consisting of nine bastions, several parapets, and escarpments completed the city's extensive system of defensive works. However, an alert British officer noticed the city was unprotected on the eastern flank of the bay, and in 1761 disembarked a small squadron of men to the east of Havana. In just one month, the British bombarded the old city, disrupted its food and water supplies, and raised the flag of the British Empire over Havana. The following year Cuba was traded for Spanish Florida,

A sugar boom following the British occupation of the 1760s brought commerce and residents to Havana, and crowding exacerbated problems within the walls. New neighborhoods sprung up outside the walled city, and the elite gradually left the walled quarters. In the early 1860s, Havana's walls were torn down, opening up a huge expanse of city blocks that were ideal for urban development. A series of civil wars between the Cuban colony and Spain, however, would delay new construction. Between 1868 and 1898, war aggravated the quality of life, deteriorated the city's infrastructure, and—except for the Albear water system completed in 1893—few public works enhanced Havana.

When the Americans occupied Cuba after the 1898 Spanish-American War under the terms of the Platt Amendment (1889–1902), they found Havana to be a lackluster place. Ripe for investment, U.S. business speculators poured into the city. Road construction, railroad expansion, banks, customs houses, sugar and cigar-factory construction, telephone services, and the newly arrived automobile offered opportunities for American capitalists. The U.S. Army Corps of Engineers lent a hand, particularly in expanding, raising, and extending the seaside promenade El Malecón—a striking seaside boulevard that graces much of Havana's northern edge (fig. 3.16).

Over the course of the 20th century, Havana would become a horizontal city in the mode of Los Angeles and less a high-density New York-like city. Unlike most Caribbean ports, Havana could expand across a wide coastal plain. It developed a series of suburban enclaves west and south of the bay. Automobile commuting for a middle-class of white-collar workers drove this suburbanization model, and led to a scattered and deeply segregated pattern of urban growth. While a streetcar network operated until the early 1950s, the automobile and bus would link Havana's new suburban and exurban developments. When the revolution of 1959 succeeded, about one in 20 residents were living in shantytowns of some sort. The socialist government imported models of high-rise prefabricated buildings like those used in Eastern Europe and the former Soviet Union (see box 1.3). While only one million residents claimed Havana as their home in 1959, the population had barely surpassed the two million mark fifty years later. Over the same period, Mexico City and Lima, Peru had increased six and three fold respectively.

This checkered pattern of urban history, warfare, and revolutions has given 21st century Havana a unique urban morphology. It is a polycentric city that has preserved distinctive architectural designs and land uses. Colonial, Republican, new government centers, and social/cultural districts characterize this panoply of urban nodes (fig. 3.17). Havana is one of the few Latin American cities where rather benign light industry (*i.e.*, cigar making) surrounds the colonial and Republican government centers.

In the first three decades of the revolution, Havana was largely a "closed" destination; few tourists came, and those who did hailed primarily from the former Soviet Union trading-bloc member states. Moreover, immigration to Havana from elsewhere is strictly controlled by a food-ration book (*la libreta*) and other

Figure 3.16 El Malecón as seen from atop the Focsa Building in Vedado, looking east, running along the Florida Straits eastward towards Centro Havana and Habana Vieja. (Photo by Joseph L. Scarpaci)

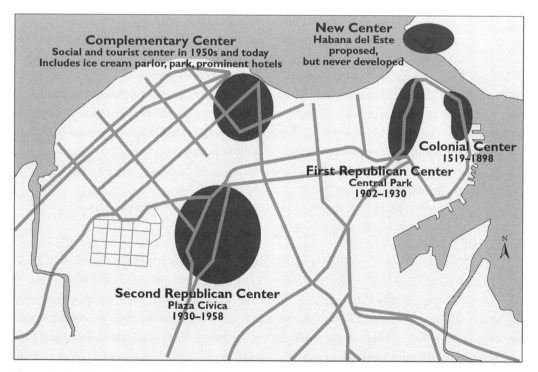

Figure 3.17 The polycentric city of Havana. Based on J. Scarpaci, R. Segre, and M. Coyula, *Havana: Two Faces of the Antillean Metropolis* (Chapel Hill, NC: University of North Carolina Press, 2002), 87.

governmental controls. However, the demise of the USSR in 1991 led to a major crisis called the "Special Period in a Time of Peace." The government tightened gasoline rations as Cuba's ability to exchange sugar for Soviet oil disappeared. Thousands of un- and under-employed Cubans have migrated illegally to Havana, mainly from the eastern provinces where the dwindling sugar economy has been devastated. In typical Cuban humor, these immigrants are called *palestinos* because they hail from the east.

Other changes are also visible in Havana. As fuel subsidies from the USSR ended, and the relative cost of gasoline soared, bus routes were scaled back to half their number and bicycling as a means of transportation boomed (from some 70,000 bicycles in 1989 to one million in 1999). Tourism was seen as a "necessary evil" to sustain the island's economy, and the city's Old Havana district (Habana Vieja), a UNESCO World Heritage Site since 1982, became a prime destination for newfound cultural tourism. In 1993, the City Historian of Havana created a moneymaking corporation to address housing, hotel construction, road paving, plaza reconstruction, and urban revitalization. This firm, Habaguanex, has become one of the most powerful state enterprises in post-Soviet Havana. The firm has embarked on an ambitious project to rehabilitate buildings and spaces in Habana Vieja. International tourism has grown significantly from about 25,000 annual visitors to Cuba in the late 1970s to just over 2.3 million visitors in 2010, comparable to the growth in tourism experienced in nearby Cancún, Mexico. In October 2010, Raúl Castro announced the downsizing of a half million state workers who will presumably be working in a newly expanded private sector. The VI Cuban Communist Party Conference held in April 2011 approved radical free-market changes such as the sale of private homes (with minimum state intervention), the downsizing of 1 million public workers, and an increase in the number of private sector jobs from 158 to 174. Time will tell whether limited these tentative steps towards entrepreneurship will change the face of a city that has been centrally planned for half a century.

Unlike many other Caribbean cities, Havana is home to a world-class biotechnology industry and boasts the third busiest airport in the region. It possesses the open space for more growth, either in the form of vacation homes for North Americans and returning Cuban-American expatriates or to accommodate an unfettered U.S. tourist market. In contrast to many Caribbean port capitals, Havana attracts only a few thousand cruise-ship passengers annually, largely because shipping companies face legal problems from the United States if they conduct business in Cuba. Nevertheless, the Caribbean manages some of the busiest maritime traffic in the world; approximately 50,000 ships navigate there and carry 14.5 million tourists annually. Havana will be on the radar of urbanists who are interested in issues of smart growth, sustainable development, and sustainable tourism as the 21st century progresses.

URBAN CHALLENGES

Some of Middle America's and the Caribbean's largest metropolises, particularly those in Mexico, reveal slowing urban rates of growth that can be attributed to a demographic transition and a decline in rural-to-urban migration. Nevertheless, these metropolitan expanses are increasingly spreading out well beyond their original boundaries. As a result, the pattern of urban settlements has become more dispersed as cities increasingly encroach on the adjacent countryside. Urbanization has spatially, economically, and socially incorporated many smaller towns and cities in this process. Guadalajara, Puebla, and Mexico City in Mexico, San José in Costa Rica, and Guatemala City reflect this process. Tools of urban and regional planning are needed to manage this level of urbanization (see box 3.3). Despite a population loss or "hollowing out" of the city center in these metropolises, the capital city continues to dominate in most countries.

Mid-size cities, on the one hand, have maintained an impressive rate of population growth. They offer new promise for job creation and enhanced quality of life as opposed to the quality of life offered in very large cities. Nevertheless, urban planners and administrators in these cities will be challenged to avoid replicating the problems that have plagued larger metropolitan areas. The viability of intermediate-sized cities will depend mainly on their economies, including the degree of integration at the global scale, the type of articulation they maintain at the national and regional level, and the extent to which they can tap into their comparative advantages.

Social and geographic segregation in the region's cities has deepened and is a serious problem. Demand for exclusive high-income communities often leads to displacing poor groups from targeted urban neighborhoods. Public housing projects concentrate at the city's edge because of lower land values. In turn, this exacerbates social and spatial segregation. High-income groups increasingly isolate themselves defensively in limited-access or gated communities that are adorned with costly houses; and are convenient to attractive retail businesses, entertainment, and recreational facilities, as well as having relatively closer proximity to work, school, and other amenities. Poor households continue to occupy precarious houses on remote and marginal lands near landfills, utility plants, factories, water-treatment plants, flood plains, and on steep terrain.

About 43 percent of the region's population was living below the poverty line in 2009, while 17 percent of the population lived in extreme poverty. Data further reveal an increase in the concentration of the poor in urban areas. Two of every three poor persons resided in cities in that same year, a greater portion than in the previous decade. Considerable disparity of income is also evident within the region. While the poor in Costa Rica amount to about one-fifth of the national population, Mexico and Panama reveal a 35 percent poverty rate, and Guatemala, Honduras and Nicaragua exceed the 60 percent mark. These latter three countries have levels of extreme poverty in excess of 30 percent. Haiti stands out, even alongside these Central American countries, as the poorest country not only in the region but also in the entire western hemisphere. Precarious shanties exist in such marginal environments as dumps, ravines, and coastal and riparian mudflats (fig. 3.18).

Natural disasters compound regional poverty. Active earthquake faults run through Mexico, Central America, and the Caribbean, and active volcanoes line these fault zones, except in the Greater Antilles. Most of the region experiences periodic intense rainfall and sometimes severe hurricanes. The steep, unstable slopes of the area's mountains encourage floods and mass landslides. Natural disasters such as these merely exacerbate these conditions.

Port-au-Prince again serves as an illustration. The city is clearly in harm's way. It lies in the middle of the Caribbean hurricane belt, on the edge of the Enriquillo-Plantain Garden Fault zone, and at the base of the steep Massif de la Selle. As the city's population exploded after World War II, shacks and poorly built cinder-block

Figure 3.18   Two aerial views of shantytowns (bidonvilles) in low-lying areas just north of the Port au Prince, Haiti, city center. Flooding occurred in these areas after Hurricane Noel struck the island of Hispaniola on October 29–31, 2007. The storm claimed at least 30 lives in the Dominican Republic and 20 in Haiti. (Photos by Joseph L. Scarpaci)

houses spread beyond the traditional boundaries of Port-au-Prince onto rapidly growing coastal mudflats, urban washes, and the slopes of Morne l'Hôpital, the mountain that looms directly behind the city. Until 2010, flooding and landslides were the most important risks; even normal storms sometimes caused deadly floods along the city's crowded ravines.

These disasters were forgotten, however, in the massive earthquake of 12 January 2010, which killed more than 200,000 people and left hundreds of thousands homeless (figs. 3.19, 3.20, 3.21, 3.22). Satellite images of Port-au-Prince in November 2010 illustrate some of the extensive destruction and displacement caused

ten months earlier by the 7.0 magnitude quake, the effect of which was made much worse by the instability of the urban infrastructure. The presidential palace—an ornate structure reminiscent of French colonial times—collapsed upon itself (Figs 3.19 and 3.20). So, too, did many other buildings, displacing thousands of people to tent camps like that at the former military airport just north of the city center (Figs 3.21 and 3.22). Such camps, conspicuous for their blue tarps, popped up anywhere space was available: vacant lots, parks, roadsides, and even on a golf course. One of the most conspicuous camps, perhaps symbolic of the continuing plight of the refugees, is right

Figure 3.19   Presidential Palace, Port-au-Prince, August 2009, five months before January 2010 earthquake. (Photo by Google-Earth)

Figure 3.20   Collapsed Presidential Palace and tent camps, Port-au-Prince, November 2010, ten months after January 2010 earthquake. (Photo by Google-Earth)

Figure 3.21   Former military airport north of Port-au-Prince city center, July 2009, six months before January 2010 earthquake. (Photo by Google-Earth)

Figure 3.22   Tent camp at former military airport north of Port-au-Prince city center, November 2010, ten months after January 2010 earthquake. (Photo by Google-Earth)

across the street from the destroyed presidential palace (Fig. 3.20).

The tragic consequences of the earthquake that struck Port-au-Prince are not unique, but rather the destruction seen in the city is an extreme example of how natural and human disasters work together in Latin and Caribbean cities. Nature presents risks, but cities amplify them by destabilizing slopes, by exacerbating flood peaks, by shunting the poor onto flood plains, and by encouraging construction of high-density homes on every available hillside. In Middle America and the Caribbean, poverty and vulnerability are intimately intertwined with natural hazard.

Early warning systems, capable management, and institutional and political development are fundamental steps in dealing with emergency preparedness and rebuilding in cities throughout Middle America and the Caribbean. Strong political will is needed to tackle the problems affecting the daily lives of people living in cities throughout the region, in keeping up with 21st-century goals of economic development through environmental sustainability.

SUGGESTED READINGS

Bolay, J.-C. and A. Rabinovich. 2004. "Intermediate Cities in Latin America: Risk and Opportunities of Coherent Urban Development," *Cities*, Vol. 21 (5): 407–421. Examines how medium-sized cities relate to other cities in their national urban systems and the wider world.

Brothers, T. S., J. Wilson, and O. Dwyer. 2008. *Caribbean Landscapes: An Interpretive Atlas.* Coconut Beach, Florida: Caribbean Studies Press. Surveys characteristic urban and rural landscapes of the Caribbean, using satellite imagery, ground photos, and essays. The last chapter focuses on Caribbean cities.

Cravey, A. 1998. *Women and Work in Mexico's Maquiladoras.* New York: Rowman & Littlefield. Examines the relationship among economic globalization, gender, and migration in Mexican piecemeal assembly-line industries. Geographer Cravey draws on economic, political, and cultural geographic resources in explaining this phenomenon.

Edge, K., H. Woofard, and J. Scarpaci. 2006. "Mapping and Designing Havana: Republican, Socialist, and Global Spaces." *Cities,* Vol. 23 (2): 85–98. Urban-historical assessment of how Havana's built environment can be read as a window to past influences from the U.S., the Soviet Union, Spain, and Cuba's own creative styles.

Scarpaci, J., R. Segre, and M. Coyula. 2002. *Havana: Two Faces of the Antillean Metropolis.* Chapel Hill and London: University of North Carolina Press. Reviews five hundred years of urbanization and Havana's spatial configuration as a mirror to periods of economic development, political control, and architectural imprint.

Scarpaci, J. and Portela, A. 2009. *Cuban Landscapes: Heritage, Memory and Place.* New York: Guilford Press. Examines the construction of sugar, slavery, heritage, and political landscapes that shape the meaning of "*cubanidad*" as seen from disciplines, including landscape architecture, art history, history, popular culture, and geography.

Scarpaci, J., Kolivras, K., and Galloway, W. 2011. Engineering Paradise: Marketing the Dominican Republic's Last Frontier. In S. Brunn, Ed., *Engineering Earth: The Impacts of Mega-engineering Projects.* Dordrecht Heidelberg London New York: Springer, pp. 1267–81. Summarizes the development of the Cap Cana project—the largest of its kind in the Caribbean—in eastern Hispaniola and examines the place-promotional strategies employed by Donald Trump, the Ritz-Carlton, and other international brands.

West, Robert C., and Augelli, John P. 1989. *Middle America: Its Lands and Peoples.* 3rd ed. Englewood Cliffs, NJ: Prentice Hall. A highly regarded text by two prominent American geographers who worked in Middle America and the Caribbean.

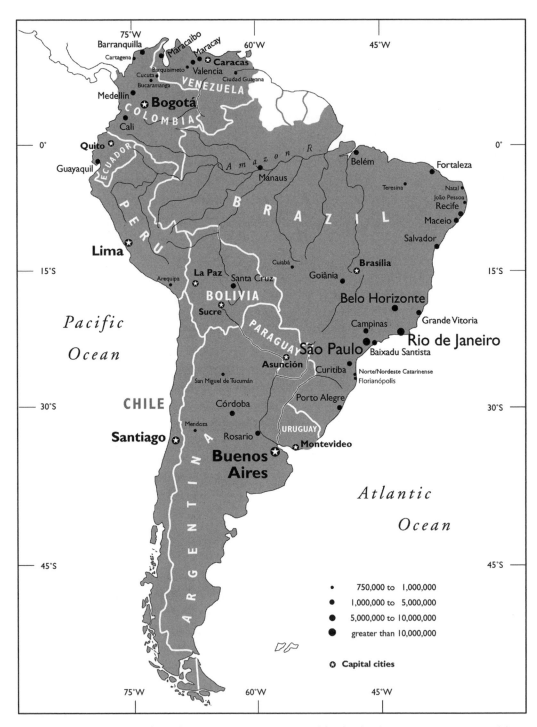

Figure 4.1 Major Cities of South America. *Source:* UN, *World Urbanization Prospects: 2009 Revision,* http://esa.un.org/unpd/wup/index.htm

4

Cities of South America

MAUREEN HAYS-MITCHELL AND BRIAN J. GODFREY

KEY URBAN FACTS

Total Population	382 million
Percent Urban Population	84%
Total Urban Population	329 million
Most Urbanized Countries	Venezuela (93.4%)
	Uruguay (92.5%)
	Argentina (92.4%)
Least Urbanized Countries	Paraguay (61.5%)
	Ecuador (67.0%)
	Bolivia (66.5%)
Annual Urban Growth Rate	1.73%
Number of Megacities	3 cities
Number of Cities of More Than 1 Million	38 cities
Three Largest Cities	São Paulo, Buenos Aires, Rio de Janeiro
World Cities	São Paulo, Buenos Aires, Rio de Janeiro

KEY CHAPTER THEMES

1. South America is highly urbanized, but its rate of urban growth has slowed in recent years.
2. The region contained three of the world's largest megacities of over ten million people and a total of 42 cities of more than 1 million in 2010.
3. Cities of Andean America includelarge indigenous and mestizo populations sharing urban space with small elite groups often of European heritage.
4. Southern Cone cities are generally heavily European in ethnic composition as well as in urban planning traditions, although recent migratory trends from Paraguay, Bolivia, and Peru have created more racial diversity.
5. Brazil's cities have a Portuguese colonial heritage and urban forms distinct from their Hispanic counterparts, including significant Afro-Brazilian cultures.

6. Most countries in South America are dominated by a primate city, often the national capital, although dynamic economic centers like Guayaquil, Medellin, and São Paulo also have arisen.

7. The cities (and countries) of South America exhibit extreme disparities in wealth, which is directly reflected in the land-use patterns and quality of life within cities.

8. Economic globalization has mainly benefitted a small segment of the urban population, despite intensifying social movements, urban protests, and governmental efforts to address inequities.

9. In recent decades a rise in urban insecurity and criminality has led to a withdrawal of elites and middle classes from many city centers, often to gated communities, shopping malls, and fortified office parks in suburban areas.

10. Rapid urbanization has caused serious environmental problems, especially air and water pollution, in many South American cities. Innovative efforts to create more inclusive and sustainable forms of urbanization have emerged in several cities, most notably such cities as Bogotá and Curtiba.

South America's cities evoke dramatic, if conflicting, mental images. The mere mention of Rio de Janeiro, Buenos Aires, Bogotá, Caracas, Lima, Quito, or Santiago conjures up scenes of spectacular natural settings, breathtaking vistas, cosmopolitan populations, picturesque colonial architecture, charming market streets, and impressive modern skylines. By contrast, their names also evoke images of squalid squatter settlements, intractable poverty, random violence, hapless street children, congested motorways, filthy air, and polluted waterways. To be sure, both images accurately portray different aspects of contemporary urban life in South America. Just as the continent is a land of great extremes, so are its cities. Despite many outward similarities, South American cities are quite diverse in form, environment, culture, economic function, political governance, and quality of life (fig. 4.1).

South America's urban centers have long participated in the world economy. Since the colonial era, cities throughout the region have served as important global producers and consumers. Today, South American cities vigorously compete for financial, manufacturing, and service-oriented multinational enterprises. Cultural currents from around the world—art, architecture, music, fashion, cuisine, athletic events, and digital technology—flow across the continent. Both advocates and critics of globalization agree that societies are being propelled in broadly similar socioeconomic, political, and cultural directions. Is it inevitable, then, that places caught up in these processes come to look and feel alike? South American cities suggest otherwise.

Collectively, South American cities contrast with those of other world regions in many ways. Various factors account for shared continental characteristics: common colonial legacies of Iberian urbanism; similar historic patterns of economic development; recent globalization of cultural tastes, production, and technology; and a growing socioeconomic polarization, spatial segregation, and informal economies. Despite such similar trends, the diversity of national and local experiences also stands out. Some cities originated with the Spanish conquest, while others derived from Portuguese colonization. Many in both categories are infused with the presence of indigenous cultures, European immigrant cultures such as German and Italian, and African cultures that derived from the slave trade. These cities exhibit disparate urban forms, contrasting levels of economic development, and varying forms of political governance, all spread across some of the most diverse natural environments on earth.

SOUTH AMERICAN URBAN PATTERNS

South America's cities may be grouped into three major cultural-ecological regions: (1) Andean America (Colombia, Venezuela, Ecuador, Peru, and Bolivia), (2) the Southern Cone (Chile, Argentina, Uruguay, and Paraguay), and (3) Portuguese America (Brazil). The cities of Guyana, Suriname, and French Guiana are more appropriately understood in the context of the Caribbean region. Despite a general adherence to many broad continental trends, there are also significant urban and regional differences.

- The cities of Andean America reveal a greater indigenous presence than do those of the other two regions. Andean cities tend to be divided by ethnicity, as large indigenous and mixed *(mestizo)* populations share urban space with small elite groups of European heritage. Rapidly growing Andean cities also are dominated by an "alternative economy" of the informal sector and popular markets. As the fastest urbanizing South American region at present, Andean cities operate under fiscal constraint and hence are experiencing severe social, political, environmental, and infrastructural crises. On the other hand, such cities as Bogotá have become world-renowned for their innovative programs of urban planning, mass transit, and sustainability programs.

- Southern Cone cities, with the exception of Paraguay, are heavily European in ethnic heritage as well as in urban traditions. Although human development indicators suggest relative prosperity, these cities contend with long-standing problems of economic stagnation and a restive middle class. Apart from Paraguay, which is similar to Andean countries in its strong indigenous presence and socioeconomic problems, the Southern Cone underwent its urban and demographic transitions by the mid-20th century. Urbanization rates in these relatively high-income countries peaked long ago. The largest cities—such as Buenos Aires, Santiago de Chile, and Montevideo—now grow relatively slowly vis-à-vis the cities of Brazil and Andean America.

- Brazilian cities have a Portuguese colonial heritage and language, a unique popular culture, and urban forms distinct from their Hispanic counterparts. The Roman name for Portugal was Lusitania, so we speak of the Luso-American cities of Brazil. In socio-cultural terms, important Afro-Brazilian populations make black-white stratification a key urban issue. With an estimated 87 percent

Table 4.1   Urbanization in South American Countries, 1850–2010

Country	Percentage of National Population in Urban Areas				
	1850	1910	1950	1970	2010
Argentina	12.0	28.4	65.3	78.9	92.4
Bolivia	4.0	9.2	33.8	39.8	66.5
Brazil	7.0	9.8	36.2	55.8	86.5
Chile	5.9	24.2	58.4	75.2	89.0
Colombia	3.0	7.3	42.1	56.6	75.1
Ecuador	6.0	12.0	28.3	39.3	66.9
Paraguay	4.0	17.7	34.6	37.1	61.5
Peru	5.9	5.4	41.0	57.4	76.9
Uruguay	13.0	26.0	77.9	82.4	92.5
Venezuela	7.0	9.0	46.8	71.6	93.4

Sources: Clawson, David L., *Latin America and the Caribbean: Lands and Peoples,* McGraw-Hill, 2006, P. 350; Population Division of the Department of Economic and Social Affairs of the United Nations Secretariat, UN, *World Urbanization Prospects: 2009 Revision,* http://esa.un.org/unpd/wup/index.html

of the 190 million Brazilians living in cities by 2010, Brazil has undergone a rapid urban transition—even vast Amazônia is now three-quarters urbanized. With over 20 million residents, Greater São Paulo is now the largest megacity in the Western Hemisphere and third-ranked in the world. Given the massive socio-economic and environmental problems of the largest metropolises, the highest rates of urbanization have now shifted to the smaller and intermediate cities.

CONTEMPORARY URBAN TRENDS

A century ago, less than 10 percent of South Americans resided in urban centers. By the middle of the 20th century, the national populations of Argentina, Chile, and Uruguay were predominantly urban. Today all countries are more than 60 percent urbanized, ranging from a low of 61.5 in Paraguay to a high of 93.4 in

Venezuela. Five countries are more than 80 percent urbanized, with Argentina, Uruguay, and Venezuela surpassing 90 percent, and Chile and Brazil not far behind. Although migration to the cities and natural population increases have declined in recent years, South America continues to face major problems stemming from a rapid urbanization (table 4.1).

Due to urban-based industrial development, South American countries today appear poised on the global semi-periphery, exhibiting characteristics of cities in the more developed "core" and the less developed "periphery." While the developing countries of Africa and Asia are urbanizing rapidly, South America's current urban levels already approximate those of North America and Europe. Yet South American cities are less affluent and more socially stratified than their northern counterparts. Sadly, urbanization and economic growth have not necessarily been synonymous: continental cities have grown more rapidly, within a highly

Table 4.2   Metropolitan Populations of South America, 1930–2010

Metropolitan Area, Ranked by 2010 Estimates	POPULATION (IN THOUSANDS)				
	1930	*1950*	*1970*	*1990*	*2010*
1. São Paulo, Brazil	1,000	2,334	7,620	14,776	20,262
2. Buenos Aires, Argentina	2,000	5,098	8,105	10,513	13,074
3. Rio de Janeiro, Brazil	1,500	2,950	6,637	9,595	11,950
4. Lima, Peru	250	973	2,927	5,825	8,941
5. Bogotá, Colombia	235	676	2,391	4,905	8,500
6. Santiago, Chile	600	1,322	2,647	4,616	5,952
7. Belo Horizonte, Brazil	350	412	1,485	3,548	5,852
8. Porto Alegre, Brazil	220	488	1,398	2,934	4,092
9. Salvador, Brazil	350	403	1,069	2,331	3,918
10. Brasília	—	36	525	1,863	3,905

Sources: Charles S. Sargent, "The Latin American city," in Brian W. Blouet and Olwyn M. Blouet, *Latin America and the Caribbean: A Systematic and Regional Survey*, pp. 188; Population Division of the Department of Economic and Social Affairs of the United Nations Secretariat, UN, *World Urbanization Prospects: 2009 Revision*, http://esa.un.org/unpd/wup/index.html

competitive global system and a regional context of poorly distributed wealth and endemic poverty. As a result, South American cities are confronted with pressing social, economic, political, and environmental issues.

CRITICAL ISSUES

Urban Primacy and the Growth of Large Cities

The urban transformation of South America has been characterized by urban primacy. Forty-two urban centers contained at least one million people by 2010, and four of them were among world's 30 largest cities: São Paulo, Buenos Aires, Rio de Janeiro, and Lima. Much of the region's urban population resides in metropolitan megacities. The disproportionate growth of primate and other large cities emerged historically, as early colonial centers became modern gateway cities appealing to foreign investors, industrialists, immigrants and internal in-migrants, transportation inno-

vations, and governmental infrastructural subsidies. Nearly every country in South America is dominated by a primate city, often the national capital. Bolivia, Ecuador, and Venezuela are each dominated by two primate cities. Brazil is dominated by a huge megalopolis anchored by two mega-cities, São Paulo and Rio de Janeiro, while the young capital of Brasília—barely fifty years old—already approaches four million residents (table 4.2).

Economic Polarization and Spatial Segregation

Although South America's megacities are centers of great wealth, this wealth is poorly distributed, a lingering impact of the hierarchical governance system that was implanted during the Spanish and Portuguese colonial period. Economic liberalization at the global scale has led to pronounced socioeconomic polarization at the national and local scales. The region's growing social divide can be read on its urban landscape. It is estimated

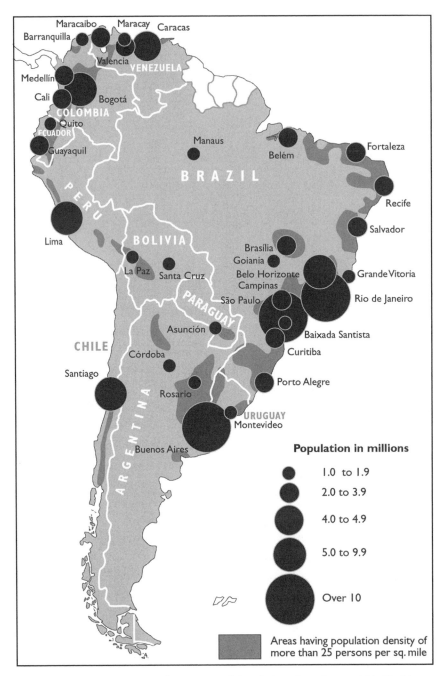

Figure 4.2   Largest Metropolitan Areas of South America. *Source:* UN, *World Urbanization Prospects: 2009 Revision*, http://esa.un.org/unpd/wup/index.html

that as many as four out of every ten urban dwellers in South America live in conditions of absolute poverty. Although many South Americans are participating in—and benefiting from—economic globalization, regular employment remains elusive for a large proportion of urban dwellers. Many are forced to cobble together meager livelihoods in the urban informal economy, laboring in such low-paying and insecure occupations as street trading, in-home manufacturing, domestic service, spot construction, itinerant transportation, and money changing. Similarly, elite and professional districts are luxurious and characterized by limited-access residential units, business and commercial centers, and entertainment facilities. Yet, a large proportion of the urban population is housed in inadequate conditions and seeks shelter in dangerous structures with poor sanitation and limited, if any, rights to the land on which their homes are built. As the gap in wealth widens in South America's cities, violence and crime are rising; personal security and political unrest are growing concerns.

Declining Infrastructures and Environmental Degradation

According to the terms of ongoing programs of economic restructuring, municipal governments have been forced to curtail expenditures, payrolls, and services. Such fiscal constraints make urban management more difficult. The gradual breakdown of urban infrastructure and service systems places enormous stress on an already strained metropolitan system. The lack of appropriate infrastructure contributes to water and air pollution, as household and industrial waste and traffic congestion degrade the urban environment. Unplanned and unregulated growth exposes vulnerable populations to environmental hazards and health risks. With basic services inaccessible for many and pervasive air pollution, the quality of urban life steadily erodes. Indeed, mortality rates in some parts of South America's largest cities are rising.

Social Movements

While many scholars interpret the contemporary problems of South America's large cities as yet another expression of the intense social and economic divisions that have characterized the region since Spanish and Portuguese colonization, others consider them a manifestation of an unfair global economy. Meanwhile, many urban dwellers are taking matters into their own hands by participating in self-help social movements in the areas of housing, health care, and service provision. Self-help movements have proliferated in recent decades, serving to fill critical vacuums as municipal budgets have contracted and city services abandoned. Yet as frustrations mount, so, too, do urban protest, social tension, and political violence. Increasing numbers of urban residents are joining broad-based calls for economic relief, human rights, and environmental justice. In a world of instantaneous communication, their causes are garnering attention far beyond the region.

HISTORICAL PERSPECTIVES ON SOUTH AMERICAN URBAN DEVELOPMENT

Pre-Columbian Urbanism

Urban settlements have long played an important role in South American societies. The

Box 4.1 Community-Based Websites and GIS in Rio's *Favelas*

Arriving in Rio de Janeiro, visitors are struck immediately by the spectacular setting of mountains and seaside districts, the warm and humid air, and the exuberant tropical vegetation. This mesmerizing vision of paradise is quickly tempered by the unmistakable presence of extensive hillside slums known as *favelas*—impoverished shantytowns precariously perched over the city, above affluent high-rise beach districts like Copacabana and Ipanema, or set in swampy, polluted, or otherwise undesirable terrain. Despite the geographic proximity, the obvious socioeconomic distance draws a powerful line between the two poles of society. Officially about 20 percent of Rio's total population (6.3 million in 2011) now lives in more than 500 *favelas*, but realistic estimates suggest that up to a third of it resides in such irregular settlements.

A similar experience unfolds for visitors to other South American cities, although local topographies do not always make the social divisions so obvious. As the region has urbanized, cities have become increasingly segregated along socioeconomic, racial, and ethnic lines. The poorest districts consist of self-constructed housing, where residents often squat without legal title to the land. Here they struggle to obtain decent livelihoods, improve their residences, and obtain urban services like water, sewerage, and electricity. Official statistics tend to underestimate the presence of these informal settlements, but they form undeniable parts of the urban landscape. They are known by a variety of local names, such as *poblaciones callampas* (Chile's "mushroom populations"), *villas miserias* (Argentina's "towns of misery"), and *pueblos jóvenes* (Perú's "young towns").

While governmental agencies often ignore the slums, community groups have developed innovative websites featuring oral histories, photography, and GIS to document their ex-

spectacular settings and monumental beauty of the Inca cities of Cuzco and Machu Picchu spring readily to mind. The Inca, however, were only the final stage in a 4,000-year history of urban development in pre-Columbian Andean America, which stretched from present-day Colombia to Chile and Argentina. Even though the urban heritage of Andean America has garnered most attention, large sedentary communities also flourished across a range of ecological settings in the Amazon region. Two notable features unite settlements in these distinctive regions: first, their successful adaptation to and strategic use of the challenges and opportunities of a diverse habitat; and second, their near total destruction by invading Europeans through violence or disease and, in some cases, their reconstruction to reflect a new and unfamiliar value system.

Colonial Cities: Spanish vs. Portuguese America

After their initial voyages of discovery and conquest in the early 16th century, both the Spanish and the Portuguese established settlements to exploit and administer their new

istence. NGOs work to change the public's prejudices, while enhancing the residents' pride of place. For example, "Viva Rio" (www.vivario.org.br) was founded in 1993 to combat urban violence, promote community development, and facilitate the education of "at risk" youth. This NGO subsequently started the "Viva Favela" (www.vivafavela.com.br) program to focus exclusively on low-income neighborhoods. A related website, "*Favelas* Have Heritage" (www.favelatemmemoria.com.br), features social histories and ongoing struggles of the communities. GIS analysis of the evolving distribution of *favelas* from 1920–2000 indicates a relative decline in the affluent central and southern locations, along with an overwhelming shift to outlying northern and western "suburbs." In addition, graphs contrast selected *favelas* with citywide standards to document glaring disparities in socio-economic status (income, employment, literacy) and service provision (trash collection, sewage, water provision).

Other NGOs take different approaches evident in their websites. Catalytic Communities, or CatComm (www.comcat.org), showcases community-generated solutions to everyday problems in poor neighborhoods. Describing itself as a tool for community support, "Catalytic Communities uses technology to link grassroots community groups so they can learn from each other's successes, and support one another's work." Since its inception in 2000, Cat-Comm has facilitated over 130 projects in nine countries. The program's founder, Theresa Williamson, traces her inspiration " . . . to direct observation of positive things that were going on in our communities here in Rio de Janeiro." This work for positive social change at the community level, despite largely negative media representations of the urban poor, suggests the importance of NGO community groups currently struggling to improve life in the *favelas*.

territories in South America. In their urban expressions, the Spanish and Portuguese colonies differed in terms of site selection, general morphological characteristics, and geopolitical strategies. In both cases, however, the enduring importance of the early colonial cities has been perpetuated in the continuing patterns of urban primacy that persist to this day. Enduring Iberian legacies are also reflected in the cultural and religious landscapes of cities, such as the dominant Roman Catholic cathedrals and parish churches in central cities and residential neighborhoods.

The main center of Spanish colonial power in South America lay in the Viceroyalty of Peru, centered on the former Inca Empire in the Andean highlands. The dramatic fall of the Inca Empire provided a rich source of labor, silver, and gold. Spain proceeded to extend this initial conquest with expeditions into other areas of the continent. Spain founded towns both on the coast and in highland areas. Port cities such as Callao on the Pacific, Buenos Aires on the Atlantic, and Cartagena on the Caribbean linked the new colonies to the Spanish homeland. In highland areas, the Spanish conquered

Figure 4.3   Spanish conquistadores built Mediterranean-style structures atop Inca stone walls in pre-Columbian cities such as Cuzco in present-day Peru. (Photo by Maureen Hays-Mitchell)

dense indigenous populations, along with their minerals, complex agricultural systems, and other natural resources. The colonial overlords rebuilt important indigenous centers to serve as new imperial cities. They forcibly concentrated the indigenous populations into arbitrarily created villages known as *reducciones,* rebuilt the Inca capital of Cuzco (fig. 4.3), and established such enduring Andean centers as Bogotá, Medellín, Quito, La Paz, and Potosí. Town founding served as a central instrument of colonization, dominating the countryside and imposing a profoundly urban civilization.

Spain did not centrally plan its earliest colonial settlements, but new towns generally adhered to a set of standards established during the late medieval Reconquista of southern Iberia and codified in the Discovery and Settlement Ordinances of 1573. The so-called "Laws of the Indies" decreed the distinctive physical form and location of Spanish settlements in the New World. The Spanish-American city adopted the distinctive feature of a right-angled gridiron of streets oriented around a central plaza. The imposed urban form of the Laws of the Indies towns essentially served as an effective instrument of social control: urban morphology and social geography were intertwined. Important institutions such as the Roman Catholic cathedral, the town hall *(cabildo),* the governor's palace, and the commercial arcade bordered the central plaza. Spanish residents clustered around the urban core, often in houses that were built with the defensive architecture such as an external wall and an enclosed inner courtyard. Indians and undesirable land uses were banished to the urban periphery, a pattern replicated in contemporary cities. These features continue to distinguish the Spanish-American city from cities founded by the Portuguese in Brazil and by the French and British in the Caribbean.

Figure 4.4   The Pelourinho historic district, named for the "pillory" formerly used to castigate slaves, reflects the strong Afro-Brazilian influence in Salvador da Bahia. The historic center of Salvador da Bahia became a UNESCO World Heritage Site in 1985. (Photo by Brian Godfrey)

In Portuguese America, the eastern coast of the continent initially proved less alluring than Spain's Andean empire, with its rich silver and gold mines, and large sedentary indigenous labor forces. Consequently, the early Luso-Brazilian settlements were somewhat smaller and less carefully planned. Early settlements in Brazil generally were located close to the coast, at convenient points of interchange between the rural areas of production and metropolitan Portugal. Except for São Paulo, all the towns established before 1600 were located directly on the coast and functioned essentially as administrative centers and military strongholds, ports and commercial entrepôts, residential and religious centers. To reinforce strategic footholds, the Portuguese crown began to designate captaincies, or land grants, in 1532. The captaincy system divided Brazil's coastal strip. It allowed Portugal to combine elements of feudalism and capitalism and to employ relatively few of the crown's funds. Yet Brazil was constantly under attack from other European powers, so Portugal created a more centralized Spanish-style system in 1549, with Salvador da Bahia as capital. From about 1530 to 1650, sugarcane cultivation on coastal plantations became enormously profitable, powered by the labor of imported African slaves. With a population of 100,000 by 1700, Salvador grew to become the most important early Portuguese settlement and the second largest city in the entire Portuguese realm, after Lisbon itself (fig. 4.4).

The coastal location of most early settlements underscored the importance of a good port and a defensible site, so settlers often favored hilly and topographically irregular terrain in the extensive Serra do Mar, the rugged mountains that stretch along much of the central Brazilian seacoast. These towns took on linear, multicentered forms. Irregular mazes of streets focused on a series of squares along the waterfront, as opposed to the more regular grid plans of the Spanish cities. Despite their apparently picturesque confusion of city streets adapted to the topography, early Portuguese settlements adhered to coherent but flexible principles of spatial order. The early colonial towns were set on defensible hilltop sites, where they prominently featured fortifications, important public buildings, churches and convents, and residential areas, all connected by a maze of winding streets and punctuated by ornate public squares. Class-segregated neighborhoods emerged, as elite mansions for rural aristocracy and urban merchant classes were set apart from slave districts. The 18th century gold and diamond boom in Minas Gerais and other areas of the interior provided new wealth and stimulated urban growth, while increasing oversight by Portuguese authorities and encouraging more centrally planned and regulated cities. Late colonial Brazilian cities also witnessed a flowering of baroque art and architecture still notable in the exquisite historic districts of Ouro Preto, Salvador da Bahia, Rio de Janeiro, and other favored cities.

Neocolonial Urbanization: Political Independence, Economic Dependence

Between 1811 and 1830 independence came to each of the countries of South Amer-

ica (except for "the Guianas"). However, throughout South America, characteristically colonial urban forms persisted, long after political independence was achieved. Until the mid-19th century, when elites embarked on campaigns of economic expansion, cities remained relatively small. Thereafter, South America became increasingly integrated into the global economy through the export of primary commodities—beef, minerals, coffee, rubber—and the import of manufactured goods. Focused on trade with North America and Europe, economic expansion fostered population growth, social change, and urban morphological adaptation. Urban growth proceeded with the creation of new transportation links, rural-urban migration, urban infrastructures, and general commercial development. First affected were mercantile cities, such as Rio de Janeiro, Montevideo, Buenos Aires, and Santiago. These leading cities in turn diffused technological innovations and capital investments to inland centers of primary-commodity production, that is, their interior hinterlands. New urban services gave the privileged cities an image of modernity and attracted migrants from the interior.

Mounting internal migration and foreign immigration contributed to South America's increasing rates of urbanization. By 1905, Buenos Aires' population surpassed one million and Rio de Janeiro's exceeded 800,000. Eight other South American cities—São Paulo, Santiago, Montevideo, Salvador, Lima, Recife, Bogotá, and Caracas—had between 100,000 and half a million inhabitants. Correspondingly, the percentage of the national population living in the largest city steadily rose. Commercial expansion and demographic growth led to widespread deficiencies in urban housing, transportation, sanitation, and public health, often the subjects of reform movements. The modern city emerged as entrepreneurs invested in new building projects and planners mounted ambitious public works projects to rationalize urban form. Architects and planners looked to European cities as their main sources of inspiration. For example, as urban renewal programs gentrified the center of Paris into an elegant residence for elites, Latin American architects and engineers similarly reformed their own *fin-de-siècle* cities. The two leading centers, Buenos Aires and Rio de Janeiro, subsequently underwent significant urban renewal programs as they competed for continental leadership. This Eurocentric focus in South American city planning paralleled the continent's political-economic and cultural dependence on distant neocolonial powers.

Twentieth Century: The Urbanizing Century

As South America moved into the 20th century, the pace of urbanization accelerated. The urban metropolis, not the rural countryside, came to define the regional landscape. The continent's neocolonial trade status subsequently shaped the course of early industrialization and urbanization well into the 20th century. The region's cities were promoted as poles of "modernization," defined in terms of urban-industrial infrastructure and industrial labor. In reality, cities became modern enclaves whose existence facilitated the extraction and basic processing of primary agricultural and mineral products for an export market. Their fate depended on the transfer of technology and expertise from more advanced trading partners, while the benefits of trade largely remained in the metropolitan regions and had little effect on the wider regional economies.

With the worldwide depression of the 1930s, demand for the region's primary products plummeted, unemployment soared, and poverty spread. By the early 1950s, a spirit of economic nationalism gripped most South American governments, as they intervened directly in the workings of their economies. The goal was to alter the pattern of producing primary products for export in favor of producing manufactured goods for domestic, and ultimately foreign, consumption. The development of domestic industry focused on major urban centers, because they offered broad access to the national market, a concentrated pool of labor, political influence, and the infrastructure of transport and communication facilities. Investment in the urban-industrial sector was generally favored over the rural-agricultural sector and life became increasingly untenable for small-scale agricultural producers. Thousands of rural dwellers were drawn to cities in the hope of finding jobs, housing, health care, educational opportunities, and social mobility for themselves and their families. Cities grew at an unprecedented rate, due to both in-migration and relatively high fertility rates.

Initially, most cities were able to accommodate their expanding populations. Rapid industrialization created manufacturing jobs as well as demand for commercial, financial, and public services. New building technologies, coupled with new forms of transportation, ensured that living conditions were at least adequate. Medical technology made cities relatively healthy places in which to live. However, as conditions of urban primacy intensified throughout the region, smaller cities languished. Rapidly growing primate cities were as dependent as ever on imported technology, in the form of modern machinery and

Figure 4.5   Carpenters in a Lima shantytown (pueblo joven) are typical of the informal economy that is so widespread in Lima and other South American cities. (Photo by Rob Crandall)

replacement parts, fostering external indebtedness and balance-of-payment deficits.

To address these shortcomings, national development shifted from an exclusive focus on nurturing domestic industries to a focus on establishing development growth poles. Growth pole development precipitated elaborate national development plans with a range of outcomes. Chile embraced this strategy, but it proved to reinforce preexisting patterns of industrialization and urban primacy. Brazil invoked this development model in its efforts to allay the vast differences in living standards between the more prosperous and industrializing coastal southeastern region and the largely agrarian and impoverished northern and northeastern regions. Although growth pole development can be credited with the expansion of Northeastern industry and large-scale mining and highway projects in Amazonia, it can also be blamed for environmental degradation and enduring socioeconomic deficiencies. The most successful example of growth pole development occurred in Venezuela, where Ciudad Guayana, founded along the Orinoco River in 1961, benefitted from hydroelectric power and mineral resources to become a center of steel production and other heavy manufacturing.

By the mid-1970s, many growth poles were perceived to be mere enclaves of foreign capital, since investment favored export industries, which were more closely linked to northern firms than to regional or national economies. Hence, most surplus capital left the region, precluding any significant spin-off of related firms and services. Development failed to trickle down the urban hierarchy and, instead, elicited massive migration to the cities and further growth of already dominant cities. Few well-paying manufacturing jobs were available to the largely underskilled rural migrants who swarmed to the cities. Most were left to seek employment at low pay and low levels of productivity, further polarizing rich and poor throughout the region (fig. 4.5).

Despite these problems, national governments continued to finance costly development—especially industrialization and infrastructure—through borrowing on foreign capital markets. Northern commercial banks aggressively courted both private and state interests in South America, as nearly every country in the region accumulated significant debt. Yet each moved steadily along the economic and social development trajectory. Primate cities remained important. They served as national headquarters for local ruling groups and multinational enterprises and as centers for the accumulation of capital and diffusion of a globalizing consumer-based lifestyle. Moreover, they provided living space for increasing numbers of working-class and marginalized peoples.

The period between 1950 and 1980 saw consistent improvement in urban living standards. Most urban centers were characterized by an expanding middle class and active government promotion of home ownership. Mortgage systems became more accessible and urban infrastructure and services improved. Water, sanitation, education, medical care, and cultural opportunities were readily accessible. Although updated motorways and increased automobile ownership facilitated the growth of elite suburban communities, cars and mortgages were largely inaccessible to lower income city dwellers. Consequently, cities underwent explosive growth in self-help housing—primarily squatter settlements—and related programs to service them.

By the early 1980s, however, the global economy had experienced a series of unanticipated shocks that would devastate urban life within the heavily indebted countries of South America. The International Monetary Fund required countries to exercise extreme fiscal restraint at every level of national life, in order to build up state revenue for debt service and eventual repayment. The debt crisis and related reforms precipitated a sustained period of deep recession and a reversal of much progress in development. While most countries transitioned from military to civilian rule by the 1990s, the dominant neoliberal model of privatization and deregulation increased socioeconomic polarization. Factories closed, public-sector employees were laid off, and social programs critical to the poor were slashed. Throughout the region, access to adequate shelter and public services worsened, and physical and social infrastructures deteriorated. Underemployment (the underutilization of one's skills or the inability to secure full-time employment) came to characterize a large portion of the economically active population in many cities.

The early 21st century has witnessed a rise of social activism and progressive democratic governments in Brazil, Argentina, Bolivia, Venezuela, and Ecuador; and more moderate-conservative tendencies have emerged in Chile, Colombia, and Peru. Creation of the South American Community of Nations (UNASUR) in December 2004 signaled increasing political-economic cooperation, despite remaining conflicts among participants. Increases in commodity trade (especially oil, minerals, soy, and other agricultural products) and rise of China's economic presence have been widely accompanied by a significant decline in poverty and broadening of domestic markets, most notably in Brazil, now the world's seventh-largest economy. On the other hand, the distribution of income remains highly uneven and slum growth continues throughout the region.

DISTINCTIVE CITIES OF SOUTH AMERICA

The spatial structure of South American cities has been an important topic for comparative urban research, given continental variations in urban form. While Spanish and Portuguese urban traditions differentiated colonial cities, subsequent post-colonial influences from France, Britain, and the United States have broadly affected the region during periods of rapid urbanization. South American cities now experience heightened degrees of socio-spatial differentiation; new areas emerged through processes of inner-city gentrification, affluent suburbanization, and peripheral commercial development of "edge cities." In the larger metropolises, functional decentralization created urban realms of varying socioeconomic levels, replete with shopping centers, office parks, and gated communities, separated from the older CBDs (central business districts).

Although the contemporary cities of South America look and feel modern and international, they are still beset by problems of poverty unparalleled in the north. It is tempting to speak of these urban landscapes as "dual cities" in which a modern, affluent, and progressive element has little to do with a poor, obsolete, and unseemly element. In reality, however, the modern, globally linked city and the impoverished, polluted city are intertwined aspects of the same metropolitan landscape. This contrasting landscape of extreme wealth and extreme poverty epitomizes the region's enduring legacy of underdevelopment, economic polarization, and social injustice. Yet each of South America's cities is unique. Rio de Janeiro and São Paulo, anchors of Brazil's megalopolis, epitomize Luso-American

urbanization, while emerging cities in the Amazon Basin have a strong regional character. Brasília deserves study as the most famous new capital of the 20th century and a bold experiment in city planning. While Lima epitomizes Spanish-American urbanization for Andean America, Buenos Aires does so for the Southern Cone. Although each city is distinct, each is representative of the evolving urban experience in South America.

Rio de Janeiro and São Paulo: Anchors of South America's Megalopolis

The vast urban region of southeastern Brazil, centered on the São Paulo-Rio de Janeiro metropolitan areas, now represents about one-quarter of the national population and one-third of the nation's GNP. This megalopolis now encompasses nearly fifty million people in an area the size of Austria. Two-thirds of the population lies in São Paulo state, including the capital region (more than 20 million in 2010) and the extended metropolitan areas of Campinas (5 million), the Santos coastal lowlands (2.5 million), and São José dos Campos (2.5 million). Rio de Janeiro's portion includes the capital region (12 million), the Paraíba Valley (2 million), and the urbanized coastal areas of the Costa Verde and Capo Frio/Búzios areas (1.5 million). The metropolis of Juiz de Fora in the neighboring state of Minas Gerais (.5 million) also forms part of this integrated urban region (fig 4.6).

Despite their increasing regional integration, São Paulo and Rio de Janeiro retain their own distinct identities. Residents of Rio (known as *Cariocas*) and those of São Paulo (known as *Paulistas* for the state or *Paulistanos* for the city) are famous for their competitive, dueling dispositions. Hackneyed images of

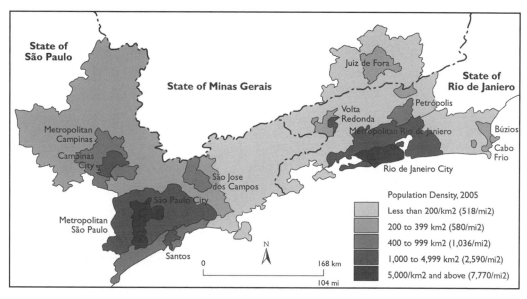

Figure 4.6   The Rio de Janeiro-São Paulo-Campinas extended metropolitan region. *Sources:* Instituto Brasileiro de Geografia e Estatística (IBGE); Centro de Informações e Dados do Rio de Janeiro (CIDE); and the Fundação Sistema Estadual de Análise de Dados (SEADE), 2007. (Map by Brian Godfrey and Laurel Walker)

the fun-loving, easy-going *Carioca* and the intense, hard-working *Paulista* are exaggerated, but like many stereotypes both reflect a particular social history. Rio de Janeiro—famous for its spectacular seaside views and popular culture of the samba, bossa nova, and carnival celebrations—has long been an international playground for the jet-set as well as a world famous beach resort. By the time that Rio lost the national capital to Brasília in 1960, rival São Paulo had taken the economic and demographic lead in this rapidly modernizing country. While Rio deindustrialized and grew increasingly dependent on tourism and other urban services, São Paulo grew through industrial, commercial, and financial dynamism to become the preferred location for multinational corporate headquarters in South America. Now considered the business capital of Mercosur—the emerging common

market centered on Brazil and Argentina—São Paulo is known as a fast-paced, resourceful, temperate metropolis with distinctly urban charms and challenges.

Rio de Janeiro: The "Marvellous City"

The Portuguese founded São Sebastião do Rio de Janeiro in 1565 on a prominent point within Guanabara Bay, one of the world's great natural harbors. With prosperous local sugar plantations and steady trade, the settlement maintained a population of several thousand, composed largely of slaves, until the discovery of gold and diamonds in Minas Gerais led to a regional spurt of growth in the 18th century. As a result, the colonial capital was transferred from Salvador da Bahia to Rio de Janeiro in 1763. After the Napoleonic invasion of Portugal, the royal family fled to Rio de Janeiro

Figure 4.7   The classic postcard view of Rio de Janeiro includes scenic Sugar-loaf Mountain (Pão de Açúcar), which guards the entrance to Guanabara Bay. (Photo by Brian Godfrey)

and the city served as capital of the Kingdoms of Portugal and Brazil from 1808–1815. Royal sponsorship after the arrival of the Portuguese court stimulated building and new institutions were founded. Extending along Guanabara Bay and scaling the surrounding hills, Rio de Janeiro acquired a linear spatial pattern oriented toward the bay (fig. 4.7).

Rio de Janeiro's status as the main seaport and capital of independent Brazil (1822–1960) secured its national primacy for over a century. As both the capital and the principal national metropolis, the city's port boomed, industry and commerce prospered, and cultural affairs flourished. Determined to compete with Buenos Aires as South America's most cosmopolitan city, Mayor Francisco Pereira Passos (1902–1906) promoted extensive urban renewal to transform Rio de Janeiro

into a "tropical Paris." Using Yellow Fever as a rallying point, municipal authorities mounted an extensive sanitation campaign and demolished thousands of buildings to make way for new boulevards and high-rise structures. The port was transferred from the downtown core to modernized facilities to the north on Guanabara Bay. The new transportation arteries encouraged real estate development in socially sorted neighborhoods during the early twentieth century. Gradually the northern zone became predominantly industrial and working-class in character, while affluent populations gravitated to fashionable districts to the south (fig. 4.8).

Even as Rio has grown, its social class barriers have remained in place. The poor are primarily non-white and the middle and upper classes remain overwhelmingly white;

Figure 4.8 Copacabana Beach, Rio's world-famous tourist playground, also is home to a dense urban neighborhood. This and other districts in the city's affluent southern zone were built up through massive public and private investments during the twentieth century. (Photo by Brian Godfrey)

and these racial disparities largely coincide with patterns of residential segregation. While a sharp north-south split plagues Rio's social geography, hillside shantytowns known as *favelas* are visible above fashionable southern seaside districts. Although *favelas* date from the turn of the twentieth century, they proliferated after World War II. By the late 1940s, as a result of the rapid rural-urban migration, the *favelas* overtook the run-down tenements of the central slums as the main form of housing for the urban poor. Today nearly one-fifth of the city population is housed in some 600 *favelas*, scattered in hillside shacks. Providing rent-free housing on government or disputed terrain close to employment, the *favelas* have become a permanent feature on the landscape, despite recurrent efforts by authorities to remove them.

By the late 20th century, long-term governmental neglect facilitated the rise of violent drug-trafficking cartels, which gained control of many *favelas*. Not surprisingly, analyses of city and metropolitan administrative districts have found strong correlations between impoverished slums afflicted with drug traffic and rates of violent death, particularly among the young male residents.

Figure 4.9   A view of the Vidigal district in Rio's southern zone indicates the informal, impro-
vised nature of Rio's favelas. (Photo by Brian Godfrey)

As a result, police forces have increasingly mounted military-style operations to rid *favelas* of drug cartels. Between 2008 and 2010, "pacification" campaigns evicted drug dealers and installed police stations in about forty of the city's *favelas*, including the "City of God" (featured in the famous film). Governmental and NGO programs to ameliorate conditions in the favelas have focused on infrastructure improvements (*e.g.,* street paving, provision of water and sewerage) and social services (*e.g.,* health center, schools, and recreational facilities). The more visible and accessible favelas of the city's southern zone have also become sites of organized "favela tours" that attract curious foreign visitors (fig. 4.9).

Rio's environmental problems have mounted along with contemporary metropolitan growth. Torrential summer storms often devastate precariously perched *favelas* and flood low-lying streets below. Fifty years ago thick hillside vegetation absorbed most of the rainfall, but most of the rainfall now runs off urbanized surfaces, dislodging unstable structures and blocking major transportation arteries. Water pollution is another major problem. Governmental agencies have made progress in curbing pollution of Guanabara Bay and the popular Atlantic beaches, but much remains to be done to conserve Rio's spectacular natural site, backdrop for the 1992 "Earth Summit," the United Nations Conference on Environment and Development. Recently Rio's prestige has rebounded with the approach of the World Cup in 2014 and the Olympics of 2016, which will be held in the city.

São Paulo: An Inland Model of Modern Urbanism

São Paulo's distinctive colonial origins began with its inland site, which contrasted with the coastal locations of most colonial centers. Jesuits founded São Paulo de Piratininga in 1554 on the gently rolling hills of an inland plateau, strategically located at a critical transportation juncture between the coast and the interior. Lacking valuable resources or lucrative plantations, the village remained small for three centuries. São Paulo's locational advantage became apparent during the mid-19th century, when the city became the center of a prosperous coffee-growing region, because of its fertile soils and generally mild subtropical climate. With railroads financed by British capital, São Paulo became the chief point of transshipment for the lucrative new cash crop. As a result, turn-of-the-century São Paulo grew rapidly, especially through Italian and Japanese immigration after the abolition of slavery in 1888 led to a shortage of labor in the coffee fields.

Profits from the coffee trade were invested in urban commerce, industry, and real-estate development. Enterprising immigrant families made fortunes in food processing, textiles, and other early industries. By the 1920s São Paulo overtook Rio de Janeiro as the principal industrial center of Brazil. Programs of import-substitution industrialization, initi-

ated under President Getúlio Vargas in the 1930s, cemented São Paulo's national industrial dominance. After President Kubitscheck designated São Paulo as the site of the nation's foreign-led automobile industry in the 1956 Development Plan, Volkswagen established the country's first automobile assembly plant there. Subsequent investments by other Brazilian and multinational firms expanded the city's industrial base.

The dizzying growth of 20th century São Paulo created successive urban layers. The modern city began to take shape in the early 20th century with the demolition of inner-city tenements to widen streets. In 1929, future mayor Prestes Maya (1938–1945) published his influential Boulevard Plan *(Plano de Avenidas)*, which provided a blueprint for opening major central avenues. Large-scale demolition, redevelopment, and new transportation lines facilitated the development of a burgeoning office and commercial district downtown; in outlying areas served by trains and streetcars, real estate speculation encouraged housing development in socially sorted districts. Working-class districts emerged in run-down central slums and near industry in the low-lying river basins and railroad corridors. Generally, the wealthy sought higher terrain in the city's southwestern districts. Along the Avenida Paulista, the townhouses of coffee barons and business leaders during the 1920s gave way to the headquarters of banks and corporations after World War II (fig 4.10).

Construction of São Paulo's modern freeway and subway systems encouraged new areas of urban expansion in peripheral areas. In fact, contemporary problems of urban transportation crystallize the city's social inequalities. Since the 1950s, metropolitan transportation policy has favored individual

Figure 4.10   Once lined by elite mansions, the Avenida Paulista became the city's corporate "Miracle Mile" after World War II. (Photo by Brian Godfrey)

automobile travel by the middle and upper classes through a massive investment in new arterial roads, while the poorer sectors of society are underserved by the city's inadequate public transportation system. Working-class areas and peripheral shantytowns often depend on tortuous, unreliable bus service.

Contemporary São Paulo is now experiencing the processes of economic restructuring, deindustrialization, and decentralization. With the transition from an industrial to a commercial and administrative-service economy, the formerly compact downtown has been split into two nodes: the traditional business center near the Praça da República and the financial district of the Avenida Paulista. Shopping malls now draw customers to the outlying areas, especially in the prosperous central-southwestern zone. The suburban industrial "ABC region"—Santo André, São Bernardo do Campo, and São Caetano do Sul—with its automobile sector and strong labor unions, faces cutbacks and job loss as industries move away to neighboring states, which have offered attractive tax breaks to lure automobile assembly plants. Meanwhile, outlying satellites beyond the official São Paulo Metropolitan Region, such as Campinas and São José dos Campos, are known for their universities and high-technology sectors (fig. 4.11).

São Paulo also now faces the problems of environmental degradation and related health

Figure 4.11　The skyline of downtown São Paulo reflects the dynamic growth of Brazil's dominant commercial center and corporate headquarters city. (Photo by Brian Godfrey)

concerns that have accumulated during years of explosive growth. Given its inland location and concentration of heavy industry, automobiles and buses, and informal peripheral growth, Greater São Paulo has endured heavy air and water pollution. Air pollution worsens particularly in the winter, when temperature inversions trap pollutants and prevent contaminants from blowing away. State agencies monitor pollution and impose penalties on offending industries. It has proven harder to regulate the more than four million cars and buses, now the main polluters, since automobile emissions are the concern of federal authorities. Sewage and waste treatment systems also remain inadequate, particularly in the peripheral informal settlements, where untreated waste often pollutes surrounding areas. Fiscal problems have hindered ambi-

tious clean-up programs in the befouled Tietê River, which snakes through the metropolitan area.

Future Prospects for the Brazilian Megalopolis

After a century of rapid growth, Brazil's two leading metropolitan areas now face the challenges of deteriorating physical and social infrastructures, traffic congestion, air and water pollution, fear of crime, housing scarcity, and saturated job markets. Industries increasingly relocate to cities in the nation's interior. Even in the face of metropolitan decentralization and economic restructuring, São Paulo and Rio de Janeiro are unlikely to lose their global and national prominence, or their key distinguishing characteristics. Both cities have witnessed growth of corporate

Box 4.2 Urban Security and Human Rights

Increasing concerns with violent crime now plague South American cities. Widespread fears of urban violence have been fed by vivid accounts in the news media, tourist guides, governmental travel advisories, and popular films. For example, such acclaimed recent films as "City of God" (Brazil, 2002) or "Our Lady of the Assassins" (Colombia, 1999) feature racy stories full of sex, drugs, and armed conflict in urban slums. Such representations sensationalize violence and serve to stigmatize the urban poor, who happen disproportionately to be of indigenous or African racial origins. The preoccupation with urban insecurity has created a culture of fear, which Brazilian anthropologist Teresa Caldeira (see Suggested Readings) relates to "the increase in violence, the failure of institutions of order (especially the police and the justice system), the privatization of security and justice, and the continuous walling and segregation of cities . . ." The widespread concern over crime has served to maintain class and racial boundaries, despite the expansion of formal democratic rights.

Official statistics often underreport crime, since distrust of the police discourages many residents from reporting incidents. Even so, studies indicate steadily increasing rates of violent crime over the last three decades. Rates of homicide (murder and manslaughter) represent the most reliable data, given compulsory death registrations. In 1980 national homicide rates in Brazil and the United States were about the same (about 10 per 100,000 population), but by the late 1990s the Brazilian rates were twice as high. Of course, violent crime tends to be worse in large cities. São Paulo, Rio de Janeiro, and Recife have been among the most violent Brazilian metropolitan regions, although rates have tended to drop more recently.

São Paulo's contemporary evolution points to widespread trends. Studies have documented a dramatic rise in the city of São Paulo's homicide rates, which more than tripled between 1980 and 2000. As in other Brazilian cities, homicides reported in São Paulo usually involve firearms, most victims are young men (15–29 years old), and there are strong local correlations with poverty and the presence of drug-trafficking activities. Other factors commonly cited include socio-spatial segregation, economic crises and high unemployment, and widening income inequality. On a positive note, São Paulo's murder rate steadily fell to 14/100,000 in 2007, which researchers have attributed to more effective policing methods and better enforcement of gun-control legislation, despite the persistence of socioeconomic problems.

Put into a social context, the rise in urban violence becomes an important issue of human rights. Community development initiatives now feature programs to prevent violence, particularly among young people in poor communities. In Rio de Janeiro, the Viva Rio non-governmental organization (NGOs) began in the 1990s to offer programs to reduce firearm injuries, promote social justice, and provide vocational training for young people in poor communities. Similarly, the Mangueira Social Project, located in one of the city's *favelas*, provides after-school programs for local youth who demonstrate regular school attendance. These and other NGOs have embarked on grassroots campaigns to change the perception of their communities through the internet, media outreach, and partnerships with the government, universities, and the private sector.

Figure 4.12 The spectacular modern architecture of Brasília, designed by Brazilian architect Oscar Niemeyer, focuses on the government ministries and the Congress buildings located along the federal district's Monumental Axis (Eixo Monumental). The "Pilot Plan" (Plano Piloto) of Brasília was declared a UNESCO World Heritage site in 1987. (Photo by Brian Godfrey)

producer services and commercial sectors, which has in large part compensated for the relative decline of manufacturing. As the twin nerve centers of a vast country and a leading emerging economy of the world, these two cities have sprawled to form the joint nuclei of an integrated megalopolis with the population of a medium-sized European country in southeastern Brazil.

Brasília: Continental Geopolitics and Planned Cities

Urbanization has now spread to South America's long-forsaken interior, particularly the Brazilian central plateau *(planalto),* the Amazon Basin, and other inland areas. The founding of new inland cities has presented a prime opportunity for modern urban planning and industrial development, as in Ciudad Guyana

of Venezuela and, in Brazil, Goiânia, Belo Horizonte and, most famous of all, Brasília. The transfer of the federal capital from Rio de Janeiro to Brasília in 1960 served dramatic notice of the determination to redistribute the population from the coast to preconceived cities of the interior. Under Juscelino Kubitschek, president of Brazil from 1956 to 1961, construction of the new capital constituted an important part of an ambitious program of national urban-industrial development. The new capital's spectacular modern design and rigorous land-use controls were meant to contrast with more spontaneous earlier cities, seen to be plagued by irregular urban growth (fig. 4.12).

Brasília's construction began in 1957 on a barren site in the state of Goiás, on the central plateau (planalto central), about 600 miles (970 kilometers) from the coast. Brazilian architect

Lúcio Costa designed the new capital's visionary plan, while his colleague Oscar Niemeyer designed the city's most impressive modernist buildings, such as the Cathedral, Senate and Chamber of Deputies complex, the Itamaraty Palace of the Foreign Relations Ministry, the Planalto Palace executive building, and the Alvorada Palace of the president. Costa's highly symbolic "Pilot Plan" of Brasília features two great intersecting axes, one governmental and the other residential, together forming the rough outline of an airplane. Federal government buildings cluster at the eastern end of the plane's body or "fuselage," around the Plaza of the Three Powers. Near the central intersection of important boulevards lie the bus terminal, stores, hotels, and cultural institutions. Farther west are the governmental complex of the Federal District, alongside a sports arena and recreational facilities. Residential areas, which extend north and south along the "wings" of the plane, comprise groups of six-story apartment buildings to house government functionaries and their families. Each "superblock" of apartments contains a school, playground, shops, theaters, and so on. On the eastern side of the Pilot Plan lies scenic Lake Paranoá, where expensive private residences have been built, especially in the exclusive Lago Sul ("South Lake") sector.

Early residents and architectural critics often found Brasília sterile and monotonous, lacking the vibrant street life of other Brazilian cities. Many government officials initially maintained homes in the former capital, Rio de Janeiro. In time, however, Brasília filled in with upscale businesses, diverse services, attractive residences and, along with these new amenities, the capital developed a certain

character. Brasília has become an effective symbol of national integration, and the central planned area—the Pilot Plan—was designated a World Heritage Site by UNESCO in 1987. The organization's International Committee on Monuments and Sites (ICOMOS) concluded that "the creation of Brasília is unquestionably a major feat in the history of urbanism," although it also cautioned that the ". . . new capital of Brazil encountered serious problems which, even today, have not been totally overcome." UNESCO's decision to recognize Brasília included a precautionary warning that "minimal guarantees of protection" must "ensure the preservation of the urban creation of Costa and Niemeyer." That the central Pilot Plan of the modernist capital of Brazil would be historically preserved, less than 30 years after its founding, reflects more than admiration of an architectural icon; it also speaks to widespread concerns over the rapid and largely unplanned urbanization of the rest of the Federal District.

Away from the central Pilot Plan of the new capital, informal settlements quickly emerged in what were called the "satellite cities"—out of sight but within commuting distance of the city center. Housing was not provided for the construction crews, other workers, and their families. So, a series of spontaneous suburbs some distance from the attractive residential "superblocks" of the city center were built by and for the migrant laborers and their kin. These unplanned communities were composed mainly of low-rise, self-constructed wooden homes and initially exhibited a ramshackle frontier atmosphere. Several of the early settlements, like Taguatinga, in time became established centers with public services, while other more recent areas are still

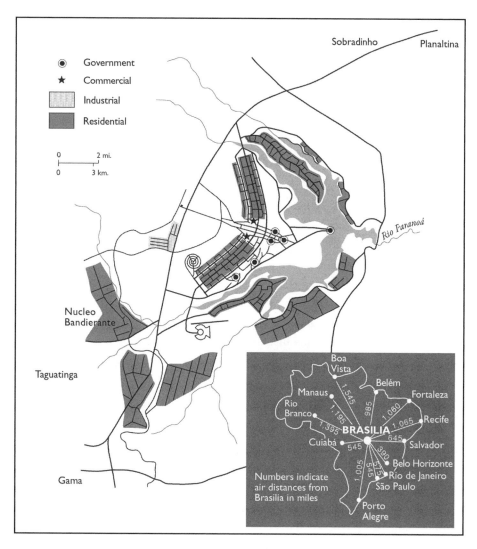

Figure 4.13   Map of Brasília. *Source:* Complied by the authors.

in rudimentary conditions. The vast majority of the population—in 2010, 2.6 million in the Federal District and 3.5 million in the metropolitan region—lives outside the Pilot Plan in what are now preferably called "surrounding cities" *(cidades do entorno)*. Despite the widespread early criticism of Brasília, the federal district's steady growth suggests a successful pole of in-migration. Yet the inability to plan

effectively the entire Federal District, the symbol of a modernizing regime, underscores the persistence of familiar social problems, such as widespread poverty, self-constructed housing, and the informal sector. The experience of Brasília speaks to the difficulty of implementing centralized planning in a developing country beset by high levels of income concentration and a dearth of basic public services.

Box 4.3 The Amazon Basin: An Urban Frontier

In the Amazon Basin, inter-regional migration and rapid urbanization now transform one of the world's last great settlement frontiers. The Amazon River and its tributaries drain a watershed of approximately 2.7 million square miles (7 million square kilometers), or 40 percent of the South American land surface. Although the Amazon region occupies large parts of Bolivia, Peru, Ecuador, Colombia, and Venezuela, two-thirds lies in Brazil. In all six Amazonian countries, contemporary regional development programs have attempted to integrate sparsely populated peripheral areas in the interests of national sovereignty. These projects have resulted in disorderly and often violent processes of frontier urbanization, which have embroiled native peoples, agricultural colonists, gold miners, forestry interests, cattle ranchers, transnational corporations, governmental agencies, and others in land conflicts. Although tropical deforestation has received the most attention, urbanization has created serious deficiencies in service provision, widespread health problems, and environmental degradation in the burgeoning towns and cities.

Brazilian Amazônia became predominantly urbanized during the late 1970s. Considering the seven primary states of the Brazilian North Region—Acre, Amapá, Amazonas, Pará, Rondônia, Roraima, and Tocantins—the urban population skyrocketed from 28 percent in 1940, to 50 percent in 1980, and to 74 percent by 2010. The Amazon Basin faces special urban problems because of its precarious infrastructure, environmental vulnerabilities, and dependence on resource extraction. Historically, boom-and-bust cycles in natural resources—rubber, gold, and diamonds, and other natural resources—generated the two primary regional metropolises of Belém (2010 population 2.2 million) and Manaus (2010 population 1.8 million). Founded during colonial times as defensive and commercial nuclei, these and other urban centers burgeoned during the height of the trade in natural rubber (circa 1870–1910). By the late 20th century, these regional metropolises became regional growth poles and tourist centers. As in-migration has mounted, new housing has come partly from residential towers in central areas, but mainly from self-constructed housing in peripheral low-lying areas, often subject to flooding by tides and rains, and ill-served by urban services.

Even as the regional metropolises of Belém and Manaus have grown to unprecedented sizes, the most significant regional transformation now occurs in the small- and medium-sized cities of the interior, which now experience the highest rates of urban growth. By

Lima: Tempering Hyperurbanization on South America's Pacific Rim

Lima and its port Callao are centrally located on South America's Pacific coast, squeezed into a narrow coastal desert between the Pacific Ocean and the Andes Mountains. The region is prone to frequent earthquakes, flooding, drought, and other disturbances. Initially serving as a point of contact between Spain and its colonial empire in South America, Lima quickly evolved into a transshipment

2010, the Brazilian North Region had sixty-three cities of over 50,000 residents, and twenty of them had populations of more than 100,000. The rates of population increase in many of the interior boomtowns have been staggering. For example, the capital of Roraima state, Boa Vista, near the border with Venezuela, grew from 17,154 in 1970 to 284,313 in 2010—an average annual increase of 38.9 percent! Similarly, over this forty-year period Porto Velho in Rondônia grew by an annual average of 19.4 percent, Rio Branco in Acre by an average 20.8 percent, and Marabá in southern Pará by an average 32.9 percent. Such dizzying rates of increase can be found in countless other Amazonian cities, where it creates severe problems of urban infrastructure, housing, and social services.

Although Brazil's Amazon expansionism attracts the most attention, the Andean countries also have vast, sparsely populated, and historically remote eastern "oriente" regions. In the 1960s, Peruvian President Fernando Belaúnde Terry envisioned a Marginal Jungle Highway (Carretera Marginal de la Selva), which would encircle the upper Amazon Basin from Venezuela to Bolivia. The pan-Andean road network would have linked the various countries and provided a unifying basis for planned urbanization of the upper Amazon settlement frontier. Although territorial suspicions among the Andean countries prevented the full implementation of the ambitious highway network and settlement program envisioned in the regional plan, the various governments subsequently all built new roads and launched colonization schemes in their own territories. Most schemes were intended to spur inland migration and thus alleviate crowding in coastal or highland cities. The discovery of oil, gold, diamonds, copper, and other minerals has fueled inland migrations into the Amazonian territories.

Regional urbanization in the eastern portions of Andean countries has been dramatic. Northeastern Ecuador became an active pioneer zone as a result of small farmer resettlement and the construction of an oil pipeline to the coast during the 1970s. Environmental groups, indigenous peoples, highland migrants, and oil companies have battled over land claims here. Similarly, with improved road access to the Peruvian interior since the 1960s, Tingo Maria has become a crossroads town in the embattled Huallaga region, a rich agricultural and coca-growing area. The formerly sleepy town of Santa Cruz de la Sierra in eastern Bolivia—with a 2010 metropolitan population of 2.1 million, now the country's largest—has become a dynamic regional metropolis as a result of agricultural colonization, exploitation of natural gas, and the improved transportation connections with neighboring countries.

point for the mineral, agricultural and textile wealth extracted from the Andean interior as well as the unrivaled capital of Spanish-American high culture.

Although Lima was founded before the Laws of the Indies, its founding anticipated them; the city was laid out in a grid pattern, with streets radiating from a central plaza in a regular east-west and north-south pattern. Urban development took hold along a set of axes, each of which had a distinctive character. The area northwestward to the port of Callao

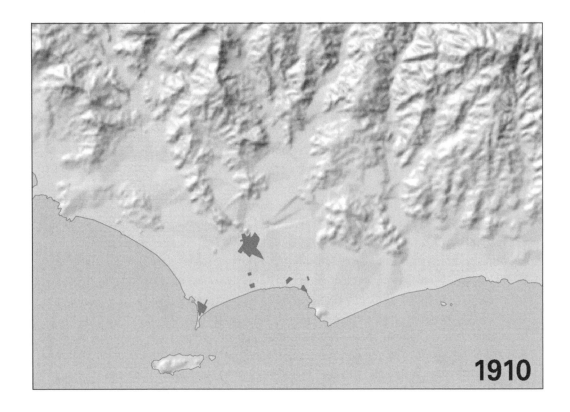

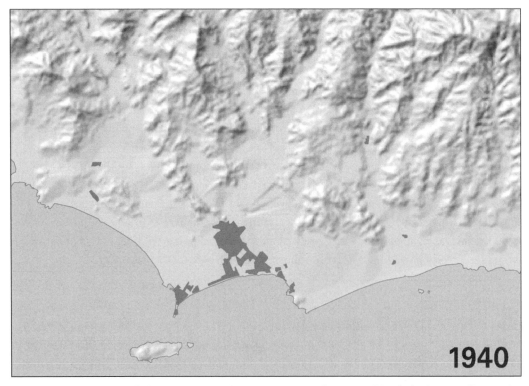

Figure 4.14 Growth of Lima, 1910–2000. *Source:* Centro de Promoción de la Cartografía en el Perú, Avda. Arequipa 2625, Lima 14, Peru. (Continued on next page.)

1970

2000

Figure 4.14 (Continued)

Figure 4.15   Lima's central plaza, known as the Plaza de Armas, dates to the city's founding and served as the central point from which streets extended in the four cardinal directions consistent with the Law of the Indies. (Photo by Maureen Hays-Mitchell)

would become the city's industrial corridor; the seacoast to the southwest would develop into an elite residential zone; and small industry would intermingle with working-class housing to the east. By the mid 20th century, the areas radiating from the old Lima center to the Pacific coast were fully urbanized. Soon, shantytowns were commonplace in the desert regions to the north and south of the city, known today as the Cono Norte and Cono Sur (fig. 4.14).

Population growth, agricultural stagnation, economic injustice, and armed violence in rural Peru have set off waves of migration to the city. Until the Second World War, mostly rural elites and people from nearby provinces migrated to Lima. Many had family contacts or skills to secure employment. In the two to three decades following the war, migration became a more generalized phenomenon, as people from all regions of the country, lured by new industry, found their way to the city. Finally, the political and economic crisis of the 1980s and 1990s brought an influx of poorly prepared and traumatized displaced persons, primarily from the southern highlands, seeking safety and refuge. In relatively short order, provincial migrants and their offspring transformed Lima from a bastion of elitist creole culture (European culture within America) into a microcosm of contemporary Peru. Today, food, music, dance, artisanry, accents, dress, and festivals from every region of Peru are found in Lima.

Demographic growth has reinforced the city's primacy. Lima dominates all aspects of national life. Seventy-seven percent of the national population currently resides in cities, with 40 percent of that population living in greater Lima. Today, upwards of nine million people live in the metropolitan region, which is more than ten times larger than the next city, Arequipa, in southern Peru. Ironically, Lima's primate status is both the cause and the effect of growth. Due to its location, Lima

Figure 4.16   Enclosed wooden balconies on colonial buildings typify the historic district in Lima, a UNESCO World Heritage Site. (Photo by Maureen Hays-Mitchell)

has long served as the gateway between the outside world and the rest of the country. The concentration of political influence, capital, industry, communications, workforce, consumers, and the most prestigious institutions of research, learning, and culture induces further concentration of all these activities and reinforces Lima's primacy. In times of economic expansion, the gap between Lima and the rest of the country grows more pronounced, with little synergy between Lima's economy and the provinces.

Today in downtown Lima, ornate colonial architecture contrasts sharply with the modern high-rise buildings that accommodate government ministries, banks, law firms, and businesses, many with global connections. Remaining colonial mansions have been subdivided into slum housing that accommodates as many as 50 families per building. The enclosed wooden balconies that typified the colonial city have become a point of interest

for preservation (fig 4.16). Although UNESCO designated much of central Lima a World Heritage site in 1991, there is little evidence of residential gentrification here. Instead, many private-sector businesses and international agencies have moved their offices to the less congested and more secure suburbs. Indeed, the most defining feature of Lima is its expansive *barriadas*, which have been euphemistically renamed *pueblos jovenes* (young towns) and most recently *asentimientos humanos* (human settlements). Shanties have been constructed on the barren slopes that rise above the red-tiled roofs of the inner suburbs and on the flat desert benches that encircle Lima (fig. 4.17). Approximately half of the city's population is estimated to reside in *asentimientos humanos,* with the Cono Norte and Cono Sur the most populous districts in the city.

The social fabric of present-day Lima is more complex than ever. Race, ethnicity, and class defy easy classification. The enduring

Figure 4.17   A shantytown (pueblo joven) outside of Lima. (Photo by Rob Crandall)

difference is between the rich and influential on the one hand and the poor and marginalized on the other. This is visible on the urban landscape. As Lima's population shifts, Andean, and to a lesser extent Amazonian, culture infuses its streets and public spaces. Pressure to assimilate is less today as migrants and their offspring assert their cultural heritage and their claim to Lima as a multicultural city. In response, wealthy Limeños pick up the process begun centuries ago of distancing themselves from the poor. Now they are moving not simply to the traditionally more elite western districts of the city, but also beyond to quasi-rural settings to the east as well as the more distant seaside communities to the north and south. Private security, gated communities, and chauffeurs are markers of their attempt to withdraw. Despite the municipal-

ity's determination to reclaim the city center for pedestrians and sightseers, *ambulantes* (street vendors) ply their trade, bedraggled children perform stunts for cash and the infirm and elderly beg for handouts. This has sparked discussion in academic and policy circles of the role that public spaces may play in enhancing the quality of urban life and creating a shared civic identity.

Today, Lima, and to a lesser extent Peru in general, is emerging from two decades of economic and political crisis. Peru has experienced five years of sustained economic growth, fueled by natural resource industries of mining, logging, oil drilling, and agribusiness; as well as a surge in tourism and construction. Poverty rates have declined, and Lima's middle class is once again expanding. Improving economic conditions have

Box 4.4 Women of the Streets

"Cómprame . . . Cómprame casera . . . Cómprame." "Buy me, buy me," they call to the potential customers who pass in the streets. These women are not selling themselves, but rather a vast array of foods and manufactured goods. They are the street vendors, or *ambulantes*, who fill the streets of South America's cities. Street vending, along with prostitution, is one of the few occupations by which women can earn a living on city streets. Whether or not it is considered "proper" is something few women street vendors can afford to consider. They are compelled to earn their livings on the streets due to the endemic poverty and limited occupations open to poor women in most South American countries.

Throughout South America, the ranks of street vendors—as all components of the informal urban economy—have swelled as the region's economic crisis has intensified. Interviews with women *ambulantes* in cities throughout Peru indicate that, although most are from humble backgrounds, an increasing proportion are relatively well-educated women who cannot find legitimate employment in their professions. Those women who depend entirely on their income from street vending tend to be heads of households; they are elderly, widowed, abandoned, single mothers, or wives of imprisoned or "disappeared" men. These women encounter the most restricted range of employment opportunities. Within their set of options, street vending is often their most viable means of employment. Increasingly, unemployed men are turning to street vending, which is heightening competition among the *ambulantes*.

In Huancayo, a bustling commercial center in the Peruvian central highlands, nearly one-half of *ambulantes* are women. As might be expected, they are rational decision makers. More than half deal in volumes of products sufficiently large to gain access to wholesale prices. Those who do not deal in sufficient volume often band together to create the requisite demand. In keeping with the traditional image of women as providers of food and domestic service, they tend to specialize in sales of fresh produce, hot meals, clothing, and household items—the most poorly capitalized product lines, with the lowest profit margins. It is likely that those who sell household items and prepared foods, rather than by selling more specialized items and filling a niche, are creating their own demand. In contrast, nearly the entire supply of fresh produce is commercialized by *ambulantes*. Although the majority of fresh produce vendors are women, they tend to deal in smaller volumes than their male counterparts, thus reaping smaller returns than men from this potentially lucrative market.

To assert their interests, many *ambulantes* have turned to political activism. Street vendors are organized into trade unions, called sindicatos, for the purpose of protecting and furthering the interests of all *ambulantes*. In Huancayo, many of the leaders of these organizations are women. Women have proven effective in organizing their unions as well as in guiding meetings away from rhetorical exhortations and toward important business matters. Despite this, they tend to be relegated to subservient roles within the organizations—an exercise that only adds to their already heavy workloads.

Like women everywhere, women *ambulantes* share worries and concerns as well as hopes and aspirations. Many, especially those who sell prohibited goods, such as raw meat, or who choose to operate without the requisite licenses, worry about harassment by the municipal police. They run the risk of losing their inventory and hence their investment. Many are bound to disadvantageous credit schemes—a cycle that ensures a lifetime of street vending. Bringing their young children to work, seen by some *ambulantes* as an advantage offered by street vending, can be a worrisome burden for many women. Children distract their mothers, who must be alert in this competitive occupation. Children are exposed to the elements and many young ones contract respiratory illnesses and die. Once they become toddlers, they play in the streets, where they often sustain injuries. Children who are school age soon become street vendors themselves. Moreover, the streets of their cities are not particularly safe. Petty thieves work them at all hours. Not even the *huachimanes* (night watchmen hired by individual *sindicatos*) can be trusted. They routinely pilfer the stands that *ambulantes* meticulously close up and leave behind.

allowed infrastructural improvements in the capital. Construction is booming, roads are being paved, public spaces illuminated, parks restored, transportation upgraded, and water and sanitation service expanded. The economic situation of some *asentamientos humanos* is improving as urban services, small businesses, and industries take root.

However, Lima confronts problems of unprecedented proportion and complexity. Opening Peru's economy to global markets has accentuated the importance of Lima as an economic center; the majority of international and national corporations operating in Peru are located in the capital. Prosperity highly concentrated in Lima, but it is unequally

Figure 4.18   A peddler ambulante selling fresh vegetables in Hauancayo, Peru. (Photo by Rob Crandall)

distributed within the city. Most poor residents do not experience the positive effects of economic globalization; and the gap between the Limeños who do and those who do not remains wide. Economic growth, however, has allowed for revitalization of the social-service sector, which had been decimated in the political-economic crisis of the 1980s and 1990s. Programs to improve the quality of life in poorer sectors of Lima focus on quality of and access to schools, health facilities, and urban technology. An experiment in wireless phone service is even underway in one district.

Lima's rapid and unplanned growth has caused severe environmental degradation, especially of the city's water and air quality. Lima is a mega-city in a desert. The rapid loss of Andean glaciers threatens Lima's sources of water. The very rivers that gave rise to human settlement here some 4,000 years ago are shrinking, as well as having been polluted by mining and agriculture runoff and residential and industrial waste. Urban sprawl has eaten away at the green space in the river valleys and has consumed wetlands, reducing biodiversity and affecting microclimates within the metropolitan region. Nearly one million people do not have water or sewer service, The UNEP (United Nations Environmental Program) has identified water as the most critical environmental problem in the Lima-Callao conurbation; and efforts are underway to improve coverage and quality of service in poorer districts.

Most Limeños, however, consider air pollution to be the most pressing environmental issue. Limeños who frequent the city center and/or reside in certain districts inhale large quantities of airborne particulates and other pollutants. A 2005 study estimated that respiratory and heart problems related to airborne particulate matter were responsible for some 6,000 deaths per year. In 2007, the Clean Air Initiative was launched and by 2010 the air quality of Lima was improving, largely due to government-led initiatives that banned the import of used diesel vehicles and sale of high-sulfur diesel, required technical revisions to vehicles, and reorganized city traffic. Catching the trend of sustainable urban planning (box 4.6), Lima is developing a high capacity transportation system of rapid buses—"El Metropolitano—that will run on natural gas and link the north and south corridors—the Cono Norte and Cono Sur—of the city.

Retrofitting automobiles and retiring old buses will not solve all Lima's air pollution problems. The countless unregulated factories, home industries, and restaurants that abound in Lima's poorer neighborhoods also contribute to the problem. Despite the health risks posed, shutting down these informal businesses is contentious because such action would impact livelihoods throughout Lima's expansive low-income communities. Not surprisingly, the very parts of Lima where life expectancy is estimated to be lower than the city's average are low-income communities with high levels of contamination. Due to its status as the national capital and primate city, Limeños have turned to the national government of Peru as well as to international actors for assistance, arguing that Lima's problems—economic, political, social, and environmental—are national problems. Recent agreements and initiatives may give Limeños cause to be cautiously optimistic.

Buenos Aires: Global City of the Southern Cone

Long regarded as one of Latin America's greatest cities, Buenos Aires stands as the visible symbol of Argentina's history and

Box 4.5 Addressing Air Quality in Latin American Cities

Rising automobile use, expanding industrial production, and increased energy generation associated with rapid urbanization in Latin America's cities exposes more than 100 million people to air contaminant levels exceeding those set by the World Health Organization. The yearly cost, according to the WHO, is thousands of premature deaths, billions of dollars in medical costs and lost productivity, and a hefty contribution to global climate change.

Air pollution in cities throughout the region affects the health and well being of hundreds of millions of people. Children, given their immature organs, are especially at risk of developing debilitating ailments. The very elderly are more susceptible to lung cancer and cardiovascular disease. And the poor, by virtue of where they live and work and how they make a living, are disproportionately exposed to the dangers associated with prolonged exposure to polluting agents known to cause cancer, cardiovascular disease, and other serious ailments. Air pollution affects the natural and built environments of cities, causing deterioration to buildings and monuments, stifling the growth and air cleansing benefits of trees and gardens, and affecting crop yields in downwind regions.

The predicted impacts of climate change in the Latin American region are severe. Temperature elevations will likely result in an increase in respiratory diseases linked to air pollution, as climate change could influence meteorological factors that impact the frequency and duration of "poor air quality" episodes in cities throughout the region. Urban water supplies, already in short supply, will likely diminish due to the loss of snow pack and glaciers in the Andes Mountains. The majority of primate cities in South America is located in coastal zones and hence these cities are vulnerable to coastal flooding due to the predicted rise in sea-levels.

As the impacts of air pollution and climate change on public health and the environment are better understood, the need to adopt strategies that recognize the importance of effectively integrating air quality and climate change considerations into social and economic development planning becomes more apparent. In September 2007 the Clean Air Institute released the draft of "The Clean Air Initiative Strategy for Latin American and Caribbean Cities 2007–2012" (CAI-LAC). The Clean Air Institute, an independent non-profit organization, was founded in 2006. It is a multi-stakeholder effort dedicated to addressing the environmental and public health concerns associated with air pollution in large cities throughout the region, as well as the region's contribution to and impacts from global climate change.

The draft initiative understands that, despite many common sources, conventional air pollutants and greenhouse gas emissions are rarely considered jointly. Jointly focusing on these factors is particularly important in Latin American cities, where resources are scarce, significant institutional and technical barriers exist, and compelling evidence suggests that the societal costs of air pollution will continue to present an enormous challenge to the countries of the region. To begin, efforts will focus on increasing the supply of clean energy and the efficiency of energy usage across the transportation industry and throughout the commercial and residential sectors. Thousands of lives are at stake.

Source: Adapted from "The Clean Air Initiative Strategy for Latin American and Caribbean Cities 2007–2012," The Clean Air Institute, draft September 10, 2007.

Figure 4.19   The colonial cabildo, or a town hall, now preserved in the historic core of Buenos Aires. (Photo by Brian Godfrey)

identity. Once a minor colonial outpost of Spain, Buenos Aires grew rapidly as a center of immigration, urban design, and modernism from roughly 1880 to 1930. While the country emerged as an agricultural and industrial power, the Argentine capital became known as the "Paris of South America," an elegant city of broad boulevards, graceful public squares, and impressive public buildings. Monumental Buenos Aires has long served as the stage for national political movements, as dramatized by the famous scenes of Juan and Eva Perón addressing the multitudes from the balcony of the Casa Rosada, the presidential palace. More recently, Mothers of Plaza de Mayo have continued to demonstrate and to protest the "disappearance" of their children during the "dirty war" of the military regime. Despite a contemporary decline in regional importance vis-à-vis São Paulo, greater Buenos Aires remains a vital metropolis with a 2010 population of about 13.1 million and a high degree of national primacy. The city's

residents, known as *porteños* (port-dwellers), continue to be trendsetters. On the other hand, the Argentine metropolis now faces growing problems of socioeconomic inequality, popular discontent and insecurity, and spatial segregation.

The city's history began in 1536, when Pedro de Mendoza led a Spanish expedition into the Río de la Plata (also known as the River Plate) and founded "Puerto de Santa María del Buen Aire" on southern shore of the Plata estuary. The Riachuelo inlet here was deep enough for anchorage by shallow-draught ships. Subsequently known as La Boca ("The Mouth"), this original port of Buenos Aires provided the best maritime landing available at the northeastern rim of the vast Argentine Pampas. Lacking precious minerals and other natural resources, and subject to continual attacks by hostile native groups, the colonists abandoned this original settlement in 1541. Given the strategic location, Spanish forces under Juan de Garay refounded Buenos Aires in 1580. Garay designed the ground plan for the fledgling town, which followed the characteristic Spanish-American urban form. The central plaza (later named the Plaza de Mayo) served as the core of the colonial settlement, surrounded by the important governmental, religious, and commercial structures. The city council, or Cabildo, sat across from the Cathedral and a commercial arcade lined much of the plaza (fig. 4.19).

The city and its port suffered from regional isolation and official neglect until the late colonial period. Andean trade routes long favored such inland cities as Tucumán, rather than the Rio de la Plata port. The Bourbon liberal reforms of the late 18th century enhanced the regional position of Buenos Aires, which became capital of the new Viceroyalty of La Plata in 1776. With relaxation of trade restrictions, the port flourished and the city's population grew to 50,000 by 1800. By this point the city became a center of agitation for independence, which aspiring local leaders proclaimed during the revolution of May 25, 1810. While colonial Buenos Aires typified a Spanish "Laws of the Indies" town, the postcolonial city's design increasingly reflected French and British influences. To resolve prolonged centralist-federalist conflicts, Buenos Aires was federalized and removed from the dominant Buenos Aires Province in 1880, and thereafter the president appointed the capital's mayor. This federal district of 78 square miles (203 km²) developed rapidly. British-financed railroads fanned out into the pampas, opening up an agricultural breadbasket to world trade, while the development of refrigeration allowed export of Argentine beef to Europe.

As Argentina became a postcolonial outpost of order and prosperity, European immigrants poured into Argentina and 30 percent of the Argentine population was foreign-born by 1914. As the federal capital, transportation hub, commercial center, cultural mecca, and immigrant port of entry, Buenos Aires experienced a high degree of urban primacy in the national city system. By 1914, the city had grown to more than 1.5 million inhabitants, fully 20 percent of the national total. The capital's population stabilized at about 2.9 million by the mid-20th century, while the metropolitan population continued to grow and now represents roughly a third of the country's total. With suburbanization, only about a fifth of the metropolitan population now resides in the capital city itself, which acquired autonomous status and began to elect its own mayors in 1994.

Rapid urbanization created a series of infrastructural problems during the late 19th century, including traffic circulation, sanitation, and housing provision. An emergent country needed a world-class capital city, graced by monuments and public buildings, worthy of Argentina's new wealth and aspirations. The Avenida de Mayo, begun in the 1870s, was torn through downtown to connect the planned capitol building with the Plaza de Mayo and the Casa Rosada. On completion in 1894, the new boulevard provided a striking visual corridor, reminiscent of the Champs-Élysées in Paris, linking the executive seat of government and the national capitol. In the early 20th century, several additional boulevards opened in central Buenos Aires, culminating with the Avenida 9 de Julio in 1936, one of the world's widest avenues, centered on an Obelisk visible from various vantage points downtown (fig. 4.20).

Unimpeded by physical barriers, districts called *barrios* covered the federal district by 1930. New immigrants first settled in central

Figure 4.20   The Diagonal Norte Northern Diagonal Boulevard, officially the Avendia Presidente Rouge Saenz Pena, highlights the imposing Obelisk monument in downtown Buenos Aires. (Photo by Brian Godfrey)

barrios near the port, such as San Telmo and La Boca: the local Italian-Spanish dialect known as "Lunfardo" emerged here, along with the Argentine "Tango" dance. The city's southeastern areas generally became industrial, working-class districts. In contrast, upper-class areas emerged on the northwestern side of Buenos Aires in such elegant neighborhoods as Recoleta, Palermo, and Belgrano, and Olivos. The two socially sorted residential sectors—generally more affluent to the northwest of downtown, more working-class toward the southeast—continued their historic trajectories in contemporary

metropolitan growth outside of the federal district. Coupled with a massive influx of impoverished migrants from the Argentine interior, Bolivia, and Paraguay—often called "Bolivianization"—poverty-stricken migrants have created extensive shantytowns or *villas miserias* ("towns of misery"). An estimated 640 *villas miserias* are home to up to a million people in the suburbs in Greater Buenos Aires, and studies suggest that these urban slums now grow ten times faster than the national population.

Argentine society has long been regarded as relatively affluent—given widespread European ancestry, middle-class living standards, and high levels of education and public health—but economic restructuring and neoliberal reforms shattered illusions of Argentine exceptionalism during the 1990s. Under President Carlos Menem (1989–1999), the country grew economically but experienced a contraction of government services, privatization of state enterprises, and widespread deindustrialization. While elites prospered, much of the population suffered from increasing unemployment and poverty. An economic recession began in 1998 and culminated in the crisis of 2001–2002, when Argentina defaulted on international debt obligations and devalued the peso. With growing public protests came new social movements, such as the *piqueteros,* unemployed workers who blocked roads, bridges, and buildings. "Unemployed Workers Movements" organized into cooperative markets and businesses during the crisis, and neighborhood-based assemblies *(asambleas populares)* arose. With the election of President Néster Kirchner in 2003, such grassroots activism declined with the return of political stability and economic growth. After Senator Cristina Fernández de Kirchner assumed the presidency in 2007, such controversies as the illfated agricultural export taxes, urban land invasions, and regularization of land titles in peri-urban areas sparked new social conflicts.

As in other metropolises, Buenos Aires has witnessed a recent proliferation of gated communities, characterized by low-density residential complexes guarded by defensive enclosures and private security. By 2000, about 350 such gated communities had developed in suburban Buenos Aires, occupying about 200 square miles (500 km²)—two and a half times the size of the federal district—and representing a residential population of approximately 100,000. These affluent enclaves cluster primarily in suburban areas with good highway access to the city center and, paradoxically, in poor localities with relaxed land-use laws. While the wealthiest municipalities tightly control land use, less affluent municipalities have relaxed building codes to attract real estate developers. The clustering of exclusive gated communities in low-income jurisdictions has deepened social polarization by juxtaposing wealthy and poor households. While social differentiation of *barrios* is not new, contemporary trends have created more pronounced and finely grained forms of residential segregation.

Despite the emergence of suburban shopping centers, office parks, and informal and gated communities, contemporary redevelopment projects suggest a continuing concern for the urban core. For example, the renovation of the abandoned downtown piers at Puerto Madero created a waterfront district of offices, restaurants, and convention facilities during the 1990s. Similar to other global cities, Buenos Aires has experienced a decline in industrial employment along with growth of commercial

and producer services sectors. While Buenos Aires retains a cosmopolitan air and cultural status, the contemporary rise of socioeconomic inequality and spatial segregation temper metropolitan prospects. Long thought to be different from other South American megacities, Buenos Aires now converges with them in terms of growing urban problems.

URBAN CHALLENGES AND PROSPECTS

The Urban Economy and Social Justice

Recent trends in economic globalization are benefiting some countries, most notably Chile and Brazil where middle classes are growing and poverty rates declining somewhat. Most countries in South America are not as fortunate. Throughout the continent, the proportion of households in poverty remains relatively high. In the cities, long-standing conditions of economic polarization and social injustice endure. Issues of employment, housing, and environmental degradation affect the poor more severely than they do other sectors of urban society.

It is not uncommon for many urban residents to spend more than half their cash income on food—only to barely meet their nutritional needs. In the absence of unemployment insurance or an adequate social security system, many South Americans cannot afford to be unemployed. The majority of urban dwellers are forced to turn to their own resourcefulness. Research on urban labor markets in South America indicates that, although participation within the paid workforce has improved, participation in the informal economic sector has increased. This is especially true among lower income groups

Figure 4.21   Recent renovation of Puerto Madero, long a deteriorated inner harbor, created a revitalized waterfront district adjacent to the downtown of Buenos aires. (Photo by Brian Godfrey)

and the more vulnerable (*e.g.*, poor women and children).

Despite indicators of stabilization at the macrolevel (*e.g.*, growth in gross national income), socioeconomic polarization persists in South American cities. When such conditions are concentrated among certain social groups or regions, they can generate restive conditions that challenge the cohesion of a society and the stability of a government. The rise of indigenous politics and social protest in Bolivian cities is a fascinating example that is playing out on the streets of La Paz, the national capital, and Santa Cruz, where a secession movement is underway.

Self-Help Housing and Defensive Urbanism

South America's cities reveal a curious sociospatial pattern of segregation that often juxtaposes those with wealth in secure high rises or gated communities alongside those without in *favelas, asentamientos humanos, villas miserias* (shantytowns). Indeed, large-scale urbanization has spawned defensive urbanism. The fear of crime has forced the urban elite to retreat into protected areas: into guarded luxury apartment buildings or suburban communities, where security is enforced by surrounding walls and armed guards, and their children are chauffeured to private schools. New security infrastructures—video surveillance, remote controlled gates, private security forces—are proliferating in cities across the continent.

Today, one- to two-thirds of the population of any given city resides in informal sector housing. Similar to its employment counterpart, the informal housing sector exists outside the bounds of "officialdom" in that it ignores building codes, zoning restrictions, property rights, and infrastructure standards. In South America, informal-sector housing is commonly known as "self-help" housing, a term that carries a double meaning. Most commonly, self-help refers to the characteristics of the homes and the process through which they are built. Self-help housing tends to be built by the inhabitants themselves, using simple—often hazardous—materials that the owner-builder-occupier has accumulated over time. Additionally, the term conjures up images of impoverished, yet well-intentioned urban dwellers "helping themselves" to unoccupied land—in the absence of a more viable option. Self-help housing communities are commonly considered shantytowns. Many settlements lack basic services, such as running water, sewerage, electricity, and garbage removal. They are constructed of scrap materials that often do not provide adequate protection from inclement weather, have limited access to services, are overcrowded, and lack the security of tenure (*i.e.,* title to the land). Shantytowns—or self-help communities—are marginal in terms of both their location on the urban periphery and the quality of the land occupied, which tends to be undesirable and often unhealthy and dangerous. They may be constructed on toxic "brown field" sites, alongside noxious landfills, on steep hillsides, or in polluted wetlands. Their overcrowded conditions are ideal for the transmission of disease. Shanties are the first structures to fall in mudslides and the first to be carried away in floods, and they easily go up in flames.

Under favorable conditions, self-help communities strengthen and improve over time. After the initial land invasion, settlements can evolve into consolidated and well-organized communities. Structures are steadily improved and basic services are addressed in one way or another. With time, municipal governments officially recognize the communities and extend urban infrastructure, supplying water and electricity, paving roads, extending public transportation lines, providing garbage removal, building schools, and staffing clinics. Despite the celebration of the self-help movement in many circles, it is nevertheless an inadequate proxy for regulated housing and urban services.

Segregation, Land Use, and Environmental Injustices

Although South American cities have long been highly segregated, the pattern of segregation is more complex today. Population expansion and variegated topography are bringing distinct social groups into closer contact. As intervening land is occupied, self-help communities and elite developments often exist side-by-side. There is little indication that residential segregation is abating. Indeed, South American cities are characterized by greater polarization in lifestyle. Glass-fronted skyscrapers and shopping malls characterize business districts and elite neighborhoods, while peripheral shantytowns are built of scrap materials and lack basic services.

Metropolitan expansion and decentralization have eroded the relative dominance of the traditional city center. Employment in the center is decreasing as industrial activity shifts to peripheral or nearby rural locations, and government and professional offices move to affluent suburbs that are less plagued by traffic congestion and crime. Although historic preservation and heritage sites in traditional downtowns have encouraged tourism, there is little evidence of residential gentrification and high-end commercial revitalization: affluent residents now prefer suburban locations with their amenities and security infrastructure. Indeed, urban elites are more likely to enjoy the advantages and to escape the disadvantages of urban living. Affluent business and residential districts tend to be better serviced with running water, sewerage, electricity, garbage service, public transportation, paved streets, sidewalks, and public parks. In contrast, low-income districts are characterized by inadequate urban services and infrastructure.

A differentiated urban landscape is also evident in environmental terms. Air pollution in some cities commonly surpasses safe levels as established by the World Health Organization. The wealthy can more readily escape these negative externalities as they listen to car stereos while waiting out traffic in air-conditioned cars.

Meanwhile, the less affluent are crowded onto hot, noisy, diesel-spewing buses. The discharge of untreated urban sewage into rivers and streams occurs more regularly in low-income districts. Children who live in shantytowns are especially vulnerable to gastrointestinal and respiratory illnesses, due to the poor water, inadequate sanitation, contaminants, open garbage, and burning refuse that characterize their living spaces. In contrast, the better off reside in less polluted areas, are more able to control some aspects of their living environment, and are more able to escape to country clubs and vacation homes. Indeed, evidence suggests that vulnerability to environmental hazards parallels income and status in South American cities.

There is evidence, however, that change may be underway as large cities endeavor to provide more inclusive, equitable and environmentally friendly forms of urbanism. Historic preservation, express buses, mass transit, pedestrian spaces, ecological restoration, biogardens, and tree planting campaigns are examples—the residents of South America's cities are paragons of creativity and resourcefulness.

Box 4.6 Planning for Sustainable Urban Development

Andrés Guhl and Brian Godfrey

Given high rates of urbanization and environmental degradation, the cities of South America have increasingly pursued policies of sustainable urban development. Two world-renowned examples are Curitiba, Brazil and Bogotá, Colombia. In these continental trendsetters, civic leaders and planners have adopted innovative policies to encourage compact, livable, and environmentally friendly urban growth. Planners have emphasized pedestrian streets and preservation of the historic center, clustered commercial corridors, parks and open spaces, ecological design, recycling of materials, educational programs, and other progressive measures. Instead of opting for costly subways, these cities have implemented public transit systems of express buses, which have proven so efficient and affordable that variations have been adopted in other cities around the world.

In Curitiba, capital of the southern state of Paraná, sustainability planning began as annual urbanization rates exceeded five percent in the 1960s. Long a regional center of agriculture and timber production, after World War II industrialization and rural-urban migration accelerated. Fears that urban growth threatened the quality of life prompted development of a 1965 Preliminary Plan by a team under architect Jaime Lerner, who later served as mayor and governor. The Institute for Urban Research and Planning of Curitiba (IPPUC, in Portuguese) developed a Master Plan *(Plano Diretor)* officially adopted in 1966. This plan proposed to minimize traffic congestion, control urban sprawl, preserve the historic city-center, provide parks and open space, and develop an efficient public transit system. Implementation of these plans began dramatically in 1972, when planners converted one of the major downtown thoroughfares, Rua XV de Novembro, into a pedestrian street. Although angry motorists initially threatened to ignore the traffic ban, they were frustrated by a famous act of public theater, as authorities unfolded large sheets of paper and invited school children to paint on the street.

Subsequently, zoning regulations have concentrated development along the arterial corridors to reduce traffic volume on the main roads, revitalize the commercial core, and maintain peripheral open space. The famous "Trinary" road system, consisting of five traffic arterials that converge downtown, separated automobile traffic in two outer lanes, each going opposing directions, from central lanes reserved for express buses. "Tube stations," which feature raised bus shelters with attendants collecting fares, and speed passengers' entry and exit to buses in the express lanes. Bus lines, operated by private companies granted concessions by public authorities, include a hierarchy of routes extending service in smaller vehicles. Beginning in the 1980s, the Integrated Transport Network (RIT in Portuguese) permitted transit between any points in the city with a unified fare. Curitiba also has promoted design-with-nature principles of urban ecology: low-lying areas subject to floods were reserved for parks, which altogether now provide about 50 m^2 of green space per resident. Despite continuing problems of poverty and service provision in peripheral shantytowns, the city's program of "Faróis de Saber" (Lighthouses of Knowledge) has offered free educational centers, including libraries, internet access, and other social and cultural resources.

These successes, however, have raised new problems. With an official city population of 1.75 million and a metropolitan population of 3.5 million in 2010, Curitiba is a major political and economic center of southern Brazil. The metropolitan demographic growth rate of 3.19 percent between 2005 and 2010 remained among the country's highest, as were per capita incomes and rates of automobile ownership. This relative prosperity has added to growth pressures. Although express buses continue to be heavily used, the development of the transit corridors has encouraged metropolitan sprawl. Whether planners can build on Curitiba's innovative record of transit-oriented development and environmental conservation to meet these new challenges remains to be seen.

Bogotá, Colombia, has faced similarly daunting growth pressures. The capital's rates of urbanization reached a dizzying 7 percent between 1950 and 1965, gradually dropping to a more manageable rate of 2.9 percent in 2005–2010. The metropolitan area grew from 630,000 in 1950 to 8.5 million residents in 2010 (table 4.2). Fortunately, Colombia's Constitution of 1991 transformed the city's institutional framework and permitted the rationalization of administration, planning, and regional management. These reforms gained momentum under a series of progressive mayors since the 1990s. As a result, *Bogotanos* have witnessed significant improvements in urban transportation, utilities, and public space. Additionally, the city's health services and libraries have expanded and about 98.7 percent of children now have access to schools.

In 2000 Bogotá adopted an urban plan (*Plan de Ordenamiento Territorial* or *POT*), revised in 2004, which has helped to prioritize resource allocation and improve the quality of life. This plan sets clear zoning patterns to regulate land use. It also recognizes the importance of the ecological assets of the city. For most of the 20th century the city filled-in wetlands, polluted waterways, and channelized streams. Now these features are recognized as key providers of ecosystem services, and efforts are proceeding to restore and manage them. Although *POT* provides clear guidelines, there are many challenges in terms of urban planning. While population densities have increased as apartment buildings replace single-family houses in the urban core, the municipalities surrounding Bogotá are suburbanizing through low-density houses and an increased reliance on automobiles.

The most important change in transportation has been the *Transmilenio*, a network of express bus lanes implemented since 2000. Based on the model of Curitiba, *Transmilenio* is part of an integrated transportation plan that will eventually include a subway system and a train system to the surrounding municipalities. So far, 84 km (51 mi) of exclusive bus lanes and 663 km (436 mi) of bus routes feed into the system; another 20 km (12 mi) are currently under construction. In 2010, *Transmilenio* supplied about 23 percent of the city's transportation needs, but some lines operated at full capacity and only 30 percent of users were satisfied with the service. Bogotá's administration has encouraged the use of bicycles by building exclusive bike lanes that provide a safe and environmentally friendly way to move about in the city (fig 4.22). As of 2010, there were 344 km (210 mi) of bike lanes in the city, and they move roughly 14 percent of the population. Main thoroughfares turn into recreational space every Sunday morning in what is called the *ciclovia* program, when about two million residents flock to the streets on their bicycles, roller skates, and other means of recreation. Unfortunately, rainy weather, safety issues, and lack of connectivity hinder this form of transportation on a daily basis. On the other hand, Bogotá has invested heavily in parks, recreational facilities, sidewalks, and public spaces. For example, Avenida Jimenez, a central boulevard in the historic city center, has been transformed into a leisurely walkway (fig 4.23). This urban intervention dechannelized one of the streams, transforming it into part of a linear park called *Eje ambiental* (environmental axis).

As in Curitiba, Bogotá has dramatically improved the quality for life of its citizens through coordination of economic development and environmental protection, but many challenges remain. Although the proportion of the population living below the poverty line dropped from 38.3 percent in 2002 to 23.8 percent in 2006, close to a quarter of the Botogá population remains poor. Despite the *Transmilenio*, transit mobility has worsened with the increasing number of cars in recent years. Air pollution remains a big problem, and largely untreated sewage is still dumped into some of the city's ecological assets. Restoring and managing the urban ecosystems has been difficult due to limited resources and lack of environmental consciousness on the part of many citizens for whom, for example, wetlands are just swamps that need to be filled in. While Bogotá has moved along the path toward sustainable urbanism, the city needs to consolidate this trend through careful planning for a more inclusive, equitable, and environmentally friendly city.

Figure 4.22   Exclusive bicycle lane in Bogotá, Colombia. (Photo by Andrés Guhl)

Figure 4.23   Eje Ambiental in historic Bogotá, where a dechannelized stream is part of a linear park along Avenida Jimenez. (Photo by Andrés Guhl)

AN UNCERTAIN YET GUARDEDLY OPTIMISTIC FUTURE

The cities of South America have long played crucial roles in a global urban network and capitalist economy. Economic, political, social, and cultural currents from around the world have flowed through the region's cities since the arrival of Iberian conquistadores. Today, as in the past, these global forces and the region's cities continue to shape and influence one another. Indeed, the escalating reach of globalization is adding new urban dimensions to long-standing problems of uneven development, regional shifts of industry, environmental degradation, social polarization, urban insecurity and violence, and spatial and environmental injustice. Such urban problems have encouraged economic decentralization and rising growth rates of small and intermediate-sized cities in South America. In cities large and small, the region's intractable social divide can be read on its urban landscape, which is at once magnificent and tragic. South America's cities contain a disproportionate concentration of regional wealth and power, as well as a disproportionate concentration of marginalized people who are undeterred in laying claim to their cities. Contemporary democratization has facilitated the rise of social movements and political activism throughout the continent, often shifting the balance of power to new groups. Although the future remains uncertain, it is being debated, contested, and acted on now.

SUGGESTED READINGS

Browder. J., and B. Godfrey. 1997. *Rainforest Cities: Urbanization, Development, and Globalization of the Brazilian Amazon.* New York: Columbia University Press. A comparative study of urbanization in Amazônia, including a general review of regional patterns and case studies of major cities and boomtowns in the states of Pará and Rondônia. The authors propose a heterogeneous model of "disarticulated urbanization" to account for the variety of settlement forms in the vast region.

Caldeira, T. 2001. *City of Walls: Crime, Segregation, and Citizenship in São Paulo.* Berkeley: University of California Press. A provocative interpretation of contemporary trends in urban segregation, defensive design, gated communities, and the widespread fear of violent crime in South America's largest metropolis. The author emphasizes how contemporary insecurities reflect and reinforce prejudices of race and class.

Cifuentes, L., Krupnick, A., O'Ryan, R., and M. Toman. 2005. *Urban Air Quality and Human Health in Latin America and the Caribbean.* Washington: Inter-American Development Bank. Detailed report on the impact of poor air quality on human health in Latin America's and the Caribbean's major cities, especially measured in hospital admissions, lost productivity, and shortened life spans.

Dangl, B. 2007. *Price of Fire: Resource Wars and Social Movements in Bolivia.* Oakland, CA: AK Press. An analysis of the rise of social movements in Bolivia over access to and control of natural resources such as water, natural gas, coca, and land, including clashes between social movements and corporate interests.

Gilbert, A. 2006. "Good urban governance: evidence from a model city?" *Bulletin of Latin American Research,* vol. 25, no. 3, pp. 392–419. A study of the transformation in urban governance in Bogotá, Colombia from 1990 to 2005, which is now considered, in certain respects, an example of "best practice."

Hays-Mitchell, M. 2002. "Globalization at the Urban Margin: Gender and Resistance in the Informal Sector of Peru." In *Globalization at the Margins,* ed. J. Short and R. Grant, 93–110. New York: Palgrave Macmillan. A study of women who labor in the informal economy in Peru, framing their work as acts of resistance to hegemonic economic policies.

Holston, J. 1989. *The Modernist City: An Anthropological Critique of Brasilia.* Chicago: University of Chicago Press. Holston emphasizes how modernist ideologies of development shaped the spatial form and social life of the Brazilian capital since 1960, often to the chagrin of the residents, who found the new city to contradict familiar relationships of buildings, the street, and public space.

Keeling, D. J. 1996. *Buenos Aires: Global Dreams, Local Crises.* New York: Wiley. This comprehensive book covers the history, urban structure and planning, political-economic development, and cultural evolution of the Argentine capital. Keeling emphasizes the growing conflicts between the city's sophisticated self-image, increasing political and economic challenges, and growing cultural "Latin Americanization" of a city long proud of its European heritage.

NACLA (North American Congress on Latin America) Report on the Americas. 2007. *Space, Security and Struggle: Urban Latin America,* vol. 40, no 4. Washington: NACLA. Issue devoted to Latin American urban spaces occupied by the poor, paying attention to how local power is exerted and how the issue of security is addressed. Special attention paid to Honduras, Brazil, Peru, Bolivia and El Salvador.

Perlman, Janice. 2010. *Favela: Four Decades of Living on the Edge in Rio de Janeiro.* Oxford University Press. This restudy of the author's classic earlier work on *The Myth of Marginality* finds that, despite improved material circumstances, most of the favela residents interviewed now felt more frustrated about the prospects for social mobility and fearful of the violence common in their favela communities.

Scarpaci, J. 2005. *Plazas and Barrios: Heritage Tourism and Globalization in the Latin American Centro Histórico.* Tucson: University of Arizona Press. A study of local response to outside demand for historic preservation and tourism in nine Latin American cities.

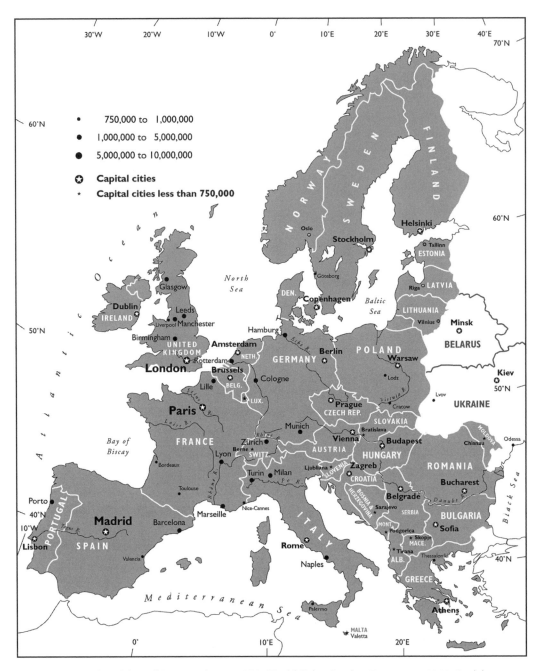

Figure 5.1   Major Cities of Europe. *Source:* UN. *World Urbanization Prospects: 2009 Revision*, http://esa.un.org/unpd/wup/index.htm

5

Cities of Europe

LINDA McCARTHY AND COREY JOHNSON

KEY URBAN FACTS

Total Population	592 million
Percent Urban Population	72.8%
Total Urban Population	430 million
Most Urbanized Countries	Belgium (97.4%)
	Iceland (93.4%)
	Denmark (86.8%)
Least Urbanized Countries	Bosnia-Herzegovina (48.6%)
	Slovenia (49.5)
	Albania (51.9%)
Annual Urban Growth Rate	0.2%
Number of Megacities	1 city
Number of Cities of More Than 1 Million	41 cities
Three Largest Cities	London, Paris, Madrid
World Cities	London, Paris, Brussels, Milan, Randstad, Vienna, Madrid, Zurich, Frankfurt, Berlin, Rome, Dublin
Global Cities	London

KEY CHAPTER THEMES

1. Europe is integral to the study of urban development because of its long history and the extraordinary impact of European urban influences worldwide.
2. Europe's urban system is dominated by the world cities of London and Paris, the result of their dominance within former vast empires.
3. Europe has quite a few cities larger than 1 million people, but in general these cities are growing more slowly than those in other world regions.

4. European cities exhibit great diversity in style and form, the result of a long history and complex mix of people and cultures.

5. The demand for low-wage labor in Western Europe has meant that immigration has gradually produced new cultural mixes in the largest cities.

6. Complex land-use patterns within European cities share certain similarities but also have important differences with cities in the United States.

7. Cities within the EU form part of an international trading bloc that contains nearly half a billion people with a combined gross national income greater than the U.S.

8. Since the end of the Cold War, Communist-era cities have undergone radical transformations, bringing them closer to their western European counterparts.

9. Europe became the birthplace of modern city planning as it reacted to the drawbacks of uncontrolled growth during the industrial period.

10. Sustainable urban management (including environmental protection and urban transportation) is an increasing priority in Europe.

Europe is a vital focus in the study of cities for a number of reasons. First, European cities are interesting in their own right; indeed, the great ones, like London, Paris, and Rome, come to mind when we think about Europe or plan a trip there. Second, because European cities are quite old, they reflect the history of many different economic, political, social, and technological systems. Third, as a hearth area of urban design, European cities are essential to understanding urban landscapes elsewhere. Fourth, a study of European urbanization is made all the more exciting by the tumultuous changes since the fall of Communism.

History, particularly recent history, has strongly conditioned the character of European cities. During the period of Soviet-style totalitarian governments after World War II, for instance, cities in much of central and Eastern Europe diverged in form and function from their western European counterparts. The late 20th century witnessed the re-integration of these socialist cities into the "new" Europe as modern democracies emerged amid the transition back to capitalism (but see box 1.3 for the lasting impressions of the Soviet era on Eastern European cities). Today, cities within the EU fall under a single economic and political framework that affects the living and working conditions of urban residents. The predominantly urban EU contains more than half a billion people with a combined gross national income greater than that of the United States. Yet, while cities across Europe share many characteristics, including bustling city centers and compact form, differences remain in land-use patterns, quality of urban infrastructure, and city planning and architectural design, to name a few.

Europe is more than 70 percent urban, but there is no standard definition for a city. National definitions range from a minimum population of 200 in Norway to 20,000 in Greece. Nevertheless, Europe's urban population of about 430 million represents just over 12 percent of the world's urban population (fig. 5.1). In fact, Europe contains over 40 cities with a population of more than a million inhabitants. Yet, wide variation is found from country to country. Moldova has the lowest level of urbanization (41 percent), while Belgium has the highest (99 percent). This variation reflects the correspondence between urbanization and factors such as level of economic development, historical circumstances, relative location, and even terrain and climate. And, as there is disagreement over what constitutes an urban population, there is even disagreement over what constitutes Europe itself (box 5.1).

HISTORICAL PERSPECTIVES ON URBAN DEVELOPMENT

One of the many exciting things about studying cities is learning to decipher the historic landscapes of bygone eras—their streets, buildings, and monuments. An historical perspective is necessary for understanding the evolution of the European urban system, because the same forces that modify the built environment of individual cities also determined where cities were initially located and how they flourished or declined (box 5.2).

Classical Period (800 B.C.E. to 450 C.E.)

In early Greek culture, independent city-states were located along coastlines, reflecting their sea-faring culture, and on easily defendable hill sites, reflecting the need for security in turbulent times. As cities like Athens, Sparta, and Corinth grew, bands of colonists left to establish cities around the Aegean and Black Seas, along the Adriatic Sea, and as far west as present day Spain.

Greek towns shared common traits. At the center was the *acropolis*, or high city, which contained temples and municipal buildings. Below the high city, in the "sub-urbs," were the *agora* (market place), more government buildings, temples, military quarters, and residential neighborhoods. Urban facilities generally were available to all citizens. These cities were laid out in a north-south grid pattern and surrounded by defensive walls. Greek cities, though, remained quite small by today's standards. Although Athens probably reached a population of about 150,000, most cities ranged from 10,000 to 15,000, while the majority had only a few thousand people.

Greek civilization was displaced during the 2nd and 1st centuries B.C.E. by the expanding Roman Empire. Although the structure of Roman cities (like Pompeii) was similar to their Greek predecessors—including the grid system, central market place (*forum*), and defensive walls—there were important differences. Roman cities were established mainly inland and operated as command-and-control centers. They were designed along hierarchical lines, reflecting the rigid Roman class system. By the 2nd century C.E., the Roman Empire extended over the southern half of Europe (fig. 5.2). Roman cities, though, remained fairly small. Although Rome's population probably reached the million mark by 100 C.E., large Roman towns contained only about 15,000 to 30,000 inhabitants, while most had 2,000 to 5,000 inhabitants.

The vacuum created by the collapse of the Roman Empire in the 5th century was filled by various tribes who greatly disrupted urban life. Most urban centers became de-populated, and their crumbling buildings were a source of building materials for rural residents. At the same time, the constant threat of attack spurred the construction of castles and other fortifications, even in some parts of Europe that had previously seen more limited urban development.

Box 5.1 Where Does Europe End and Asia Begin?

Perhaps the world's ultimate mountaineering challenge is the "Seven Summits." An elite group of globe-trekking climbers has achieved this milestone, which entails summiting the highest peak on each of the "seven continents." The location of Europe's tallest peak, at least for this challenge, illustrates the difficulties in determining what constitutes "Europe" and where it ends. Though few geographers could be pinned down to an easy definition of Europe's eastern border, for many, the Caucasus Mountains, along with the Ural Mountains and the Bosporus and Dardanelles straits, are seen as convenient physical dividing lines between Europe and Asia. Consequently Elbrus, the tallest mountain in the Caucasus, is commonly called "Europe's tallest mountain." But is it? Elbrus is located in a sparsely populated mountain range surrounded by one of the most ethnically complicated regions in the world. Turkic languages are spoken on the Russian, or so-called European, northern side of the mountain. A neat cultural dividing line based on religion is similarly hard to pin down. The population on the Russian side of Elbrus is majority Muslim, while the population on the Georgian to the south is majority Christian and has been part of the Christian world for some seventeen centuries. Armenia, which in this book is included in the Greater Middle East, has been Christian for even longer. So where does Europe end? Should these climbers be making their ascent some 1,700 miles west at Mont Blanc in the Alps?

An important, if somewhat tricky, issue for geography is that precise definitions of regions are elusive. A look at the overview map at the beginning of this book shows that what is labeled as "Europe" is, physically speaking, a peninsula of a much larger landmass, Eurasia. As even more sensitive issues than Seven Summits illustrate—including political debates in Brussels about the accession of Turkey into the European Union, and potential deliberations about including Azerbaijan and other Caucasus-bordering countries in North Atlantic Treaty Organization—a definition of Europe is not settled. Geographic labels such as those in the chapter titles of this book are more convenience rather than objective truth. Indeed, Europe is much more a cultural idea than a neatly bounded region on a map.

In many cases, the definition of world regions depends on the theme being addressed. For cities, Russian ones have fairly distinct characteristics compared with those in more westerly parts of Europe, even though textbook definitions of Europe often include the western part of Russia (where Moscow, St. Petersburg, and other important cities are located). Consequently you can understand the editorial decision to separate "Europe" from "Russia" in the table of contents of this book.

Medieval Period (450–1300 c.e.)

Feudalism curtailed the development of cities during the early medieval period because its highly structured nature favored the self-sufficient country manor as the basic building block of settlement. The only urban places to thrive or even survive were religious, trade, or defensive centers. With the resumption of long distance trade after 1000 c.e., many medieval towns grew along commercial routes that crisscrossed Europe (fig. 5.3).

Box 5.2 Digging into the Past Using Geospatial Technologies

In the not so distant past, looking at a reconstruction of an historical urban landscape often involved a visit to that city's museum. There, in one of the first rooms, you would find a glass case containing a scaled model of, for example, the Roman city of Colonia (Cologne or *Köln*), complete with forum, temples, and city walls.

Museums are still excellent entryways into how life in cities might have been and how urban morphology has changed over time. But recent changes in geospatial technologies have made accessing the urban past more interesting and entertaining. As a complement to traditional archeological methods of excavation and systematic surveys, architects, geographers, archeologists, and historians have used GIS and computer-aided design software, as well as remote sensing and ground-penetrating radar, to create more accurate visualizations that bring to life the way European cities looked in previous historical times.

Any number of data layers can be incorporated into re-creating past urban landscapes. In addition to historical maps, which can be digitized, many governments and empires kept accurate cadastral records. By entering this information into a GIS, and incorporating a digital elevation model, a dynamic spatial-temporal map of the past can be created. Interactive web-based interfaces allow people the world over to explore virtually a city's past. Even where little material evidence remains, it is possible to visualize in multiple dimensions how a city might have looked.

Other than historical curiosity, how are these virtual models useful today? One of Europe's most important industries is tourism, and cities are popular destinations. Cultural and historical tourism is a major part of this industry. Consequently, the use of new geospatial technologies helps visitors envision city walls and other parts of the built environment that no longer exist. Thematic urban itineraries—downloaded to a smart phone or a tablet computer—range from identifying residential areas of the Greek polis in Naples, Italy, to charting the location of various stalls in a medieval market in Lyon, France. For every construction crane you see in any city, you can be certain that some inventory has been taken to understand what layers of archeology still exist. In the post-socialist parts of Europe, especially, where massive efforts have been undertaken to rebuild historical urban centers in recent years—in Krakow, Bratislava, and Sofia, as examples—virtual models have been enormously useful in historical preservation and new construction, where the planners want to preserve the integrity of a city's historic morphology.

At the center of the typical medieval city was the market square. In larger cities, the market square was surrounded by the main church or cathedral, the town hall, guildhalls, palaces, and the houses of prominent citizens. Close to the center were streets or districts that specialized in particular functions, such as banking, furniture, or metalwork. The streets and alleys were quite narrow. The enclosing walls often had water-filled moats to enhance defensive capability. Finally, medieval towns were decidedly unhygienic. Given

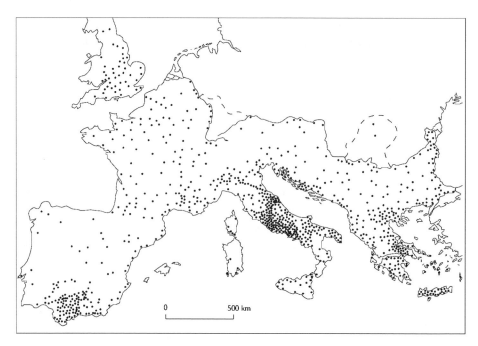

Figure 5.2   Roman Cities in Europe, 2nd century C.E. *Source:* Adapted from N. J. G. Pounds, *An Historical Geography of Europe* (Cambridge: Cambridge University Press, 1990), 56. Reprinted with permission

the cramped conditions, lack of air circulation, poor sanitation, and absence of waste treatment, it is little wonder that the Black Death (1347–1351) progressed so easily, killing one-third of the people in the towns.

Most development during the medieval period was in the western and southern parts of Europe that had a Roman heritage of city building. Urban development was impeded in southeastern Europe where the Byzantine Empire was in control, whereas much of eastern and northern Europe remained in a pre-urban state. Conversely, the Moors, who spread into Iberia in the early 700s, founded or restored many cities and elevated urban culture in what would become Spain. At the close of the medieval period, Europe had about 3,000 cities, most of which had fewer than 2,000 people; only Milan, Venice, Genoa,

Florence, Paris, Córdoba, and Constantinople had more than 50,000.

Renaissance and Baroque Periods (1300–1760 C.E.)

The Renaissance (1300–1550 C.E.) was marked by significant changes: in the economy (from feudalism to merchant capitalism), in politics (rise of the nation-state), and in art and philosophy. Beginning in Florence in the 1300s, these changes spread throughout western Europe; conversely, feudalism was still strong in eastern parts of Europe; southeastern Europe fell within the Ottoman Empire; while much of northern Europe remained outside the progressive influences of the Renaissance.

Spurred by heightened demand for such luxury goods as spices and silks introduced dur-

Figure 5.3   Ljubljana, Slovenia, has taken advantage of the collapse of Communist rule to bring out the medieval elements of the city's center, including the Dragon Bridge and St. Nicholas cathedral. (Photo by Donald Zeigler)

ing the time of the Crusades (1095–1291 c.e.), merchants greatly expanded the trade functions of Mediterranean cities; later, the economic center of gravity shifted to the port towns along the North and Baltic Seas in conjunction with the Hanseatic League, an association of towns with the goal of promoting trade.

Changes in the political system, in particular the growth of nation-states, had an impact on European urbanization. Best exemplified by Paris and Madrid, the central location of these capitals aided the process of political consolidation; in turn, both cities were given further impetus for growth by their administrative functions and enhanced status at the vortex of social, economic, and political change. Similarly, regional centers and seats of county government emerged to fill-out expanding national urban networks.

The overall appearance and structure of cities changed because of the new forms of art, architecture, and urban planning. Especially in capital cities, flourishing artistic and architectural expression brought about greater use of sculpture in public areas. Other urban beautification such as fountains and embellishment on monumental buildings reached a peak during the baroque period (1550–1760).

After the introduction of gunpowder, massive city walls became obsolete. In many cities, the walls were removed to make space for wide boulevards that were becoming fashionable. Also, the accumulation of great wealth by the nobility led to opulent palaces being built in many cities, notably Vienna and Paris, and to replanning parts of cities. Beginning in Paris, many districts containing narrow medieval streets were torn down to make way for wide boulevards that radiated outward and connected the various palaces and formal gardens laid out for the nobility. The emphasis on the control of visual perspective and the rediscovery of classical models of design marked a significant departure from medieval times. Overall, the urban network remained largely unchanged.

Industrial Period (1760–1945 C.E.)

Large-scale manufacturing began in the English Midlands in the mid-1700s and spread to Belgium, France, and Germany, reaching Hungary by the 1870s. New factories, making a range of products from textiles to machine tools, changed the structure of cities and led to massive rural-to-urban-migration.

In many cities, whole districts of factories emerged, easily identified by their belching smokestacks, deafening machinery, and general hustle-and-bustle of industrial activity. By the mid-19th century, trains transported much of the industrial inputs and products, so new tracks, stations, and rail traffic began to play a significant role in urban development. Public transportation—trolleys and subway systems—also modified the look and functioning of cities. Large tracts of often cramped worker housing were constructed. The industrial period also heralded the development of the central business district ("CBD") with its office buildings and corporate headquarters.

The growth of cities closely mirrored the spread of industrialization. By the mid-1800s industrial towns in the English Midlands and Scotland, such as Birmingham and Glasgow, had grown to more than 100,000 inhabitants each and the proportion of the population living in cities larger than 10,000 had risen to 30 percent. The growth of industrial cities in France, Belgium, and Germany reflected the same pattern. In contrast, expansion of the industrial sectors in southeastern Europe did not occur until the early- or mid-20th century.

URBAN PATTERNS ACROSS EUROPE

A glance at the map of Europe shows the impact of central place theory on the size and spacing of urban places. In southern Germany, the largest metropolitan areas—Frankfurt, Munich, and Stuttgart—are spaced about the same distance apart. In Hungary, centrally located Budapest is ringed by the regional centers of Debrecen, Miskolc, Szeged, and Pécs. Of course, political, economic, cultural, environmental, technological, and other changes can alter the role and rank of a place within an urban hierarchy. Still, the empirical observation of rank-size distribution holds for Belgium, Germany, Italy, Norway, and Switzerland. In other national urban systems, a deviation occurs at the top of the hierarchy to create primacy. Primate cities, which are often national capitals, include Athens, Budapest, Dublin, London, Paris, Reykjavik, Sofia, and Vienna.

Historically, rural-to-urban migration was the most important component of urban growth, especially during the industrial period. This form of migration within Europe, though, has largely ceased. And, as birth rates have fallen considerably in recent decades, European

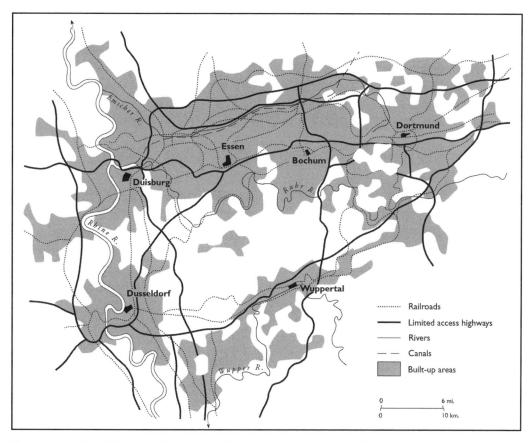

Figure 5.4   The Rhine-Ruhr Conurbation in Germany. *Source:* Compiled from various sources

cities are among the slowest growing in the world, averaging just 0.18 percent a year.

The cities of Europe, however, have begun to expand outward and coalesce into conurbations in tandem with the transportation and communications infrastructure. Europe now contains about 50 conurbations with more than a million inhabitants. The metropolitan regions between London and Newcastle form an area of extensive urbanization in England. Germany's Rhine-Ruhr conurbation has a diameter of about 70 miles (113 km) and runs from Düsseldorf and Duisburg in the west to Dortmund in the east (fig. 5.4). Of similar diameter is the Randstad, a densely populated horseshoe-

shaped region in the Netherlands that runs from Utrecht and Amsterdam in the north through The Hague and Rotterdam in the west, to Dordrecht in the southeast (fig. 5.5). Only 60 miles (97 km) apart, these two conurbations may eventually coalesce to become a dominant European metropolitan core.

Post-War Divergence and Convergence

Western Europe

After World War II, separate urban systems developed on either side of the Iron Curtain—the boundary that divided Europe into a capitalist

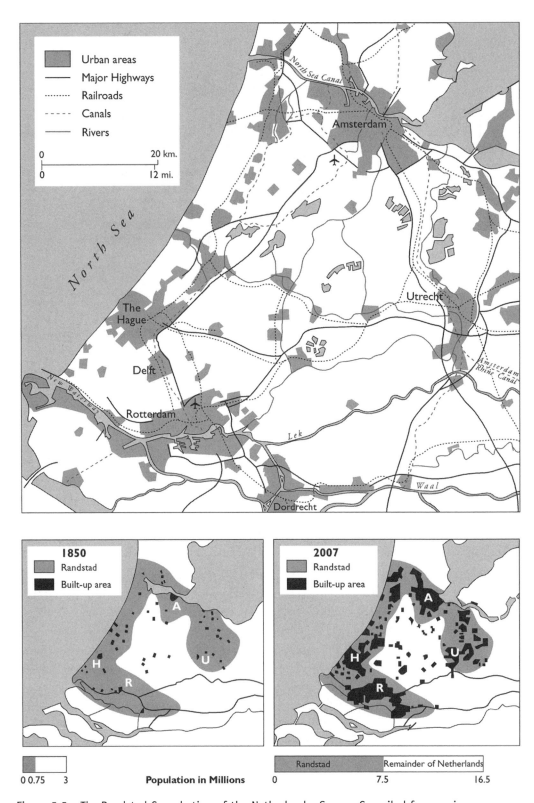

Figure 5.5   The Randstad Conurbation of the Netherlands. *Source:* Compiled from various sources

Figure 5.6   Impacted adversely by metropolitan decentralization, this derelict inner-city area in Dublin contained vacant and deteriorating residential and industrial properties. (Photo by Linda McCarthy)

west and communist east. Cities in Western Europe cultivated connections with the capitalist world, especially the United States, whose Marshall Plan funded rebuilding in cities that had suffered appalling wartime destruction. The reconstruction effort was seen as an opportunity to replan bombed-out urban areas. Some of the most heavily damaged cities, like Rotterdam and Dortmund, completely redesigned their street systems for new commercial and industrial buildings. Most cities, including Cologne and Stuttgart, incorporated the surviving historic structures and medieval street patterns into their reconstructed city centers. Rouen and Nuremberg went so far as to rebuild exact replicas of their destroyed historic buildings.

Rapid economic and demographic growth fueled remarkable urban growth. Cities of all sizes grew in residents as well as in extent due to suburbanization. But this period of remarkable urban growth began to slow by the early 1970s as widespread economic recessions followed the end of the Baby Boom. In addition, counter-

urbanization (metropolitan decentralization) promoted development in nearby towns and rural areas, while growth slowed toward the urban core (fig. 5.6). Peripheral areas attracted residents and businesses looking for more space and less pollution and crime. Most city centers lost retail and office employment to outlying areas. Medium-sized cities attracted employment in expanding sectors of the economy like information services, high-tech industries, or modern distribution activities. These smaller cities at the periphery of major metropolitan centers had lower rents and congestion while enjoying nearby transportation routes, airports, universities, and skilled workers.

Deindustrialization and corporate restructuring contributed to massive job losses and urban decline in traditional centers of industry. The jobs created by the relocation of labor-intensive manufacturing benefited some urban areas in Ireland, Spain, Portugal, and Greece. Branch plant operations, however, are vulnerable to decisions made outside the area and to company

Figure 5.7   Communism brought extensive industrial development (evident in the background) and isolation to Plovdiv, but post-Communist cell phone networks now connect a new generation of Bulgarians to the world. (Photo by Donald Zeigler)

relocations when government tax incentives expire. Since the early 1990s, competition for investment has come from central and eastern Europe where production costs are lower.

During the last few decades, the city centers have seen quite significant changes as employment has shifted to professional and business services such as banking and insurance. New developments include shiny new high-rise offices, luxury condos and apartments, and gentrified neighborhoods with expensive restaurants, bars, and boutique stores. In world cities like London and Paris, the most visible group of people on the streets of the CBDs are young professionals chatting on their cell phones and wearing the latest fashions.

Socialist Urbanization

Following World War II, cities behind the Iron Curtain developed independently of their western counterparts. Totalitarian governments engaged in sweeping reforms that led to considerable changes to their national urban systems, which evolved in response to centralized planning rather than market forces.

Communist governments had to contend with the pressing need to rebuild cities left in ruins after the war. The damage sustained by these cities, particularly Dresden, Berlin, and Warsaw, was more severe than in Western Europe, but subsidies were not available through the Marshall Plan. The earliest stage of postwar economic development involved rapid expansion of heavy industry, particularly iron and steel, chemicals, and machinery. Coupled with collectivization and increased mechanization in agriculture, this *extensive* industrial development soon led to unprecedented rural-to-urban migration (fig. 5.7). Levels of urbanization rose quickly in conjunction with rising levels of primacy, severe housing shortages, insufficient social infrastructure and basic services, and *environmental degradation.*

By the late 1960s, for example, Budapest contained 44 percent of Hungary's urban population. The numbers masked a new reality: *underurbanization.* The number of industrial jobs grew faster than the number of housing units, forcing workers to commute long distances. As a result, beginning in the mid-1970s, Communist governments set out to erase the difference between city and village life by emphasizing light industry and services, decentralizing production from capital and larger cities to smaller ones, developing transportation networks, and increasing levels of public infrastructure and housing in cities. Rural-to-urban migration slowed significantly. Despite these efforts, by the collapse of Communism, the urban network in much of Central and Eastern Europe was still less developed than in the West. Only seven cities, all national capitals, contained more than a million people: Budapest (2.1), Bucharest (2.0), Warsaw (1.7), Belgrade (1.5), Prague (1.2), East Berlin (1.2), and Sofia (1.1).

Post-Socialist Changes

Since the fall of the Iron Curtain in the late 1980s, Communist governments have been replaced, Germany has been re-united, and Czechoslovakia, Yugoslavia, and the USSR have broken up. Central economic planning was abandoned in favor of democratization and transformation from socialist to market economies. The demise of the Soviet Union opened the way for western investment to move in and people to move out. These changes impacted the cities and urban systems in a number of ways. First, city names

that were inspired by revolutionaries, such as Leninváros (Lenin City) in Hungary (now Tiszaujvaros) and Karl-Marx-Stadt in Germany (now Chemnitz) were changed back to their pre-war names or to honor individuals or events associated with the 1989 velvet revolutions. Statues of Communist and Soviet leaders were removed and some were later put on display in statue parks or museums. Second, foreign direct investment flooded in, targeted mainly at capital cities. This boosted the transition to capitalism and fueled speculative construction booms and gleaming, Western-style commercial and residential developments. Third, decentralization gave more authority to city planners.

The transition to a market economy, however, did not come without hardship. In addition to the loss of state subsidies, some cities lost revenues that had been generated by the Soviet military presence. Other towns lost economic status because of the re-orientation of trade westward. Industrial towns suffered severely due to the closure of outdated plants. Rapid, large-scale privatization brought skyrocketing unemployment, unknown under socialism. The number of industrial workers dropped sharply compared to those in services. These processes particularly affected large and industrial cities and resulted in dilapidated infrastructure, loss of green space, and overconstruction in resort towns (see box 1.3). In the former Yugoslavia, post-Communist hardship was accentuated by damaging wars.

More color and neon lights now characterize the cities of Central and Eastern Europe. Advertising has replaced communist slogans on billboards. Shabby, old department stores have been renovated or replaced by boutiques and shopping malls. Beggars have appeared, as well

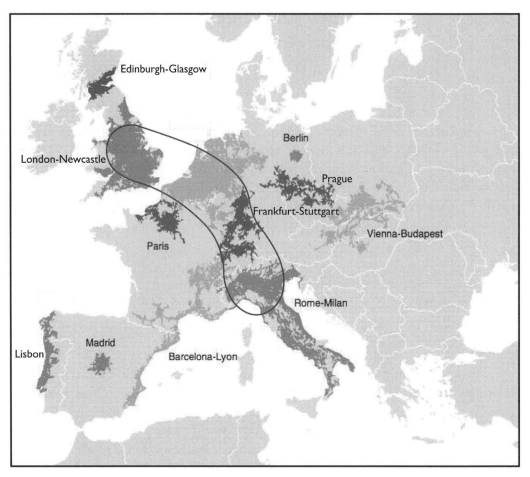

Figure 5.8   Europe's conurbations within the context of Europe's "Blue Banana." *Source:* Linda McCarthy

as casinos and nightclubs; crime has risen; and congestion is ubiquitous. Social differentiation in housing has increased tremendously. Democratization and market economies are erasing the communist legacy and bringing these cities closer to their western European counterparts.

Core-Periphery Model

A core-periphery model is often used to describe urban patterns in Europe (fig. 5.8). The dominance of cities and conurbations at the European core is based on their superior endowment of factors influencing the location of economic activity, such as accessibility to markets. The largest cities are connected by the most advanced transportation and communications systems. Labor force quality and government policies make the core the most attractive area for modern companies. French geographer, Roger Brunet, identified the "Blue Banana," a curving urban corridor of high-tech industry and services that includes core cities such as London and Frankfurt.

Cities in the core and periphery are linked in a symbiotic if unequal relationship. Core cities prosper and maintain their economic dominance at the expense of the periphery by capturing flows of migrants, taxes, and investment in cutting-edge industries such as high-tech manufacturing and in command-and-control functions like the headquarters of transnational corporations ("TNCs"). Peripheral cities have more limited potential for economic development, and attract tourists and investment in branch plants from core locations. In between are the semi-peripheral cities that have economic links with both the core and the periphery.

The core is shifting to the south and east, to areas of high-tech industrial growth. New core cities include Munich, Zürich, Milan, and Lyon. This southeastward shift intensified after the fall of the Iron Curtain; people and companies have been attracted to cities such as Bratislava and Warsaw because of the surge in economic activity, as well as to the region as a whole due to its relatively low production costs. London and Paris, however, have retained their historic importance because of their size and established positions as major national and international cities. The continued economic strength of the core is reinforced by the considerable political control that comes with the role of the largest cities as major centers of international decision-making.

IMMIGRATION, GLOBALIZATION, AND PLANNING

Integrating Immigrants: A Challenge of European Urbanization

The rebuilding of western Europe's urban infrastructure and industry after World War II generated strong demand for labor, especially in the more prosperous countries. In the 1950s and 1960s, rural-to-urban migration fueled urban growth. In addition, foreign guestworkers were brought in to fill low-wage assembly line and service-sector jobs that the more skilled domestic labor force would not take. Guestworkers came from Mediterranean Europe and former colonies. West Germany attracted immigrants from Turkey and Yugoslavia; France brought in workers from northern and western Africa; and Britain drew on Commonwealth citizens from the India, Pakistan, and the Caribbean.

In the EU today, 4 percent of the people—about 20 million—were born outside its member countries. More than one-third of the immigrants in France are concentrated in the Paris region where they represent over 15 percent of the population. Foreign-born residents comprise about a fifth of the population in German cities like Frankfurt, Stuttgart, and Munich. More than half the population of Amsterdam is now non-Dutch, with large immigrant streams from Morocco, Turkey, and Ghana. Even Finland's immigrant community is growing (fig. 5.9). Despite the lack of comparable data for immigrants in cities across Europe, their presence is evident in the cultural diversity reflected in the names of stores and restaurants in some neighborhoods and even in the list of most popular baby names for cities like Brussels (table 5.1).

Immigrants typically live in poor quality suburban high-rise apartments or inner city enclaves left vacant through suburbanization. Each enclave in the city is dominated by a particular ethnic group. The enclaves in Frankfurt and Vienna contain mostly Turks, while in Paris and Marseille, the different enclaves house Algerians or Tunisians. In large British

Figure 5.9   Kurds, with flag unfurled, assemble in front of the Parliament building in Helsinki. Even though Finland has one of Europe's smallest immigrant
communities, the numbers are growing. (Photo by Donald Zeigler)

Table 5.1   Top 10 Boys' and Girls' names in the Brussels Metropolitan Area.

	Boys' Names	Girls' Names
1.	Mohamed	Lina
2.	Adam	Sarah
3.	Rayan	Aya
4.	Nathan	Yasmine
5.	Gabriel	Rania
6.	Amine	Sara
7.	Ayoub	Salma
8.	Mehdi	Imane
9.	Lucas	Ines
10.	Anas	Clara

Source: The Telegraph newspaper, 2007, http://www.telegraph
.co.uk

and West Indians, the foreign-born population represents only 15–20 percent of the
population within most neighborhoods.

In addition to outright discrimination, the
labor and housing markets help create inner
city enclaves. Low wages force immigrants
to rent lodgings in deteriorating inner city
locations. The internal cohesiveness of the
ethnic groups also contributes to residential
segregation. Existing residents are more likely
to share information about vacancies in their
neighborhood with members of their own
ethnic group.

European and Global Linkages

European cities are part of urban networks
that operate at different spatial scales. Since
1989, cities on either side of the former Iron
Curtain have become more interconnected.
Increasing EU economic and political integration has influenced the development of

cities, in contrast, there is significant mix-
ing of different ethnic groups. Within each
neighborhood, however, the ethnic groups
are highly segregated from each other. And
although there are large numbers of Asians

the European urban system. For example, the removal of national barriers to trade within the EU, with the internationalization of the European economy, has encouraged population increase along certain border regions. Urban growth zones straddle the boundaries between the Netherlands and Germany, Italy and Switzerland, and the southern Rhine regions of France, Germany, and Switzerland.

A select group of cities contain the headquarters of major international agencies, many of which were founded after World War II to promote economic, political, or military cooperation. Geneva is the main European center for the United Nations. Paris is the headquarters for the Organization for Economic Cooperation and Development (OECD) and the European Space Agency. Vienna is the headquarters for the Organization of Petroleum Exporting Countries (OPEC).

Important decision-making functions are located in the EU's "capital cities": Brussels (both the Council and the European Commission), Strasbourg (Parliament), and Luxembourg (Court of Justice). Brussels is also the headquarters of NATO and Strasbourg additionally serves as the headquarters of the Council of Europe, an organization of nearly 50 countries that promotes European unity, human rights, and social and economic progress.

The major centers of international banking and finance in Europe historically have been London and Paris, but now include Frankfurt, Zürich, and Luxembourg. Frankfurt hosts the Bundesbank, Germany's influential central bank, as well as the European Central Bank that manages the euro, making it the financial capital of the EU.

London and Paris rank among the select number of world cities that contain the headquarters of some of the most powerful TNCs

Figure 5.10 Milan is one of the fashion capitals of the world, a fact in evidence on the Piazza del Duomo. (Photo by Donald Zeigler)

in the world. London contains about 17 of the 500 largest global companies (58 percent of the U.K.'s total), including BP, HSBC, and Lloyds. Paris has even more—25 of these companies (64 percent of France's total), including AXA, Christian Dior, and Vivendi. In addition to housing 4 of the 500 largest global companies, Rome contains Vatican City, the seat of the Roman Catholic Church. Milan and Paris are major centers of fashion and design, while London is the premier insurance center (fig. 5.10).

Accessibility via the latest transportation and communications technologies allows some cities to strengthen their international positions. The high-speed train network reinforces the dominance of London, Paris, Brussels, Amsterdam, and Cologne. The cities with the busiest airports are London, Paris, Frankfurt, and Amsterdam.

At the mouth of the Rhine, Rotterdam is Europe's largest port and one of the largest in the world. Its annual turnover of about 400 million metric tons of cargo is third only to Shanghai and Singapore. Rotterdam's water and pipeline connections with the Ruhr in

Germany make it the main oil distribution and refining center in Europe. Antwerp, Marseille, and Hamburg are also major ports.

Trucking is the most important mode of ground transportation for freight. Nearly 2,000 billion tons of goods are transported by road annually in the EU alone compared to only about 450 billion by rail. There are close to 500 passenger cars for every 1,000 people in the EU. With nearly 600 cars per 1,000 inhabitants, German and French levels already match those of Canada, but even Italy's higher rate of more than 650 is still far below the U.S. level of more than 800. Steadily increasing automobile ownership, with the distance traveled tripling in the last 30 years, has overwhelmed existing and new highway capacity and led to traffic congestion within and between cities throughout Europe.

Formerly Communist areas of Europe still lag behind in terms of the extent and efficiency of their transportation systems, though this is changing. In Germany, rail and road links abandoned during 40 years of Communism that divided the country into West and East have been rebuilt. Even though Communist states were allied with each other and the Soviet Union, this did not ensure a high degree of infrastructure interconnections. Traveling between Budapest and Warsaw by land, for example, entailed multiple border crossings, transit visa requirements, and long waits. Travel was also slowed by narrow, dangerous, often pre-war roads. In part due to large investments by the EU, new multi-lane highways are being built, but it will be years before transport linkages resemble those in the West.

Urban Policy and Planning

Europe became the birthplace of modern city planning as it reacted to uncontrolled growth during the industrial period. Planning now pervades European city life. In Western Europe, postwar planning for growth ended in the early 1970s. Declining population growth rates and widespread economic recessions forced governments to reconsider large-scale publicly funded projects. Given dissatisfaction with alienating high-rise buildings and open spaces, policy shifted to planning for conservation and restructuring, and combating urban decline.

This reappraisal has had two, often conflicting, components: budgetary constraints forcing governments to seek private-sector investment in revitalization projects, and growing concern for social equity, citizen participation, environmental protection, and aesthetic quality. These competing elements are pitted against each other at three levels of government: local, national, and international.

Local Policy and Planning

Since the early 1970s, urban revitalization policies in the older industrial cities of western Europe have promoted economic restructuring away from traditional manufacturing. Cities use local, national, and EU funds to attract private-sector investment in high-tech and service industries. The transition toward K-economies (knowledge economies) has favored metropolitan areas with diversified economies, highly skilled workforces, major universities, and good quality of life. Until the global economic downturn in the late 2000s, national government policies in Ireland, such as low corporate taxes and investment in education, combined with EU funding, helped fuel the period of phenomenal economic growth in Ireland that came to be know as the "Celtic Tiger," and allowed the Dublin region to attract global information technology companies like Dell, Intel, and Microsoft.

Traditionally in southern Europe, cities in lower-cost production areas attracted labor-intensive branch plant industries. More recently, cities like Montpellier in France, Bari in Italy, and Valencia in Spain have focused on providing attractive environments for high-tech industries. Since the early 1990s, Communist-era cities like Dresden, Budapest, and Warsaw, have worked to attract new commercial and industrial investment.

In recent years, most countries in Western Europe have decentralized power and responsibility for urban planning to local governments. In addition, smaller units of local government have been consolidated into larger regional ones to achieve economies of scale. These changes have set the scene for more coordinated regional planning. The Dutch "compact city" policy in the Randstad endeavors to curb counterurbanization by concentrating new development within existing major cities in an effort to maintain their economic competitiveness.

National Policy and Planning

In Western Europe, national policies promoted regional decentralization after World War II. Industry, commercial activity, and population were redirected from the large congested cities to new towns, as in the case of Abercrombie's Plan for Greater London. In the 1970s, declining population growth and widespread economic recessions forced national governments to reassess the need for new towns.

By the 1980s, a national policy shift reflected factors such as the severity of decline in the central parts of larger cities and their importance as national engines of growth in a global economy. The U.K. government established Urban Development Corporations to attract businesses to declining industrial and port areas in cities like London and Liverpool. In the 1990s, government policy shifted yet again—away from expensive nationally devised strategies to "community empowerment" initiatives in which communities carry out revitalization programs and projects that are sensitive to local challenges.

In Communist parts of central and eastern Europe after World War II, government planning was guided by the basic tenets of Marxist-Leninist ideology: to remove the "contradiction" between living standards in urban and rural areas and to create a classless society. Urban planners sought to avoid excessive population concentration in large cities and to achieve a balanced urban infrastructure. These social goals, however, often clashed with economic directives, especially the development of heavy industry.

In an attempt to increase overall industrial capacity and provide urban functions to underserved areas, governments implemented a program of new town construction away from existing cities. These towns were developed around a large industrial facility, typically an iron and steel mill or chemical processing plant. New towns included Eisenhüttenstadt in East Germany and Nova-Huta in Poland. By the 1970s and 1980s, Communist planners had turned their attention away from promoting large-scale industry and new towns to developing light industries and filling-out the national urban systems. In many countries, central place theory became an explicit guide as planners tried to create multi-tiered urban hierarchies that provided goods and services to particular regions according to their size and function.

Since the early 1990s, central and eastern European cities have experienced dramatic changes because the transition to a market economy involved rapid, large-scale privatization of state-owned housing, industry, and services. National policies have evolved to address urban problems that were unknown in the former socialist states, such as unemployment, crime, poverty, and homelessness. The poor Roma (Gypsy) quarters in larger cities were disproportionally affected. EU integration has helped alleviate some of the problems through a number of urban redevelopment projects, though funds are being gradually reduced.

International Policy and Planning

Europe is the scene of significant international urban planning and management initiatives. The Council of Europe and the EU, for example, celebrate Europe's cultural heritage through European Heritage Days. Cultural events are planned in cities and towns across the EU that are aimed at bringing European citizens together through highlighting local traditions, skills, and works of art and architecture. The EU also selects two cities every year as European Capitals of Culture, with Turku in Finland and Tallinn in Estonia selected for 2011.

The EU established the European Green Capital Award to promote local government efforts to improve their urban environment and to showcase best practice. The first award winner in 2010 was Stockholm, followed in 2011 by Hamburg. In terms of more measureable environmental quality, Stockholm also ranked highly, second behind Copenhagen, on the European Green City Index (table 5.2).

EU integration efforts have led to unprecedented achievements in international policy and planning. The publication of the "Green Paper on the Urban Environment" in 1990 reflected the need for EU policies to address specifically urban issues. Certainly, European cities are generally pre-disposed to being

Table 5.2   European Green City Index 2009: Top 10 Cities

Rank	City	Score*
1	Copenhagen	87.31
2	Stockholm	86.65
3	Oslo	83.98
4	Vienna	83.34
5	Amsterdam	83.03
6	Zurich	82.31
7	Helsinki	79.29
8	Berlin	79.01
9	Brussels	78.01
10	Paris	73.21

* Out of a possible 100, based on eight categories (CO_2, energy, buildings, transport, water, waste and land use, air quality, environmental governance) using 30 indicators, conducted by the Economist Intelligence Unit, sponsored by Siemens.

"green" because of their high density, compact form, and associated walkability and high usage of public mass transit. Even so, European cities are not exempt from environmental problems associated with issues like traffic congestion. Over the years, the EU has adopted policies that have provided funding for innovative environmental management projects, including the Sustainable Cities project involving research, information exchange, and networking in support of the implementation of Local Agenda 21, and Eurocities, a network of more than 140 major cities across 30 countries that provides a platform for best practice exchange.

CHARACTERISTIC FEATURES WITHIN CITIES

"Our cities are like historical monuments to which every generation, every century, every

Figure 5.11   The House of Science in Bucharest is typical of Soviet-inspired architecture throughout central and eastern European cities. Note the pedestal in front of the building, which formerly supported a large statue of Lenin. (Photo by Darrick Danta)

civilization has contributed a stone" (Ildefons Cerdà, Spanish town planner, 1867). The landscape of European cities today represents an incomplete catalog of urban development and redevelopment over time. Typical historic and contemporary features include:

Market Squares

The town square, the heart of Greek, Roman, and medieval towns, has often survived as an important open space in contemporary European cities. Some medieval town squares boast a continuous tradition of open-air markets. In central and eastern Europe, the large open square, typical of socialist cities, was used for political rallies. Today, like their western European counterparts, many central squares and their historic buildings contain modern commercial functions, such as tourist offices and fashionable restaurants and cafés.

Major Landmarks

Historic landmarks in western European city centers have become symbols of religious, political, military, educational, and cultural identity. Many cathedrals, churches, and statues often continue to dominate the skyline. Town halls, royal palaces, and artisan guild-halls have been converted into libraries, art galleries, and museums. Medieval castles and city walls are tourist attractions. Today, of course, the major landmarks are expressions of economic power the offices of TNCs and sports stadiums, for instance.

In Central and Eastern Europe, in addition to pre-war landmarks, the hallmarks of socialist cities included massive buildings in "wedding cake" style (fig. 5.11), red stars, and "heroic" statues. Since the late 1980s socialist political symbols have been replaced by billboards advertising the trappings of consumer culture.

Figure 5.12   Here on Ludgate Hill in the City of London, a new immigrant from Bangladesh directs people to the nearest McDonalds. In medieval times, this area would have been a shadowy tangle of narrow alleys that passed for streets. (Photo by Donald Zeigler)

Figure 5.13   A busy pedestrianized shopping street in the heart of Dublin. (Photo by Linda McCarthy)

Complex Street Pattern

The narrow streets and alleys of the medieval core developed in the pre-automobile era (fig. 5.12). During the medieval period, suburban areas grew around long distance roads that radiated outward from the city gates. In the 19th century, cities like Munich, Marseille, and Madrid made radial or tangential boulevards the axes of their planned suburbs.

High Density and Compact Form

The constraints of city walls kept population density high during medieval times. Several factors maintained the compact form that is now characteristic of large cities in Europe. A long tradition of planning that restricts low-density urban sprawl dates back to strict city building regulations in the earliest suburbs. Compact urban form also reflects the relatively late introduction of the automobile and high gasoline prices.

Bustling City Centers

Their high density and compact nature create city centers that bustle with activity (fig. 5.13). Heavily used public transportation systems of buses, subways, and trains converge on the core, and central train stations figure prominently.

In larger cities, distinct functions dominate particular districts. Institutional districts house government offices and universities. Financial and office districts contain banks and insurance companies. A pedestrianized retail zone often leads to the train station. Cultural districts offer museums, art galleries, and theaters.

Many buildings in the city center have multiple uses. Apartments are found above shops, offices, and restaurants. Large department stores, such as Harrods in London, Printemps in Paris, and Kaufhaus des Westens in Berlin, are prominent features in most city centers. Modern centrally located malls include Westfield London (255 stores) and Prague's Palace Flora, both accessible by mass transit.

Suburban malls are becoming prevalent. Many coastal or riverine cities have also refurbished old port and industrial buildings to house mixed-use waterfront developments like Ķīpsala in Riga and HafenCity in Hamburg. Other cities have renovated obsolete historic structures, such as London's Covent Garden, as festival marketplaces with specialized shops, restaurants, and street performers.

Low-Rise Skylines

For North American visitors, the most striking aspect of the older parts of European cities is the general absence of skyscraper offices and high-rise apartments. City centers were developed long before reinforced steel construction and the elevator made high-rises feasible. Building codes designed to minimize the spread of fire maintained building heights between three and five stories during the industrial period. Paris fixed the building height at 65 ft in 1795, while other large cities introduced height restrictions in the 19th century. Still regulated today, high rises are found only in redevelopment areas or on land at the periphery of the city, like La Défense in Paris. Skyscrapers have also been built in the central financial districts of some of the very largest cities, including London.

In socialist parts of Europe, there was no private ownership of land, and so no urban land market. Until recently, the tallest buildings were usually Communist Party and state administrative buildings, massive "Houses of the People," and TV towers.

Neighborhood Stability

Western European cities enjoy remarkable neighborhood stability. Europeans move houses much less frequently than North Americans. As a result, older neighborhoods at or near the center of large cities enjoy remarkably long lives, despite suburbanization.

The districts of handsome mansions built by speculative developers for wealthy families in the 17th and 18th centuries remain stable, high-income neighborhoods, including Belgravia and Mayfair in central London. High-income suburban neighborhoods developed in the western parts of older industrial cities, upwind of factory smokestacks and residential chimneys.

Wealthy residents, in fact, have remained at or near the city center in Western Europe since before the Industrial Revolution. Higher taxes on city land until the late 19th century kept the poorest people outside the city walls. Beginning in Paris in the mid-19th century, this tradition was strengthened by the replacement of slums and former city walls with wide boulevards and imposing apartments.

Since the 18th century, however, urban growth has also spread to suburban areas and even enveloped freestanding villages and towns. Yet these separate urban centers became distinct quarters within the expanding city as they maintained their long-established social and economic characteristics and major landmarks and links to the city center by public transit. During the second half of the 19th century, annexations of these suburban areas produced distinctive city districts with their own shopping streets and government institutions.

In the past few decades, city governments funded urban renewal projects designed to attract higher-income residents to the revitalized parts of central areas. The success of these large-scale redevelopments has given rise to gentrification in the surrounding area. Demand for housing that can be renovated for higher-income occupants, however, has raised property values in certain areas and pushed out lower-income residents.

Housing

Apartment living is common in Europe. Apartments are a good land use choice when space is at a premium and land values are high. Instead of growing outward, cities grew upward, to the limit of the height regulations.

The multistory apartment building originated in northern Italy to accommodate the wealthy during the Renaissance. By the early 18th century, apartment buildings had spread to the larger cities in continental Europe and Scotland. Until the invention of the elevator, social stratification within individual buildings was vertical: wealthier families occupied the lower floors; poorer residents lived in smaller units above. Horizontal social stratification also developed within apartment blocks. Larger expensive units faced the front; small low-rent units faced the rear. By the late 18th century, as the Industrial Revolution spurred increasing urbanization, apart-

ment blocks had spread to medium-size cities. Speculators built large-scale standardized tenements for middle-income occupants and barracks for low-income residents.

The two-story, single-family row house with small garden is distinctive to England, Wales, and Ireland (box 5.3). This tradition can be traced back to efforts to restrict congestion in London in the late 1500s that made it illegal for more than one family to rent a new building.

The serious housing shortage that started with the economic recession of the 1930s was exacerbated by the lack of construction and significant destruction during both world wars. The public housing programs that began in Vienna in the early 1920s were stepped up after World War II across Western Europe. Modern architecture and urban design were combined with low-cost factory production. Many war-damaged historic houses and dilapidated 19th century tenements were replaced by monotonous highrise apartments after World War II.

In the 1950s and 1960s, most governments adopted a policy of metropolitan decentralization. Massive modern high-rise apartment blocks were concentrated in large peripheral housing estates known by their French name—*grands ensembles*. The amount of public housing was highest in cities with serious housing shortages and liberal municipal governments, such as Edinburgh and Glasgow in Scotland, where the number of public units grew to well over half the housing stock. Traditionally, public housing comprised 25 percent of the total in England, France, and Germany, and 10 percent in Italy. Public housing represents only 5 percent or less of the housing in the more affluent and conservative Swiss cities. Since the 1970s, however, dependence on public housing has declined due to

---

Box 5.3 Growing Power: Urban Agriculture in Europe

Cities around the world have witnessed a revival in urban agriculture during the last decade or so. Liberalized laws mean that apartment dwellers in London and Rome, like those in New York and other global cities, can keep chickens. Rooftops and garden patios now double as vegetable gardens, while old allotments are seeing new life as people want food security and quality assurance about the food they eat.

Urban agriculture is not new. Rapidly industrializing Germany in the 19th century was home to a number of firsts in urban agriculture, as socially conscious lords and later city governments sought to offer relief to migrants who lived in squalid conditions in cities such as Berlin, Munich, and Leipzig. Urban community gardens had their origins as part of larger projects designed to provide low-income residents with a means to feed themselves. Allotment gardens, in Germany called *Kleingärten* (small gardens) or *Schrebergärten* (after a physician who promoted gardening as a means for urbanites to escape the ills of city life), were usually located on low value land, such as railroad rights-of-way. In Scotland, the 1892 Allotments Act provided a legal means for working-class people to petition for an allotment garden. During wartime in the 20th century, allotments served a crucial role as sources of food. "Dig for Victory" was a rallying cry for Britons during World War II in the same way as victory gardens sprouted across the U.S. After the war, when critical shortages of food in Central Europe caused widespread malnutrition in cities, small plots provided much needed vegetables.

Urban agriculture's recent strong comeback across Europe is not about fear of starvation, but about a trend toward a "return to roots." The slow food movement, which originated in Italy and spread throughout the world, emphasizes that overall physical and psychological health is tied to healthy eating and locally produced foods.

The geography of urban agriculture reflects general shifts in urban morphology during the last century. Where market gardens once occupied peripheral lands around European cities, those areas have long been overtaken by suburbanization. Since urban land commands premium prices, gardeners have found novel locations for farming. Old tourist boats that once plied Amsterdam's canals have found new life as floating greenhouse gardens. Sections of the former "no-man's land" of the Berlin Wall are now community gardens. Paris rooftops hum with beehives, providing honey to kitchens and restaurants. And the 19th century allotment gardens are abuzz with a rejuvenated agricultural economy centered on the sustainable local provision of food.

---

government cost-cutting and privatization programs.

In contrast, cities that developed under socialism were less spatially segregated. Certainly mansions, the prewar residences of the social elites, were used for political purposes to house party officials, foreign delegations, or institutes. But housing was viewed as a right, not a commodity, and each family was entitled to its own apartment at reasonable cost.

In the face of the tremendous housing shortfalls following World War II, as well as the needs of rapid industrialization, Communist governments built massive housing estates. Prefabricated multistory apartment blocks were constructed in groups to form a *neighborhood unit*, with shops, green space, and play areas for children at the center. Individual apartments were small (460 to 650 sq ft). The ability to get an apartment in socialist countries was often linked to family status; so it was common to marry and have children much earlier than in Western Europe. The housing estates were typically built in large clusters, forming massive concrete curtains, on land near the edge of cities. As a result, urban population densities could actually *increase* near the urban periphery.

MODELS OF THE EUROPEAN CITY

The concentric zone model, with concentric circles of increasing socioeconomic status with distance from the center, is most applicable to British cities. In contrast, Mediterranean cities, as in Latin America, exhibit an inverse concentric zone pattern. There, the elite concentrate in central areas near major transportation arteries, while the poor live in inadequately serviced parts of the periphery. In Europe, the number of people per household usually increases with distance from the city center.

The sector model explains the pattern of socioeconomic status in which different income groups congregate in sectors radiating outward from the city center. The wealthy may prefer to locate along monumental boulevards or upwind of pollution sources. Poorer residents are left with unattractive sectors along railway lines or strips of heavy industry.

Finally, the multiple nuclei model describes the pattern of ethnic differentiation in which different groups are concentrated in ethnic neighborhoods within the inner city or in high-rise public housing near the periphery.

Northwestern European City Structure

The pre-industrial city center contains the market square and historic structures such as a medieval cathedral and town hall (fig. 5.14). Apartment buildings host upper- and middle-income residents above shops and offices. Narrow, winding streets extend out about a third of a mile. Some wider streets may radiate out from the square to form a pedestrianized corridor that runs to the train station and contains major department stores, restaurants, and hotels. Skyscrapers are concentrated in the commercial and financial district. There are downtown shopping malls or festival marketplaces in refurbished historic buildings. Some old industrial and port areas may have been recycled into new retail, commercial, and residential waterfront developments.

Encircling the core are some zones in transition. The area of the former wall is a circular zone of 19th century redevelopment. Some of the deteriorated middle-income housing has been gentrified, while other sections provide low-rent accommodation for students and poor immigrants.

Surrounding this area is another zone in transition—an old industrial zone with disused railway lines. In the 1950s and 1960s, new industrial plants (*e.g.,* light engineering, food processing) replaced many of the derelict old factories and warehouses. Low-income renters and owners live in run-down 19th century housing. Some houses have been

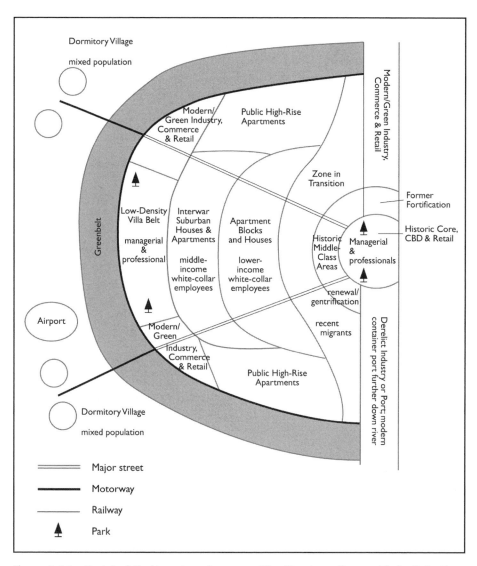

Figure 5.14   Model of Northwestern European City Structure. *Source:* Linda McCarthy

refurbished or replaced. Certain neighborhoods are quite distinctive because they house foreign immigrants who often live above their exotically painted stores and restaurants.

Beyond this inner area is a zone of "workingmen's homes," a stable, lower middle-income zone dating from the early 20th century. These *streetcar suburbs* contain apartment blocks and houses without garages, and are typically anchored by small shopping areas, community centers, libraries, and schools. Beyond these areas are middle-income automobile suburbs containing apartments and single-family homes with garages that correspond with the zone of better residences in the concentric zone model.

Further out are clusters of the most exclusive neighborhoods.

The multiple nuclei model explains the estates of public high-rise apartments and new middle-income "starter" homes at the urban periphery that lack basic amenities like shops and banks. The periphery also contains commercial and industrial activities, such as shopping malls, business and science parks, and high-tech manufacturing.

Beginning in the early 20th century, cities like London established a greenbelt at the edge of the built-up area where development was prohibited. The greenbelt was intended to prevent urban sprawl and provide recreational space. Commuters live outside the greenbelt in dormitory villages and small towns that correspond with the commuters' zone in the concentric zone model. Airports and related activities, such as hotels and modern factories, are located further out on major freeways.

Mediterranean City Structure

In the Mediterranean, pre-industrial urban cores reflect the unique history of each city (fig. 5.15). In Greece and Italy, the historic core exhibits traces of the grid pattern of streets from the first walled enclosure of Greek or Roman origin. In Iberia, remnants of narrow alleys of the Arab quarters date back to Moorish control. The central square is home to markets and festivals; and in Spain, bullfights. The area around the town square contains the cathedral, town hall, and the narrow streets of the walled medieval city. Lower income residents live at high densities above street-level shops and offices. A retail corridor runs from this old commercial core to the train station. The high-rise offices of the modern CBD are nearby. As in the multiple

nuclei and sector models, new industries are found in former old industrial sites and in locations well served by the Mediterranean region's generally more limited transportation infrastructure.

Until the 19th century, urban growth was absorbed in increasing densities within the medieval city. Larger cities like Barcelona that removed their medieval walls in the 19th century laid out new monumental districts. A grand new thoroughfare lined with public works such as statues and fountains was extended out from the city. This area attracted commercial development and wealthy residents, as suggested by the sector model. These elite residential areas of parks and tree-lined boulevards were flanked by middle-income neighborhoods.

In the early 20th century, suburban sprawl became a problem, especially in cities experiencing rapid growth due to industrialization and rural-to-urban migration. Squatter settlements encircled the outskirts of cities. After World War II, these were replaced with low-cost, high-rise public housing that today contains low-income households. Further out, near a natural resource or industrial plant, are the remote, poorly serviced satellite communities for low-income residents and recent immigrants.

Central and Eastern European City Structure

Prior to World War II, the internal structure of cities in Central and Eastern Europe was much the same as in Western Europe. Beginning in the late 1940s, however, the imposition of socialist planning set "Eastern Bloc" cities on a different trajectory, resulting in a set of features that typified Communist-era cities.

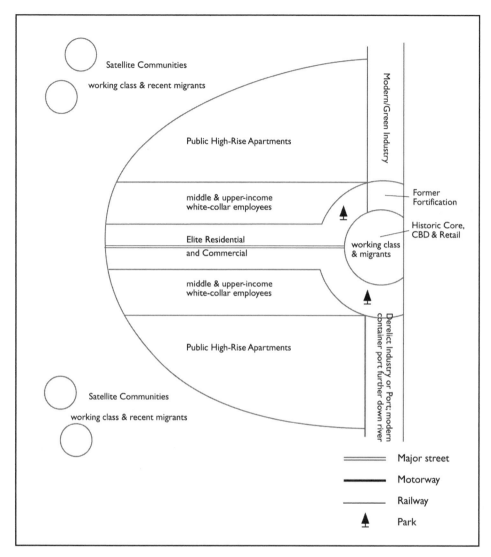

Figure 5.15   Model of Mediterranean City Structure. *Source:* Linda McCarthy

A typical socialist city contained a central square for political gatherings. Following the imposition of socialism, former mansions were converted to government use, religious establishments were used for other purposes, and statues of revolutionary heroes dotted the cityscape. Clusters of housing estates and neighborhood units were interspersed with factories, transportation hubs, and retail establishments. These cities did not conform to Western models of urban structure because land use was based more on government decisions than economic forces. Not all cities exhibited these features to the same degree. Few socialist elements are evident in central Prague, which escaped major destruction during World War II. The socialist city model was most clearly achieved in cities that had been severely damaged during the war (such as Warsaw), in the industrial new towns, and in countries where socialist ideology was especially strong.

One of the first changes to occur in the structure of Communist-era cities after 1989 was an increase in tourist facilities—hotels, restaurants, and entertainment—to cater to foreign visitors. A building boom, especially in the capitals, larger cities, and tourist centers, resulted in foreign-financed office buildings, trade centers, and shopping malls becoming a common feature of cities like Berlin, Budapest, and Prague (fig. 5.16). These are also found in brownfield redevelopments. Corporate logos and billboards have become very visible signs of change. Suburbanization has increased dramatically, but relatively strict traditions of planning regulation often mean that a suburban development will have a core of a historic village and be linked to the city center by public transit. Nevertheless, many cities are becoming increasingly oriented toward the automobile, as witnessed by large commercial and office park developments at the periphery along ring roads.

DISTINCTIVE CITIES

London: Europe's Global City

As hub of the British Empire, London became the center of global economic and political power in the 19th century. Today, London enjoys global city status shared only by New York and Tokyo. Greater London has a population of over 7.5 million, and, including its metropolitan fringe, boasts as many as 19 million people. At the head of navigation on the Thames River, London dominates the U.K. from southeast England. It is the seat of national government, core of the English legal system, headquarters for TNCs, and a leading center for banking, insurance, advertising, and publishing.

In 1666, a fire destroyed virtually the entire city, setting off an immediate building boom. Many historic structures survived until bombed in World War II. London's old nickname, "The Smoke," recalls the days when a haze of pollution from industrial and domestic chimneys hung over the city. London has since undergone deindustrialization and a shift to services and modern manufacturing. Despite these changes, London remains firmly rooted in its historic past.

Central London grew around two core areas, both along the Thames: the City of London (port and commercial hub) and the City of Westminster (government and religious hub). The former developed from a Roman fort, Londinium, which became the fifth-largest city north of the Alps with trade networks extending as far as the Baltic and Mediterranean. In the medieval period, London's protected inland site and strategic location for North Sea and Baltic trade allowed port and commercial activities to thrive. The docks spread from the Tower of London into the East End. Specialized market areas developed near St. Paul's Cathedral in the original square mile of the Roman city. This "City of London" is now the financial precinct, containing the offices of the world's largest banks and insurance companies. The City also houses powerful institutions such as the Stock Exchange and the Bank of England.

About 2 miles (3.2 km) upriver, the "City of Westminster" developed around Westminster Abbey to become a second core during

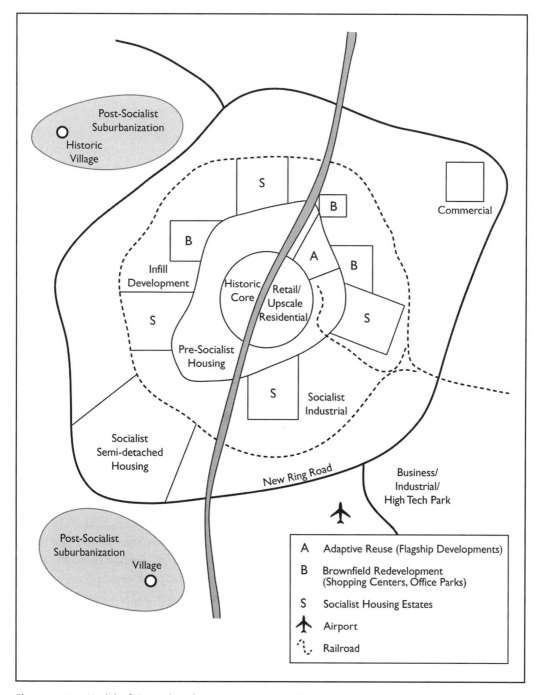

Figure 5.16   Model of Central and Eastern European City Structure. *Source:* Corey Johnson

Figure 5.17   The Houses of Parliament dominate the City of Westminster. The Clock Tower (often called Big Ben) and the London Underground (subway) are landmarks that symbolize London to the world. (Photo by Donald Zeigler)

the medieval period. The present Houses of Parliament were built in the mid-19th century (fig. 5.17) and Queen Victoria made Buckingham Palace the monarch's residence in 1837. This institutional core grew eastward toward the commercial core along Whitehall, where government offices include 10 Downing Street, the prime minister's residence. The royal hunting grounds in the west became St. James's, Green, Hyde, and Regent's Parks. The area attracted mansions of the nobility, centers of culture such as the National Gallery, and exclusive shops. In the 17th and 18th centuries, large-scale townhouse developments were speculatively built for the aristocracy. Belgravia, the last of these West End developments, has survived as an affluent neighborhood.

In the 19th century, major retailing axes developed along Oxford and Regent Streets. In addition to the West End and The City, the inner city (thirteen of London's 32 boroughs) comprises a ring of 19th- and early-20th-century suburbanization. From the early 1840s, the railways allowed wealthier families to move further out. The higher density Victorian and Edwardian housing nearer the center included middle-income detached and row houses such as those in Islington and laborers' cottages in the East End. With increasing industrialization and the incredible growth of new docks, the East End became home to the poorest immigrants.

Much of the original housing in the East End is gone—destroyed in World War II air raids or replaced by high-rise public housing, now deteriorating. Other housing, dispersed among old factories, warehouses, docks, and railway yards, is in poor condition too. Many of the decaying middle-income residences have been subdivided into low-rent apartments. Within the inner city, however, residents are differentiated into neighborhoods, each with its own high street, socioeconomic and ethnic mix, and political and sporting allegiances.

Since the early 1980s, an extensive area of London, the Docklands, has been revitalized through public and private investment. The British government established an Urban Development Corporation that used public funds to stimulate private development. The city's tallest building, a fifty-story tower containing offices and specialty stores, was built at Canary Wharf. Dockland revitalization projects now extend as far west as upscale Saint Katharine Docks (just east of the Tower of London). These dockland developments have attracted higher income occupants and promoted gentrification.

Outer London is a lower density belt of interwar housing with some shopping streets and industrial parks. These outer suburbs comprise the remaining 19 boroughs. Between

1918 and 1939, the expansion of the London Underground (subway) and private automobile use promoted suburbanization. Middle-income residents live in well-maintained houses with gardens. Neighborhood stability is strong. Second-generation immigrants have moved into pockets of older housing. An innovative approach to London's traffic congestion was the introduction of a daily congestion charge for motorists driving into the most heavily congested zone of the city. The estimated benefits have been a more than 20 percent reduction in traffic entering the zone, and funds to reinvest in public transportation.

The outer suburbs end abruptly at a 5–10 mile (8–16 km) wide greenbelt within which development is restricted to prevent sprawl and provide recreational space. Villages and small market towns remain much as they were when the greenbelt was established in 1939. Growth pressures are evident only in the rural dwellings that have been gentrified by new wealthy residents. Prohibiting development within the greenbelt has forced growth into either the existing built-up area or further out. Eight new towns were built beyond the greenbelt to house London's overspill population and migrants from the rest of the U.K. This metropolitan fringe extends up to 50 miles (80 km) from the city center and includes large towns like Guildford, Reading, and Luton. The relatively strong economy, the removal of trade barriers within the EU, and business from the Channel Tunnel have put pressures on housing, government infrastructure, and transportation services.

Paris: Primate City Par Excellence

With a population of as many as 16 million people including the metropolitan fringe,

Paris is Europe's second-largest metropolitan area. As France's primate city, Paris dominates the national urban system and the country's economy, politics, and culture. In the wake of deindustrialization, Paris has become a major international center for modern industry and finance. The outer suburbs contain high-tech plants and research and development companies. Inner-city workshops produce exclusive items such as fashion clothing and jewelry.

Since World War II, Paris has grown almost continuously due to migration from the rest of France and the former French empire, and the city's high proportion of young adults of childbearing age. Much of this growth has been concentrated in the outer suburbs. The city's center and inner suburbs are losing population.

The original site of Paris was an island in the Seine, today called *Île de la Cité*. The Romans seized the island in 52 C.E. from the Parisii, a Gallic tribe. They built a temple and a palace for the city's governor, and the island settlement attracted convents and churches. The magnificent Gothic cathedral of Notre Dame was begun in the 12th century and took more than 170 years to complete.

As a royal center, the grandeur of its architecture and planning made Paris an intensely monumental city. The "Royal Axis" is the imposing entry to the city. It runs from the Louvre (a royal palace, now national art gallery) and Tuileries Gardens across the Place de la Concorde, along the Champs-Élysées, to the Arc de Triomphe. The nearby Eiffel Tower was erected for the Paris exposition of 1889. The tallest structure in Paris, the Eiffel Tower is one of the most recognizable monuments in the world. Paris still produces imposing architecture. Recent, initially controversial structures include the sleek glass pyramid in

Figure 5.18   The sleek glass pyramid, designed by I. M. Pei, is the new, famous entrance to the Louvre, the national art gallery in a former royal palace, in Paris. (Photo by Linda McCarthy)

the 19th-century forecourt of the Louvre (fig. 5.18) and the Pompidou Center, the National Museum of Modern Art, nicknamed the "arty oil refinery" for its multicolored exterior ventilation and steel-and-glass escalators.

The Paris region, *Île de France*, comprises eight administrative units or *Départements*, of which the innermost coincides with the historic City of Paris. This high-density area developed within the confines of the medieval wall. Its distinct quarters include *Île de la Cité*. Facing downstream, the "right bank" of the Seine has become the economic heart of Paris. It contains offices, fashionable shops, hotels, restaurants, and high- and middle-income apartments. The "left bank," the seat of intellectual and cultural life, is dominated by its oldest part, the Latin Quarter, with the Sorbonne University, bookshops, theaters, and middle- and low-income apartments. Unlike London, there are few large parks. Paris gets its feeling of openness and greenery from the wide boulevards and tree-lined river walkways.

The outer parts of Paris include the "little ring" (*petite couronne*) of inner suburbs that extends out about 15 miles (24 km) from the center. It developed between the late 1800s and World War II. Interwar speculative developments of single-family homes were built on prime sites. Public high-rise apartments were erected later on the less marketable land. The big ring (*grande couronne*) of outer suburbs spreads out another 10–15 miles (16–24 km) and contains the postwar *grands ensembles* of poorly serviced public high-rise apartments.

Beginning in the late 1940s after the publication of Jean-François Gravier's book, *Paris and the French Desert*, planners began to focus on counteracting the extraordinary economic and demographic primacy of Paris. National decentralization policies attempted to limit growth and congestion problems within the Paris region, while promoting development in the eight *métropoles d'équilibre* of Lille-Roubaix-Tourcoing, Metz-Nancy, Strasbourg, Lyon, Marseille, Toulouse, Bordeaux, and Nantes-St. Nazaire. Five new towns (St.

Figure 5.19   Looking northwest from the Eiffel Tower, Paris. The Palais du Chaillot in the foreground was built for the International Exposition of 1937. In the distance, beyond the Bois de Boulogne, is La Défense, a planned suburban business district. (Photo by Linda McCarthy)

Quentin-en-Yvelines, Evry, Melun-Senart, Cergy-Pontoise, and Marne-la-Vallée) were built along two east-west axes of growth to the north and south of Paris. These new towns grew as extensions to the city, however, and became middle-income dormitory commu-

nities for some of the more than one million daily commuters to central Paris.

Complementing the new towns are four suburban employment centers. The largest and most successful is La Défense (fig. 5.19). It boasts high-rise offices containing the head-

Figure 5.20   Throughout Catalonia, signs of Catalan nationalism—and separatism—are to be found. This banner, in Girona, speaks to the world in English. (Photo by Donald Zeigler)

quarters of TNCs, shops, public buildings, and housing. Its modern Grand Arche is visible from the Arc de Triomphe along Avenue Charles-de-Gaulle—the modern extension of the "Royal Axis."

Barcelona: Capital of Catalonia

With more than 1.6 million people, Barcelona is Spain's second largest city after the capital, Madrid. On the northeastern coast of Spain, Barcelona is the country's largest port and leading industrial, commercial, and cultural center. Housing the seat of the Catalan government, this regional capital of Catalonia is a bilingual city: Spanish and Catalan are widely spoken official languages (fig. 5.20).

The Phoenicians founded Barcelona more than 2,000 years ago. The street plan reflects its three main phases of growth—its ancient and medieval origins, 19th-century additions, and late-20th and early 21st century suburbs. The old town is the symbolic and administrative center of the city. Remnants of the Roman wall and grid pattern of streets are overlain by the narrow streets of the medieval core. Here, residents and tourists alike stroll along the famous *Ramblas*. Barcelona is the most popular tourist port in the Mediterranean with more than 6 million annual visits, including 2 million from cruise ships.

In 1859, Ildefons Cerdà drew up a plan of expansion into the area of the former medieval wall. His pioneering design was based on a grid pattern with wide, straight boulevards and unique 8-sided city blocks containing parks surrounded by apartment houses. Largely ignored during the 19th- and early-20th-century era of speculative growth, Cerdà's plan was fully realized only in the *Eixample* district, a new precinct that was built just north of the old city. This new development also contains the high-rise offices and apartments of the modern CBD.

At the end of the Spanish Civil War in 1939, under Francisco Franco's dictatorship,

Barcelona's Catalan culture was suppressed and the city experienced uncontrolled speculative development without adequate public infrastructure and services. Massive rural-to-urban migration fueled rapid population growth. Tens of thousands of illegal squatters ended up in shantytowns at the sprawling edge of the city. In the 1960s and 1970s, several hundred thousand poorly designed and serviced peripheral high-rise public apartments were built to address the acute housing shortage.

Since the mid-1970s and the establishment of Spain's parliamentary democracy, the increased autonomy of Barcelona's elected local governments has contributed to a rebirth of planning as well as growing prosperity. Barcelona's urban renewal program benefited from funding for infrastructure from the EU. The construction of the 1992 Olympic village helped rejuvenate an area of derelict docks into waterfront redevelopments. Popular World Heritage sites include a park by Antoni Gaudí—Park Güell—and his unfinished church—Sagrada Familia—financed by private donations since 1882!

At the same time, a continued influx of poor residents has put pressure on housing, infrastructure, and services. These poor migrants become socially, economically, and locationally polarized in the poorest inner-city neighborhoods and peripheral public apartment blocks. In contrast, higher-income residents live in nicer central districts or well-serviced lower density parts of the suburbs.

Oslo: Low-Key Capital of Norway

Oslo is the largest urban center in Norway as measured by both the nearly 600,000 people that inhabit the city and its metropolitan area

population of about 1.4 million. At the mouth of the Oslofjorden, this city is Norway's capital, main port, and leading commercial, communications, and manufacturing center.

Oslo was founded around 1,000 C.E. and became the national capital in 1299. After a devastating fire in 1624, King Christian IV of Denmark designated another site for the town nearer Akershus Castle on the east side of the bay. The new town was named Christiania (later Kristiania). It was planned with a grid system of spacious streets, a square located between the town and castle, and ramparts protecting its northern flanks. For fire resistance, buildings were required to be constructed of brick or stone; soon, however, extensive tracts of wooden houses were built on the outskirts of the built-up area. The town grew slowly: in 1661 only around 5,000 residents had made Christiania their home; by 1800 the population had risen to only 10,000.

During the mid-1800s, the administrative function of the city was augmented by industry, based mainly on textiles and wood processing. Many landmarks, such as the university, royal palace, parliament, national theatre, and stock exchange, were built. The city expanded in a largely unplanned manner as the population swelled to 28,000 by 1850 and 228,000 by 1900. In 1925, the city reverted to its original name, Oslo. After World War II, Oslo's outward expansion continued, largely as a result of public policies that subsidized owner-occupied housing.

Oslo has 40 islands and 343 lakes; about two-thirds of the city comprises protected natural areas, which give it a picturesque appearance. While most of the surrounding forests and lakes are private, the public is strongly against developing them. As is common throughout northern Europe, the city extends around the port, is flanked by the centrally located train station, has a royal palace overlooking the historic core, and pedestrianized shopping streets. Oslo is where the Nobel Peace Prize is awarded because Alfred Nobel decided that this prize—awarded to Barack Obama, the International Panel on Climate Change (IPCC) and Al Gore, and Doctors without Borders, to name a few—was to be awarded by a Norwegian committee (while the other four prizes were to be awarded by Swedish committees). Despite being Scandinavia's oldest capital, Oslo today is a modern, though low-key, city.

Berlin: The Past Is Always Present in Germany's Capital

Around every corner, across every bridge, and in nearly every *U-Bahn* (subway) station, Berlin offers titillating morsels from its fascinating past. Relatively unimportant for most of its 800-year history, Berlin found itself in the middle, figuratively and literally, of many of the important political struggles of the 19th and 20th centuries. More recently, as the capital of reunified Germany and a key political node in the EU, Berlin is undergoing another period of profound changes.

Berlin had rather humble beginnings in the 13th century on the flat, glaciated marshlands of the North European Plain at a convenient crossing point on the Spree River. The growth of Berlin reflected its political fortunes as the center of what would become the Kingdom of Prussia. The architectural heyday of the city came after 1701 as the official capital of Prussia. The Prussian royalty sought to give the capital an impressive built environment worthy of long-established capital cities such as Paris, Vienna, and London. Walking along

Unter den Linden, the city's most important axis, you see the bravado of the Prussian ruling family, the Hohenzollerns, who built up the boulevard during the early 19th century to project their pride after (helping) defeat Napoleon's armies. The street culminates at the Brandenburg Gate, a monument symbolic of Prussian power, German Imperial pretension, cold war division, and since 1990, a reunified Germany. As people walk through the Brandenburg Gate they are confronted by more history: to the south, the Memorial to the Murdered Jews of Europe; to the north, the center of Germany's government in the renovated Reichstag building with its huge glass dome symbolizing the transparency of the Federal Parliament. Perhaps more than any other spot in the city, modern and old, painful and joyous, co-exist in an almost surreal urban assemblage.

The growing power of Prussia during the 1800s was accompanied by industrialization. Large companies such as Siemens and AEG were founded, while Berlin-based insurance firms (*e.g.,* Allianz) and banks (*e.g.,* Deutsche Bank) served the booming industrial economy. Reflecting shifting political fortunes, none of these TNCs remain. But the evidence of Berlin's status as one of Europe's major industrial cities can still be seen. Working-class apartment houses from its industrial heyday are found in neighborhoods such as Wedding and Kreuzberg, which encircle the historic core. The city's old industrial breweries have found new life in post-unification Germany as cultural and arts centers. The Schultheiss brewery in Prenzlauer Berg is now a cultural center and shopping area called *Kulturbrauerei*, and the Kindl brewery in Neukölln houses artist studios and apartments. Berlin also became a major transportation hub during the 19th and early 20th centuries. The major trunk rail line winds its way through the city center, stopping at iconic stations like Zoologischer Garten, Lehrter Station (the main station, the largest in Europe), Friedrichstrasse, and Alexanderplatz.

Berlin's economic and demographic peak came in about 1940 when it was the industrial, transportation, and governmental center of the Third Reich. Its population was over 4 million in 1939 on the eve of Germany's invasion of Poland, compared with just over 3 million today. The Nazis left their indelible mark on Berlin, although not entirely as planned. Hitler's planned Germania, a megalomaniacal rebuilding of Berlin's core as the capital of the German Empire, was never realized. The most readily apparent legacy of the war is destruction. The core of the city was up to 95 percent destroyed during bombing raids by the Allies and the bloody campaign of the Soviet Army during the last days of the war.

After World War II, when Berlin was divided into four sectors by the Allies, the priority was constructing housing for the remaining people of the city and the large refugee population in both East and West Berlin. The results were often more functional than architecturally appealing. Though bland housing blocks are still apparent, some cold war prestige projects remain. The Stalinallee (later Karl-Marx-Allee) in East Berlin was designed as a showcase of communist architecture, and it offers remarkable insights into the aesthetic ideals of the communist regime. In West Berlin, projects such as the Cultural Forum, home to the concert hall of the Berlin Philharmonic Orchestra, is a modernist icon, which was designed to show off the merits of West Germany's social market economic system.

Since reunification, Berlin has been the site of one of the largest urban reconstruction efforts of all time. Potsdamer Platz, a pre-war buzz of activity, was in the area of the Berlin Wall during 1961–1989. It was not until the 1990s that it was rebuilt in the then popular steel and glass styles. Friedrichstrasse, the major north-south axis that the Wall once cut in two, has blossomed as Berlin's most fashionable shopping and office address. Meanwhile, Kurfürstendamm, an icon of West German consumerism, appears frozen in the 1980s.

Berlin's fascinating history is also the subject of frequent debates over appropriate land uses, and nearly every major building decision is accompanied by sensitive emotional excavations of the more sordid periods of that past. In recent years, this is perhaps best illustrated by the reconstruction of the *Stadtschlos* (city palace). The damaged 19th century original was destroyed after the war by the East German regime, and on its site a modernistic Palace of the Republic was constructed, a building intended to distance the site from a militaristic Prussian history. Demolished in 2008, the Parliament voted to rebuild an exact replica of the city palace that would be called the Humboldtforum and would house the Humboldt collection and gallery of non-European art. In 2010, the project—a casualty of Germany's and the world's economic downturn—was put on hold.

Bucharest: A New Paris of the East?

On the Romanian Plain between the Carpathian Mountains to the north and the Danube River lowlands to the south, Bucharest is by far Romania's largest city, with a population of around 2 million. In addition to being the capital, it is the country's most important economic and industrial city. Like Berlin, Bucharest bears the marks of its communist past, in the form of monumentalist Stalinist architecture, large communist-era apartment blocks, and a somewhat run-down pre-war built environment. Prior to World War II and the post-war communist era, however, its elegant architecture and high culture elite made it the "Paris of the East." Since Romania's accession to the EU in 2007, there are signs that Bucharest would like to reclaim that title.

Bucharest is relatively young by European standards: the first references to the city date to 1459. During the 1800s, Bucharest became an important transportation hub, acquired a manufacturing base, and became Romania's capital when the country was formed in 1862. By the end of the century, Bucharest boasted a tram system and the world's first electric streetlights. The city enjoyed continued growth up to World War II. Population totals for Bucharest rose from about 60,000 in 1830 to slightly more than one million at the end of World War II. During the 1930s a master plan for the city sought to vastly change the compact medieval core and envisioned expansion like Berlin and Paris with their wide boulevards, parks, and grand public buildings. This laid the groundwork for what would come later under the post-war communist leader, Nicolae Ceaucescu.

Bucharest suffered heavy damage during World War II from Allied and Nazi bombing. After the war, socialist planning guided development: industrial capacity was greatly expanded; new housing was constructed; and former villas were converted into government offices and foreign embassies. The population increased to 1.4 million by 1966. Perhaps no city in Europe bears the personal imprint of an individual to the same extent as Bucharest at the hands of this deposed leader,

who was executed along with his wife after a hasty trial in 1989 amid a democratic revolution. Romania's capital allows fascinating insights into the impact of a megalomaniac personality. Ceaucescu's program for urban redevelopment, systematization, led to single family houses in some suburbs being replaced with apartment blocks. He greatly expanded housing estate construction within existing districts and redesigned certain boulevards as impressive entryways to represent the revolutionary aesthetics of socialism.

After a 1977 earthquake caused significant damage, Ceaucescu turned his attention to central Bucharest. By 1989, approximately 25 percent of the historic central area had been bulldozed for a new civic center. An estimated 40,000 families were evicted practically overnight, and those with dogs were forced to let them go (giving rise to Bucharest's intractable stray problem). At the heart of the construction scheme was The House of the Republic, which was to be the new seat of government. It became one of the world's largest buildings. Over a thousand acres of neighborhoods were torn down to make room for the complex. This building consumed virtually all the country's marble during construction and featured hand-carved wood paneling and crystal chandeliers. The second component of the scheme was the construction of the Victory of Socialism Boulevard. This finely appointed ceremonial route was intended to be longer and grander than the Champs-Élysées in Paris and eventually was adorned by a lavish fountain and lined by Bucharest's finest apartments. A rather bizarre element of Ceaucescu's modifications involved churches. He simply did not like them; but rather than being accused of ordering their destruction, he had many moved behind other buildings.

For much of the 1990s, Bucharest suffered the aftermath of Ceauçescu's experimental planning. The city was rundown and known to visitors more for its stray dogs than as the "Paris of the East." The completed House of the Republic became the seat of Romania's democratic parliament, symbolizing a coming to terms with the past and a more forward orientation for the country and its capital city. More recently, as Romania has benefitted economically from its integration into Europe, including accession to NATO and the EU, Bucharest is again blossoming. Wide boulevards that were once used to showcase a regime's hold on power are now monuments to consumerism. Bucharest still has its stray dogs—100,000 by recent estimates—but like so many of its neighbors, Bucharest seems happy to continue to move toward becoming a European-oriented, sophisticated, bustling, economic and cultural hub.

URBAN CHALLENGES

Compared with the problems faced by cities in many other regions of the world, European cities are fairly well off. At the same time, cities in Europe suffer from problems similar to those in other more developed parts of the world, including terrorist threats (box 5.4).

As the earliest place to industrialize, Western Europe was also the first to suffer deindustrialization. Rising long-term unemployment among inner-city residents has concentrated poverty and a wide range of social problems in some neighborhoods of older industrial cities. These neighborhoods also contain the city's oldest and most deteriorated housing and urban infrastructure. Privatization of public housing by governments attempting to cut back on expenditures has exacerbated the shortage of decent affordable housing in many cities.

Beginning in the early 1990s, governments across Central and Eastern Europe faced the daunting task of transforming to market-based economies. Under socialism, workers were publicly employed and all businesses—from huge factories to small bread shops—were operated by the state. Many economic activities under socialism were run inefficiently, so privatization resulted in reduced production, plant closures, and unemployment, something previously unknown.

Privatization had a particularly troubling impact on the housing sector. People were accustomed to paying very little for housing. With privatization, a burgeoning land market and general atmosphere of lawlessness arose, and some investors reaped huge profits; most individuals, though, suffered hardship.

As private automobile ownership has risen, traffic congestion and air pollution, especially in the medieval cores, have reached critical levels (fig. 5.23). Transportation policies in Western Europe typically shifted from investment in freeways and central parking facilities to transportation demand management involving ride sharing and public transit. Many cities built or extended their subway and light rail systems.

The inadequate road system throughout central and eastern parts of Europe, particularly given the dramatic increase in car ownership rates since 1990, placed considerable strain on roads and parking facilities. Previously, most people could not afford a car, waiting periods for car orders were long, and restrictions on ownership applied in some countries. A major challenge continues to be to improve the transportation infrastructure

Box 5.4 Tackling Terrorism in London

Cities—especially world cities like London—are a preferred location for terrorist attacks for several reasons. First, cities have symbolic value. They are not only dense concentrations of people and buildings but also symbols of national prestige and military, political, and financial power. A bomb in London's Underground (subway) arouses international alarm and is communicated instantly to a world audience. Second, the assets of cities—densely-packed with a large mix of industrial and commercial infrastructure—make them rich targets for terrorists. Third, cities are nodes in vast international networks of communications—reflecting not only their power but also their vulnerability. A well-placed explosion can cause enormous reverberations by triggering fear and economic dislocation. Finally, word gets around quickly in high-density localities. These kinds of environments can be a source of recruits for terrorist organizations.

Central London has attempted to reduce the real and perceived threat of terrorist attacks. Physical and increasingly technological approaches to security have been adopted at ever more expanded scales. In 1989, the Prime Minister installed iron security gates at the entrance to Downing Street to control public access (fig. 5.21).

Figure 5.21   The iron security gates at the entrance to Downing Street in London prevent the public from getting close to the official residence of the Prime Minister. (Photo by Linda McCarthy)

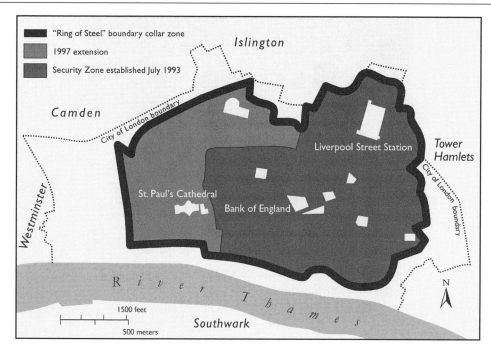

Figure 5.22   Since the 1990s, terrorist threats have increased and so has the security zone in London's financial district, "The City." (Adapted from J. Coaffee, "Rings of Steel, Rings of Concrete and Rings of Confidence: Designing out Terrorism in Central London pre and post September 11th," *International Journal of Urban and Regional Research* 28, 2004, 204)

In 1993, a security cordon was set up to secure all entrances to the financial zone of the City of London (the "Square Mile"). The 30 entrances to the City were reduced to seven, with road-checks manned by armed police. Over time the scale of this security cordon was increased to cover 75 percent of the "Square Mile" (fig. 5.22).

As a territorial approach to security, this cordon was augmented by retrofitting the closed circuit TV (CCTV) system. The police, through its "CameraWatch" partnership effort, encouraged private companies to install CCTV. At the seven entrances to the security cordon, 24-hour Automated Number Plate Recording (ANPR) cameras, linked to police databases, were installed. The City of London is now the most surveilled space in the U.K., and perhaps the world, with more than 1500 cameras.

Journal of Urban and Regional Research 28 (2004), 201–11; H. V. Savitch with G. Ardashev, "Does Terror Have an Urban Future?" *Urban Studies* 38 (2001) 2515–33.

Figure 5.23   European "smart cars," such as this one parked along a street in London, are a practical response to narrow streets and high gas prices in Europe. (Photo by Linda McCarthy)

as part of a European-wide transportation network.

The presence of significant numbers of foreign workers and their families has generated problems in Western Europe. Language differences create difficulties for the educational system in countries with large numbers of children born to foreign workers. These students represent more than 10 percent of the school population in some French, German, and Swiss cities. During times of recession and rising unemployment, existing prejudices can be intensified based on differences in language, culture, or race. Xenophobia, or anti-foreigner sentiment, has contributed to some governments, as in France, banning headscarves worn in public schools and universities. Vicious attacks on immigrants by violent elements such as skinheads have occurred, especially in some German and French cities. Discrimination against African immigrants,

as well as high unemployment and lack of opportunities in France's poorest immigrant suburbs (*banlieue*), have sparked riots.

More recently, the considerable economic and social changes in Central and Eastern Europe have increased the opportunities for criminal activities. The incidence of petty crime, such as pickpocketing and graffiti, has increased (box 5.5). Organized crime has also grown. Besides the more typical criminal pursuits of drugs, gambling, and prostitution, "Mafia" type crime organizations are common in some cities.

Pollution is a problem in nearly every city of the world, though western European cities rate more favorably than most. Indeed, Swiss, Austrian, and Scandinavian cities are practically sanitized daily and most citizens are conscientious and do not litter. Many streets and squares in Italy and Spain are routinely washed down. Levels of pollution in most

Box 5.5 Urban Graffiti in Europe: Is the Writing on the Wall?

Art or eyesore? Free expression or vandalism? European cities bear the marks of graffiti. On a stroll through just about any neighborhood, you see spray paint on buildings, signs, buses, and trains. An aesthetic issue for many, graffiti is also often a political issue tied to ethnic, socioeconomic, or generational conflict. Imagine the events of 1989 in Germany, when the Berlin Wall came down, without the iconic images of the spray-painted wall being chipped away by gleeful Germans. A painted portion of the wall recently was sold for nearly $10,000, while other parts are displayed in museums around the world.

Urban graffiti is nothing new. Ancient Greek and Roman cities had graffiti, and you can see a caricature of a politician etched on an outdoor wall in the excavated Roman city of Pompeii from around 79 c.e. More recently, graffiti has played an important role in some of Europe's ethno-political struggles. The Basque separatist group ETA and sympathizers with its cause have used graffiti to protest their lack of autonomy within Spain. During Spain's forty-year dictatorship until 1975 under Francisco Franco, graffiti was one of the few forms of protest the ETA could get away with, while in the last thirty years graffiti as political statement has continued alongside more publicized acts of violence. Similarly, Irish Republican Army (IRA) markings were common during the height of "The Troubles" and survive in Northern Ireland cities like Belfast. Likewise politically motivated graffiti by Bosnians and Serbians can be found in Sarajevo two decades after the outbreak of war and the breakup of Yugoslavia.

Graffiti has also come to be associated with counter cultural movements, such as hip hop and punk as well as anarchists, fascists, and anti-fascists. European graffiti artists, often using aliases to avoid being caught by police, achieve notoriety in the underground world of their respective movements as well as occasionally more widespread recognition. The British artist, Banksy, whose anti-war and anti-establishment work appears across Europe, has produced a widely acclaimed documentary film about graffiti, and individual pieces of his work have sold for more than $500,000. Most graffiti artists are content to ply their trade in the warehouse, residential, and commercial districts of cities, and are more likely to be arrested for vandalism than to receive a premium price for their work.

Graffiti is, unsurprisingly, not welcomed by all Europeans. Very expensive targets, such as trains, government buildings, and public artworks, have become prized canvasses for graffiti artists. But there are also significant risks—not to mention clean-up costs—associated with this activity. Many cities have undertaken programs to address graffiti, from expensive clean-up efforts to the use of CCTV surveillance to catch taggers. In spite of this, graffiti, a practice dating back millennia, continues to color the European urban landscape.

of the former heavy industrial regions have fallen. Indeed, air quality in the English Midlands has improved since polluting industries have closed or relocated, and even Germany's Ruhr area boasts clear skies and clean lakes.

Formerly communist parts of Europe are still tackling the legacy of weak Soviet-era environmental standards, use of higher risk industrial processes, and greater reliance on aging Soviet-designed nuclear reactors. The EU's stricter environmental regulations have closed most of Central and Eastern Europe's iconic smoke-belching plants.

Since the early 1990s, improved air, rail, and road transportation linkages connecting the major urban centers across Europe have been laying the foundation for the complete reintegration of the urban system. Businesses and city governments in the eastern European countries that joined the EU in the 2000s (Poland, the Czech Republic, Hungary, Slovenia, Estonia, Slovakia, Latvia, Lithuania, Romania, and Bulgaria) have already developed stronger ties with their counterparts in the preexisting member states.

Membership in the EU has enhanced the opportunities for former socialist cities to address their pressing social and economic problems. Certainly, the future and prosperity of Europe as a whole in the global economy depend on creating a more economically and socially equitable situation for all European urban residents.

As the common currency, the Euro, is gradually extended to more countries, the urban system will undergo more profound changes as governments, businesses, and people in cities across Europe and elsewhere reorient their activities to take advantage of the changing economic environment.

At some point in the future, serious consideration will be given to moving some of the administrative functions of the EU to cities further east. While London, Paris, Berlin, Frankfurt, Brussels, and Milan will continue to dominate as major financial, political, and cultural centers, qualitative improvements in cities like Warsaw, Prague, Budapest, and Sofia will surely shift the center of gravity of the core-periphery model further east as the 21st century progresses.

SUGGESTED READINGS

Beatley, Timothy. 2000. *Green Urbanism: Learning from European Cities.* Washington, D.C.: Island Press. Examines policies in 25 innovative cities to identify best practices for sustainable urban development.

Hamilton, F. E. Ian, Kaliopa Dimitrovska Andrews, and Nadasa Pichler-Milanovic, eds. 2005. *Transformation of Cities in Central and Eastern Europe: Towards Globalization.* New York: United Nations University Press. An overview with rich examples of the experiences of major cities on the road to globalization and European integration.

Kazepov, Yuri, ed. 2005. *Cities of Europe: Changing Contexts, Local Arrangements, and the Challenge to Urban Cohesion.* Malden, MA: Blackwell. Chapters focus on important issues such as segregation, gentrification, and poverty.

Kresl, Peter K. 2007. *Planning Cities for the Future: The Successes and Failures of Urban Economic Strategies in Europe.* Cheltenham, UK: Edward Elgar. Examines the relationship between urban competitiveness and economic-strategic planning for 10 internationally networked cities.

Florida, Richard, Tim Gulden, and Charlotta Mellander. 2007. *The Rise of the Mega Region.* Toronto: University of Toronto. A study that uses datasets of nighttime light emissions, GDP, and other indicators to identify global megaregions, including in Europe.

Moulaert, Frank, Arantxa Rodriguez, and Erik Swyngedouw, eds. 2003. *The Globalized City: Economic Restructuring and Social Polarization in European Cities.* Oxford: Oxford University Press. Case studies of large-scale redevelopment projects and their social implications, including the Olympic Village in Athens.

Murphy, Alexander B., Terry G. Joran-Bychkov, and Bella Bychkova. 2008. *The European Culture Area*, 5th ed. Latham, MD: Rowman and Littlefield. An updated version of a major European text with chapters on cities, culture, the EU, and the environment.

Ostergren, Robert, C and Mathias Le Bossé. 2011. *The Europeans: A Geography of People, Culture, and Environment*, 2nd ed. New York: Guildford.

An updated comprehensive view of Europe with two chapters on towns and cities.

Penninx, Rinus, Karen Kraal, Marco Martiniello, and Steven Vertovec, eds. 2004. *Citizenship in European Cities: Immigrants, Local Politics, and Integration Policies.* Aldershot, UK: Ashgate. Examines citizenship in European cities with a focus on immigration policies and immigrant participation in local civil society.

van den Berg, Leo, Peter M. J. Pol, Willem van Winden, and Paulus Woets. 2005. *European Cities in the Knowledge Economy.* Aldershot, UK: Ashgate. Examines local dimensions of the knowledge economy and policy options using case studies of Amsterdam, Dortmund, Eindhoven, Helsinki, Manchester, Munich, Munster, Rotterdam, and Zaragoza.

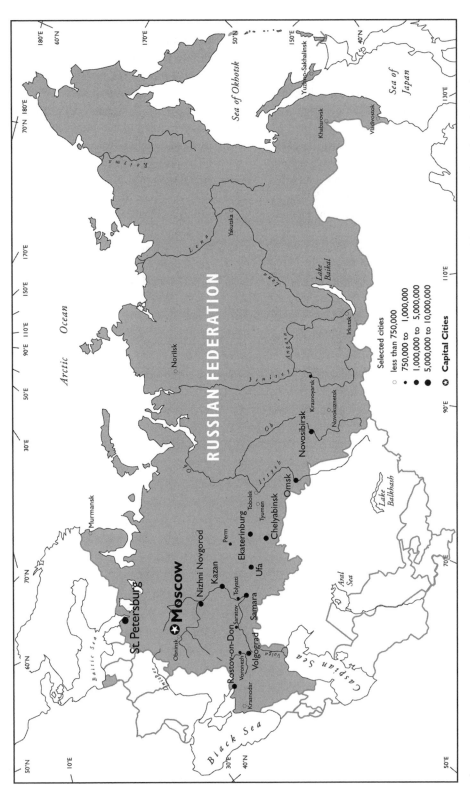

Figure 6.1 Major Cities of Russia. *Source:* UN, *World Urbanization Prospects: 2005 Revision,* http://www.un.org/esa/population/publications/WUP2005/2005wup.htm

6

Cities of Russia
JESSICA K. GRAYBILL AND MEGAN DIXON

KEY URBAN FACTS

Total Population	140 million
Percent Urban Population	73.2%
Total Urban Population	103 million
Annual Urban Growth Rate	−0.33%
Number of Megacities	1 city
Number of Cities of More Than 1 Million	12 cities
Three Largest Cities	Moscow, St. Petersburg, Novosibirsk
World Cities	Moscow

KEY CHAPTER THEMES

1. Russia's urban development reflects the impact of three distinct eras in the country's history: tsarist, Soviet (communist), and post-Soviet.
2. Russia's cities experienced two reconstruction phases in the 20th century: one after the creation of the Soviet Union in 1917 and the other when the Soviet Union collapsed in 1991.
3. The main pattern of the urban system, with its strong reflections of European urban planning characteristics, was established in the tsarist era.
4. Russia's rapid urbanization in the early 20th century accelerated the country's historic patterns of urban growth and contraction over the last thousand years.
5. As a result of the disintegration of the Soviet-era socialist support system, crime and corruption have hindered the emergence of a democratic post-Soviet governance and the development of a civil society, especially in cities with increasing in-migration.
6. Environmental issues in Russia's urban centers are increasingly recognized as severe and have become an important issue requiring attention from post-Soviet city leaders.
7. The need to overhaul and redesign urban places and urban governance raises new questions about the roles of government and citizens in the post-Soviet era.

8. Changing demographics and shifting cultural and religious identities have reinvigorated questions about tolerance and acceptance of multiculturalism in post-Soviet cities.

9. In the post-Soviet period, cities are no longer subsidized by the central government; many have experienced economic recession, significant population loss, and at least seasonal de-urbanization or ruralization.

10. Cities that are prospering are those with superior locations, strong historic roots, or attractive environments for foreign investment and economic growth.

The urban landscape of the Russian Federation, commonly known as Russia, is today characterized by ornate tsarist-era buildings and monuments (palaces, churches, museums) standing alongside the utilitarian, concrete-and-steel structures of the Soviet era (office buildings, communal apartments, community centers) and the newly erected European-style, elite apartments and shopping centers of the post-Soviet era. This landscape reflects the impacts of urban development during three distinct periods in the country's history: tsarist, Soviet (communist), and post-Soviet.

Diverse ethnic groups have inhabited this region of Eurasia for at least a thousand years, continually contributing to Russia's diverse population. A blending of many cultures, religions, and histories across the European and Asian realms of Russia for several centuries resulted in multicultural settlements that eventually grew into towns and cities during the Tsarist Russian Empire (1721–1917). Under the Union of Soviet Socialist Republics (the USSR, or the Soviet Union) from 1917–1991, that multiculturalism was celebrated but also tempered by communist universalism; standardized Soviet urban forms spread across a Eurasian territory, either rebuilding existing cities or creating new ones (box 1.3). Across Russia but especially in cities, the collapse of the Soviet Union was symbolically marked on December 25, 1991, as the Soviet hammer-and-sickle flag was lowered and replaced by the red, white, and blue flag of the Russian Federation, now one of 15 independent post-Soviet nations.

Russia began the 20th century with less than one-fifth of its population classified as urban; by 1989, 74 percent of Russia's population lived in urban places. The average percentage leveled off in the late 1990s at about 73 percent, a figure that remains stable today. Although it has been two decades since the end of the Soviet period, the imprint of Soviet-era urban policy and form still profoundly affects the larger territory—urban and rural—of the Russian Federation (fig. 6.1) and other post-Soviet states. The Soviet attempt to provide greater cultural, educational, employment, and housing opportunities in cities produced specific urban landscapes; many of these remain as Russia continues to undergo a series of economic, political, and social transformations, shaping urban trends that contradict some expectations common in the West. For example, the severity of the economic collapse following the end of the Soviet Union and Russia's abrupt confrontation with the global economy prompted many urban workers to fall back on *dacha* settlements and villages (rural areas established in the Soviet-era) to practice subsistence farming, a process known as *ruralization*.

Not surprisingly, in a country nearly twice the size of the United States, the effects of "wild" capitalism, and its visual imprint on the landscape are unevenly distributed across

Figure 6.2   New construction in cities around Russia relegates Soviet urban landscapes to the background as new commercial and residential buildings vie for valuable real estate locations. (Photo by Jessica Graybill)

Russian cities. In some Russian cities, the built environment has changed so dramatically since 1991 that many cities are nearly unrecognizable to those accustomed to quiet, somber Soviet landscapes. Commercial retailers, private transportation, and new housing construction have altered Russian social and cultural urban landscapes (fig. 6.2). For example, Moscow's Red Square is no longer a nearly deserted public space awaiting military parades; instead, this central city landmark area abuts a bustling high-end retail and tourist space (fig. 6.3). This new socioeconomic landscape changes how people use the built environment, creating new cultural spaces and practices. Luxury shopping and lingering in cafes have become daily activities for Muscovites and tourists alike.

Construction of Tsarist and Soviet cities emphasized urban planning principles such as pedestrian walkways and mass transit, thus ignoring or minimizing the needs of automobile drivers. Indeed, across the territory of the former USSR, over 20 metros (subway systems) were developed, far more than any other country in the world. The Soviet vision of accessible transportation for all urban citizens thus predates many of the sustainability-minded mass transit projects now becoming popular in the West. However, an exponential increase in the number of private and commercial vehicles has occurred in post-Soviet Russia; and cities were neither built nor have they been re-developed to accommodate them. The concept of rush hour has great meaning now, and for many Russian cities, especially Moscow, rush hour

Figure 6.3   Renovations in GUM shopping center on Red Square make it a top destination for tourists and Russia's elite seeking high-end shopping experiences. (Photo by Jessica Graybill)

begins in mid-afternoon and extends through the evening (fig. 6.4).

The typical Soviet rings of monolithic apartment complexes on city outskirts are increasingly mixed: new elite apartment buildings are constructed alongside Western-style suburban developments of "cottages" and gated communities (fig. 6.5). Soviet neighborhoods were often ethnically and socioeconomically intermixed; but today, depending on the city, stratification by socioeconomic class and sometimes by ethnicity is beginning to occur, as upwardly mobile residents choose to live in newly constructed, high-security apartments or McMansions in suburban or ex-urban locations. Cultural and social change is reflected in the replacement of Communist Party billboards (formerly present in every city and town) with brilliant neon and banner-type commercial advertisements for consumer goods and services along major urban thoroughfares.

Post-Soviet Russia still grapples with the legacy of the spatial framework created by Soviet urban development. The Soviet planning regime often located settlements near natural resources in isolated, inhospitable environments, resulting in far-flung, potentially unsustainable urban growth. In the post-Soviet period, capitalist notions of efficiency have made such cities' locations and industrial operations unprofitable, resulting in economic decline and population outflow. This type of development has been called *archipelago urbanization*, because Soviet cities arose like urban islands in a vast rural Eurasian hinterland that remained—and remains—seemingly unchanged culturally or economically for centuries. These Soviet era cities' positions as

Figure 6.4   Since the fall of communism, automobile ownership in Moscow has soared, and with it has come urban gridlock. (Photo by Alexei Domashenko)

Figure 6.5   New microrayon developments, with varied architectural styles, are rapidly changing the face of Russia's suburbs. This picture is from Yuzhno-Sakhalinsk. (Photo by Jessica Graybill)

islands within Russia's vast territory persist due to a lack of transportation infrastructure connecting them or, due to the lack of affordable transportation to any destination but Moscow. For example, the price of a round-trip plane ticket from Petropavlovsk-Kamchatsky to Magadan (a 1-hour flight) is usually greater than a round-trip ticket from either of these cities to Moscow (a 9-hour flight), indicating the centralized power that Moscow still wields over individual Russian regions.

In both tsarist and Soviet Russia, many cities owed their existence and location to the needs of national security; but military-industrial complexes now play a lesser role in determining the location of urban investment and growth. Today, cities previously favored by Soviet urban and economic policies are undergoing economic restructuring processes not dissimilar to the restructuring experienced by major North American and European cities during the period of deindustrialization that began in the 1970s. For example, cities such as Ekaterinberg in western Siberia are becoming service-based transportation and corporate centers for European businesses, and other gateway cities near the Chinese border (*e.g.*, Vladivostok and Khabarovsk in eastern Siberia) are transforming Russian–Asian business relations.

The early post-Soviet period exposed many cities to increasing poverty, economic collapse, and restructuring, as well as to large inflows of refugees from more troubled parts of the former Soviet Union (table 6.1). Today, rather than buying subsidized goods and services from the state, urban governments are challenged to transform into self-sufficient capitalist entities responsible for self-promotion in an economic climate marked by rapid and widespread changes in the distribution of development both within cities and between them. Urban in-migration and development has led to increased growth of larger cities (*e.g.*, 100,000 or more) in western and southern Russia (fig 6.6). Just as throughout Russian history, harsh climate, a poorly developed (and frequently impassable) network of roads, and immense distances still exacerbate the fragmentation of the Russian urban system. Russian cities will continue seeking successful solutions to these challenges well into the 21st century.

Table 6.1   Percent Urban Population in Each Federal Okrug

Federal Okrug	1926	1939	1970	1989	2002	2010	Percent change, 1926–1989	Percent change, 1989–2010
Central	19	34.2	64.3	78	79.1	81.3	310.5	4.3
Northwest	29.2	48	73.3	82.2	81.9	83.5	181.5	1.6
Southern	19.2	31	52.1	60	57.3	62.4	212.5	4.1
North Caucasus[1]	-	-	-	-	-	49.1	-	-
Privolzhskaya	12.1	23.8	56.1	70.8	70.8	70.8	485.1	0.0
Ural	21	45.4	71.3	80.2	80.2	79.9	281.9	-0.3
Siberia	13.3	32.6	62.5	72.9	70.5	72.0	448.1	-1.3
Far East	23.4	46.5	71.5	75.8	76	74.8	223.9	-1.4
Russian Federation Total	17.7	33.5	62.3	73.6	73	73.7	315.8	0.1

[1]   The North Caucasian Federal District was split from Southern Federal District on January 19, 2010.
Source: 2010 Census Data; www.perepis-2010.ru

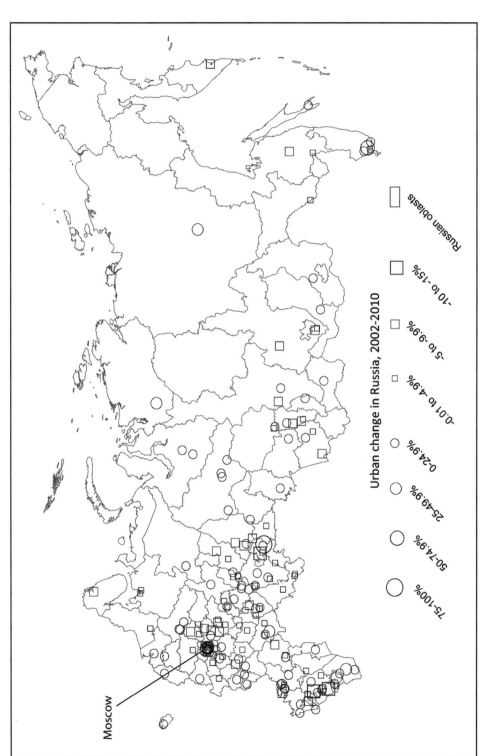

Figure 6.6   Population Change in Russian Cities, 2002–2010. Map by Jessica Graybill using Russian Census 2010 data: www.perepis.ru.

HISTORICAL EVOLUTION OF THE RUSSIAN URBAN SYSTEM

The Pre-Soviet Period: Birth of the Urban System

Historical settlement patterns have depended on access to water, transportation, and the location of military and economic outposts. The eastward spread of Russia's urban population dates to the first Slavic cities that appeared in the Valdai Highlands of the Russian plain at the end of the 9th century. A vast river network provided connectivity through this region, often called Rus, creating vital trade routes between Scandinavia, Russia, and the eastern Mediterranean regions. The Vikings established a set of city-principalities, at once both military outposts and trading centers, where they collected tolls from merchants traveling through the region. Kiev (or Kyiv, now the capital of independent Ukraine), Novgorod, and Smolensk were among the earliest urban settlements of this period.

The region gradually began to function independently from the Viking settlers, and Kiev became the focal point for Slavic political and economic development because of its location on the navigable Dnieper River with access to the Black Sea and Constantinople. In 988, Orthodox Christianity became the official religion, constraining the open practice and tolerance of other religious beliefs (*e.g.,* Islam, Judaism, paganism, pantheism). Most cities in Kievan Rus were located along rivers and were originally established as *kremlins*, or forts, because of constant conflict among the settlements as well as for protection against raids by Mongols and, later, Mongolian Tatars. The importance of hills for defense and rivers for communication lines during this period explains common features of many city centers. Kremlins were always located on high riverbanks and streets were radially planned, to facilitate rapid dispatch of troops. Many of these cities have survived today with their kremlins still intact. The famous Golden Ring cities around Moscow (*e.g.,* Yaroslavl, Suzdal, Vladimir) have origins in Kievan Rus and remain important centers of Russian culture.

After an important victory against the Tatars in 1480, a new polity arose, called Muscovy, aided in the development of a new type of urban network that developed east of Rus. The growing city of Moscow dominated this new region of settlement from its location at the center of another river system, allowing cultural and economic growth in new directions. Access to the Volga and its tributaries aided eastward expansion; the Western Dvina led to the Baltic; and the Don and Dnieper rivers led to the Black Sea. Theologians who envisioned Moscow as the "Third Rome" provided Muscovy with a vision and a self-proclaimed mission to build a new Russian empire firmly rooted in Christian missionary traditions with expansionist intentions.

Indeed, Russian settlements expanded eastward across the Ural Mountains into Siberia, encountering little resistance after final defeat of the Tatars at Kazan in the middle of the 16th century by Tsar Ivan the Terrible. New settlements, such as Tobolsk and Yakutsk, began as military outposts; they remained isolated frontier towns until Soviet expansion. Trappers plundered Siberia for furs; explorers and scientists who sought to map the territories to the East and South also brought back tales of ethnic groups and raw resources in Siberia, the Far East, and Central Asia.

As the 17th century ended, Russia's network of cities had become landlocked. Seeking access to the sea, Tsar Peter the Great founded St. Petersburg in 1703, touching off spectacular urban transformations within Russia. Consistent with Russian urban history, St. Petersburg was built for economic and security reasons, to be a showcase naval and commercial port with crucial access to maritime routes. New to Russian urban development, however, was the cultural purpose of becoming the country's "window on the West;" the city was designed according to European planning principles. Peter the Great also expected adoption of western cultural norms. For example, he required men in cities to cut off their long beards or pay an annual beard tax, thus hoping to shear away old Muscovite customs and traditions in favor of new styles and habits of living.

As the new national capital, St. Petersburg quickly supplanted Moscow. The urban focus moved westward physically and culturally. Reforms undertaken by Tsar Peter revitalized both local and long-distance trade and encouraged growth in new market centers as well as in more established ones. The creation of this new, Western-oriented city fueled social and spatial tension between those who believed in modernizing the country and those who emphasized Russia's traditional Slavic origins. Current debates about Russia's direction of development mirror these earlier ones; some look to the West or East for support, and some look inward for purely Russian inspiration and solutions. The development of St. Petersburg in originally inhospitable swampy land was also a precursor to the Soviet belief that humans can conquer nature in the name of economic progress. By the end of the 19th century, about 16 percent of Russia's population lived in urban areas. Factories in the region around Moscow and several nearby centers (*e.g.,* Tver, Vladimir,

Ivanovo, Kostroma) fueled economic and urban development.

The Soviet Period: New Urban Patterns

After the Russian Revolution (1917) and the ensuing civil war, the Communist Party took steps to consolidate its political power and reshape the economy, establishing a political-economic and urban system unlike any other in the world. In 1918, the leadership moved the capital from St. Petersburg (renamed Leningrad in 1924) back to Moscow. The move was both symbolic and strategic: the relatively recent capital built by the tsars as a "window on the West" was replaced by an older capital (Moscow) in the country's heartland, which would be easier to defend. It also made a statement that the country's gaze was no longer to the West but to the East and within the empire.

Soviet doctrine privileged urban life over rural life as the proper environment for communist individual, drawing on both British and French models and experiments in worker housing. Urbanization was seen as necessary to create an industrialized working class that would embrace communist ideals, and thus became synonymous with the construction of communism. Russia rapidly urbanized after the communist era began in 1917; and significant levels of urbanization continued in every region of Russia throughout the Soviet period, bringing electricity and indoor plumbing to many regions. Even in predominantly agricultural regions, more than half of the population lived in urban places by 1979.

The Communist Party established a new economic system guided by communist and socialist principles instead of market forces; it was called the command economy because a group of central planners located in Moscow made all decisions. Central planners allocated all investment resources and set standards for urban development, prioritizing needs national over local needs. This meant that cities had little influence over local economic development, urban growth, and internal city structure.

Private property was abolished. To provide immediate housing for the crush of population moving into cities and to supply industrial labor, private apartments were appropriated and subdivided to create *kommunalka*, or communal apartments, which provided immediate housing in the era of urban industrialization. Multiple households had to share spaces that formerly accommodated a single family. At first, such lack of privacy was tolerated as part of the excitement of building a new communist society. Later, however, *kommunalka* became slum-like dwellings for the urban poor where sharing was largely practiced out of economic necessity, not out of idealism about a better future.

Obeying another ideological principle, Soviet planners attempted to distribute urban settlements evenly across the Soviet expanse, even into harsh, inhospitable regions, obeying the injunction of Friedrich Engels to distribute large-scale industry equally across the country. Using algorithms based on European and North American urbanization models, Soviet planners chose so-called optimal locations for industrial development and built cities around them. This led to the construction of new cities with predetermined sizes (*e.g.*, less than 50,000 people, more than 100,000 people) in previously lesser developed and less-populated regions. This approach created a seemingly irrational pattern of economic flows between quite distant cities; the locations of suppliers, intermediate producers, markets, and managerial bureaucracies were

of little concern in a system where transportation and energy costs were state subsidized and therefore perceived to be virtually free.

The artificiality of this urban planning policy is especially visible in the rapid settlement and urbanization of Siberia, the Far East, and the Far North. Soviet planners regarded mastery of these regions both as an ideological necessity and as a challenge to their technological ability to tame harsh environments, such as permafrost or the steppe regions. Prior to the Soviet period, small autonomous villages and indigenous settlements dotted the vast territory of the Russian Far North and East. Modernization of lifestyles in these regions in the early Soviet period was achieved by pushing people off their native lands and into regional towns and collective farms (*kolkhozi*). Ostensibly undertaken to ease central management and regional planning for cities and towns, resettlement of nomadic and semi-nomadic indigenous peoples from small villages across the former Soviet territories ultimately enlarged the Soviet industrial workforce but greatly altered residential patterns and traditional ways of life.

High rates of urbanization in the Arctic and Siberian regions existed as early as 1959 and continue today (table 6.1). Even into the late 20th century, the population of the Far North (the Arctic region) remained nearly 80 percent urban, well above the Russian average of 73 percent. The ideologically based subsidies that enabled this process included higher salaries, offered to persuade people to join the social and physical construction of communism in the new industrial settlements. This practice highlights a significant mismatch between the location of labor resources, markets, and urban-industrial power in the western portion of the country, and the location of natural resources, including energy, in the eastern and northern portions of the country.

After WWII, national security needs also prompted the creation of a vast urban network connected to the military-industrial complex (MIC). Closed to outside visitors, these cities grew in economic importance and population only because of their attachment to the MIC. Decades of defense-related investment in these cities' industrial bases, housing stocks, roads, schools, and other urban infrastructure influenced their urban geographies in ways that would have been impossible in capitalist economies (in fig 6.20, note the enclave of Zelenograd near Moscow but distinct from it).

Subsidization of transportation included spectacular megaprojects that aimed to increase connectivity in the urban system. Joseph Stalin envisioned the construction of a canal network across northern European Russia to promote trade. While never completed, it is remembered for the use of prisoners to construct it, especially the White Sea-Baltic Canal (Belomorkanal) portion. In the 1970s, Soviet planners began constructing a second Siberian rail route, the Baykal-Amur Mainline (BAM), to supplement the capacity of the Trans-Siberian Railroad. The BAM facilitated new natural resource exploitation and the transportation of goods across Russia's vast expanse, providing lifelines to cities located thousands of miles from central Russia. In the Far North, cities primarily depended on boats using the Northern Sea Route along the Arctic coastline or Siberian rivers. Even today, frozen rivers are used as winter roads until the ice breaks. Crucially, however, the USSR never developed a network of highways such as those in North America or Europe, thus greatly hampering circulation between cities.

To help carry out political and economic agendas, as well as to reflect the new ideology, the communist leadership established a hierarchical urban administrative system that located all power in Moscow. Administrative centers in the *oblasts* (political units comparable to states or provinces) were subordinate to the central planners in Moscow, who controlled resource allocation and use in each region. Not surprisingly, administrative centers benefited disproportionately from those central investment decisions. *Oblast* centers became the locations for massive industrial investment and grew rapidly. Historic industrial centers such as Moscow, Yaroslavl, and Kazan were joined by administrative/industrial centers in Siberia and the Far East (*e.g.,* Omsk, Novosibirsk, Krasnoyarsk, Irkutsk, and Vladivostok). Many *oblast* centers still function like primate cities in other world regions, where investment, services, and labor are concentrated in one city, creating uneven regional development.

Planners also used investments to develop a system of secondary industrial cities focused on heavy industry (*e.g.,* the automotive industry in Togliatti, aluminum production in Bratsk) or natural resource exploitation (*e.g.,* nickel in Norilsk, oil near Surgut). Thus, a new system of large cities (*i.e.,* cities of more than 50,000 people) developed in Russia. In a country where bigger was seen as better, planners and politicians spoke glowingly of cities with more than 1 million inhabitants. Indeed, many Russians have a cultural urban bias; they consider it more prestigious and advantageous to live in cities and, despite urban hardships, and would prefer not to leave them for more rural settings.

While planners directed investment resources to specific cities, they simultaneously pursued a contradictory policy: limiting population growth in many of the same cities through formal control mechanisms such as the *propiska* (legal permission to live in a specific city). Many individuals found legal ways around the system, such as marrying someone who had a *propiska* to live in the city or finding employment and having the employer secure a *propiska*. Ultimately, pressure for ever greater production made investments in established sites more economically rational. But additional production created demand for increased labor. This had the dual effect of increasing city sizes beyond intended targets and intensifying industrial production (and thus industrial pollution and waste) inside city boundaries.

Urban and Regional Planning under the Soviet System

Central planners also influenced the internal spatial structure of Soviet cities. To create cities consistent with socialist ideals, planners adopted specific principles to guide urban planning that included adopting urban growth boundaries in order to constrain city sizes, distributing consumer and cultural goods and services equitably to the population, minimizing travel distances to work and providing public transportation for spatial mobility, and segregating urban land uses. Interestingly, some of these Soviet principles, such as urban growth boundaries and reduced commuting, have been abandoned in Russia today; but they are now being propounded in the West as "smart" growth.

The basic building block of Soviet cities was the *microrayon*. Constructed near industry and other places of work to minimize journeys to work, *microrayons* housed 8,000–12,000 people in living areas designed as integrated units of high-rise apartment buildings, stores, and schools, providing residents with cultural and educational services required by Soviet norms. In this urban planning scheme, all daily life activities (*e.g.,* education, shopping, use of city services like post offices and utilities payments) could be conducted without leaving the *microrayon*, thus influencing how children, workers, the elderly, and others moved through the city. People living in close proximity in the same city, but not in the same *microrayon*, might never meet on the street because of the highly structured nature of urban life in these neighborhoods.

Built using standardized plans irrespective of local environmental conditions, *microrayons* are numbingly similar whether in Novosibirsk, Vorkuta, or Moscow (fig. 6.7), and large tracts of identical or similar multistory apartment buildings still ring Russian cities. For example, similar construction materials and designs were used to build *microrayons* located in diverse physical geographic regions found across the former Soviet Union, such as in earthquake hazard zones (*e.g.,* Almaty in Kazakhstan); in cold, damp climates (*e.g.,* Petropavlovsk-Kamchatsky); in flood hazard regions (*e.g.,* St. Petersburg); or on the semi-arid steppe (*e.g.,* Barnaul). *Microrayon* locations were set by the so-called General Plans, which determined the location of *microrayons* within cities, as well as all other land uses. General plans were so detailed that a milk store could not be built legally on a site designated for a bread store. General plans were intended to complement the shorter term five-year economic plans, which determined what would be made, how it would be made, who would make it, who would receive the final product, and at what price.

Figure 6.7   Close proximity of apartments, small retail and service buildings typify a Russian microrayon, also often built near an industrial center (top, left). In these satellite images of Norilsk, note the regularity of built living spaces in the city at macro (top, right) and meso (bottom; corresponds to white box on top image) scales. *Source:* Google Earth

The Urban Environment

Absent from Soviet planning principles were concerns about the impact of industrial or urban development on the environment or about enforcing environmental constraints. Planners, and the Soviet system in general, believed in and practiced technological control over nature. This practice, combined with the zeal to reach economic goals, resulted in

Figure 6.8   On-shore Russian-owned oil fields are found on the outskirts of the city of Okha on Sakhalin Island. Decades-long oil spills and seepages run into the Sea of Okhotsk, only 10 miles (16 km) away. (Photo by Jessica Graybill)

almost complete disregard for the ecology in and near Russia's cities. Teams of planners, geo-engineers, and economic geographers choreographed large-scale development projects to modernize society, especially in large urban areas. For example, dams, hydroelectric power stations, and industrial complexes were constructed in and near cities. "Progress" was narrowly conceptualized as industrialization at all costs; and nature was society's tool to create the new socialist reality.

For example, near the city of Okha on Sakhalin Island, onshore oil deposits have been exploited since the early 1900s. Exploitation increased in the Soviet period, and the evidence of poor environmental standards for extraction remains today. On a road out of town, adjacent to local residents' summer homes (*dacha*s), and within sight of high-rise apartment buildings, numerous rusting and leaking oil pumps stand in pools of stagnant water mixed with leaked oil. Although signs posted in this suburban oil field warn pedestrians of the toxins in the area, they are often illegible or half-buried in oil muck (fig. 6.8). This mixture runs into local creeks, which in turn empty into the Sea of Okhotsk, where discharge from runoff pipes disrupts ecologies in bays and coastlines. This environmental and human health hazard was—and remains—less important than the economic bottom line.

Examples like these abound in and around Russian cities and can be understood as examples of environmental and social injustice.

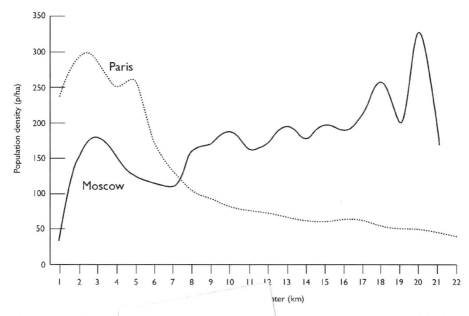

Figure 6.9　Compar̶　　　　　　　̶uilt-up areas of Moscow and Paris.
Source: Beth Mitchn̶

Although many urban res̶　　　　　　　air pollution to rebound. Disregard
of urban environmental iss̶　　　　　　impact of economic and industrial
papers and scientific engin̶　　　　　̶ent on ecology shaped investment
remained silent about the g̶　　　̶̶̶̶actices that continue today.
mental problems across m̶ ̶̶ ̶ of Russia's　　The Soviet history of urban and regional
industry-driven cities until the late Soviet　planning has left an indelible mark on Rus-
period, when the extent of *environmental deg-*　sia's built and natural environments precisely
radation began to be publicized. Only in the　because buildings and industries were located
late 1980s did people openly begin to express　without reference to market forces and envi-
concern about environmental health issues　ronmental conditions. In the command econ-
after nationwide reporting of air, water, and　omy, land was not bought and sold in Soviet
land pollution; environmental degradation　cities, but allocated roughly in accordance
with economic consequences (*e.g.,* decreased　with the socialist ideology and planning prin-
fishing catches in lakes and rivers); and human　ciples outlined above. The absence of a free
health issues (*e.g.,* asthma, kidney diseases,　market meant land was not recycled for other
lung diseases). Many urban dwellers found　purposes, as would have been the case in mar-
a silver lining in the industrial decline of the　ket economies. As a result, new construction
1990s—the spiraling decline of urban envi-　tended to continue moving outward from the
ronments was temporarily disrupted causing　city center. Compare, for instance, the popula-
a brief decline in environmental degradation　tion density of Paris and Moscow as it varies
until massive increases in automobile use　with distance from the city center (fig. 6.9). In

Figure 6.10   The crumbling historic center of Balakovo (founded 1762) is surrounded by microrayons on all sides. (Photo by Jessica Graybill)

Paris, market forces mean that valuable land near the city center is more densely populated than less valuable land on the city outskirts. In Moscow, just the opposite is true. The most densely populated parts of the city are located on less valuable land far from the city center.

Clearly visible historic rings of development remain. Beginning in the 1930s, huge factories were erected outside tsarist-era city cores as part of the industrialization drive. In subsequent years, especially after the late 1950s, a near catastrophic housing shortage and renewed determination to improve people's living conditions began decades of construction of *microrayons* on city outskirts, while historic buildings crumbled in city centers (fig 6.10). Instead of developing technology to build skyscrapers on valuable land in the city center, central planners focused on building high-rise apartment buildings on the outskirts, creating a characteristic "bowl" skyline in many cities. Ironically, this under-valuing of centrally located land preserved many older buildings up until the post-Soviet period.

Late Soviet Period: The Beginning of Change

Noticeable urban restructuring preceded the final economic and political collapse of the Soviet system in 1991. Starting in the 1980s, a phenomenon of "disappearing cities" revealed

Figure 6.11 A submarine in Kaliningrad, a former secret military city in the former Soviet Union, is now used as a tourist attraction. (Photo by Annina Ala-Outinen)

a pattern of manufacturing and industrial decline. Towns in European Russia lost so many people that they no longer appeared in Soviet statistical accounts of urban places. While distributed throughout Russia, more than half of these towns were in the industrial core regions around Moscow and St. Petersburg and in the Urals. By the mid-1990s, however, Siberian cities accounted for a larger proportion of shrinking urban areas. The exodus of residents from shrinking towns, combined with new westward migration patterns, contributed to the worsening situation of overburdened and decaying urban infrastructure, including housing and utilities, as population became concentrated in many large Russian cities. Thus, despite natural population decline (a greater number of deaths than births), many Russian cities—especially in the south and west—are growing rapidly even as others shrink (fig. 6.6).

Analysis of the profile of disappearing and shrinking towns reveals the impact of economic restructuring on the urban system. Because of the boom-bust economies surrounding mineral and other resource exploitation, mining towns account for a large proportion of declining towns (roughly a quarter by 2000). Urban centers related to the MIC (military-industrial complex) comprise another large proportion of declining cities. Previously, MIC cities received special attention from the central government, including better-than-average access to goods and services as well as higher salaries. This special treatment does not continue today in most MICs. For example, in the border zone of Kamchatka in the Far East, MICs have become near ghost towns due to a perceived end in need for border control in this region.

An interesting counterpoint to the disappearance of many towns is the appearance of previously unacknowledged cities, known as "secret cities" or, in Soviet parlance, "closed administrative-territorial formations." Secret cities never appeared on maps, and some estimates place the number of secret cities around 40. Located throughout Russia, but with clusters in Murmansk Oblast, the Far East, the Urals, and Moscow Oblast, employment in these cities was focused on highly classified military production including nuclear research and missile production (fig. 6.11).

CONTEMPORARY RUSSIA: RECONFIGURING THE URBAN SYSTEM

In the post-Soviet period (1992–present), heavy industry was especially hard hit by the reintroduction of market forces and the reorientation of the economy towards a consumer oriented capitalist market model and away from defense. As Russia's borders opened, it was flooded with cheaper and better consumer goods, and demand for locally produced goods—everything from steel to planes—dried up. Cities experienced economic restructuring processes not dissimilar to those in major North American and European cities during deindustrialization in the 1970s. At the same time, regions began scrambling for new investment capital to replace the funds, which used to flow in from Moscow. Economic restructuring away from manufacturing to increased extraction of raw resource and export means that cities dependent on manufacturing struggle to cope with high unemployment rates and few opportunities for new economic development. It is precisely these de-industrializing cities that are losing population in the current period.

In the post-Soviet period, the places with the highest rates of gross regional product today are found in just a few natural resource-rich regions in Siberia, the Far East, and in Moscow. As a direct result of the change from Soviet central planning to market-driven economic processes, there has been significant movement out of Siberian and Arctic cities (as well as from cities in the newly independent countries in Central Asia and the Caucasus) back to European Russia. The introduction of market forces in cities located in harsh and inaccessible places, such as Norilsk and Surgut, has caused rapid increases in costs for energy and transportation, food, housing, and industrial production. As subsidies dropped sharply, urban industrial complexes closed and unemployment surged. Many who could simply pulled up stakes and moved.

The uneven spatial distribution of the benefits of economic reform has created vast differences among individual cities in Russia's urban network. Some cities are clearly thriving in the transition to a market economy, but those whose geographic locations are not conducive to taking part in transformation are struggling. However, many people (including the elderly and the poor) stay in these cities because of strong social or kinship ties forged in the urban archipelago, or a strong belief in the Soviet system. For those who stay behind, the 'new' traditional economy depends on activities like hunting, gathering, fishing, and domestic agricultural production. For example, people have intensified agricultural crops and livestock husbandry on household plots and at *dachas* (fig 6.12). In some places, foresters report increased gathering of communal forest resources such as mushrooms, berries, or herbs. In this way, traditional agricultural activities are woven into everyday life in shrinking urban places across Russia.

Cities in regions with growing economies and growing populations (like Moscow,

Figure 6.12   Space around many Russian apartment buildings and houses is devoted to agriculture during the short summer season. (Photo by Jessica Graybill)

Yakutsk, Yuzhno-Sakhalinsk, Nakhodka, and Kazan) have become attractive destinations for migrants from more depressed areas of the country. This phenomenon has rearranged hierarchical relationships in the urban system by increasing the population and the economic clout of previously minor cities. For example, it is not only migrants from across Russia, but migrants from across the former Soviet republics, who are attracted to jobs associated with oil and gas development on Sakhalin Island. Cities in the Southern Federal Okrug, such as Krasnodar, Stavropol, Vladikavkaz, and Novorossisk, are also growing, but mostly as a result of large influxes of migrants from conflict-ridden areas of the North Caucasus and the former Soviet republics.

Significant population losses nationwide and depopulation of many urbanized regions in the post-Soviet period (table 6.1) are clear signs of the failure of the communist approach to city planning. In the early 21st century, Russia is experiencing a dramatic restructuring of urban centers. Whereas the centralized economic authorities used to designate the political and economic importance of any particular city as well as the location of residences, individual Russians can now choose from a widening array of housing options. A growing trend is suburbanization and new housing has developed outside city limits (fig. 6.5). Many new housing developments are video-monitored, guarded by security officers, and approachable only by vehicular transportation. Although this contributes to increased traffic woes, it satisfies a desire for private space for those who can afford this lifestyle. While there are no systematic data on suburbanization processes across Russia, it is clear that this trend is especially prevalent in the European portion of the country. Yet like past urban processes, it is spreading eastward.

Figure 6.13   Opened in 2010, "City Mall" in Yuzhno-Sakhalinsk is the largest shopping mall in the Russian Far East and boasts a microbrewery for beer and loudspeaker announcements in Russian and English. (Photo by Jessica Graybill)

Political Urban Transformation

Democratization and political decentralization have also influenced post-Soviet urban geographies. The first democratic elections took place in cities throughout Russia shortly before the end of the Soviet Union. For the first time, local politicians, at least in theory, became accountable to local populations instead of to higher-level government officials. During the 1990s, this accountability had important new implications for the spatial structure of cities, as urban geographies began to reflect local economic needs and social desires instead of national ones.

For example, one result of local autonomy in the 1990s was the popularity of renaming cities and streets. Names associated with prominent Soviet leaders have been replaced with historic names of the tsarist past, for example, Leningrad reverted to St. Petersburg and Sverdlovsk (named after a local communist leader) reverted to Ekaterinburg (literally, Catherine's City, after Tsarina Catherine the Great). Similarly, streets were renamed: Moscow's Gorky Street, named after the Soviet writer, was renamed Tverskaya Street. The changes contributed to a feeling among Russians of reclaiming their cities and neighborhoods.

Unlike during the Soviet era, it is now possible to freely express political visions for cities and the state that may not toe the party line. For example, the rise in cities of visible extremist and nationalist groups, especially those targeting non-Slavic populations (*e.g.*, Africans, African-Russians, ethnic peoples of the former Soviet Union especially from the southwest), creates new conflict in many urban settings that is often violent with occasional acts of armed violence (box 6.1).

Changing Urban Structure and Function

Notable changes to urban structure include new kinds of infill within the city, suburbanization, and "slumification." Important changes to urban function include increased finance and retail commerce at multiple scales (fig. 6.14). Processes absent during most of the 20th century govern these changes—including market forces and the active participation of municipal and regional governments.

New infill appears in city cores as old factories on land surrounding historic city centers are increasingly torn down and the land is reused for other purposes, such as housing (apartment buildings) or retail. Existing buildings in poor condition but in good locations are purchased, upgraded, and converted into office space or upscale, gated apartments (fig. 6.15). In Moscow and some other Russian cities, this has led to gentrification and displacement of long-time residents from city cores.

Suburbanization is another visible change in urban form that results from the development of real estate markets in cities where they had been prohibited for most of the 20th century (fig 6.5). Carefully guarded single-family housing has appeared seemingly

overnight in what has become a new ring of housing developments referred to as cottages (*kottedgi*) surrounding the older Soviet and new post-Soviet multifamily high rises.

"Slumification" of parts of Russian cities is a result of transition from a command to a market economy. Run-down high-rise apartment buildings far from the city center are located on nearly worthless and often polluted land near former industrial production sites. These high rises, and sometimes entire *microrayons*, are deteriorating rapidly as better-off tenants move to superior locations, leaving only poorer residents behind in what will likely become vertical slums (fig 6.16).

Finance and banking—particularly international banking—is increasingly a feature of larger post-Soviet Russian cities. Just as in de-industrializing North American and European cities, the economic function of Russian cities and the new labor market are more oriented toward services in general, and toward financial and retail services in particular. Members of the international financial and banking sector have added Russian locations near manufacturing or natural resources. For example, European and Japanese banks can be found in Yuzhno-Sakhalinsk today, which marks a change in the function of this city from a small regional capital to a globalizing city involved in oil and other natural resource exploitation. Other Russian cities actively seek foreign investment to restructure their cities through partnerships among government actors and local and international businesses.

Retail commerce, powered by market forces, has also visibly changed the economic geography of Russian cities. Previously, retail trade occurred in state-owned stores or in a

Box 6.1 Hate Crimes in Russian Cities

Imagine commuting to work on Moscow's metro system. With over 6 million people traveling Moscow's metro daily, it is the world's second busiest transit system; only Tokyo's is bigger. You reach your destination, the train doors open, and you step onto the platform. Suddenly, an explosion rips through the platform; the lights go out, causing mass panic. You and hundreds of fellow travelers run for the escalators, but are stopped. Panic subsides but fear develops as people process the impact of the explosion, aid the injured, and mourn the dead. Immediate police response reveals that yet another suicide bomber from the restive Caucasus region has targeted Moscow, causing injury, death, and urban infrastructural damage. Hours later, commuters are released, riding escalators up into the fresh air and renewed cell phone service linking them to family and friends who have been desperate for knowledge of their status.

Since 1995, 23 terrorist attacts have plagued Russian cities and transportation networks, largely in Moscow and in cities around the Caucasus region (*e.g.*, Vladikavkaz, Beslan, Stavropol), killing over 1500 people and injuring over 5000. Common terror tactics include suicide bombing and hostage taking. Allegedly, terrorist acts are conducted almost exclusively by Chechen separatists intent on building a new ideological identity–only partially based in Islamic faith–and an autonomous territory in the North Caucasus region. By using violence against Russian civilians, they hope to advance their political or ideological objectives by creating fear in the Russian populace and government. Although the probability of experiencing a terrorist attact remains low, citizens and tourists traveling through and between cities are protected by armed police who patrol the city and the metro system, demanding inspection of documents (*e.g.*, passport, visa) seemingly without cause, particularly targeting people with darker skin thought to be from the southern regions near the Caucasus or the Democratic Republic of Georgia. Critics call such racial profiling a human rights violation, while others find it justified given the frequency of attacks in post-Soviet cities. In a twist on the story, the writer and human rights activist, Anna Politkovskaya, who opposed the Russian wars in Chechnya, was assassinated on October 7, 2006, ostensibly for her criticism of then-president Vladimir Putin.

More isolated, smaller hate crimes against ethnic groups—especially people with darker skin or from minority religious groups—are also increasing in the post-Soviet era in Russia. For example, people with multiple ethnic backgrounds who were once embraced by the Soviet Union's multiculturalism now find themselves targets of attacks, including beatings, torture, and mutilation, by radical ultranationalist groups who practice violence against non-Russians residing in Russian territory. This radical ultranationalism is often explained as a result of resurgent national pride combined with unemployment, inflation, and decreased educational opportunities after the fall of the Soviet Union. Twenty years after the collapse, young adults who never knew the Soviet Union comprise a new generation with new understandings of cultural and political allegiances. These young Russians, competing with migrant laborers from neighboring states for low-paying jobs, do not share a collective cultural or economic future, as their parents did in the Soviet past, and instead find the slogan "Russia is for Russians" more fitting for today's reality.

Figure 6.14 At a small scale, city centers are revitalized with first-floor shopping venues bringing brand names into the rapidly-growing consumer market. (Photo by Jessica Graybill)

Figure 6.15   Two existing Soviet-era buildings are joined together with new technology and architectural design as elite apartments are developed in the core of Yuzhno-Sakhalinsk. (Photo by Jessica Graybill)

limited number of farmers' markets. Now, the spatial structure of urban retail has been altered dramatically. Transportation hubs (subways, rail stations) are multiscalar centers of retail trade where peddlers vend wares and where retail centers, such as malls, have been built (fig. 6.17). Some prerevolutionary shopping centers have regained their functions; for example, Moscow's famous GUM department store was remodeled as a high-end shopping mall. The urban periphery has turned into a new retail environment; megastores, such as IKEA, are opening outside of traditional retail centers, extending urban retail spaces (fig 6.18).

Sociocultural Urban Transformation

Notable social transformations include changing labor and leisure structures. There is a growing class with extra money to spend, resulting in increased consumerism and more availability of goods. A revived entertainment sector has also sparked a service industry catering to twenty- and thirty-year olds who have cash to spend. However, goods, services, and entertainment remain expensive. Many people cannot afford a high quality of life in the new Russia, or cannot afford it after leaving home or college, as the costs of buying and furnishing apartments are out of reach for many.

Figure 6.16 Vertical slums develop in Petropavlovsk-Kamchatksy as people migrate out of this run-down neighborhood to other parts of the city and suburbia or for other regions of Russia. (Photo by Jessica Graybill)

Although the *propiska* no longer exists, the current registration system exasperates people hoping to move around the country searching for jobs. Nonpermanent residents working away from their hometowns must purchase temporary urban registration in a semi-legal system, thus increasing their cost of living and jeopardizing their ability to succeed in Russia's new spatial economy. This type of compulsory registration system relegates them to second-class citizenship in their city of employment, often putting them and their children last in line to receive other government services, such as socialized health care, education, or employment services.

Transition from the Soviet to the post-Soviet era was difficult for those who were raised, trained, and already employed in the Soviet system, as it led to the disappearance of many jobs, the depletion of pension funds, and an unknown future. Many turned to the countryside to survive (ruralization); but others (considered "victims" of economic transition) turned to alcohol, theft, and/or prostitution. Because of Russia's slow response to providing social services to people in need, a new and growing class of the very poor, the homeless, and the disenfranchised exists in cities today.

Urban governments have meager funds to allocate for social services; in Moscow, only

Figure 6.17 Street peddlers hawk a variety of goods outside a Moscow metro station. (Photo by Jessica Graybill)

1,600 beds in 2005 gave shelter to the entire homeless population (with homeless population estimates ranging from 30,000 to 1 million). Official sources reported in 2006 up to 55,000 homeless children in Moscow alone and 16,000 in St. Petersburg, but the problem occurs across Russia. These figures include children—many teenagers but some younger—living without parents on the streets in Russian cities.

Many of Russia's homeless children leave home to escape domestic situations that include extreme poverty, alcoholism, domestic violence, and neglect. Research suggests that these social ills result from the post-Soviet economic restructuring and are experienced at the household level. Other causes include forced migration as a result of civil unrest in particular regions of Russia and a migration process in the 1990s that brought people

Figure 6.18   One of three IKEA stores in Moscow, this one is attached to a megashopping mall just outside the city limits, where people travel by minibus to purchase goods. (Photo by Jessica Graybill)

from former Soviet republics to Russian cities, where they often failed to find housing or employment. These forms of displacement especially affect children. Street children often do not attend school, experience harmful health impacts (*e.g.*, contraction of infectious diseases) and they frequently become targets for illegal activities (box 6.2). Some work in slave-like conditions, whereas others engage in child prostitution or drug trading.

21st Century Environmental Concerns

First raised openly in the late 1980s, environmental concerns are now growing across Russia. Increasingly, publications relate the harmful environmental legacy of Soviet urban development, suggesting an energized engagement with socio-environmental issues today. For example, one prominent issue plaguing urban areas is garbage. Soviet goods were often wrapped only in paper and string; larger amounts of waste in the post-Soviet era from imported packaged goods have not led to increased infrastructure to contain or remove garbage from urban centers. This results in garbage accumulating in public spaces (fig. 6.19), increasing environmentally hazardous conditions and contention between citizens and city governments.

Remedying urban environmental concerns, such as motor vehicle emissions and

Box 6.2 Human Security Concerns: HIV/AIDS and Human Trafficking

As of 2009, HIV/AIDS prevalence among Russian adults was 1 percent, well above Europe's highest rate (Portugal, 0.6 percent) although near other post-Soviet Eastern European nations (Ukraine, 1.1 percent and Estonia, 1.2 percent). A 2008 UNAIDS report notes that in 2007, 75 percent of newly infected people in Russia were aged 15 to 30. In 2001, the International Crisis Group suggested that Russian GDP might eventually be reduced as much as 1 percent by the effects of AIDS and other diseases on worker productivity, and several international charities currently run programs to help slow the spread of the disease in Russia.

According to UNAIDS (2006), Russia approves the growth of social programs such as HIV testing, treatment for pregnant women, and activities promoting increased societal tolerance of people living with HIV. Despite high-level support and the fact that the AIDS epidemic in Russia has primarily spread through intravenous drug use, criminalization of drug use and addiction prevents infected people from seeking medical attention or taking part in harm-reduction programs such as needle exchange (only 70 such programs exist for an estimated intravenous drug user population of nearly 2 million). Rates of HIV/AIDS are becoming significant among young women, suggesting a spread from high-risk groups into the general heterosexual population.

Considered an urban phenomenon in Russia, HIV/AIDS is associated with economic and social dislocation, drug use, and the illegal drug trade. In 2006, St. Petersburg reported 22,000 infected people. In Kaliningrad, where trends often presage events in the rest of Russia, sexual contact accounted for an estimated 30 percent of new cases in 2001. Some regions in Siberia now have

chemical poisoning (*e.g.*, lead) from contaminated water supplies remain largely unaddressed in many cities. Urban environmental problems are largely understudied and misunderstood in the post-Soviet era. Many people feel that solving environmental woes is the government's responsibility and that they cannot respond individually because socioeconomic and political issues are currently more pressing. In some cases, where urban and regional growth infringes upon land valued by environmentalists, conflict arises when citizens join forces to criticize or block urban growth (box 6.3). Such instances, however, are not the norm in today's Russia, where most citizens are not well versed in opposing the government and instead see the state as responsible for protecting people and the environment.

At the federal and international levels, however, environmental discussions since the mid-1990s have included the concept of sustainability. In creating new environmental policy directives, Russian policymakers invoke tsarist-era Russian and Western ideas about living in harmony with the biosphere as a foundation for creating sustainable development. Indeed, recent legislation that recognizes the need to address anthropogenic climate change provides hope that achieving economic growth while

higher infection rates than Kaliningrad; but officials in western Russia fear a new wave of HIV infections linked to prostitution, particularly along the Polish and Lithuanian borders.

Russian cities are a primary destination for women trafficked for sex work or escaping poverty. Since the beginning of market reforms, privatization, and withdrawal of state subsidies have forced women to pursue a wider range of income-generating activities—including prostitution. Many traffickers pose as employment agents, targeting cities with high unemployment. Attempts to migrate for legal work often ensnare women in sex slavery from which escape is difficult. One survey in Moscow found that 88.5 percent of prostitutes were not native Muscovites and less than half had lived in Moscow for over a year.

Even when law enforcement officials are committed to stopping the drug or sex trade operations, understaffed and underfunded local authorities have few resources to fight powerful trafficking organizations. Government-funded social programs to help victims are sorely lacking in Russia. For example, as of 2005, no women's shelters existed in eastern Russia, and those in Moscow did not have enough space to provide women with effective refuge from their traffickers.

Human security issues such as HIV/AIDS and human trafficking are serious issues that Russia struggles to contend with economically, politically, and socially in the post-Soviet era. Economically, many Russian regions and individuals have experienced crushing poverty since 1991, from which they are only beginning to recover. Politically and socially, addressing drug use, sexual activity, and the implied connections to infectious diseases has proven difficult, as many individuals deny having or cannot find help solving—at home, among friends, or through organizations—drug or sexual abuse situations.

---

balancing social and environmental goals is a real concern among politicians and will increasingly become so among citizens in Russia today.

DISTINCTIVE CITIES

Moscow: Russia's Past Meets Russia's Future

Perhaps no city captures Russia's long history as vividly as Moscow. Modern Moscow is a chaotic blend of brash and unfettered capitalism, seen in its casino lights and chic boutiques; monolithic apartment blocks from the Soviet period (where most residents live);

new construction of glass skyscrapers and gated communities; and buildings renovated in the old Russian style. New Russian Orthodox churches join places of worship of many faiths (including other forms of Christianity and Islam) on the urban landscape. Museums and theaters have undergone much revitalization and reconstruction in the post-Soviet era, bringing new cultural capital to the city.

In Moscow, Russia's past lives alongside its future. Founded over 850 years ago in the declining years of Kievan Rus, the city grew rapidly in importance until Peter the Great moved the capital to St. Petersburg. The 1917 Russian Revolution returned the seat of power to Moscow. On December 25, 1991,

Figure 6.19   Increasing consumption and lagging public services are reflected in the garbage-strewn landscapes surrounding many Russian apartment buildings. (Photo by Jessica Graybill)

resignation of the last Soviet leader, Mikhail Gorbachev, once again brought Moscow into the international limelight and launched massive socioeconomic and cultural changes.

After the collapse of the Soviet Union, Moscow exploded with the signs of capitalism. Foreign investment flooded the city, and new foreign and Russian capital created business centers, real estate companies, and a new retail sector. Once-empty avenues filled with cars—seemingly overnight. In the early 1990s, kiosks appeared everywhere stocked with an improbable mix of everything from candy bars to vodka, socks, and toys. Now, more permanent stores selling food, liquor, clothing, and toys, as well as every other possible consumer good, largely replace kiosks. Megascale shopping malls in the city outskirts also have replaced hastily built and remodeled stores. Restructuring of Moscow's new economy (historically the political, economic, media, educational and cultural center of the country), has added importance for the city's existing functions, especially as it rises to be the nation's banking and consumer capital.

Moscow is vastly richer than most other parts of Russia, partially as a result of inheriting immensely valuable real estate from the Soviet government and Communist Party. But the introduction of capitalism has still resulted in a highly fractured city. City residents experience life in vastly disparate ways. For the expatriate community, continuously growing as a result of foreign investment, Moscow is ranked as one of the most expensive cities in the world in terms of the

Box 6.3 Showdown in Khimki: Defense of Suburban Forest as a Form of New Protest in Russia

Mikhail S. Blinnikov

On July 23, 2010 at 5 A.M. unknown men wearing white masks suddenly attacked a camp of environmental activists inside Khimki Forest. The camp had been in place for a few days to protest construction of a new tollway through one of the last remaining old-growth forest fragments near Moscow City near the Sheremetyevo airport. The tollway would have bisected the forest and effectively destroyed its rare oak ecosystem and numerous wetlands. The activists knew that the company doing the logging did not have official permits to conduct the clearing. What they did not anticipate was the brutality of the anonymous thugs who were deployed to break-up the loggers' camp. When the riot police arrived after being called by the activists, the policemen attacked the defenders, not the loggers. The activist camp was destroyed, and seven activists and two journalists were arrested. The police seriously beat two women. The logging was called off in late August after rallies in Moscow caused President Medvedev to intervene. While the project was put on hold temporarily in the fall of 2010, activists prepared for a protracted physical and legal battle to resume in 2011. In April of 2011, the logging did resume after a governmental panel supposedly found that the only viable route for the tollway was through the forest.

Khimki Forest is located within the protected green belt established in 1935 to allow for recreational opportunities and air quality protection near Moscow (fig 6.20). The belt has seen destruction of 8,200 hectares (20,254 acres) of forest between 1992 and 2008–5 percent of the total area—to commercial and residential development. Located northwest of Moscow and east of the city of Khimki, the forest is a magnificent 1,118-hectare (2,762-acre) fragment of mixed oak and pine left intact since the times of Catherine the Great, and within a 30 minute walk for 300,000 people.

While Russia needs more roads to accommodate increasing suburban traffic, building the tollway through the forest is not imperative and clearly would serve only the interests of a few well-heeled businesspeople presumably connected to the governmental elite. The forest's defenders explain the routing choice as the result of the personal ambitions of a few corrupt officials with connections to the international airport, the city of Khimki, and Moscow *oblast* government. Putting a new road inside the forest will promote the rapid privatization of land in the highly valuable green zone. Also, a substantial amount of government-backed financing is involved—over $2 billion—which will allow for the quick embezzlement of funds allocated for road construction.

The protests in Khimki are nationally significant because they were only one of a handful of protests in Putin-Medvedev-ruled Russia in which local people have been able, at least temporarily, to halt an environmentally destructive project and reach a national audience on the national news and obtain a hearing in local courts. Many activists are local residents who believe that a cleaner suburban environment belongs to all, and that a corrupt development scheme should not be permitted to proceed at the expense of non-elite people.

available choices for housing, dining, and other forms of consumption. For "New Russians," the city is a 24-hour shopping, dining, and business extravaganza.

For the vast majority of Muscovites, however, life in the city is, as they say, *normal'no* (normal). In addition to the choking traffic jams, the overcrowded metro commutes during morning and evening rush hours, and a decrease in the environmental quality along major roadways due to the explosion of automobile ownership in the city, *normal'no* also includes a dizzying array of consumer services, fulfilling every desire, such as 24-hour gyms and sushi and coffee bars, businesses offering extended educational opportunities (*e.g.*, computer science, foreign language training), rental agencies, and much more. In Moscow, it is no longer a question of seeking out something new to do or become; rather the challenge is to choose among the many options.

Along with growth in consumer options, the city is slowly expanding beyond its urban growth boundary, a road that rings the city 20 km (12.4 miles) outside the city center. Delineated by urban planners in the Soviet era, land outside the boundary was meant to be a green zone for leisure purposes (*e.g.*, hiking, picnicking, camping). With the development of megashopping centers and the single-family housing market, the green zone is slowly diminishing as the city creeps outwards radially, especially along major transportation routes (fig 6.20).

Always a multiethnic city, Moscow is now increasingly so because of a large, constant flow of labor migrants from other parts of Russia and the former Soviet republics. While migrants arrive from all regions of Russia—urban and rural alike—perceived increasing numbers of ethnic migrants from the south (*e.g.*, Geor-

gia, Chechnya, Kazakhstan) causes inter-ethnic conflict in Moscow and other major Russian cities (see box 6.1). While some ethnic migrants are actually Russian citizens, some ethnically Slavic portions of the population feel that increased numbers of non-ethnically Russians threatens the economic and cultural futures of ethnic Russians in Russian cities. Largely, however, Moscow's long-standing tradition of being a multicultural locale remains, and the city is highly cosmopolitan and vibrant. In the post-Soviet era, Moscow now also ranks as a world city, the first in Russia.

St. Petersburg: Window on the West—Again?

Peter the Great founded St. Petersburg in 1703 as Russia's "window on the West," making it Russia's capital city, full of ornate palaces and bridges over winding canals, although thousands of Russians lost their lives while building the city. In 1917, the battleship *Aurora*, now a museum, fired shots at the tsar's winter palace, thus signaling the start of the Russian Revolution. When victorious Soviet leaders centralized economic and political resources in Moscow, again making it the capital, St. Petersburg was freed of its tsarist-era bureaucratic atmosphere and was renamed Leningrad in 1924. Aggressive neglect by Stalin ironically preserved the city's green expanses and architectural heritage.

While Leningrad did not grow as quickly as Moscow, the population doubled from 1917 to reach a high of about 5 million in 1989. In the post-Soviet period, its official population size declined to about 4.6 million in 2006 but has risen again to 4.8 million in 2010. It remains Russia's second largest city. Well-known worldwide as the home of a unique cultural legacy, including the renowned Hermitage Museum,

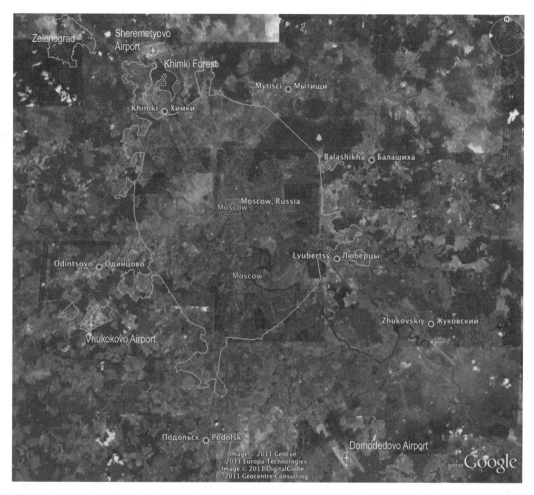

Figure 6.20 Retail and residential urban growth outwards creeps up against Moscow's Urban Growth Boundary. *Source:* Google Earth

Leningrad gained its prestige in Soviet times from its many prestigious universities and research institutions. The city's economy depended on educational and research activities, particularly defense-related industries in the Soviet military-industrial complex.

After 1991, the city's original name was restored, and it moved to capitalize on its distant past. A deteriorating economic and financial situation caused by the near collapse of government support for education, culture, and especially defense signaled a need for significant restructuring. In the mid-1990s, as a result of leadership from Mayor Anatoly Sobchak, a former lawyer active in the political transition, the city embarked on a strategic planning process—the first city in Russia to do so. In contrast to top-down Soviet central planning, this plan aimed to be participatory, building on a partnership among local government, private businesses and citizens to analyze possible scenarios for the city's development. Extensive discussions with private businesses, residents, and local organizations in St. Petersburg pinpointed the city's highly educated population; its ranking as the largest-capacity port in Russia; and its favorable location, not far from Finland and the rest of Europe, with excellent access to major railroads and highways. An initial concept document laid the foundation for the subsequent Master Plan and new legislation to standardize the construction and development process as well as modernize city zoning. Under City Governor Valentina Matvienko (2003–2010), the city embarked on numerous high-profile projects to improve its appearance and infrastructure, including a ring road to divert industrial traffic away from the city center. It remains to be seen whether the government of St. Petersburg can improve living standards and promote long-term restructuring. Halting attempts to increase hotel capacity and urban amenities suggest that tourism has not grown as expected. Meanwhile, road congestion and air pollution have increased dramatically because per capita car ownership has more than doubled recently; the city struggles to provide adequate traffic flow and parking throughout the historic area.

St. Petersburg has been at least modestly successful at attracting investment, including plants for Toyota, Ford, and Hewlett-Packard. However, thanks to its culture of participatory urban development, it has experienced vigorous public outcry over recent land use choices. For example, from 2006 to 2010, a struggle took place over plans by the state-owned natural gas corporation, Gazprom, to build a new skyscraper in St. Petersburg, called Okhta-Center. Fueled by residents' pride in the traditional low-rise city skyline punctuated by cathedral spires, resistance to construction of a central skyscraper became an international cause célèbre. Since the entire historic city center is on the list of World Heritage sites, UNESCO asked the city of St. Petersburg to halt the project in order to study potential impact on the city's historical monuments. Protest against the plan was a central theme in a series of unprecedented street demonstrations in 2007 and 2008; several prominent cultural figures in St. Petersburg and eventually Moscow joined the opposition. While plans for Okhta-Center were eventually scrapped, other less high-profile buildings are evidence of the power that business interests wield in the city administration; it is these interests that are quietly effecting the transition to a new skyline. These include the Stock Exchange building on Vasilievsky Island, erected in 2007, and the new Stockmann shopping center behind historic Ploshchad Vosstaniya (fig 6.21).

Vladivostok: Russia's Pacific Capital?

Vladivostok, home to the Russian Navy's Pacific Fleet from 1958 to 1991, was a closed city where even Soviet citizens needed permission to enter. Before this time, Vladivostok had been an international city. Valuable for its port facilities and proximity to Asian markets, the city drew a diverse international population, including Chinese, Koreans, Japanese, and Americans. Since 1991, these populations are back, as the city is now open to foreign tourists and businesses. Indeed, both the United States and Japan maintain consulates in Vladivostok.

Founded in 1860, Vladivostok is the capital of Primorsky Krai and is the Russian Far East's largest city, with nearly 600,000

Figure 6.21   Opened in 2010, the new Stockmann retail shopping center shocked urban conservationists in St. Petersburg. (Photo by Nathaniel Trumbull)

residents. It is the largest Russian port on the Pacific Ocean, and historically it has been an important regional industrial center for shipping and fishing. Located 3,800 miles (6,430 km) from Moscow, it is the eastern terminus of the Trans-Siberian Railroad. The great distance to European Russia feeds the image of Vladivostok as a gateway to exotic, Asian Russia. Historically, this self-image led Vladivostok residents to be self-reliant and to expect little from Moscow; indeed, the first Soviet leader to visit Vladivostok was Nikita Khrushchev in 1954. The city hosts the headquarters of the Far Eastern Division of the Russian Academy of Sciences, an academy of 14 academic and research institutions. These factors have fueled hopes that Vladivostok could become an urban economic hub for a new range of businesses. For example, entrepreneurs dream of Vladivostok-based regional eco-tourism, and the port facilities are a strong asset in rebuilding a strong import–export city on the Pacific. Views of the bay from the hills around the city also provide incentives for restoration of the historic center and its tsarist-era buildings and monuments, many of which could be refurbished to rival those in St. Petersburg. As an intellectual capital of the Russian Far East, Vladivostok has much to offer post-Soviet Russia, despite the need to crack down on polluters and numerous illegal activities.

Unfortunately, Vladivostok has been plagued by two negative economic developments since 1991: the rise of organized crime and the legacy of environmental degradation.

The rise of informal and mafia-driven economies has hindered the growth of legitimate, tax-paying businesses. Illegal Chinese immigrants who have moved into the city largely control the retail-oriented black market. The city is also an aviation gateway to Asian cities, since its airport is one of the few that can handle international flights. When Russians from across the Far Eastern region return from destinations across Asia via Vladivostok, they generally bring commodities to sell in Russia.

Because international trade with Asia is funneled through this Russian city, organized crime in Vladivostok and Primorsky Krai since 1991 purportedly involves not only Russian mafia but numerous mafia-like groups from the former Soviet republics in Central Asia. Many elected government officials work with, instead of against, organized crime. Illegal trade includes marine resources from China, cars from Japan, heroin from Central Asia, and timber exports from Russia to China and Republic of Korea.

In addition to economic difficulties, Vladivostok suffers from severe air, water, and soil pollution. Ecologists consider much of the city to be hazardously polluted by heavy metals and industrial (cadmium, mercury, arsenic) and agricultural (nitrates, phosphates) waste. Despite the city's location adjacent to the Pacific Ocean, local wind and water circulation patterns in Amursky Bay do not remove pollutants from the densely populated or intra-urban industrial areas. Unchecked, pollutants have built up over time, and their detrimental effects on human health are only slowly being recognized.

Despite its distance from European Russia and from the federal center, Vladivostok (and other Siberian and Far Eastern cities) remains in the cultural, political, and economic orbit of Moscow. While the rich economies of Japan and the Republic of Korea and the rapidly developing economy of China are far closer neighbors than Western Russia, Vladivostok is a city settled by ethnic Slavs (largely Russian and Ukrainian), who remain the largest proportion of the population today. Generally, Russians are suspicious of the Chinese and their interests in Russia's vast raw materials and land resources. The fact that the Russia-Chinese border is largely unmanned and many Chinese enter Russia illegally to reap resource rewards is not unnoticed by Russians; indeed, many in this region deeply dislike and distrust the Chinese. Thus, cultural allegiance to European Russia is stronger than their new Asian economic ties. The loyalty of Vladivostok's political elites to the nation's center is being rewarded by massive capital investment in Vladivostok by the federal government. Today the city is a giant construction zone, where new city infrastructure (*e.g.*, sewage treatment plants, underground freshwater reservoirs), highways, bridges, dams, and airport terminals are under construction. Vladivostok is thus being positioned to become a center of international cooperation in the Asian Pacific region, which will aid Russia in stabilizing the population and growing the economy in the Far Eastern region.

Yuzhno-Sakhalinsk:
The International Power of Oil

Yuzhno-Sakhalinsk, located on Sakhalin Island in Russia's Far East, is an oil boomtown fueled by multinational investment. This city, long a small urban hub for military and natural resource exports (coal, oil, fish, timber), is ushering in the era of globalization in the Russian Far East because of its proximity to offshore oil and gas reserves in the Sea of Okhotsk. The *oblast* is a leading destination for foreign direct investment—second only to the city of Moscow—indicating the importance of natural resources and regional centers to Russia as a whole. Located 4,000 miles (6,500 km) from Moscow and only 110 miles (175 km) from Japan, Yuzhno-Sakhalinsk is sited on a tsarist-era settlement for exiled prisoners (Vladimirovka) that later became a Japanese village (Toyohara) during Japanese rule of southern Sakhalin Island from 1905–1945. After WWII, the USSR reclaimed the entire island, and Yuzhno-Sakhalinsk was created as the new *oblast* capital for Sakhalin and the Kuril Islands.

Yuzhno-Sakhalinsk embodies the multiethnic character of Soviet-era cities. Home to about 175,000 people, the urban population is comprised of ethnic Russians, Ukrainians, and other Slavs; native peoples of Sakhalin (Nivkh, Evenk, Orok); and ethnic Koreans. Koreans, the last "newcomers," were brought to Sakhalin during WWII to work the coal mines. Expatriates associated with the hydrocarbon industry have arrived since 1995; and it is common to find workers from Europe, North America, and Russia s neighboring states residing semi-permanently in city hotels. The city's multiethnic history is noted in the mélange of architectural styles in the city. The few remaining traditional Japanese structures stand next to tsarist-era frontier houses (turn-of-the-20th-century wooden multifamily dwellings with rudimentary utilities), Soviet-era five-story apartment buildings, post-Soviet suburban *kottedgi* on the city's outskirts, and gleaming Western-style offices and houses occupied by expatriate executives. Previously a small urban center connected to the MIC, the large and increasing presence of foreigners in Yuzhno-Sakhalinsk is a big change for this formerly closed city. Offshore hydrocarbon sites lie further north than Yuzhno-Sakhalinsk, but smaller settlements there lack the infrastructure and political capital necessary to accommodate international companies. Hence, as the *oblast* center, Yuzhno-Sakhalinsk has become the bustling urban hub for the hydrocarbon industry. New service industries associated with hydrocarbons are changing the labor market for the city, but not for the entire island. Current spatial patterns of economic growth mimic the Soviet urban settlement pattern where large urban centers were favored. The economic boom occurring in this city provides hope for future regional economic growth, and it is a refreshing change from the Soviet period when the government had to send people to work on Sakhalin using the *propiska* system. Today, it is a destination city both for younger generations from the Russian Far East and for international migrants associated with the hydrocarbon industry, each of which seek promising jobs.

While Yuzhno-Sakhalinsk enjoys international investment and a relatively high standard of living, many residents wonder when they will see benefits from oil extraction promised to them by the regional government and multinational companies. Many fear that promises made in the mid-1990s to develop the island will remain unfulfilled. For example, infrastructure projects (*e.g.*, transportation, education, health facilities) promised in return for allowing hydrocarbon extraction have not actualized, and residents remain saddled with decrepit Soviet-era dwellings, transportation systems, and services. Gated and gleaming buildings for expatriate workers taunt neighboring buildings lacking decently operating heat, hot water, or electricity. This disparity raises questions about the strength of Yuzhno-Sakhalinsk as an emerging hub in Russia's globalizing economy, as well as the preparedness of urban and regional governance to secure even economic growth for all citizens in the post-Soviet era. Many residents of Yuzhno-Sakhalinsk struggle to understand their rapidly changing socioeconomic and cultural place in the post-Soviet era.

Norilsk: The Legacy of Heavy Industry

Planned, developed, and federally subsidized to house over 100,000 people above the Arctic Circle, Norilsk is the northernmost large city in the world (population 175,300). Temperatures in the city can reach -58°C, and snow cover lasts 70 percent of the year. Gulag laborers (prisoners in the Soviet penal system of forced labor camps) worked from the 1920s until the mid-1950s to construct many of the city's buildings, mines, and smelting facilities. Norilsk remains a closed city, ostensibly maintaining security of nationally valued metallurgic operations by restricting travel and residency of both non-resident Russians and foreigners. It is the world's largest producer of nickel and palladium and one of the world's largest producers of platinum, rhodium, copper, and cobalt.

The city was carved onto the tundra homeland of the indigenous peoples of the Taimyr Autonomous Okrug, bringing modernization and traditionalism into close spatial con-

tact; it is not surprising to see native Evenk driving caribou-drawn sleighs through the city, vividly juxtaposed with the city's modern transportation and high-rise buildings. As a Soviet creation, Norilsk's urban landscape consists of high-rise, concrete-panel *microrayons* intermixed with and surrounded by mining and metallurgical industrial facilities (see fig. 6.7). At 69° N, the city lies in the continuous permafrost (soil at or below the freezing point of water) zone. As a result of the urban heat island effect and anthropogenic climate change, the permafrost warms under the city's foundations, compromising the city's transportation networks and building foundations.

Although Norilsk's history as an "urban gulag" is legendary, the longest lasting legacy of Norilsk may be as Russia's most polluted city. One percent of global emissions of sulfur dioxide are estimated to come from Norilsk. Air, soil, and water pollution degrade the physical environment and the health of all living inhabitants. For example, in the process of ore smelting, sulfur dioxide (SO_2) is emitted. Acid rain forms and precipitates back on the city, releasing contaminants (acid, heavy metals) into the soil and water supply. The legacy of environmental damage to vegetation and waters is noticeable at the regional scale (fig 6.22); but is highlighted in the city, where the surfaces of buildings and monuments decay where acid rain eats away at the stone or cement. Some businesses propose mining Norilsk's urban soils, because the proportion of metals in them is economically viable. As in many industrial Russian cities, illnesses (*e.g.,* lung cancer, asthma,) affect the young and old alike. Despite the risks, many residents remain because of the importance of resource extraction and

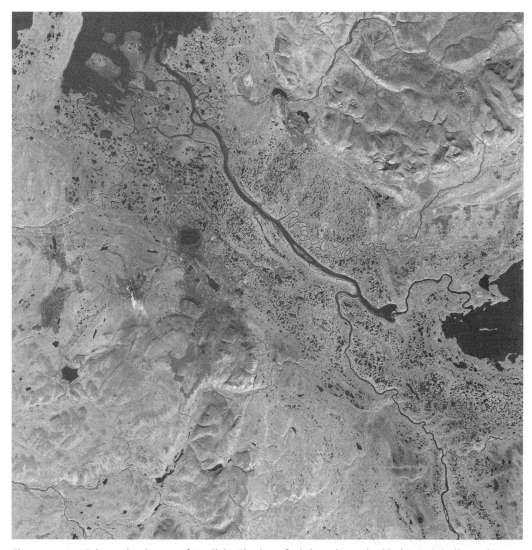

Figure 6.22   False-color image of Norilsk. Shades of pink and purple (dark gray) indicate bare ground (e.g., rock formations, cities, quarries) where vegetation is damaged from heavy pollution. Brilliant greens (light gray) show mostly healthy tundra-boreal forest. South and southwest of the city are moderately to severely damaged ecosystems; those northeast of the river and away from the city and industrial centers are healthier. *Source:* NASA.

industry in Russia, reflected in high wages (up to four times the national average), and because of strong historical roots or social networks linking people to place. Once state-owned but now in private hands, the smelting operations of Norilsk Nickel, the primary regional employer, drive the continued existence of the city. Ongoing urban pollution problems remain largely unchecked by Norilsk Nickel's corporate headquarters in Moscow (a remnant of command economic planning).

The combination of being situated physically in the environmentally sensitive far north and being economically crucial to Russia's resource industry makes addressing Norilsk's environmental issues both urgent and necessary. As awareness of Norilsk's capacity to alter regional and possible global environments (through global climate change) grows in Russia and abroad, industrial managers and the government are beginning to implement environmental management systems, which include statements of the environmental and social impact of urban industrial activity. Some industry leaders also wish to streamline Russian environmental legislation to meet international environmental safety standards, which could work to reduce regional and global pollution derived from industrial activity in Norilsk. In this way, Norilsk is poised to become a leader in addressing urban environmental concerns of the north, especially with regard to industrial pollution and global climate change.

Kazan: Volga Port in Tatarstan

Kazan, the capital of the Republic of Tatarstan, is the seventh largest city in Russia (1.1 million) and one of the largest outside of European Russia. Kazan is distinctive for many reasons, not least of which is the political struggle of the 1990s that resulted in the establishment of the republic as one of the leaders of the independence movement within the Russian Federation. Two recent events have drawn worldwide attention to this city along the Volga River. In 2000, UNESCO included the historical city center on its World Heritage List, which denotes places of universal value to the world community. And in 2005, the city celebrated its 1,000th anniversary.

While most cities in Russia are largely multiethnic, few have as large and powerful a non-Russian population as this city and republic. Tatars, whose origins are in the Central Asian steppes, make up the largest ethnic group in the republic. Russian settlement and domination of the city, however, goes back to the 16th century when Ivan the Terrible invaded the region. Even today, despite a plurality of the Tatar population in the republic, the city has a slight majority of Russians—about 50 percent relative to about 42 percent Tatar. Other non-Russian populations from the Volga region also live in the city including the Chuvash, a Turkic group, and the Maris. Migration in the 1990s brought new ethnic groups to the city, namely those from the Caucasus and Central Asia.

The Tatar population is traditionally Islamic. Kazan's history as a prominent Muslim city extends to the 14th century. Today, the city is home to a new school that trains Russians in the Islamic religion, the Islamic University. Since the early 1990s, at least 40 new mosques have been built in the city. The city government helped construct a new mosque on the grounds of the historic kremlin. The political and social significance of both the site and the leading role of the city government should be recognized as a symbol of cultural as well as political independence from Russia. The city's cultural independence from Russia is also seen in the historic forms of architecture that make up the urban built environment. Buildings in the city combine many architectural styles, ranging from Baroque to Moorish. Bas-reliefs created by traditional Tatar stone workers embellish buildings in the city, and minarets dot the skyline.

Kazan and the region hold major economic significance for the Russian economy. The city is a major port on the Volga River, the main water route through European Russia. The city and the region have been a transportation gateway for centuries. Currently, the European Union is helping to modernize its port facilities; and in 2005 it became one of only a few cities in Russia to receive a direct loan from the World Bank. The city's economy is also strongly tied to the production of transportation equipment. It produces military transport equipment, including helicopters, and is home to KamAZ, a colossal automotive production firm.

PROSPECTS

Cities in Russia today are the products of tsarist, Soviet, and post-Soviet Russian societies, as well as the many different ethnic groups and cultures that have inhabited the region for at least 1,000 years. The blending of this diverse set of cultures and histories results in cities with varied built and social landscapes. The natural environment, however, has always wielded an important influence over the location of urban settlements, irrespective of the time period or the dominant ethnic group. The harsh Siberian landscape originally posed barriers to Russian expansion and settlement. However, widespread urban settlement of Siberia became a great accomplishment of Soviet central planners—in spite of environmental degradation wrought by settlement, the immense social and cultural costs of stranding people in isolated and inhospitable places, and the immeasurable financial

implications. Cities in the post-Soviet period continue to struggle with the consequences of attempts to harness nature.

Municipal and the federal governments are attempting to integrate economically, politically, and geographically disparate cities into a larger geopolitical and economic framework, with implications for Russian transportation and communications systems. For example, should cities that are closer to Tokyo or Beijing than to Moscow rely primarily upon trade within Russia for economic direction, or should they look to Asia for new markets and influence? What will happen to cities constructed within the Soviet system but now functioning under another system? How will existing structures such as factories, housing, roads, schools, and other buildings be adapted for new uses in a market economy? How (and by whom) should pollution in urban environments be addressed? How sustainable are cities in extreme environments, such as the Arctic and Siberia? How can cities that are quite distant from one another remain connected both economically and politically? Perhaps most importantly, what will happen to the people who live and work in Russian cities? Should the Russian government promote integration into the European Union? How far down the path of destroying the Soviet housing system should Russian cities go? How best should Russian cities manage land-use change? Will the nascent increase in birth rates stem the overall population decline in Russia? If not, what will a nationally declining population mean for the regeneration of many Russian cities? Answers to these and many other questions confront the people of Russia today as they continue the process of reinventing their cities.

SUGGESTED READINGS

Axenov, Konstantin, Isolde Brade, and Evgenij Bond-archuk. 2006. *The Transformation of Urban Space in Post-Soviet Russia.* London, New York: Routledge. Focuses on the transition from socialism and communism to democracy and capitalism in urban areas of post-Soviet Russia and Eastern Europe.

Bater, James H. 1996. *Russia and the Post-Soviet Scene: A Geographical Perspective.* New York: John Wiley. Surveys the human geography of the former Soviet Union.

Chernetsky, Vitaly. 2007. *Mapping Postcommunist Cultures: Russia and Ukraine in the Context of Globalization.* Montreal: McGill-Queen's University Press. Focuses on post-Soviet cultural developments, puts them in a global context, and suggests that Russia and Ukraine form the basis of post-Soviet culture.

Dienes, Leslie. 2002. "Reflections on a geographic dichotomy: Archipelago Russia." *Eurasian Geography and Economics* 43(6): 443–458. Conceptualizes the economy of metropolitan areas as surrounded by dead space, invoking the concept of nodes and networks to describe the dichotomous development and decline of cities and the countryside in the post-Soviet era.

Feshbach, Murray. 1995. *Ecological Disaster: Cleaning Up the Hidden Legacy of the Soviet Regime: A Twentieth Century Fund Report.* New York: Twentieth Century Foundation. An early report on the underestimation of the former Soviet Union's health and environmental problems, with interesting coverage of satellite monitoring.

Figes, Orlando. *Natasha's Dance: A Cultural History of Russia.* New York: Picador Press, 2002. A survey of European Russian culture from the beginning of the Tsarist era through the Soviet era, especially focusing on the roles of multiculturalism, Europe, peasant society, and expansionism in the creation of the arts and lives of citizens in the Tsarist and Soviet empires.

French, R. Antony. 1995. *Plans, Pragmatism and People: The Legacy of Soviet Planning for Today's Cities.* London: University College London Press. Examines the assumption that cities in the Soviet Union exemplified Marxist social planning.

Hill, Fiona, and Clifford G. Gaddy. 2003. *The Siberian Curse: How Communist Planners Left Russia Out in the Cold.* Washington, D.C.: Brookings Institution Press. Traces the failed attempt to establish an industrial base in Siberia, and argues for abandoning the eastern territories because of their economic instability.

Hiro, Dilip. 2009. *Inside Central Asia: A Political and Cultural History of Uzbekistan, Turkmenistan, Kazakhstan, Kyrgyzstan, Tajikistan, Turkey, and Iran.* New York: Overlook Duckworth. Looks at the historical events shaping Central Asia, including ancient history, the Soviet period and the rise of cultural and political movements across this diverse region since 1991.

Ioffe, Grigorii, and Tatyana Nefedova. 2000. *The Environs of Russian Cities.* Lewiston, NY: Edwin Mellen Press. Looks at the outlying areas of Russian cities, and examines how urban and rural areas interact; presents case studies of Moscow and Yaroslavl through the 1990s.

Stoecker, Sally, and Louise Shelley. 2005. *Human Traffic and Transnational Crime: Eurasian and American Perspectives.* Lanham: Rowman and Littlefield. An insightful collection of social scientific writings by scholars from European, Siberian and Far Eastern Russia about the social, political and economic issues surrounding human trafficking in Russia and Ukraine.

Turnock, David. 2001. *Eastern Europe and the Former Soviet Union: Environment and Society.* New York: Oxford University Press. Analyses the transition from centralized totalitarian government within the former Soviet Union and Eastern Europe.

Utekhin, Ilya, Alice Nakhimovsky, Slava Paperno, and Nancy Ries. *Communal Living in Russia: A Virtual Museum of Soviet Everyday Life.* http://kommunalka.colgate.edu. An online ethnographic museum of the social phenomenon of Soviet communal apartments, describing social and built environment living conditions in Russia's large cities during the Soviet era.

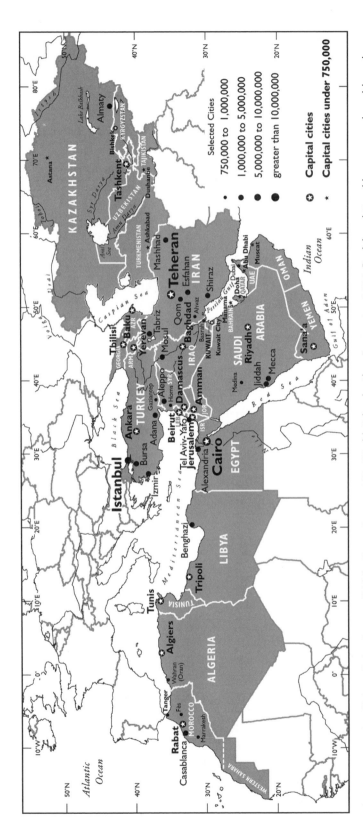

Figure 7.1   Major Cities of the Greater Middle East. *Source:* UN. *World Urbanization Prospects: 2009 Revision*, http://esa.un.org/unpd/wup/
index.htm

7

Cities of the Greater Middle East

DONALD J. ZEIGLER, DONA J. STEWART, AND AMAL K. ALI

KEY URBAN FACTS

Total Population	539 million
Percent Urban Population	60.3%
Total Urban Population	325 million
Most Urbanized Countries	Kuwait (98.4%)
	Qatar (95.8%)
	Israel (91.9%)
Least Urbanized Countries	Tajikistan (26.35)
	Yemen (31.8%)
	Kyrgyzstan (34.5%)
Annual Urban Growth Rate	2.34%
Number of Megacities	3 cities
Number of Cities of More Than 1 Million	37 cities
Three Largest Cities	Cairo, Istanbul, Teheran
World Cities	Istanbul, Cairo, Jerusalem

KEY CHAPTER THEMES

1. Urban landscapes of the Greater Middle East have been shaped by the natural environment and religion (particularly Islam, but also Judaism and Christianity).
2. The location of cities has been strongly influenced by the availability of water in the form of rivers, springs, and underground aquifers.
3. The world's first cities grew up in the Fertile Crescent, along the Nile, and on the Anatolian Plateau.
4. Traditional city cores are, or were, walled and dominated by a citadel or kasbah.
5. Urban economic geography has traditionally been shaped by the commerce that coursed across the region, a result of its relative location at a tricontinental junction.

6. Some states have a primate city, some have two or more competing large cities, and a few have fully developed urban hierarchies.

7. The "urban triangle" that defines the region's core has a foothold on all three continents and in three different culture realms (Arab, Turkic, Persian).

8. During the 20th century, oil and gas revenues have turned some of the least urbanized countries into some of the most urbanized.

9. The urban population geography of the oil-rich states has been transformed by the millions of "guest workers," particularly from South and Southeast Asia.

10. The domino effects of the Arab Spring revolutions in Egypt, Libya, Tunisia, and Bahrain are shown in the current uprisings in Syria and Yemen and in the protest movements in other Arab countries. These will change city life as cities become centers of reforms and hopes.

11. Major urban problems range from rapid population growth and unemployment to the preservation of heritage resources.

Urbanism as a way of life seems to have originated in the region we call the Middle East (fig. 7.1). Solidly anchored in this tri-continental junction are the roots of the Western city. Perhaps it is no accident that our word *urban* still carries the name of the world's first truly urban places, Ur and Uruk, in southern Mesopotamia. These early agglomerated settlements, dating from at least five millennia BC, offered protection, security, and the ability to control resources (such as water) and trade. Ideas about urban development and planning originated here and spread as byproducts of commerce and conquest.

Over the course of the 20th century, the "Middle East" as a regional appellation replaced the older "Near East." As a vernacular term, it has also expanded in geographical coverage. For the sake of brevity, the term *Middle East* is used in this chapter to refer to the "Greater Middle East," that crescent of territory stretching from Morocco, eastward across North Africa through the lands of southwest Asia, to the steppes of Kazakhstan. This great sweep of territory is united by a basic core of similar geographical characteristics and shared history (fig. 7.2). First, in its physical attributes, the Middle East is predominantly arid, though river systems and oases mitigate the region's dryness. Second, in its cultural attributes, the Middle East is marked by the shared history experienced through successive Islamic empires and their interaction with the wider world. Third, in its relative location, the Middle East really is in the "middle"—the middle of the Eastern Hemisphere, a frontier between the civilizations of Europe, eastern Asia, and Africa. In short, cities here have been designed around access to water, God (or gods), and trade.

Dry and seasonally dry, arid and semi-arid, desert, and steppe—these are descriptors of the Middle East's physical geography. Most of the region suffers a fresh water deficiency. Available water comes from winter showers, orographic precipitation, exotic streams, rivers,

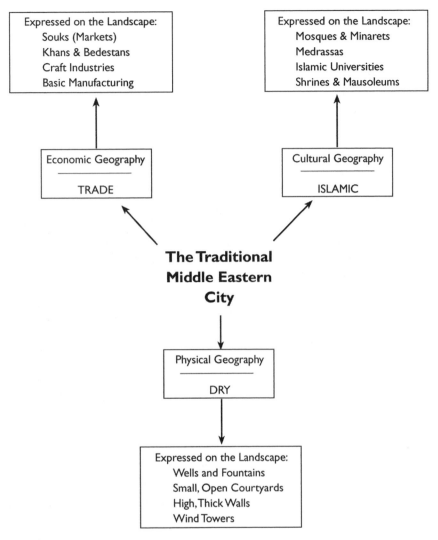

Figure 7.2 The Traditional Middle Eastern City. *Source:* Donald Zeigler

natural springs, and shallow aquifers. In arid areas, the location of water has historically determined the location of population centers and the growth of towns and cities. In turn, the dry environment presented a set of natural obstacles, which these urban centers had to overcome: scorching sun, high daytime temperatures, desert winds, dusty air, and scarcity of water. The same dry environment—the desert—offered the earliest of cities a set of natural frontiers. The surrounding desert served as a buffer, providing protection from potential invaders.

Contemporary Middle Eastern cities also share a common element of cultural geography, that of religion. This region is the birthplace of the three largest monotheistic faiths: Judaism, Christianity, and Islam. All three have left their enduring stamp on the region; but it is Islam that is most widely associated with the Middle East, having begun here in the early 7th century. Within the region, cultural geography varies from place to place—Arab, Persian, Turk, Kurd—but the cultural matrix is webbed together by the historical and current impact of Islam and the religious, social, and political systems it created. Christians and Jews (and some smaller groups like the Bahai, Druze, and Zoroastrians) are found living within the Islamic matrix; but they are the exception, not the rule. Only in a few select areas were the urban landscapes historically punctuated by anything but the minarets of mosques. In the Middle East, the city has always been a center of spiritual and intellectual life and a generator of new ideas about people's relationships with God and each other. Islam itself first developed in the urban centers of Mecca and Medina.

The relative location of the Middle East has given its cities a third set of common characteristics: they are centers of trade and commerce. Prior to the 15th -century discovery of water routes around Africa and around the world, trade between the great civilizations had to pass through the dry-land crescent separating Europe from eastern Asia and sub-Saharan Africa. Trade among these three regions—by geographical necessity—passed through the Middle East. In fact, the configuration of land and water—the interpenetration of seas (*e.g.,* the Red Sea), land bridges (*e.g.,* Anatolia and Persia), and peninsulas (*e.g.,* Arabia), suggested a multitude of routes through this dry-world barrier. Trade in food and fabrics, gold and copper, spices and perfumes, frankincense and myrrh, all helped build the cities of the Middle East. In fact, new ideas about how to create and expand wealth, perhaps even capitalism itself, were born in cities here. Their market places—called *bazaars* in the

Persian language, *pazars* in Turkish, *souks* in Arabic, and *shuks* in Hebrew—are among the oldest in the world.

In other words, cities in the Middle East have evolved around water, the house of worship, and the marketplace (fig. 7.3). Until the late 20th century and the genesis of oil economies, relative city size was, in fact, a corollary of the above: larger cities evolved in direct proportion to the availability of fresh water, the abundance of "spiritual capital," and the bounty of trade. Powerful countries were, and are, built around powerful cities. As the concentration of political power in cities increased, it gave rise to a fourth essential element of the urban landscape, the palace. It was the symbol of secular power, the rival of the temple precinct. Today, the palace has evolved into the buildings of governmental administration. Governments, in fact, have shaped cities past and present in the Middle East under regimes that were both monarchies and republics. For example, Uzbekistan and Turkmenistan were parts of the former Soviet Union until 1991. Their cities were heavily influenced by the socialist city model in which their governments intervened in housing, economic activities, and services. Also, the French colonization of most of the Arab Maghreb influenced city cultures, especially in terms of language in Morocco, Tunisia, and Algeria. This history makes *Dependency Theory* very relevant to the Middle East as most Arab countries are still connected to their former colonizers (*e.g.,* Libya and Italy, Tunisia and France). In addition, the Palestinian-Israeli conflict since 1948 has led to wars and unrest in cities of the Middle East such as Jerusalem, Ramallah, and Gaza.

It was physical geography, however, that historically commended sites for urban devel-

Figure 7.3   A street vendor does business as it must have been done for centuries, in front of a shuttered Greek Orthodox church in the Istanbul neighborhood of Karakoy. (Photo by Donald Zeigler)

opment: a hill, a natural spring or oasis, a small headland or peninsula, a harbor or river mouth, a confluence of streams, or a point able to command passage through a gap or across a river. A peninsula, for instance, that set the stage for the evolution of Istanbul, an oasis for Damascus, a hilltop for Aleppo, a spring for Tehran, and a harbor for Beirut.

While site characteristics may be responsible for the founding of a city, location relative to routes of commerce and seats of power more often determines whether a settlement grows and prospers or withers and dies. Some cities have the potential to be trading hubs or imperial capitals; others do not. Furthermore,

a good location in one age may be a bad location in another. Baghdad replaced Damascus as the seat of the Islamic caliphate in 750 AD. The opening of the Suez Canal drew trade away from the cities of the Fertile Crescent after 1869. Iskenderun (originally Alexandretta) on the Mediterranean was cut off from its natural Syrian hinterland in 1939 when the French transferred the territory to Turkey. During Lebanon's civil war (1975–1990) much of Beirut's economic activity, especially insurance and banking, relocated to Manama, Bahrain, and then Dubai. All across the Middle East, "dead cities" and archaeological *tells* (hills created as one city was built on top of its predecessors) offer examples of how changing relative location can influence the geography of growth and decline.

THE ORIGINS OF THE MIDDLE EASTERN CITY

The Middle East was home to the world's first large, agglomerated settlements. They evolved as centers of production, worship, defense, and trade. They distinguished themselves from the countryside by offering the best life had to offer, at least for powerful elites and their clients. The oldest cities in the world, though most often classified as proto-urban, were born during the Neolithic period. They are all associated with the beginnings of agriculture. Their locations form a triangle with one vertex in Iraq, one in Palestine, and one in Turkey. In lower Mesopotamia (the "land between the rivers") were the cities of Ur, Uruk, Eridu, Kish, and others. As truly urban places, they emerged in the 4th millennium BC. They were the largest cities in the world until the rise of Babylon. Only archaeological tells remain, but these forerunners of the modern Middle Eastern city set off a chain reaction in urban innovation that continues to this day. Earlier than this, however, in Palestine, on the other side of the Fertile Crescent, ancient Jericho (now in the West Bank) boasted a wall and watchtower as early as nine millennia BC. One has only to look at the natural endowments of Jericho's site to understand the impetus for prolific agriculture here. Deep within an arid rift valley, this "city of palms" is located next to a gushing spring, and not far from the *wadi* (river bed), which brings runoff from the Judean Hills into the sun-drenched Jordan River valley.

On the Anatolian plateau, just to the north of the Fertile Crescent, stands the most recent-ly discovered prototypical city, Çatal Höyük, which dates to about 6500 BC. It was located on a small river in the Konya Plain, today a rather desolate area of Turkey. It had an estimated 50,000 inhabitants, and it was probably one of the largest and most sophisticated settlements in the world in its day. Why? This was because of the successful domestication of wheat and other staples. Just as its size proved the viability of large-ness, its form proved to be a pacesetter in the development of urban landscapes. Çatal Höyük, Jericho, and Ur all illustrate the principle that civilization and urbanization evolve hand in hand.

The Iranian Plateau and the Mediterranean basin gave rise to some of the world's earliest empires. Each conquest opened a new chapter in urban history. Persian culture transformed Babylonian, Median, and Phrygian cities. Then, Greek culture, under Alexander and his successor generals, blended with various "eastern" elements to create a new Hellenistic city. Later, Roman culture transformed Phoenician trading posts and gave rise to a new set of cities in northern Africa and southwest Asia, some built on their Hellenistic predecessors. Empires built by the Persians, Greeks, and Romans succeeded because they were successfully administered (sometimes brutally) and promoted trade, not just in goods but also in ideas. Innovations in urban form and function rolled across the region's urban landscapes, converging around the institutions of government, commerce, and religion. In the Roman city, for instance, forums, basilicas, coliseums, amphitheaters, public baths, and temples provided many elements which linger in the landscapes of the Middle East today. El Djem, Tunisia, is no longer a major city, but it still has its coliseum, second in size only to Rome's. Remnants of the Temple of Jupiter guard the western entrance to Damascus' grand souk. And, Istanbul, Turkey, still has its hippodrome, a racetrack for chariots, dating from 330 AD, when Constantinople was built as a new capital for the crumbling Roman Empire.

The Roman Empire became officially Christian in the 4th century A.D. Thereafter, the followers of Jesus transformed the cultural landscape and social geography of the empire's cities. Its successor, the Byzantine Empire, became the guardian of Christianity in the eastern Mediterranean. Nevertheless, a new religion, Islam, born on the Arabian Peninsula, was to conquer the Byzantine lands of southwest Asia and North Africa. Between the 7th and 10th centuries, Islam created the unique urban landscapes we know today as "the Islamic city." It was a city of mosques, madrassahs (religious schools), and universities, a city built on free thinking and scientific progress, a city of honest trade, tolerance, and justice. Its daily routines, seasonal rhythms, architectural appearance, and governing system were all heavily influenced by Islam. The term *Islamic city* is often used to refer to the historic urban areas, originally constructed during the Islamic empires. These Islamic cities are characterized by several common elements: mosques, markets, forts, palaces, and city walls. The Ottoman city continued to be an Islamic city.

After World War I, when the Ottoman Empire was defeated, much of the Middle East came under the control of European powers. In European style, the colonizers added new sections onto the traditional city. Independence came to the region after World War II, and new governments built skyscrapers in the global style. Today, the "Middle Eastern" city often has a historic core composed of the original Islamic city, and new sectors reflecting European and global architectural influences. And yet, the essence of the distinct

Islamic matrix continues to characterize cities throughout the region.

Many countries of the Middle East have an urban tradition that transcends not just centuries but millennia: Iraq, greater Syria, Turkey, Egypt, Iran, and Uzbekistan. Yet, some countries of the region entered the 20th century without an urban tradition at all. The emirates (principalities) of the Persian Gulf (known as the Arabian Gulf in Arab lands) knew nothing of city life—until recently (box 7.1). The Arab side of the Gulf was punctuated by small fishing and pearling ports. Arabia's urban population was almost entirely *hajj*-related, with Mecca, Medina, and Jeddah surfacing at the top of the urban pyramid. The ancient cities of the frankincense trade, cities like Ubar, had been reclaimed by the desert, and the few large seaports, Aden and Musqat, served offshore interests more than their hinterlands. Today, whether a country's cities date back to antiquity or to the recent past, the Middle Eastern population tends to be decidedly urban, a response to the declining viability of nomadism, the rapidly growing numbers of people, the restricted range of arable land, the rise of prosperous fossil fuel economies, increased educational opportunities, the accessibility to international economic networks, and the political decisions of powerful elites.

CITIES AND URBAN REGIONS

The Middle East entered the 21st century as a majority-urban region. Today, over 60 percent of its population lives in cities. National statistics range widely, however. In some countries, more than 9 out of 10 inhabitants live in urban settings. The most urbanized are the small states of the Persian Gulf, the oil-rich/

Box 7.1   Monitoring the Growth of a Desert City in the UAE

M. M. Yagoub

People have found Al Ain an attractive place to settle for thousands of years due to the availability of ground water, oases (dates), low humidity, and its location at a transit point between inland areas and the Arabian Gulf. In just over thirty years Al Ain has gone from a *desert oasis* to a thriving modern city. Its population has increased from 51,000 in 1975 to an expected half a million by the year 2010. The city has received two international awards for development and landscaping, the first was from Spain in 1996 and the second from the United States in 2000.

Figure 7.4   The oasis of Al Ain in 1976. *Source:* M. M. Yagoub

Figure 7.5   The oasis of Al Ain in 2004. *Source:* M. M. Yagoub

Those awards were due in part to the use of geographic tools in monitoring and planning the growth of the city: maps, color aerial photographs, satellite images, the Global Positioning System (GPS), and Geographical Information Systems (GIS). Change and development in the city is evident from a comparison of old maps, aerial photographs, and satellite images (fig. 7.4 and fig. 7.5). Since 1976, Al Ain has exhibited a tendency for major expansion to the west and southwest, in the direction of Abu Dhabi, the UAE's capital. Expansion has followed the road network, the water pipelines, and the power transmission lines that connect the two. Similar factors lead to the expansion of Al Ain to the north, towards Dubai. Al Ain's expansion in both directions has been governed by the gravity model, the availability of utilities, and economic activities along roads. It has been shaped and limited by geographical constraints such as valleys, sand dunes, and mountains, and by legal constraints such as the boundary with Sultanate of Oman and domestic planning, and institutional ordinances. Revenue from oil has been the main driver of this development.

In the absence of accurate socioeconomic data, remote sensing can be used to chart broad dimensions of change over time in socioeconomic indicators such as population, water consumption, electricity, and solid waste. The policy of the UAE has been to encourage farming and conservation of the environment while permitting urban development. In Al Ain, geospatial research has concluded that urban development is tied to conservation of agricultural areas (oases) and reclamation of the desert. The reasons for conservation of the oases are historical, social, and recently associated with eco-tourism. Thus, while many (if not most) cities around the world expand at the expense of agriculture, the urban expansion of Al Ain has resulted in the expansion of agricultural land—77 percent increase between 1990 and 2000 according to satellite imagery.

Sources: Dona J. Stewart, 2001, New tricks with old maps: urban landscape change, GIS, and historic preservation in the less developed world, *The Professional Geographer*, 53(3), 361–373; M. M. Yagoub, 2004, Monitoring of urban growth of a desert city through remote sensing: Al-Ain (UAE) between 1976 and 2000, *International Journal of Remote Sensing*, 25 (6): 1063–1076.

rain-poor country of Libya, and Israel. Lebanon is not far behind, at 87 percent urban. The most urban countries are all among the most economically developed in the region. The least urban countries are the least economically developed: Yemen and Tajikistan have fewer than 3 out of 10 inhabitants in cities. Egypt and Syria, both with enviable agricultural resources, have large rural populations because the ability to make a living off the land holds people in the primary sector of the economy. Egypt's ratio of urban to rural, at 43 percent vs. 57 percent, is one of the most stable in the region with very little change over the past three decades. In Central Asia, the rate of urbanization has been consistently low: only Kazakhstan has a majority of its population living in cities. In the region, it is oil wealth that seems to have been the most effective stimulator of urban growth: Oman was 5 percent urban in 1980, 72 percent today; Saudi Arabia was 24 percent then and 81 per-

cent today; and Kuwait's population increased from 56 percent to almost 100 percent urban over the same time period.

Why has urbanization transformed the Middle East over the last century? From a demographic perspective two forces are at work: natural increase and migration. As death rates have declined faster than birth rates since World War II, the population size of every country has soared. Some countries in this region (*e.g.*, Syria, Iraq, Yemen, Palestine) continue to have among the highest birth rates in the world. In addition, two types of migrants have swollen the urban ranks: migrants from rural areas and, in oil-rich states and Israel, and immigrants from abroad. For instance, one-quarter of Saudi Arabia's population and 80 percent of the UAE's population is foreign-born. Israel received a million Russian Jews between the mid-1980s and the end of the century. Israel has also received major immigrant streams from Ethiopia, Argentina, France, the United Kingdom, and the United States.

Rapid city growth in the Middle East is related to developments that have transformed life in urban settings faster than life in the countryside. Technology has created jobs in cities; governments have provided services in cities; and people have greater access to health care and education in cities. Growth in the urban population may be related to political developments, as well. For instance, Amman, Jordan, a village of 2000 people in 1950 has grown into a major regional metropolis. Much of that growth has been the result of refugees moving into Jordan from neighboring Palestine in the wake of wars between Arabs and Israelis. Later, the arrival of Iraqi refugees, fleeing violence in their own country, propelled the population of Amman to beyond the 2 million mark. Refugee communities are also prominent in the cities of Syria and Lebanon.

The rapid growth of cities in the Middle East has created numerous environmental challenges. Not only does demand exceed the available water resources in many countries, but water quality is also suffering. In many places, sewage systems are inadequate or non-existent, waste and refuse is dumped into nearby canals or drainage ditches. Often these canals are used for bathing, dish and clothes washing, and their water may be used for drinking. Unregulated industrial activities, especially small-scale enterprises such as tanning, add further pollutants to water sources. Moreover, the number of cars in Middle Eastern cities has increased dramatically with growth in population and prosperity. Not only do these cars clog ancient narrow streets never designed to accommodate them, they also contribute to air pollution. Car exhaust and industrial effluent, containing lead and other heavy metals, pose a health hazard to urban dwellers.

The core of the Middle East is defined by a decagon formed by five seas (Mediterranean, Black, Caspian, Persian Gulf, Red Sea) and five land bridges (Anatolia, Caucasus, Iran, Arabia, Suez). On the perimeter of this core, lie three of the world's twenty most populous metropolises. They anchor an intercontinental, international, and intercultural urban triangle (fig. 7.6):

- Cairo, Egypt, on the continent of Africa, has 16 million people. It is the largest city in the Arab region.
- Istanbul, Turkey, on the continent of Europe, has 13 million people. It is the largest in the Turkic realm.
- Tehran, Iran, on the continent of Asia, has 13 million people. It is the largest in the Persian realm.

Yet, in 1950, only one city in the entire Middle East, Cairo, had more than a million inhabitants. In 1900, there were none. Today, there are 43 cities over one million inhabitants, eight in Iran, six in Turkey, and four in Saudi Arabia. Except for their historical centers, all are products of the late 20th century—they are new cities, not old. Virtually all of their growth has post-dated World War II. Yet, in the countries which anchor the Middle East's triangular urban core, all have a less developed frontier side, too. Eastern Iran, eastern Turkey, and most of upper (southern) Egypt remind us that urbanization has not entirely transformed even the most urbanized countries in the region.

Beyond the Cairo-Istanbul-Tehran triangle and its immediate environs, there are important regional metropolises, all of which fall

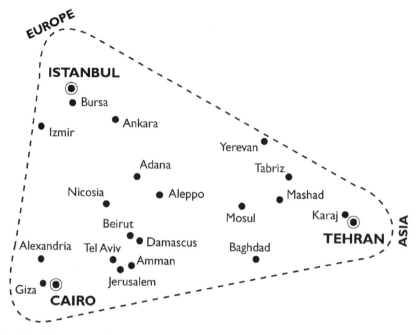

Figure 7.6   The Urban Triangle of the Middle East shows the relative locations of major cities. These cities are in their correct geographical locations, but shown without the base map underneath. *Source:* Donald Zeigler

into the one-million to two-million population range. Only two exceed two million inhabitants: Casablanca, Morocco, and Tashkent, Uzbekistan. One anchors the far west, called Al-Maghreb by Arab geographers. The other anchors Central Asia or, in its former incarnation, Turkestan.

Some states of the region are anchored by a single primate city (*e.g.,* many Arab states on the Gulf and all of Central Asia); others have two rival urban cores, a condition which has the potential to set into motion strong centrifugal forces. Perhaps the best example of this is Syria. With over two million inhabitants each, Aleppo (now the larger) and Damascus are quiet rivals. Other countries with dual anchors and inter-city rivalries include: Yemen (Sana'a and Aden), mostly as a result of its colonial history; Libya (Tripoli

and Benghazi), as it was cobbled together by the Italians; Israel (Tel Aviv and Jerusalem), split by its secular and religious axes; and Kazakhstan (Almaty and Karaganda), with its Turkic and Slavic flanks. Iraq, on the other hand, has one clearly dominant city, Baghdad; but two smaller anchors, Mosul in the north and Basra in the south, have such different cultural geographies that Iraq effectively is a country divided.

In between these two extremes are the states with complex urban hierarchies. Turkey and Iran are both punctuated by booming metropolises, regional centers, small towns, and villages. Morocco, likewise, is a country of cities large and small, many of which have assumed a highly specialized role in the Moroccan urban system. Morocco has a political capital, Rabat, but three other

Figure 7.7   Here in Bursa, a "young Turk" takes over the family kiosk while his father chats with some customers outside. (Photo by Donald Zeigler)

cities—Marrakesh, Fès, and Meknes—have historically served in that role. In addition, Morocco has an unofficial economic capital, Casablanca. Its former diplomatic capital (as far as ambassadors were permitted to venture into the country), Tangier, now serves as a bridgehead to Europe. A city in the south, Ourzazate, has even surfaced as one of the movie-making capitals of the world, starring in films such as *Lawrence of Arabia* and *Star Wars*, and providing the venue for one episode of *The Amazing Race*.

Not only have cities in the region increased in population and territorial extent, they have also begun to coalesce in a fashion reminiscent of Jean Gottmann's Megalopolis. In these urban mega-regions, life is at its most intense, a characteristic symbolized by the pace and volume of traffic along connect-ing thoroughfares. There are perhaps seven megalopolises developing on the 21st century map of the Middle East. The first four are rather well defined linear regions. The most populous stretches from Cairo across the Nile to Giza then to Alexandria. Now that the superhighway has been completed across the desert, development is even more rapidly filling in this 100 mi (161 km) corridor. The second is the staple that holds together the continents of Europe and Asia. It is Turkey's Marmara megalopolis, anchored by Istanbul on the European side and Bursa, an old Silk Road city in Asia Minor (fig. 7.7). It virtually engulfs the entirety of the Sea of Marmara, whose ports attract industry and whose vistas attract residential and recreational developments. Third is coastal Morocco from Casablanca to Rabat, only about 50 mi (80 km) apart. Fourth is the Israeli megalopolis, which begins in Haifa and Akko, stretches along the coast to Tel Aviv-Jaffa, and then up to Jerusalem. Although punctuated by some Arab enclaves, this crescent is essentially a belt of intense Jewish settlement, tied together by highways and expanding settlement.

The next three megalopolises are best understood as nests of coalescing cities. The Mid-Mesopotamian megalopolis is anchored by Baghdad but stretches between the Tigris and the Euphrates Rivers. The Tehran-Karaj megalopolis stretches along the base of the Zagros Mountains. The only international megalopolis is the one stretching from coastal Saudi Arabia, beginning with the oil-engorged cities of Dhahran (headquarters of Saudi Aramco, the world's largest oil company) and Dammam, and then across the four-lane causeway to Bahrain and its capital, Manama.

Capital Cities of the Middle East

In the Greater Middle East today, there are 27 independent states, hence, 27 national capitals. Indeed the very concept of a capital city originates with the early states and empires that grew up in the river valleys of Egypt and Mesopotamia. Memphis, in Egypt, was one of the world's first real capitals. Its location served as a hinge between Lower Egypt (the delta) and Upper Egypt (the river valley). It was able to command the entire Nile from the Mediterranean to the first cataract (rapids). A later Egyptian capital, Thebes (today's small city of Luxor) served pharaonic Egypt, with only brief interruptions, for over 1500 years, beginning in the 21st century B.C., one of the longest runs in history. Its monumental architecture affirmed one of the rules of capital city development—nothing is too large, too beau-

tiful, or too expensive for the head city of a proud and prosperous state. The power of the sovereign is expressed in powerful landscapes of control, where respect for and obedience to the ruler, or the law, is demanded by the symbolism of the built environment. Thebes is a case study in itself. It is still intimidating and inspiring. The feeling of grandeur, even today, is not unlike the 21st century's world capitals—Washington, London, Moscow, and others.

The Middle East may have the largest collection of historical capitals in the entire world. Some, such as Memphis, Carthage (Tunisia), and Persepolis (Iran), are well known; others are now but evocative ruins of the geopolitical past—Sardis, Susa, Ashur, Tuspa. The history of the Middle East is the history of major states and empires, amoeba-like in their evolution, and all anchored by capital cities. The 19th and 20th centuries, however, presented a major break with the past. These were eras of extraterritorial rule when the capital cities of the Middle East, the decision-making centers of the region, became Paris, London, Moscow, and Rome. Even earlier, Ottoman Istanbul became an extra-territorial ruler of the Arab lands of Southwest Asia and most of North Africa. With the collapse of the Ottoman Empire, the rule of Europe became even more widespread. Only a few states–Saudi Arabia, Oman, much of Yemen, modern Turkey, and, arguably, Iran–avoided colonial domination or protectorate status.

With independence, which came to most states of the region after World War II, a new set of capitals was born. For some states, the choice was simple. Cairo was already serving as the national capital of Egypt. In Syria, Lebanon, and Jordan, the administrative centers that served French and British interests,

Figure 7.8   Inner-city apartments with balconies are part of the residential landscape of Almaty, former capital of Kazakhstan. (Photo by Stanley Brunn)

Damascus, Beirut, and Amman, respectively, became the national capitals. Libya had to decide between two capitals, Tripoli and Benghazi, one in command of Libya's west and the other of Libya's east. Only with the creation of a highly centralized, unitary state, was Tripoli, the larger of the two, designated as the sole capital. Turkey, born out of the Ottoman Empire after World War I, had an obvious choice of cities for its capital: Constantinople, now called Istanbul. It had served as head of an empire for 1600 years under Romans, Byzantines, and Ottomans. In 1923, however, Atatürk, as one of his first acts as President, made a not-so-obvious choice: Ankara, a city of the interior. Why make such a dramatic change? To symbolize the death of an old state, the Ottoman Empire, and the birth of a new one, the Republic of Turkey. Turkey was not to be the Ottoman state with a new name, it was to be a new creation rooted in the civilizations of Anatolia, the plateau of Asia Minor. By choosing Ankara, a city of some antiquity,

the modern state was able to emphasize its roots in a pre-Ottoman past. Except for the architecture of its mosques, Ankara bears almost no trace of the Ottoman Period.

Kazakhstan provides the most recent case study of a country which has chosen to relocate its capital. As a modern independent state, it began life in 1991 with Alma Ata, renamed Almaty, as the center of government. Alma Ata, the country's most populous city by far, had served as capital during the period of Soviet rule when Kazakhstan was a Union Republic (fig. 7.8). Within a few years, however, the President of Kazakhstan made known his intention to move the seat of government to Akmola, which was eventually renamed Astana, a word which means "capital" in the Kazakh language. Real and symbolic factors were at work in the decision. Almaty was not centrally located. To boot, it was perilously close to the Chinese border. The choice of Astana solved both of these problems. Moreover, Astana is five meridians closer to Europe, and Kazakhstan has taken pride in calling itself a "Euro-Asian" nation, overlapping the continental boundary as it does. Outside the city center, however, the urban form remains similar to other post-Soviet cities (box 1.3). Yet, while flirting with the European connection, Kazakhstan draws its national identity from the steppes, the grasslands of Central Asia. Astana, the new capital, presides over those steppes, whereas the previous capital was in the mountainous foothills of Kazakhstan's east.

The urban geographer Jean Gottmann compares a capital city to a hinge, and the choice of Astana fits that analogy almost perfectly. First, the new capital serves as a hinge between the pre-Soviet past and the nation-building present. Second, it serves as a hinge (or hopes to) between Asia and Europe. Third, it serves as a hinge between the two sociocultural parts of the nation, the Russified north, inhabited mostly by transmigrant Slavs, and the Turkic south, inhabited by Kazakhs and their Uzbek populations. By moving the capital city onto the Slavic frontier, Kazakhstan proclaimed its intention to maintain territorial integrity and to continence no irredentist claims by Russia.

The Middle East also boasts a set of very specialized capitals that no other part of the world can equal: the "head cities" of the three great monotheistic faiths. Mecca, in Saudi Arabia, is the religious capital of the Islamic world, a pilgrimage destination for two million *hajjis* (pilgrims) annually, but a destination reserved strictly for Muslims. Jeddah serves as the gateway to Mecca for those arriving by boat and by air. Those who make the hajj visit Islam's second holiest city as well, Al-Madinah Al-Munawarah (the "Illuminated City") or Medinat al-Nabi, the "City of The Prophet," commonly called just

Medina. It was to this city that Mohammed fled in 622 AD, marking the beginning of the Muslim calendar. For a short time under the first caliphs, Medina served as the capital of the expanding Islamic empire. The third holiest city to Muslims is Al-Quds ("the holy") in Palestine—it is known to the rest of the world as Jerusalem. These three cities are important to the entire Islamic world, but Shiah Islam has an additional pair of holy cities, which are also pilgrimage sites: Najaf and Karbala in southern Iraq. The "capital" of Islam's predecessors, Judaism and Christianity, is also in the Middle East (fig. 7.9). The ancient walled city of Jerusalem is the holiest in the world to both Jews and Christians, and a focus of world pilgrimage for both. Medieval European T-in-O maps, in fact, showed Jerusalem to be at the center of the world, exactly where it remains today, but only in a religious sense.

Port Cities of the Middle East

Port cities are gateways to the world. Egypt's major port of Alexandria was founded by Alexander the Great in 331 BC on a natural harbor created by an offshore island. Its hinterland comprises the most fertile and productive land in the Mediterranean basin, the delta and valley of the Nile River. Its earliest maritime links connected the port to the entire Hellenistic world, particularly the Greek Aegean. Alexandria hinged together Greek and Egyptian civilizations. The symbol of its commercial prosperity was its lighthouse, "the Pharos," one of the seven wonders of the ancient world. The symbol of its intellectual prowess, stimulated by the cultural encounter, was the library of Alexandria, the greatest of antiquity. Neither has survived, but the lighthouse remains a symbol of port

Figure 7.9   Landscapes of three faiths are seen from the rooftops of old Jerusalem. These young Orthodox Jews are silhouetted by Islam's Dome of the Rock, and in the background is what Christians know as the Mount of Olives. (Photo by Donald Zeigler)

cities everywhere, and in 2002 Egypt unveiled its magnificent new library, the Bibliotheca Alexandrina, as an attempt to revive the city's standing in the intellectual world.

Just as some of the world's oldest capital cities are located in the Middle East, so are some of the world's oldest port cities: Tyre, Sidon, Acre, Jaffa, Ashquelon, and Gaza on the eastern Mediterranean coast, for instance. Older ports, however, just cannot meet the needs of modern countries. If a port was mentioned in the Bible, it is almost surely inadequate for today's commerce. Acre has been superseded by nearby Haifa, and Jaffa by neighboring Tel Aviv. Beirut and Alexandria have built new ports beyond their natural harbors. Antioch (modern Antakya, Turkey), the premier city of the eastern Mediterranean in Roman times, is a port no more. Gaza, which is now challenged to become a modern port for an emergent Palestine, illustrates an important rule of port geography: every country (unless landlocked) must have a seaport of its own. Where good natural harbors are not available for port development, artificial harbors must be built. The configuration of international boundaries, virtually all products of extraterritorial decision making, heavily influenced the port geography of the Middle East. Syria (historically the entire western Fertile Crescent), for instance, was cut off from its natural port cities, historic Antioch, Turkey, and modern Tripoli and Beirut, Lebanon, by European-drawn boundaries. The result has been a substantial investment in enlarging the small natural harbor at Tartous and building a large new port at Latakia.

Likewise, the impact of international boundaries on port geography is evident at the head of two arms of the sea: on the Persian Gulf where Iraq, Iran, and Kuwait converge; and on the Gulf of Aqaba where Jordan, Israel, Egypt, and Saudi Arabia converge. In both places one port, operating at economies of scale, would make more sense, but political boundaries dictate otherwise. Ditto, the UAE, the only federation in the Middle East. Here, there is lively competition among the seven emirates to capture a share of maritime trade and in-transit manufacturing. Consequently, a country of eight million people has sixteen commercial seaports, the most important of which are in the emirate of Dubai.

In the Middle East, seaports are not the only port cities. Important historically have been "ports" on the edge of the desert, cities that have dispatched caravans to distant places and opened their gates to merchant traders arriving from afar. The seas traversed by the caravan trade were desert seas—"seas of sand" (even though most are not sandy). Sijilmassa, in southern Morocco, now in ruins and only recently excavated, was one such desert port. It linked the Atlas oases with the Sahel—the southern "shore" of the Sahara—before coastal shipping proved more economical. Sijilmassa served the same function as a seaport, and so did its southern counterpart, Timbuktu. Likewise, Damascus grew up around an oasis

on the edge of the Syrian Desert. It attracted caravans coming from Mesopotamia. In central Asia, Samarkand, at the foot of the Tien Shan, was also such a desert port. Thus, on the edges of the Sahara, the Syrian Desert, and the Karakum, ports developed to act as hinges between civilizations.

MODELS OF THE URBAN LANDSCAPE

At the heart of every traditional Islamic city was a fortress. It may be called the citadel, *al-qalat*, or, in the Maghreb, the *kasbah* (fig. 7.10). It usually covered only a few acres and occupied the most defensible site, often on a hilltop. Typically surrounded by a wall, it might also have been protected by water. In the past, it would have served as the administrative heart of the city, the site of the palace. Today, the citadel is most likely to be a preserve of history, valuable as a visual reminder of the past, important in building national identity, and part of the historic core within the modern city. Its landscape usually exudes a powerful spirit of place. It may or may not continue to be inhabited.

Surrounding the citadel is the old city itself. It, too, was usually walled, and its often-ornate gates gave access to the world beyond. In the Maghreb, the old Islamic city is called the *medina* (Arabic for "city"). It is the city of antiquity, at least as it has survived conquest, disasters, and well-meaning modernization attempts. For most people in the world, its landscape—compact, congested, cellular, and fortified—provides the stereotype for the region's cities. Walls, along with their watchtowers, until the 20th century differentiated quite sharply between city and country. Outside the walls were olive groves, grazing lands,

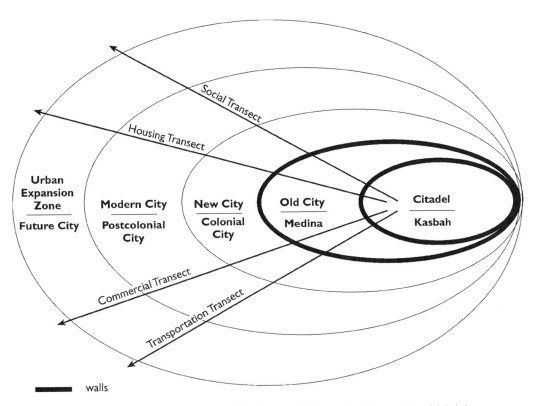

Figure 7.10   Internal Structure of the Middle Eastern Metropolis. *Source:* Donald Zeigler

cemeteries, quarries, and periodic markets, not to mention potential enemies (fig. 7.11). At most, the medinas of modern Middle Eastern cities contain no more than 3 percent of the urban population, and residents of modern neighborhoods rarely patronize the old city cores.

Beginning with the colonial era (British, French, Italian, or Russian/Soviet), a new city typically developed outside the old city walls. In the Maghreb, it was called, in fact, *la nouvelle ville* ("the new city"). It was a city between two worlds, traditional and modern. The traditional elements of urban form, from mosques to bakeries to public baths, were incorporated into the new city, but they were supplemented by modern ameni-

ties and architectural styles, including larger stores and hotels, wider boulevards and traffic circles, European-style churches, new government buildings, and corporate offices, all plastered with the language of the colonizers (fig. 7.12).

Gradually, the new city came to be enveloped by the modern, post-colonial city. Courtyard homes all but disappeared from the landscape, their place taken by apartment blocks, with high-income flats and single-family units becoming increasingly common. The postcolonial city became the zone of international hotels, corporate headquarters, and modern universities. It also became the zone of squatter settlements (which can invade even the inner zones if there is any unoccupied space)

Figure 7.11   In the old part of the Middle Eastern city was the citadel, or old fort, as seen in this remnant in the United Arab Emirates. (Photo by Donald Zeigler)

where recent in-migrants from the villages find lodging while they work their way up the urban social pyramid. In some countries, such as Morocco, these squatter settlements remain poor, under-served by the urban infrastructure, and socially marginalized. In other countries, such as Turkey, squatter settlements are quickly legitimized by the government, and the process of incumbent upgrading begins.

Beyond the modern city is the urban expansion zone. Here, small villages find themselves undergoing urbanization *in situ*. Here, also, may be new industrial estates, modest new housing tracts, or, in the case of Egypt and Saudi Arabia, new cities, built from scratch.

For the richest countries, those able to afford automobiles and the gasoline to fuel them, it can also be a zone of raging urban sprawl. Most often, the international airport is located here as well.

The Middle Eastern city today, therefore, can be seen as an interlocking set of concentric zones, patterned in time: citadel, old city (Islamic), new city (European), modern city, and urban expansion zone (box 7.2 and fig. 7.13). As one moves from the innermost zone to the outer-most, changes in urban form and function are evident. What do transects through these zones reveal?

- A transect along the social axis: The old city is becoming increasingly marginalized by society, particularly as the well-to-do move out. The modern city is attracting the lion's share of new neighborhood investment and the best of social services. Even tourists typically stay in the modern city and depend on air conditioned buses to drop them off at a city gate for a brief sojourn into the past.
- A transect along the housing axis: The old city (except in Turkey and Yemen) is a zone of traditional two or three story courtyard houses. The modern city is a zone of mid-rise and high-rise apartments. In fact, the multi-story apartment block is now the most typical component of the Middle Eastern city's residential landscape (and a reminder of how architecture has pulled away from the physical environment and its indigenous roots).
- A transect along the commercial axis: The old city still displays fully functional souks, traditional industries, and small-

Figure 7.12   To ensure freshness, bread is baked where the people live. These young men are picking up bread at the bakery for delivery to stores and hotels in the Marrakesh medina. (Photo by Donald Zeigler)

scale, family-owned artisanal enterprises. The modern city displays its logo-laden landscape of chain stores, international (and national) franchises, and ever-more snippets of English signage (fig. 7.14).

- A transect along the transportation axis: The old city reveals narrow streets clogged with pedestrians, taxis, and even donkey carts. The new city reveals the proliferation of privately owned automobiles and the amenities, like gas stations and parking spaces, they require.

These transects of Middle Eastern cities virtually recount the history of cities everywhere. Middle Eastern cities are the places where the 21st century BC meets the 21st century AD.

It is not the periphery that defines the personality of a city, however. It is the center, the old city, the city of history. The pre-industrial core also offers a glimpse of the prototypical Islamic City, for Middle Eastern cities seem to be held together by their mosques and minarets. Therefore, in the Islamic city, no one could live or work beyond earshot of a minaret. Without amplification as an aid to the human voice, this meant the network of mosques had to be very dense, indeed. In the Islamic city, the skyline is punctuated by minarets, occasionally joined by church spires in the Christian quarters. The so-called Friday mosques stand out as the city's architectural masterpieces. The religious landscape of the Middle Eastern city is typically supplemented by Koranic schools, traditional universities, shrines, and mausoleums.

Trade dominated the region long before Muhammad, himself a merchant, revealed a

Box 7.2   The Middle Eastern City Model: Exploring Mecca Using Google Earth

Zia Salim

While visiting cities around the world is an exciting and educational experience, not every-one has the time or the means to travel to the dozens of interesting places discussed in this book. Using free applications such as Google Earth and similar programs to explore some aspects of the broad patterns of the form and morphology of cities can be extremely useful. In this example, we will use Google Earth imagery to examine the zones of the Middle Eastern city model. Of the four transects, two (transportation and housing) are easily visible in the imagery. In addition to the factors of distance and time, Google Earth is useful for another reason—access to Mecca is limited only to Muslims because it is the holiest site in Islam for prayer and pilgrimage, the *Masjid al Haram* or Grand Mosque, is located there (fig. 7.13). It should be noted that all of the images in this box are taken from the same "height" or "eye altitude" (4000 feet [1219m], in this case) to facilitate comparison.

In the Middle Eastern city model, the *medina* or old city surrounds the citadel. In Mecca, the core is dominated by the mosque instead of a fortress. Parts of the old city can still be seen in the area near the mosque, but the expansion of the mosque and associated de-velopment have greatly reduced the imprint of the old city. This area of the city is tightly organized into residential zones or quarters with small, densely packed houses. It has nar-row, winding lanes, and is organically laid out and not comprehensively planned. Use Google Earth to zoom out just a bit from the Grand Mosque and you will get a bird's eye perspective of the old city. The rest of the city has grown to a point that the remaining old city area is very small in comparison. Some neighborhoods have had wide streets cut through them and have been gradually modernized. The fringes of this zone are developed, with newer and taller buildings, and the little of the old city that survives clings to the less accessible mountain slopes, safe for now from the bulldozers and cranes.

The next part of the Middle Eastern city model reflected in Mecca is that of the colonial city. While (non-Muslim) Europeans had no colonial presence in Mecca, the (Muslim) Ot-toman Empire colonized part of the Arabian Peninsula. In Mecca, only small and isolated structures, such as a solitary complex of government offices, remain to bear testament to the Ottoman presence. Finding these structures using Google Earth is difficult since it requires a fairly detailed knowledge of the city; in addition, the individual structures are small and not necessarily distinguishable using Google Earth.

In comparison, the next zone, that of the modern post-colonial city, is easily distinguish-able using Google Earth. It represents the everyday Mecca that nobody hears about—this is where the city's million-plus residents live, work, and play. The modern city has sprawled into the valleys away from the core mosque area in all directions. Following the Middle East-

Figure 7.13   The top photograph shows the Grand Mosque at the center of Mecca; the bottom photograph shows the urban expansion zone on Mecca's periphery. *Source:* Google Earth

ern city model, this large zone has single-family homes, apartment blocks, schools, small parks, and offices and commercial buildings. See if you can find the planned layout of wide main streets and narrower side streets. They are clearly visible from Google Earth's vantage point. A more regular street pattern can sometimes be seen. The modern city's urban form is a mix of single-family, two- to three-floor, walled villas, and mid-rise apartment buildings. Again, the larger size and more planned layout of the houses can be seen in the image. The Saudi government has established a Real Estate Development Fund that provides long-term interest-free loans to private sector builders. As a result of this financing model, most of the buildings in the residential neighborhoods are built by families and individuals tract houses do not exist in the modern city.

The final zone in the Middle Eastern city model is that of the urban expansion zone, which in the case of countries like Saudi Arabia, contains new housing tracts. Numerous examples of large areas of government-built master-planned tract housing exist on Mecca's outskirts. The planned nature of the street layout and housing can be clearly seen at the bottom of fig. 7.13. Spaces for neighborhood mosques and parks and open space have also been included in the design. This zone is the newest and smallest zone in Mecca.

The case study of Mecca thus provides a useful example of the Middle Eastern city model. Although models are generalizations, they offer very functional ways of examining urban structure and form. The use of Google Earth or other imagery sources illustrates how cities, even ones as unique as Mecca, can reflect broad regional trends. This kind of imagery enables the study of patterns of urban form and morphology and provides insight into urban function. Actual personal experience in the locations being studied is a time-honored method of studying cities; but the use of newer tools and technology such as Google Earth can be very informative, as well as an effective way to reduce the effects of time, distance, cost, and access in studying cities around the world.

new religion, so it is little wonder that *souks* (markets) are another typical part of the traditional city landscape. Often, souks envelop the neighborhood mosques. Guided by the religious duty of zakat (alms), the surrounding businesses and residents help support the mosques and the social services they provide. The central souks usually evolved around the city's "grand mosque." These souks are specialized affairs: shoes in one area, produce in another, fast-food (traditional style) available throughout. Merchants dealing in the same product compete with each other, and the concept of fixed prices does not exist. Shoppers know the techniques of hard bargaining and expect low prices. In the past, the souks were supplied by two institu-

tions: (1) the city's artisans who made pottery, jewelry, carpets, and everything else, and (2) the city's *khans* (or caravanserais) which served as wholesale centers for goods coming in from distant lands (originally via camel caravans). The khans were essentially inns for merchant traders and their pack animals; many now stand abandoned or have been turned into historical landmarks. In today's city, specialized outdoor souks (*e.g.*, vegetables, fruits, fish, and clothes) spread across central areas and traditional neighborhoods. However, mass produced goods are as common as locally produced products. In other realms of the traditional (albeit disappearing) city: bread tends to be purchased daily at the bakery itself, most public baths (hammams)

Figure 7.14   The landscape of Amman, Jordan, shows the signs of global commercialization in the form of this bilingual advertisement for Subway. (Photo by Donald Zeigler)

have become historical sites since houses are served by water pipes, modern beauty salons are located in almost every neighborhood, coffee houses take the place of bars for men (no alcohol permitted by the Koran), and private life takes place in very private places, like the home (box 7.3).

In fact, the Islamic city is set up to assure privacy, particularly for women and families. Houses are not recognizable as discrete entities; they appear as doors (often brightly decorated) in thick earthen walls (whitewashed at most). Along a typically narrow street (often a dead end), doors are staggered and never give casual passers-by a view into the house; windows are well above eye-level; and second-story windows are enclosed by a wooden lattice so one can see out but not in. The lattice also wards off the sun and ushers in the breeze. The privacy of the house is carried into the street by a veil or simple head scarf. The traditional house is built around an open-air courtyard with covered, roomy bays on each side. Rarely does anyone but family make it deeper into

the home. Courtyards tend to be small for the same reasons that streets tend to be narrow: to provide shade and to keep distances walkable. The middle of the courtyard is often anchored by a well for water or a fountain to humidify and cool the air. In houses of the Persian Gulf and Iran, wind towers were added for the same reason: they channel air downward creating a breeze. Along the edges of the courtyard, vines and arbors provided shade and produced fruit as a bonus. Mosques, khans, and palaces also included courtyards, making "cellular" one of the words most often used to describe the traditional Middle Eastern city. Supplementing courtyards are flat rooftops (little rain), often enclosed by low walls over which carpets are flung for cleaning. Rising from the rooftops are laundry lines, cisterns, antennas, satellite dishes, and solar water heaters. In the summer, roofs may be cluttered with mattress pads, for use at night, and ubiquitous plastic chairs.

The most coveted space in the pre-industrial Islamic city was at the center, close to the seats of economic power and social interaction.

Box 7.3   Home Space in Tehran

Farhang Rouhani

Since 1979, the Islamic Republic of Iran's efforts to supress modernization along Western lines have imposed strict policies on Iranian citizens. In urban centers, most notably Tehran, these include the state policing of public spaces such as parks and commercial streets. What is particularly striking, however, is how these politics invade even the most private of spaces: the home.

Home spaces in Tehran were significantly rearranged over the course of the 20th century. The traditional home was divided into separate male (*birun*, outer, public) and female (*andarun*, inner, private) sections. A lack of street-facing windows accentuated the sense of privacy; indeed, internal courtyards served as the home's central focus. Modernization under the Pahlavi Shahs (1920s to 1979) and the demands of rapid population growth ushered in western high-rise apartment-style living, without the physically gendered division of space, and more public street-facing windows. The importance placed on familial privacy, though, has been maintained in different ways, including the prevalence of walls around the complexes and buzzers to screen visitors.

On top of this urban transformation, the politics of satellite television viewing has brought the middle-class Tehran household to the forefront of Iranian state politics. The clerical government first instituted a ban on the sale, import, and use of satellite dishes in 1994 because of the proclaimed polluting effect of Western media products on Iranian society and particularly upon the youth.

Under Iran's more liberal-democratic government, the late 1990s witnessed a relaxing of the ban, including the requirement that police have a warrant for home searches. The current, more conservative-theocratic government, however, has recommended the policing of public and private morality. While the ban on satellite dishes still exists and periodic raids to confiscate them are made by the morals police, a majority of the population of Tehran has access to the air waves, whether directly in their homes or through family and friends. Ultimately, police regulation and infiltration of people's homes have intensified fears in an already politically insecure society.

The practice of policing the home has led to dynamic conflicts within the Iranian state and society over the role of privacy in a democracy, the importance of the home as a space of refuge, and the moral and social effects of the global media. It is within this context, that the infiltration of American television shows such as *The Simpsons* and, *American Idol* is viewed as having transformed the living room furniture into a realm of resistance.

Figure 7.15   In the classic medina, undesirable activities such as leather tanning were relegated to the margins of the city, as seen here in Fès, Morocco. (Photo by Donald Zeigler)

Palaces and merchants' houses were central elements of the urban landscape, so were the central souks. Noxious enterprises like tanneries were relegated to the periphery. (fig. 7.15) The residential population of cities, in the traditional city model, tended to be concentrated in "quarters": sections for Jews, for Europeans, for different Christian sects, different ethnic groups, and people of different village or regional origin. In the past, many of these quarters had gates of their own. Residential space was highly segregated, yet it also mimicked the security and social cohesiveness of the village by providing a scale of life to

which people were accustomed and a set of community institutions which people of like mind could control. Within each quarter lived both rich and poor.

As they have survived, the narrow streets of old residential and commercial quarters tend to be penetrable only by foot traffic and donkey carts. Still, this does not stop the occasional taxi or service vehicle from squeezing through. The cramped feeling is heightened by second and third stories jutting out overhead, sometimes joining and creating tunnels. The largest streets lead into the city from gates in the walls, but a large public square dominating the center of town is a rarity (except in Iran). A grid pattern manifests itself (sort of) in towns with a Roman or Byzantine heritage, and it is common for elements of the pre-Islamic landscape to become parts of the working city, usually unrecognized for their historic value. Roman pavements may lie deep underneath the contemporary street, and what was a wide Byzantine thoroughfare may now be subdivided into three or four narrow, parallel alleys, each serving as a souk of its own. Periodically spaced along the most well-traversed routes, are richly decorated drinking fountains (or their remnants), often given to the public by wealthy merchants. Otherwise, water vendors (dressed like neon signs) wander around high-traffic areas selling drinks to the thirsty (fig. 7.16).

The preceding generalizations apply to the old cities of the Middle East. An entirely new set of cities has grown up on the 20th century's oil and natural gas fields. They are best exemplified by the cities of Kuwait, Saudi Arabia, Bahrain, Qatar, and the UAE (though oil revenues have transformed cities in Iran, Iraq, Libya, and Algeria, as well). Urban landscapes from Kuwait to Dubai have been built

Figure 7.16   Water is precious in the dry environments of most Middle Eastern cities. From these public water urns in Cairo, everyone drinks from the same cup. Note the political ads in the background. (Photo by Donald Zeigler)

on revenues from the world's largest oil fields. While lacking the historical depth of a city like Aleppo, they do offer a glimpse of what a modern Middle Eastern city can be. These are cities built on petrodollars, largely since the 1960s. Generally, they consist of two zones patterned in time. First, in the center of each is a tiny core anchored by a fort and perhaps the remnants of a traditional wharf area or well-to-do neighborhood. Although small, these cores are the source of local identity, particularly in layout and architectural styles. Second, around each core is a post-industrial city of high-rise office buildings, shopping malls, gardens and golf courses, apartment complexes, sprawling suburbs, and mosques. With oil wealth, the world's most creative architects are not out of range, and the cities they have built blend modern structures with traditional themes (box 7.4). One of the surprising elements of these new urban landscapes is how green they are. Turning

itself into a Garden City, in fact, has been one of Dubai's urban planning objectives. Figuratively, oil is turned into water, and water into green space. Oil is also turned into new human geographies. Petro-economies have, by governmental design as well as economic magnetism, virtually eliminated nomadism as a way of life. The rural population has become urbanized. In addition, the faces of the oil-engorged boomtowns have also changed. The economic magnetism of Kuwait, Dammam, Abu Dhabi, and others, draws unskilled workers from as far away as India, Pakistan, Sri Lanka, and the Philippines, and skilled workers from Europe and the United States, as well as other parts of the Arab world.

What is missing from the landscapes of Arab cities on the Gulf is the industrial era. Manufacturing is limited to local craft industries and manufacturing-in-transit at the free ports of the region, most notably those in the Emirate of Dubai. As banking and trading

---

Box 7.4   Cities of Spectacle: Abu Dhabi and Dubai

Dona Stewart

Renowned urban geographer, David Harvey, notes the role of spectacle on the urban land-scape, in creating a city's identity and offering an experience for its inhabitants. The creation of spectacular urban landscapes is common in the post-industrial capitalist urban system of North America and Europe, but it remains less common in economically less developed countries.

In the Middle East, the oil rich United Arab Emirates is creating spectacular urban land-scapes, designed to bring global attention, and investment, to the tiny emirate. Costing billions of dollars and designed by architects of international stature, the skylines of Dubai and Abu Dhabi, the country's two largest cities, compete to be the most spectacular.

At present Dubai is completing the tallest skyscraper in the world. The Dubai Tower (Burj Dubai) will surpass the Sears Tower in number of floors and Taipei 101 and the Petronas Towers (Kuala Lumpur) in overall height. Though the final height of Burj Dubai has long been kept secret, it is expected to exceed 2,275 ft (693 m). The interior of the tower will be decorated by Giorgi Armani. The tower is part of a large-scale development that includes 30,000 homes, malls, and a human-made lake, at a cost of $20 billion.

While Dubai is the commercial hub of the UAE, Abu Dhabi is its capital, and has focused its attention on being the UAE's cultural center. Presently it is constructing a major cultural development on Saadiyat Island ("Island of Happiness") off its coast. It will contain branches of The Louvre Museum, estimated to cost $1 billion, and the largest museum ever constructed by the Guggenheim Foundation. The Guggenheim collection will focus on modern and contemporary art. Abu Dhabi's cultural enterprises will also include sports, such as Formula One racing and a golf tournament on the prestigious PGA tour.

---

centers of the Middle East, however, some of these cities—Manama, Dubai, Abu Dhabi—have been thriving as increasingly important transactional nodes in the economic systems reshaping the region. The center of Arab World banking has shifted (in part) from Beirut to Manama, Bahrain, and Dubai as a result of Lebanon's civil war. Doha, Qatar, is now the headquarters of Al Jazeera, the most popular satellite television channel in the Middle East. Since the 1990s, the port cities of the Persian Gulf have been the entrepôts supplying

Central Asia with cars, electronics, and other high-end goods. While built on oil, the cities of the Gulf are likely to continue thriving only if they diversify and lay the groundwork for post-petroleum economies.

With the growth of the global Internet, all cities in the Middle East are developing into *cyber cities* (fig. 7.17). The region's universities, manufacturing establishments, and traders are increasingly tied to constant flows of information that arrives by waves, wires, and photo-optic cables. Every computer terminal

Figure 7.17   Wi-fi is free and increasingly available in central Istanbul's public spaces such as here in the gardens that frame Emperor Constantine's Hagia Sophia cathedral, now a museum. (Photo by Donald Zeigler)

becomes its own harbor in the post-industrial landscape. The public is demanding frontage on these "harbors." Internet cafés, born in 1984, numbered fewer than 100 worldwide by the early 1990s. Today, there are thousands throughout the Middle East alone. Turkey, Israel, and the emirates of the Persian Gulf lead the region in Internet connectivity, but Iran and Egypt are catching up. The UAE and Dubai have the highest Internet bandwidth of any countries in the region (followed by Egypt and Morocco). Kuwait is also developing into a major cybercity, even as it lags in the development of a petrochemical industry.

ARAB CITIES AS SEATS OF UPRISING

Since December 2010 cities of the Middle East have witnessed uprisings and revolutions to demand economic and political reforms. In the republics of the Arab world, most governments have been in power for over 25 years. During that time, patrons of those regimes took control of the country's resources, while freedom of speech was repressed, income inequality grew more extreme, and corruption became intolerable. According to the 2010 report of Transparency International, Qatar, ranking 19th in the Corruption Perception

Index (CPI), is the least corrupt Middle Eastern state, while Iraq, ranking 175th , is the most corrupt, making it one of the four most corrupt countries in the world. One of the striking examples of corruption was the plan of most Arab leaders in non-monarchy countries such as Egypt, Libya, and Yemen to make their sons heirs to the presidency in order to keep power in their families. That plan emerged after the success of President Bashar Al-Assad's ascent to rule Syria after the death of his father President Hafez Al-Assad. These succession plans left no hope for the Arab people to free their countries from dictatorial presidencies rooted in powerful families, corrupt supporters, and brutal police. Therefore, uprising started in Tunisia, Egypt, Libya, Yemen, Syria, Bahrain, and other Arab countries.

Tunisia: The Jasmine Revolution

For decades Tunisians suffered from corruption, income inequality, unemployment, and rising prices. However, a specific event started the flame of the Tunisia's Jasmine Revolution. In December 2010 a policewoman slapped Mohamed Bouazizi, a university graduate who was selling vegetables in the street to generate income since he could not get a job in Sidi Bouzid, his home town. Bouazizi's humiliation, loss of dignity, and loss of income, led him to commit suicide by burning himself publicly. Shortly thereafter, the uprising started in Sidi Bouzid, then spread to the rest of the country. The police used excessive force in an attempt to stop it. After 29 days of continuous protests, the Tunisian president, who had ruled for more than 23 years, fled to Saudi Arabia. Tunisia, free from dictatorship, began moving toward democracy by reforming its government, political parties, and police systems. The success of the Jasmine Revolution sparked uprisings in other Arab countries.

Egypt: The Youth Revolution

Egyptian activists were inspired by the Tunisians; but the roots of the Egyptian Revolution actually go back to 2005 when President Mubarak began his fifth term by preparing his son Gamal to be the next president of Egypt. Mubarak used the Emergency Law to give his security forces broad powers while demanding little accountability. Therefore, political movements such as Kifaia ("Enough"), April 6th , and Youths for Change emerged to call for political and economic reforms. However, the police succeeded in brutally controlling these movements until an event similar to Bouaziz's death in Tunisia took place in Egypt.

On June 6, 2010, two policemen beat and killed Khalid Said, a young businessman living in the City of Alexandria. Protests in the streets and online called for serious investigations and immediate actions against the brutal policemen. Wael Ghonim, an Egyptian activist, created a Facebook page called "We are all Khalid Said" and used it to organize protests. On January 25, 2011, protests started in Tahrir ("Liberation") Square, Cairo, and spread to Alexandria, Suez, and several of Egypt's other major cities to demand freedom, justice, and economic reforms; later, protesters asked Mubarak to step down. The police initially used excessive force to end the revolution; but to everyone's surprise, the police disappeared within three days. After 18 days of protests, President Mubarak resigned. The success of Egypt's popular uprising could not have been achieved without the Egyptian Military; they supported the protesters' demands and kept order after the police security apparatus collapsed. During the revolution, "people's committees" were created in every neighborhood to guard properties and protect residents' lives. The committees worked with the military to keep order, which was demonstrated by the revolution's slogan "the people and the military are one hand." Free and fair elections lie ahead. The "Hollywood" ending of the Egyptian Revolution revealed to people around the world the power of peaceful protest.

Libya: From Uprising to Civil War

The domino effects of the Egyptian and Tunisian revolutions spread across the Middle East. In Libya, protesters demanded regime change and free elections. As a response, Colonel Muammar Gaddafi, after ruling the country for 42 years, ordered his military, police, and private security forces to end the uprising by force. As more protesters were killed, some military units, governmental officials and politicians joined the opposition movement. Within three weeks, protests succeeded in controlling most of eastern Libya, including Benghazi, the second largest city. Rebel forces were successful in capturing Tripoli and other cities from Gaddafi's forces in August.

Libya's uprising was transformed into a civil war as Gaddafi made plans to fight in every city, town, neighborhood, and alley. One of Gaddafi's infamous speeches, in fact was turned by the opposition into a hip-hop song called *Zanga-Zanga*—"Alley-Alley" —to make fun of Gaddafi's brutality. Gaddafi , meanwhile, claimed that protesters were misled and drugged by al Qaeda, the west, and other Arabs. In the summer of 2011, the antigovernment rebels formed an Interim Transitional National Council that was recognized by France and other EU states and the United

States as Libya's legitimate government. As the death toll of protesters increased, the international community put pressures on Gaddafi to leave power. The International Criminal Court announced that it would investigate Gaddafi for crimes against humanity. The United Nations Security Council adopted resolution 1973 to demand an immediate ceasefire and stop violence against civilians. The resolution also authorized the international coalition to impose a no-fly zone over Libya and to use all means to protect civilians. Later, NATO (the North Atlantic Treaty Organization) led the international forces to enforce the no-fly zone to defeat Gaddafi's forces.

Uprisings in Other States

As of 2011, there were major protests in Yemen, Syria, and Bahrain, and to a lesser degree in other countries. In Yemen, several weeks of constant protests in Sana'a, Aden and other cities made President Ali Abdullah Saleh, who had ruled the country for 32 years, announce he was willing to resign and leave power to "safe hands," something which the Gulf Co-Operation Council threw its weight behind. In Syria, protests started in the city of Deraa after the arrest of 15 children who wrote anti-government slogans on walls. Afterwards, protests spread to Damascus, the capital, Hama, Latakia, Homs, Banias and other towns. President Bashar Al-Assad blamed outsiders for the current unrest in Syria and used elements of the armed forces that were loyal to him to end the protests. However, the uprising continued to strengthen. It seemed as of President Al-Assad was following the steps of Mubarak and Ben Ali, the former Presidents of Egypt and Tunisia. So, will he step down like them or will he succeed in staying in power despite all these challenges? At the time of this writing, the fate of the Syrian uprising remains unclear.

Protesters in Manama, Bahrain, demanded governmental reforms, the institution of constitutional checks on the monarchy, and, for some, the end of monarchy altogether. As in other countries, unemployment and corruption were major reasons for the uprising, but Shi'a Muslim protesters complained about discrimination in addition to economic hardships. Protests in the Bahrain's Pearl Roundabout were met with violence by the King's security forces. As protests increased, Saudi Arabia sent military forces to support the Bahraini government. Interestingly, less significant protests took place in Saudi Arabia to demand a constitutional monarchy and other reforms. Nevertheless, King Abdullah responded by announcing measures to relieve economic hardships and granting women the right to vote, while the government warned people from protesting. Sporadic demonstrations also took place in Algeria, Jordan, Morocco, Oman, Bahrain, Kuwait, Iran, and the Palestinian territories.

In the cities of the Middle East, people struggle for freedom, democracy, and social justice. They have witnessed how peaceful protests, with the support of the military, is able to remove dictators (as in Tunisia, Libya, and Egypt) and how dictators can create civil wars to stay in power (as in Yemen and Syria). In the Middle East's cities modern history is still being written.

DISTINCTIVE CITIES

Cairo: Al-Qahirah, *"The Victorious"*

Al Qahirah means "victorious," and Cairo has emerged victorious as the most populous city in the Arab World, in the Middle East, and on

the continent of Africa. Greater Cairo with about 19 million people consists of governorates of Cairo, Giza, Helwan, 6th of October, and Kalyobia, which makes it one of the largest metropolitan areas in the world. It is known as "the City of 1000 Minarets" since mosques spread across all city neighborhoods. With its movie industry and annual International Film Festival, Cairo is known as the "Hollywood" of the Middle East, as well. The Arabic-language cinema and popular Arab music have made Cairo—along with Beirut, Lebanon—one of the cultural epicenters of the Arab universe. *Al-Ahram*, a Cairo daily newspaper owned by the government, has the largest readership in the Arab world, and its English-language edition circulates globally. Several private newspapers such as *Al-Wafed*, *Al Masry Al-Youm*, and *Al-Shorouk* are also published in Cairo to represent critical views of governmental policies. As the capital of Egypt and the headquarters of the League of Arab States, Cairo is also the head city of pan-Arab politics, a role facilitated not only by the city's size, but also by its relative location. It is the home of foreign embassies and cultural centers and to a major educational center where Cairo University, Ain Shams University, and other public and private universities are located. It has Al-Azhar University, the world's oldest Islamic university and center of Sunni Islamic education, to which students from around the globe come to study Islam and other subjects. Cairo is positioned between the western Arab world of North Africa, and the eastern Arab world of Asia.

Cairo, over a thousand years old, is a multi-layered city; its buildings and neighborhoods reflect the impact of various historical periods. At its dense core lies the "Islamic" city. To the east of the Nile River this city began as a military encampment and grew to include a citadel, mosque, and city walls. Today most walls are gone, torn down as the city's boundaries expanded. However, three of the original city gates remain to draw tourists, who mingle with the area's residents. In medieval times Cairo was an epicenter of world trade; caravans brought luxuries and necessities to the city's famed markets. Today, tourists crowd the Khan al-Khalili to buy souvenirs, most often trinkets reflecting Egypt's pharaonic history, or to sip tea in traditional coffee shops formerly frequented by Naguib Mahfouz, Egypt's Nobel laureate.

As dams tamed the Nile floods in the 19th century, the city expanded onto its banks. At the same time, Europe was expanding its interests overseas, establishing colonies and political control over countries in Asia and Africa. In Cairo, new neighborhoods were constructed in European design. Houses were built that looked like Italianate villas. Parks were constructed. Indeed, a "new" downtown was created to mimic the design of Paris, and upper-class Egyptians enjoyed performances in the new opera house. In 1871, an opera premiered there that has become one of the world's most popular: *Aida* by Giuseppe Verdi.

Cairo's explosive population growth occurred in the era after World War II, when Egypt became an independent republic. Migrants from the rural areas flooded the capital in search of jobs and opportunities; they created enormous economic challenges for the young government. Massive high-density apartment blocks, with bleak architectural designs and poor quality construction, were built to accommodate the influx. At the same time the government became concerned about Cairo's massive size, and its military vulnerability.

In an effort to stem the tide of urban expansion onto valuable farmland, the Egyptian government began to re-direct growth into the desert. The result was government-built, industry-based cities distant from Cairo. The 10th of Ramadan City, for instance, located on the way to the Suez Canal, was built to have an industrial base anchored by several thousand factories. Since the 1970s it has offered jobs, housing, and some services. The new towns, however, never met their target population goals and did little to relieve the population pressure on Cairo. Also, new settlements were built along the ring road surrounding Cairo to redistribute the population. Many of these settlements became homes for the middle and upper classes. Since the 1990s, Cairo's landscape has increasingly reflected the impact of globalization. Chili's, TGIF, Hardee's and other fast-food chains with global reach are ubiquitous. Massive malls, office towers, and new hotels now line desirable locations along the Nile. In addition to the American University in Cairo (AUC) found in 1919, new international universities such as the German University in Cairo (GUC) were established to internationalize the city's educational opportunities. Foreign banks (*e.g.,* CitiBank, HSBC, and Scotia bank) are symbols of Cairo's integration into the global economy.

As a megacity, Cairo is really a product of the 20th century. As it has expanded, however, it has engulfed dozens of predecessor settlements and unique historical landscapes (fig. 7.18). These visual reminders of the past, numbering in the hundreds, make Cairo a vast open-air museum. Their distribution most closely resembles the multiple nuclei model. Large tracts of 20th century blandness separate such historical nucleations as the following:

Figure 7.18   The vast metropolis of Cairo sprawls to the horizon from the base of the old Citadel. (Photo by Jack Williams)

- The great pyramids (and the sphinx) of Giza, on the west bank of the Nile, date back to the Old Kingdom, but they have been encroached upon by an expanding city, deflating some of the excitement of first-time visitors who may be disappointed to find a Pizza Hut practically at the pyramids' base.

- Heliopolis (on the way to the airport) was one of the ancient world's cult centers, but only a single obelisk remains—now in the middle of an urban park.

- Babylon-in-Egypt, now known as Coptic Cairo (because of its Coptic Christian inhabitants), has a history associated with the world's most famous refugee family—Mary, Joseph, and Jesus. It is now engulfed by the modern suburb of Ma'adi.

- Cairo's Citadel was built by Saladin in the 12th century; it was transformed by the Mamluks and then the Ottomans.

Recently, Tharir Square has become a major landmark in Cairo. It was the focal point of the Egyptian Revolution that started on January 25, 2011, to demand freedom, social justice, and economic reforms. With the success of the revolution, Tahrir Square has become a place where Egyptians go to demand more economic and political reforms. It is becoming the Egyptian equivalent of London's "Hyde Park Speakers' Corner" in which people openly voice their opinions and debate political issues.

Whether counting people or cars, the growth of Cairo has been meteoric. Today, the city's traffic snarls are of world renown. Not until the late 1980s was Cairo's long-anticipated metro rail system opened to help move the city's millions. It has since expanded to include a subway line tunneling under the Nile to the west bank. The new metro stations are conveniently sited, brilliantly lighted, and immaculately clean. The Metro links some outer suburbs (but not yet the satellite cities) with the center of the metropolis, and at least one car on every train is reserved for women. Trips that at one time took three hours by car, now can take as little as half an hour. As the city's transportation system continues to become more efficient, so will its economy.

Figure 7.19   The Umayyad Mosque was built in 705 AD in Damascus, Syria. It took about 10 years to complete this magnificent structure. (Photo by Amal Ali)

Damascus and Aleppo: Fraternal Twins

The Arabs call them Al-Sham and Halab. Outsiders know them as Damascus and Aleppo. With over two million people each, they are the most populous cities in Syria. Each one claims the same superlative: "the oldest, continuously-inhabited city on earth." It was Damascus that Mohammed deliberately decided never to visit because he wanted to enter Paradise only once. But it was Aleppo where Abraham (the father of monotheism) stopped to milk his cow on the journey south from Haran. Such stories tell more about the pride each city takes in its heritage than the cold facts of history. Like fraternal twins, their looks and personalities differ, but you can tell they are from the same family.

The location of each city is impressive enough to give credence to their storied histories. To appreciate the site of Damascus, one needs to look no further than the mosaics of the Umayyad Mosque, built at a time when the city was the capital of the first Muslim empire, the only time in history when the entire Islamic world was politically united (fig. 7.19). In green and gold tiles, the mosaics portray a lush oasis where life was at its best. Next to the northwest corner of the Umayyad Mosque is the mausoleum of Saladin who successfully led the Arabs against the Crusaders. To honor Saladin and mark his great victory, the Saladin monument was placed at the entrance of Souk el-Hamidiyeh (fig. 7.20).

Thanks to the Barada River, Damascus is one of the world's premier oasis cities. The Barada comes tumbling down the slopes of Mount Hermon. As its velocity diminishes, the channel splits into a handful of distributaries, and water sinks into the ground, forming an oasis on the edge of the Syrian Desert. Ancient Damascus, as if to avoid occupying precious agricultural land, lies to one side where growth has pushed it up the last fold in the mountains of Lebanon. Mountains to the west and the desert to the east protected the ancient oasis city.

Damascus' site is dramatic when seen from the air; Aleppo's site is dramatic when seen a ground level. Rising from the surrounding

Figure 7.20   The monument of Saladin reminds visitors to Souk Al-Hamidiyya, Damascus, of the glory days of Saladin. (Photo by Amal Ali)

plain is Aleppo's citadel, a fortified hilltop that identifies the original city. The north Syrian plain is a steppe (grassland), which stretches southward from the Anatolian plateau until it grades off into the Syrian Desert. In northern Syria, nature has set the stage for dry farming, and the Syrians have made it even more productive with irrigation. The region is a center of wheat, olive, almond, and pistachio production. Aleppo's potential as a breadbasket holds even more promise for the future, a promise affirmed by the United Nations when it located the International Center for Agricultural Research in Dry Areas here.

No matter how productive the oasis nor how impregnable the hill, neither Damascus nor Aleppo would have occupied pivotal roles in Middle East history were they not situated at the crossroads of trade. The oldest trade routes followed the parabolic shape of the Fertile Crescent: from the Tigris and Euphrates Valleys, along the spring line at the base of the Anatolian Plateau, and south along the mountainous littoral of the eastern Mediterranean, thence on to Egypt and beyond. Between the two arms of the Fertile Crescent is the Syrian Desert. The desert provided a series of east-west shortcuts: the northernmost shortcut focused on Aleppo, half way between the Euphrates and the coast. A longer shortcut passed through the mid-desert oasis of Palmyra (today's Tadmour) and focused on Damascus. Both Aleppo and Damascus were also on the north-south axis of trade and thus became urban growth poles early in their history.

Damascus and Aleppo are modern cities today, but not as modern as they might be were it not for the protectionist, almost isolationist, stance of the Syrian government. Only in the early 1990s were strict rules of socialist planning and government ownership eased in Syria. The result was a mini-renaissance in economic activity in both cities, but too many restrictions remained in place and the renaissance was short-lived. One positive result is that authentic Arab traditions, not the artificial niceties contrived for tourists, remain triumphant in both cities, and the global economy has yet to overtake either one. Each city still has a character of its own. One feels a greater reverence for the party line (the autocratic Baath Party) in Damascus. Aleppo, farther from the corridors of power, has a more freewheeling, freethinking air about it; yet more of its women are still fully veiled, and the largest stores are no bigger than an American 7–11. As Syria's capital city, Damascus is the almost-monopolistic gateway to the entire country. And, Aleppo seems to yearn for a greater role in Syria and the world beyond.

Jerusalem: Clash of Religion and Politics

Jerusalem occupies neither an attractive site nor a strategic location. It is not central in a geographical sense and lies astride no major

Figure 7.21   The Dome of the Rock and the Western Wall are symbols of a religiously divided Jerusalem. (Photo by Donald Zeigler)

trade routes. It is a city that should have been bypassed by time. Instead, Jerusalem has become an epicenter of religious veneration and conflict. Three religions regard it as a holy city: Judaism, Christianity, and Islam. Muslims rank it behind only Mecca and Medina in importance. For them, it is the place from which Mohammed made his "night journey" to heaven to talk personally with God. To mark the place of his ascension, Muslims built the Dome of the Rock in 691 AD (fig. 7.21). It is one of the oldest Islamic structures in the world. To Jews, Jerusalem is the city of David's kingship and Solomon's Temple. The wall of the platform on which the Temple stood is all that remains; it is called the Western Wall and is the focal point of Jewish prayers. The Dome of the Rock may be located on the very site of

Solomon's temple. To Christians, Jerusalem is the city where Jesus of Nazareth revealed himself to be the Messiah. Since Byzantine times, the Church of the Holy Sepulchre has sheltered the place of the crucifixion, entombment, and resurrection of Jesus. It is easily visible from the Temple Mount and Dome of the Rock.

One always speaks of "going up" to Jerusalem. It began as a hill town at the very southern tip of the western Fertile Crescent, but it did not have a commanding position. The land enclosed by the "old city" walls is lower in elevation than the land to either the north or the east. The only feature of the physical environment that commended the site was a spring, now known as the Gihon. The Jebusite village at the site of the spring was destined for prominence, however, largely because its rela-

Figure 7.22   The Jewish Quarter of Jerusalem's old city has been entirely rebuilt since 1967. The Israeli flag was adopted when Israel received independence in 1948. (Photo by Donald Zeigler)

tive location made a difference 3,000 years ago. The village was conquered by the Hebrew king, David, who needed a centrally located capital between the northern and southern tribes of the Hebrew people. Jerusalem fit the bill. With the decision to move the Arc of the Covenant (containing Moses' tablets of stone) to Jerusalem, the city began to acquire the religious capital needed to sustain its spiritual centrality for three millennia. Jerusalem became the place where the God of Moses and Abraham permanently resided; and before the first Muslims prayed facing Mecca, they prayed facing Jerusalem. Religion endowed Jerusalem with elements of centrality that geography could not.

The current walls of the old city of Jerusalem date to the Ottoman period. Within the walls, Jerusalem is divided into four so-called quarters: Muslim, Jewish, Christian, and Armenian. The mental map conjured up by such a description, however, belies the reality of the city's cultural geography. In fact, almost the entire old city has an Arab feel about it, save for the Jewish Quarter. Furthermore, the boundaries of the quarters no longer (probably never did) define the cultural divisions of the city. Residential patterns and movement into and out of the four quarters challenge the idea that they are homogenous neighborhood groupings of like-minded souls. In the "old city," Muslims and Jews are the primary actors, Christians are diminishing in numbers, and Armenians are doing what they have done best for over 1500 years, surviving as a culturally distinct Christian minority. Muslim Arabs are expanding into the Christian quarter, the traditional niche of Christian Arabs. Jews are solidly in control of the Jewish quarter; but they are also acquiring property in the other three quarters, which they conspicuously mark with signs, synagogues, and Israeli flags (fig. 7.22). Jews moving into the old city are more likely to be extremely religious, and those leaving it are more likely to be secular. Armenians, especially seminarians, flow through the Armenian quarter from all over the world, their identity bolstered since 1991 when Armenia reappeared on the map of sovereign states.

The "old city" is only one of two Jerusalems. The other is the sprawling modern metropolis. While the walled city is no larger than a college campus, metropolitan Jerusalem covers at least 50 square miles (129 sq km). Despite being governed as a single municipality, the metropolitan area is bisected by a cultural fault line. West Jerusalem is thoroughly Jewish and provides the site for Israel's parliament, the Knesset. East Jerusalem is primarily Arab (including both Muslim and Christian Arabs), a collection of Arab villages, one of which may someday become the capital of a Palestinian state. "Occupied" East Jerusalem, however, is not as homogeneous as West Jerusalem. In the east, Jewish settlements occupy a dozen hilltop sites, all of them new (post-1967), wealthy, and strategically positioned to maintain control of greater Jerusalem for the Israelis. To the north, south, and east of Jerusalem, where border crossings are patrolled by Israeli soldiers, the Palestinian West Bank begins. The city's relative location between Israel proper and the West Bank gives it a frontier feel, of being along a tension-ridden international boundary with a more-developed country on one side and a less-developed country on the other.

The future of Jerusalem will be determined by the ability of the Israelis and the Palestinians to negotiate a peaceful resolution to their conflicting ambitions. In the meantime, repeated conflict between the Israelis and the Arabs has destroyed the infrastructure of many West Bank cities, such as Ramallah and Jenin. The construction of a separation barrier (also known as the separation wall or fence) by the Israelis further complicates efforts to resolve the conflict. Fifty-six miles (90 km) of the barrier is in Jerusalem, much of it thrusting deep inside the pre-1967 border. The barrier separates Jerusalem from the West Bank. Although it was constructed to increase Israeli security, the barrier also restricts the ability of thousands of Palestinians to reach their jobs, fields, and medical services.

Istanbul: Bi-Continental Hinge

Istanbul has existed for almost 27 centuries (since 657 BC), for 16 of them as an imperial capital. Its original name was Byzantium, but it was re-christened the New Rome in 330 AD. Almost immediately the people began calling it Constantinopolis, Emperor Constantine's city. Today, it appears on the map as Istanbul. It has been the dominant city of the eastern Mediterranean realm for more than a millennium, having surpassed a million inhabitants by 1000 AD. Something about this location just begs for an urban place.

Istanbul's location makes it a hinge between continents. From its situation on the European side of the Bosporus, it is positioned to control overland trade between Europe and Asia, and the shipping lanes between the Mediterranean and Black Seas. The huge empires which Istanbul commanded—Roman, Byzantine, Ottoman—also gave it the ability to control overland access to Arabia, the Indian Ocean, and Eastern Asia. Until the sea route round-Africa was fully opened in the 16th century, Istanbul was able to control trade

Figure 7.23 Istanbul is a city of thriving neighborhoods. Here men gather in the late afternoon under a bust of Atatürk, the revered father of modern Turkey. (Photo by Donald Zeigler)

between north and south, east and west. As the imperial capital of the Ottoman realm since 1453, it reached its peak in the 1500s when the emperor, Suleyman, commanded so much wealth that he was known to the world as "The Magnificent." His city was at that time larger in population than London, Paris, Vienna, or Cairo. During that century, however, the power of Constantinople began to wane. No longer did Ottoman subjects hold a monopoly on the ancient silk routes across Asia or on the Fertile Crescent caravan trade from the Eastern Mediterranean to the Persian Gulf. As technology enabled mastery of the sea, caravels replaced camels as the most reliable and economical modes of transport. Well before the end of the

19th century, the Ottoman Empire was the "sick man of Europe," and "Stamboul" was a city in decline. Modern Turkey was born out of the Ottoman Empire thanks chiefly to the secular nationalism inspired by Mustafa Kemal Atatürk (fig. 7.23).

The core of Istanbul, the historical city, occupies a peninsular site with deep water on three sides. Crowning the peninsula, and visually dominant from the sea, are seven hills, just like Rome. The peninsula is bordered on the south by the Sea of Marmara, on the east by the Bosporus, and on the north by the Golden Horn, the large, sheltered harbor that enabled the city to dominate the shipping trade. The Bosporus and its companion strait, the Dardanelles, enabled ocean-going vessels to penetrate central Eurasia. On the Black Sea's northern shore, the ancient Greeks implanted colonies. The fertile hinterlands of these colonies became a breadbasket, producing wheat for the Aegean core of the Hellenic world, wheat that went to market via the Bosporus. The first "world class" city to dominate these straits goes back to the Bronze Age. Its name was Troy, and it was located at the southern end of the Dardanelles. Troy was the Istanbul of its day.

In the 20th century, the Bosporus provided one of the Soviet Union's few outlets to the world ocean and was consequently a point of strategic significance during the Cold War. By controlling the strait at Istanbul, NATO could deprive Moscow of dominating one of the world's most strategic locations. Even now, the Bosporus continues to be important to the Ukraine and the Russian Federation. It has also taken on a new strategic significance in ensuring the steady flow of crude oil from the landlocked Caspian Basin fields, much of which transits the Bosporus, already one of the world's busiest straits. At Istanbul, the Bosporus is at its narrowest, assuring full and busy roadsteads and shipping lanes, the ingredients of potential collisions, particularly given the swift currents that flow out of the Black Sea.

Oil is not the only commodity important to Istanbul's economy, however. During most of the 20th century, Istanbul's European hinterland was all but severed by the Iron Curtain and the animosity of neighboring Greece. Now, however, Eastern European nations have opened their borders. The routes of commerce between Europe and Asia are once again funneling traffic across the Bosporus, and Turkey finds itself a possible candidate for European Union membership. The growth of trade is nowhere more powerfully symbolized than by the growing volume of truck traffic navigating the trans-continental Bosporus bridges, completed in 1974 and 1988. Now, a new rail line designed to carry passengers and freight under the Bosporus is in the offing, and Turkey sometimes calls itself a European nation. At the same time, it is emerging as a gateway to Central Asia, where the Turkish people originated and where most of the languages spoken are Turkic in origin.

Fès and Marrakesh: Imperial Rivals

In Morocco it was Fès and Marrakesh that grew up as anchors of the country's urban system. Neither city functions as the capital today, but both have in the past. Nor is either located in the Rabat-Casablanca megalopolis, or even along the coast. Instead, Fès faces east towards Morocco's Arab Islamic heritage; and Marrakesh faces south towards the Atlas Mountains and Africa's desert oases beyond.

The seeds of Fès, and with it the seeds of the Moroccan state, were planted shortly after "the Arab conquest," about 790 AD. The Arabs arrived via the Taza Gap, the only natural corridor between the Maghreb and the rest of North Africa. An Arab village joined a Berber village on a tributary of the largest river in Morocco. This city, Fès al Bali, commanded passage through the Taza gap to the east. It was also located near well-watered and forested mountains to the north and south, and had a rich agricultural hinterland to the west. As trade developed, the souks of Fez became the best in the Maghreb.

Marrakesh was born in the 11th century when the indigenous peoples of the Sahara established a trading post just north of the snowcapped High Atlas Mountains. Their rulers were so successful in consolidating power and trade routes that the city soon became their capital. As a result, Marrakesh and Fès both claim to be the historical core of the Moroccan state, with Fès contributing its Arab and Andalusian culture (Arab culture transplanted from Spain) and Marrakesh contributing the culture of the African Sahara.

A comparative analysis of the cultural landscapes of Fès and Marrakesh illustrates their roles in building the national urban system. Like cities everywhere, the personality of each city is expressed most clearly in its center, within the walls of the old city, the *medina*. All Moroccan cities have their medinas; but the finest in the entire Maghreb, and arguably the best-preserved pre-industrial city in the world, is Morocco's first imperial capital, Fès. The second finest medina may be in Marrakesh, the imperial capital of Morocco's south. So authentic are the landscapes of intramural Fès and Marrakesh, that they were among the first cities to be placed on the United Nations' World Heritage List. Fès and Marrakesh have shaped the culture that is uniquely Moroccan. Common to the cultural landscape of both Fès and Marrakesh are the souks. They are the rabbit warrens of shops and craft merchants competing to provide not only the necessities but also the finer goods and higher-level services that mark the differences between urban and rural life.

Both Fès and Marrakesh served as focal points of commerce, political power, religious life, and public celebrations. But why have these ancient walled cores been able to stall the passage of time so well? Their survival is related to the survival of Morocco itself. More than most other places in the Middle East, Morocco held outside conquest at bay,

turning back both the expanding European kingdoms to the north and the Ottoman Empire to the east. Few Westerners even laid eyes on the interior of Morocco until the French established a protectorate in 1912. For an "infidel," penetrating Morocco was the equivalent of penetrating Tibet–one did it only under cover. Many of Morocco's medinas, therefore, survived intact. Fortunately, when the colonizing French did arrive, they chose not to modernize them, but to build new cities outside the old walls, thus saving medieval urban landscapes from disfiguring modernization.

The most well-known symbols of Fès are the Kiraouan and Andalusian mosques. The first is among the most revered in the entire Muslim world. It stands not alone, but with its university, among the oldest in the world, and with the tomb of the founder of Fès. The spiritual and intellectual life of Fès set the tone for the evolution of Moroccan culture. Like Fès, the most recognizable site in the Marrakesh medina is a mosque, the Koutoubia. It was built by a Berber dynasty that re-centralized power in the south. The Koutoubia's minaret, its most conspicuous feature, arose as a statement of political power, as a mirror image of the mosque built in Cordoba (Spain), where Andalusian culture thrived. In Marrakesh commercial power blended with political power, whereas in Fès commercial and political power blended with the power of Islam. It was probably that religious anchor that kept the capital city function returning to Fès.

The physical settings of Fès and Marrakesh have influenced each city's urban landscape. The Fès medina, wedged into a narrow, V-shaped valley, epitomizes pre-industrial urban crowding. Every square inch of space is accounted for. As a result of its more northerly location and higher altitude, Fès' climate is cooler and rainier. The Marrakesh medina, on the other hand, seems to epitomize the desert, which begins just to the south of the High Atlas. The spaciousness of the desert is reflected in the public square in the center of the Marrakesh medina. Such an expanse of open space in a Middle Eastern city is a rarity.

Tashkent: Anchor of Central Asia

With over two million people, Tashkent is Central Asia's largest city. Since independence from the Soviet Union in 1991, it has come to see itself as the region's anchor and gateway, and as a trans-continental bridge linking east and west. It is also the capital city of Uzbekistan, the most central state of Central Asia. Uzbekistan borders every other country in the region (but not Russia), giving it a pivotal geopolitical role in the 21st century's "great game" for influence over people, trade corridors, and resources (natural gas, gold, uranium, and cotton).

Almost every city of Central Asia originated as a trading post on the "silk roads." Tashkent is no exception. It was a port on the edge of the desert, a "stone town" on the northern route from Kashgar (now in China) to the Caspian Sea, thence on to Europe. Its founding goes back at least 2000 years to the place where the Syr Darya's tributaries back up to the water-laden Tien Shan (mountains). With such enviable access to the region's most precious resource, water, Tashkent offered unique opportunities for modern growth. The city is also situated near the mouth of the Fergana Valley, one of the choice spots for fruit and vegetable production in the region. It is also

in a commanding position relative to the cotton fields snaking across the irrigated desert to the Aral Sea. These economic attractions, combined with the region's gold resources, resulted in the Russian conquest. Tashkent was in the first khanate to be absorbed, in 1865, during a time when Europe had been cut off from American supplies of cotton by the Civil War.

During the Soviet period, Tashkent was selected to be the capital of the Uzbek SSR, replacing Samarkand in 1930. Before World War II its population was only half a million, but during the war the Soviets mounted a massive effort and protectively located many strategic industrial operations beyond the Urals, thus swelling the city's population. In fact, Tashkent's most renowned product is still airplanes, and Seattle, Washington, home of Boeing Aircraft, is one of Tashkent's "sister cities." When the Soviet Union collapsed, the city was the country's fourth largest. It was also the most Soviet city in Central Asia thanks to two factors—Communism's emphasis on centralized state planning, and the forces of nature. Tashkent was almost completely destroyed by an earthquake in 1966. The Soviets seized upon the opportunity to re-fashion the landscape in a socialist mode: "a new Soviet city" for "a new Soviet man."

Tashkent's ancient core survives only in the four historic M's: Market, Mosque, Medrassa, and Mausoleum. Since independence, a fifth M has been added, a new Museum dedicated to the life of Tamerlane. Although Tashkent was not Tamerlane's capital, the city has seized upon Amir Tamur as the quintessential symbol of its independence and the quest for respect. Like many states that cast off Communist regimes, Uzbekistan has mined the archives for inspiration drawn from its pre-Soviet and pre-Russian past. The opening of the Tamerlane Museum symbolizes that transition. Tashkent was a regional city beholden to Moscow. Today it is the capital city of an independent state seeking to play the same role in the modern world as did Tamerlane's empire when it stretched from the borders of Europe to the borders of China.

It is iconography rooted in Turkic culture that is now remaking the landscape of Tashkent (even though Tamerlane was probably Mongol, and his capital was Samarkand). Lenin and Marx are gone; heroes on horseback and other images of the steppes have taken their place. Government buildings fly flags that are Turkic blue not Soviet red. New structures are conglomerates of steppe-colored brick, not piles of cinder blocks. One out of every five cars is a Daewoo, from Korea. Traditional domes and arches (bedecked with blue tiles) are replacing socialist perpendiculars. Mosques are being built again, after decades of anti-religious propaganda. Uzbek, a Turkic tongue, has replaced Russian as the official state language (in 1989); and it is being written in the Roman alphabet, not Cyrillic, a fact of life easily read in the city's linguistic landscape.

Meanwhile, Tashkent still has a predominantly Soviet feel and appearance (see box 1.3). About a third of the population is Slavic in origin and speaks Russian. Nine-story apartment blocks dominate residential neighborhoods. European-style parks and open space, a positive contribution of socialist planning, were incorporated into the landscape after the earthquake. Opera, ballet, and the puppet theater still dominate the performing arts. The only subway system in Central Asia offers Tashkent the same efficient service and rich artistry as Moscow's metro. Flights to Moscow still far outnumber flights to any other destination. Finally, a layer of transnational landscape elements has appeared: Western products in the stores, logos of multi-national corporations downtown, American fast-food chains, international hotels, copious commercial advertising, and the international tourist trade that has come to exploit the lure of Asia's "old silk road."

KEY PROBLEMS

Middle Eastern cities do many things well. First, they reflect the hospitality of their inhabitants, people who easily talk to visitors, who are anxious to communicate despite linguistic barriers, and who have time to spend in casual conversation on the street (fig. 7.24). Arabs, Turks, and Iranians are among the friendliest people in the world, and their cities make you feel at home. Second, Middle Eastern cities, with few exceptions, are safe day and night. There are "eyes upon the street" all the time, whether you can see them or not; and family networks, under-girded by strict codes of conduct, hold family members accountable. Plus, alcohol consumption among Muslims is prohibited, so drunkenness is uncommon. Third, the generations mix freely. Neither the old nor the young are warehoused; parents are seen with children; teenagers use the same streets as the elderly; and young apprentices are common in the city's businesses. Households are often multi-generational. Fourth, homelessness, although it does exist, is less common than in Western cities. It is taken for granted that some people will not be able

Figure 7.24   The coffee urn is a symbol of hospitality throughout the Arab world, as seen here in Jordan's capital, Amman. (Photo by Donald Zeigler)

to live self-sufficient lives, so families compensate for personal inadequacies, and many social needs are taken care of by the Islamic emphasis on required almsgiving and charity. Fifth, almost every city takes pride in its food, whether served in sit-down restaurants or on the street. Middle Eastern cuisine helps to define both national cultures and urban life. Furthermore, it is healthy food, not over-processed, and rarely fried. Sixth, cities are well served by a variety of transportation. Cars are not required; and in the old cities they may be a hindrance. City buses, taxis, service taxis (often 12-passenger vans), and fixed-rail

lines (in a few cities) always make it possible to get around at very low cost. Besides, cities are compactly organized, so walking is always a possibility.

Urban life in the Middle East is not utopian, however. Cities have their problems, some seemingly intractable, just as in other world regions. These troubles include: too rapid population growth, pollution, and transportation chaos (much of it caused by the impact of the automobile). Five problems acutely felt in the Greater Middle East, however, are the lack of fresh water, unemployment and underemployment, housing affordability and infrastructure, cultural homogenization, and preserving the region's heritage.

Fresh Water

There is a shortage of water in the Middle East. As cities expand, they can accommodate growth only by developing water resources for new homes, businesses, and industries. Surface and groundwater resources are being utilized to the maximum throughout most of the region. Even the "fossil waters" of the Sahara are being used by the Libyans to supply their urban populations in the north; and the Arab states of the Persian/ Arabian Gulf have all gone into desalination in a big way. Nevertheless, most cities do not use their water resources efficiently. What water they do have could go further if leaky pipes were repaired, if water-conserving technologies were used, and if irrigation systems could make do with less. Plus, the water problem is not simply a problem of quantity; it is also a problem of quality. Virtually every city must concentrate on upgrading its water treatment operations so that tap water is safe to drink.

Unemployment and Underemployment

Unemployment is the most serious challenge facing cities of the Middle East since it affects people's ability to obtain food, decent housing, education, and social services. Except for Gulf Countries, unemployment rates range between 8 percent and 15 percent in countries of the Middle East, but reach 26 percent in the Occupied Palestinian Territories. With the increase of unemployment, underemployment grows as educated young people accept jobs that do not meet their qualifications or financial needs. Consequently, poverty has increased in all but the major oil-producing countries. Some Middle Eastern countries instituted economic reforms that actually widened the gap between the rich and the poor and increased income inequality. Multinational corporations operate in the Middle East to take advantage of cheap labor. Indeed, they provide jobs and incomes for people, but in many cases they pay low wages that do not adequately compensate people for the long work hours.

Housing Affordability and Infrastructure

Housing is an important element of cities of the Middle East. Natural population increase combined with the growing rural-urban and urban-urban migration has raised demands for housing in major cities like Cairo and Beirut among others. The reduction of public housing projects has made housing less affordable for low-income groups. In fact, housing has been a challenge for the middle class, as increasing housing prices were not accompanied with a similar increase in wages. Therefore, squatter settlements or shantytowns have emerged as alternative shelters for the poor who cannot afford living in formal settlements. Unique names have been developed to describe squatter settlements: geçekondu ("built overnight") in Turkey, bidonvilles (after the tin "oil cans," *bidon à pétrole*, used to build them) in the Maghreb and other former French colonial territories. These settlements usually lack basic infrastructure and public services such as clean water and sanitation and waste disposal systems. As the number of squatter settlements grows, governments attempt to upgrade them by expanding water pipes and electricity lines and providing community services to improve the living conditions for the residents. Infrastructure upgrading is also needed in old neighborhoods in core cities such as Cairo, Baghdad, and Damascus to assure the adequacy and efficiency of infrastructure and public facilities.

Cultural Homogenization

The trend throughout the Middle East is toward more culturally homogenous populations, a concomitant of European-style nation building. Istanbul is more ethnically Turkish and more thoroughly Muslim than ever before in history; Alexandria and Cairo are more Arab. Slavs have left Tashkent by the thousands since independence from the Soviet Union. Many Israelis see Jerusalem as becoming ever more thoroughly Jewish. Religious minorities have been squeezed out of virtually all cities in Iraq and Iran. Europeans, seen as "the colonizers," have abandoned the cities of the Maghreb. The Jewish population of every Middle Eastern city outside of Israel has decreased substantially or disappeared since 1948. There seems to be a desire for "ethnic purity," and some see this as a problem of urban cultural geography.

Heritage

In all cities of the Middle East, modernization threatens heritage resources. New roads are cut through walled cites, suburban expansion buries Roman villages, and property owners upgrade without considering historical preservation. In the architectural realm, urban expansion zones are looking more and more alike throughout the world. Meanwhile, the integrity of old urban landscapes is scarred by deterioration or unregulated incumbent upgrading. The new post-war landscape of central Beirut, for instance, has been re-built in an "international style," quite in contrast to Sana'a, the capital of Yemen, which has used United Nations aid to maintain the architectural unity of its historic core despite rapid population growth.

These problems, along with others too numerous to mention, are not unique to Middle Eastern cities. They are worldwide problems. As a majority of the world's population has become urban, the problems of cities are increasingly the problems of all humanity. Having invented the city in the 4th millennium BC, the people of the Middle East are now challenged to see if they can perfect it in the 21st century AD.

SUGGESTED READINGS

Abu Lughod, Janet L. 1971. *Cairo: 1001 Years of the City Victorious.* Princeton, N.J., Princeton University Press. A chronicle of Cairo from 969 to 1970, and a glimpse of how to make sense of any urban landscape.

Benvenisti, Meron. 1996. *City of Stone: The Hidden History of Jerusalem.* Berkeley: University of California Press. Offers a balanced view of Jerusalem's urban landscapes, boundaries, and demographics.

Bonine, Michael, ed. 1997. *Population, Poverty, and Politics in Middle East Cities.* Gainesville: University Press of Florida. Profiles of various cities and articles on political, historical, and gender-related themes.

Elsheshtawy, Yasser. 2004. *Planning Middle Eastern Cities: An Urban Kaleidoscope.* London and New York: Routledge. Presents urban planning in the context of globalization, with separate chapters on Cairo, Dubai, and Algiers.

Hitti, Philip K. 1973. *Capital Cities of Arab Islam.* Minneapolis: University of Minnesota Press. Thoughtful profiles of historical capitals: Mecca, Medina, Damascus, Baghdad, Cairo, and Cordova.

Hourani, A.H., and S. M. Stern. 1970. *The Islamic City.* Philadelphia: University of Pennsylvania Press. Delves into the question of whether there is an "Islamic" city, with specific reference to Damascus, Samarra, and Baghdad.

Kheirabadi, Masoud. 2001. *Iranian Cities: Form and Development.* Syracuse, N.Y.: Syracuse University Press. A thorough treatment of the spatial structure and physical form of Iranian cities.

Messier, Ronald. *Jesus: One Man, Two Faiths: A Dialogue between Christians and Muslims.* 2010. Murfreesboro, Tenn.: Twin Oaks Press. Examines the similarities in the Christian and Muslim vision of Jesus in an attempt to bridge the interfaith divide.

Salamandra, Christa. 2004. *A New Old Damascus.* Bloomington: Indiana University Press. Presents a portrait of Damascus' cultural anthropology, with considerable attention focused on the problems of historical preservation.

Serageldim, Ismail, and Samir El-Sadek, eds. 1982. *The Arab City: Its Character and Islamic Cultural Heritage.* Riyadh, Saudi Arabia: Arab Urban Development Institute. Photographs, drawings, and readable text on city form and urban planning.

SUGGESTED WEBSITES

Middle East Online
http://www.middle-east-online.com/english/
News from and about the Arab World

Transparency International
http://www.transparency.org./
Covers corruption everywhere in the world

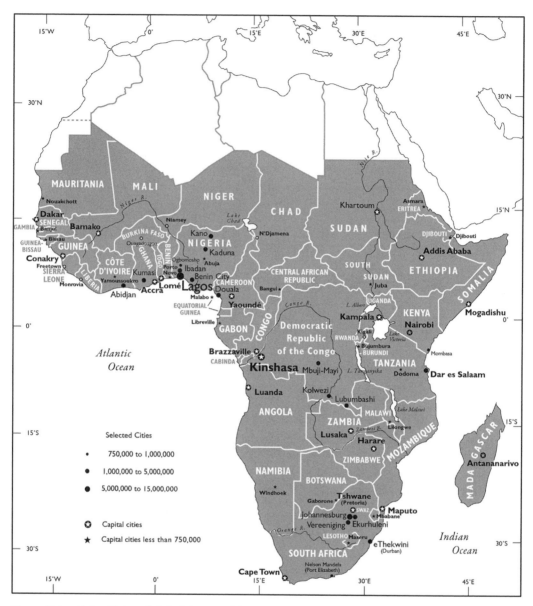

Figure 8.1 Major Cities of Sub-Saharan Africa. *Source:* UN, *World Urbanization Prospects: 2009 Revision,* http://esa.un.org/unpd/wup/index.htm

8

Cities of Sub-Saharan Africa

GARTH MYERS, FRANCIS OWUSU, AND ANGELA GRAY SUBULWA

KEY URBAN FACTS

Total Population	980 million
Percent Urban Population	41.9%
Total Urban Population	411 million
Most Urbanized Countries	Gabon (86.0%)
	Djibouti (76.2%)
	Botswana (64.6%)
Least Urbanized Countries	Burundi (11.0%)
	Uganda (13.3%)
	Ethiopia (16.7%)
Annual Urban Growth Rate	3.39%
Number of Megacities	1 city
Number of Cities of More Than 1 Million	43 cities
Three Largest Cities	Lagos, Kinshasa, Khartoum
World Cities	Johannesburg

KEY CHAPTER THEMES

1. Sub-Saharan Africa (SSA) is among the least urbanized of the world's regions, but it has some of the world's most rapidly urbanizing countries.
2. A rich urban tradition preceded the arrival of colonialism in several parts of Sub-Saharan Africa.
3. Colonialism had profound impacts on urban development, particularly in the creation of what would become primate cities along the coast.
4. Rates of urban primacy are generally high across the region, with a few exceptions, and economic production and political power are concentrated in the primate cities.

5. Many, though not all, primate cities are also capital cities.
6. Many SSA cities have experienced major impacts from cultural globalization, as in changing patterns of consumption, but minimum impacts from economic globalization, in terms of production and investment.
7. Most SSA urban land use patterns and urban economies develop outside of formal regulation, but with significant overlap of the "formal" and the "informal" urban structures and economies, both of which are highly gendered spaces.
8. Many Sub-Saharan African cities are characterized by spatial, socio-economic, and gender inequalities and high rates of urban poverty.
9. Great cultural diversity and creativity help shape very dynamic urban life experiences for residents of the region's cities.

SSA's interlocking urban environmental problems are magnified by shortcomings in management and oversight by both governments and the private sector. Patrick lives in Dar es Salaam, Tanzania, where he works as a chef at a Chinese restaurant. Patrick is a mixed-race South African, born in Cape Town. He worked for many years as a cook on oil tankers, where many of the crew members were Bangladeshi, Philippino, or Tanzanian, the latter often from the Zanzibar islands or from Dar es Salaam. After a stint cooking for an offshore oil rig in Cabinda, Angola, he took up an offer from a Tanzanian friend to come start a new high-end restaurant in the rapidly gentrifying inner city Kariokoo neighborhood in Dar es Salaam. Patrick has found it exciting to learn KiSwahili, which will be his sixth language once he conquers it, but he emails his South African friends around the world in Afrikaans or English. He loves the mix of foods and cuisines available in Dar es Salaam, but favors Chinese food, which is rapidly gaining a foothold in Tanzania and much of eastern and southern Africa along with Chinese investments in the region's cities in the early 21st century. He hopes the restaurant will take off, with an eclectic mix of Tanzanian African, Asian, and European customers.

Meanwhile, across the world in Houston, Jamila, a Nigerian-born software engineer, receives a text message from her father in Calabar asking her to call home. She knows what this means, but hesitates, because she wants to have good news for him before she calls. She sends an email to the secretary of the local hometown association for southeastern Nigerians in Texas, and asks for an update on her plea for help in raising funds—because her father will be telling her, she knows, to come home for her mother's funeral. Jamila has the money for her own plane ticket, but she knows the family will expect her to pay all of the funeral costs and to bring her twin daughters with her. She is conflicted, since she knows that the whole extended family feels that they have invested in her education and emigration with the expectation that she will provide support through remunerations. She has succeeded for many years in sending enough money home to build her parents the nicest house in their neighborhood in Calabar; but her husband's recent

Figure 8.2   Cape Town's Black African township of Khayelisha is experiencing sky-rocketing growth. Its population may exceed one million with many living in densely packed and poverty-stricken developments such as this. (Photo by Stanley Brunn)

death has put a major strain on her household financially, to say nothing of her sadness. The hometown association secretary tells her the news she has been waiting for: in just two days, the large southeastern Nigerian community in Texas has raised more than $10,000 on her behalf. She does not know how she will ever thank these people, many of whom she does not know and only a handful of whom would ever have met her mother. She would do the same for them, she tells herself, because "all of us in what we call The Remote Lands have to stick together to help our Motherland." She calls her father in Calabar, Skype-to-Skype since it is free, with the good news, and on her laptop video camera box through the Skype software, she sees tears on her father's face, for the first time in her life.

AFRICAN URBANIZATION

Sub-Saharan Africa (SSA) has long been among the least urbanized world regions. But many Sub-Saharan countries have been urbanizing rapidly since the 1960s (fig. 8.1). This rapid urban growth has come with limited opportunities for employment in the formal economy or for effective governance. African cities also suffer from a lack of decent and affordable housing, failing infrastructure and basic urban services, alongside increasing inequalities (fig. 8.2). But negative views of contemporary cities are overly simplistic and pervasive. African cities are also creative engines of cultural change and dynamic centers of political and associational life. Many accounts of cities in SSA miss the resourcefulness, inventiveness,

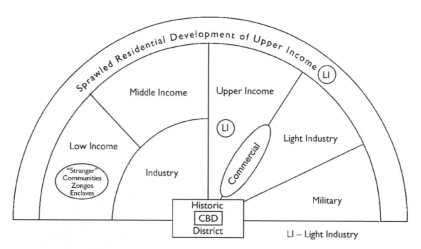

Figure 8.3   A model of the Sub-Saharan African City. *Source:* Samuel Aryeetey-Attoh, "Urban Geography of Sub-Saharan Africa," in *Geography of Sub-Saharan Africa*, edited by Samuel Aryeety-Attoh (Upper Saddle River, N.J.: Prentice-Hall, 1997). 193. Reprinted with permission

and determination of millions of ordinary people who manage to negotiate the perils of everyday life, to make something out of nothing.

Cities in Sub-Saharan Africa are diverse and heterogeneous. Scholarly efforts to construct an ideal model of a generic "African city" in terms of urban structure have failed to find a single profile that fits all cases. Anthony O'Connor, for example, tried to fashion such a general scheme 25 years ago, but his effort led him toward not one but six possible types. O'Connor identified and diagrammed city morphologies that he classified as the indigenous city, the Islamic city, the European city, the colonial city, and the dual city, with examples across the continent. His sixth category—what he termed the hybrid city—actually functions as a kind of catch-all for cities with multiple morphological characteristics. Over time, more cities in Africa seem to have become hybrid cities.

The paths to such African hybrid cities are complex and sometimes contradictory. While many cities came into existence as overseas extensions of European colonial powers seeking to establish beachheads on the African continent, their subsequent growth and development did not conform to one pattern. City-building processes that took place under the dominance of European colonialism often left an indelible imprint on the original spatial layout, built environment, and architectural styles of cities in Africa. Yet with time, these features have sometimes been modified beyond recognition. Thus cities that were built specifically for Europeans, such as Cape Town or Nairobi, still have clear European influences, but these have been overwhelmed by African and even Asian urbanisms. The colonial urbanisms, too, have been dramatically transformed by what amounts to fifty years of independence for most cities. Likewise indigenous, Islamic, and dual cities have in nearly all cases witnessed the steady overlay and erasure of their original forms through colonial and postcolonial impacts (fig. 8.3).

In SSA cities, previous patterns of government dominance in urban centers have been replaced by more reliance on private sector and/or non-governmental institutions. Municipal authorities have not kept up with the demand for infrastructure, social services, or access to resources. Many urban residents have looked outside the formal economy and conventional administrative channels to gain access to income, shelter, land, or social services (box 8.1).

Box 8.1   Multiple Livelihood Strategies

The economic crisis that spread across Africa in the 1970s and 1980s and the structural adjustment programs that were introduced in the 1980s have caused major upheavals in the livelihood strategies of millions of people in African cities, including formal sector employees. For instance, while the introduction of cost-recovery measures has escalated the prices of critical urban services, the real salaries of formal sector employees, especially public sector employees have remained stagnant or, in some cases, declined. Moreover, the limited job creation potential of the private sector in Africa, combined with employment freezes, and retrenchment in the public sector, have reduced avenues for employment in the formal sector. At the same time the policies have led to increased poverty in urban areas. The combined effect of the economic reforms and the urbanization of poverty is that many formal sector employees, either by necessity or by choice, have joined the informal sector in an effort to increase their income earning opportunities and diversify their sources of income.

Since the early 1990s, an increasing number of studies from selected African countries and cities have documented how people of various socioeconomic backgrounds and status seek additional income by engaging in multiple economic activities. Overall, this literature shows that participation in multiple activities is not limited to the urban poor; but it also includes other social classes such as the middle and professional classes that were previously assumed to be immune from the pressures of economic change. Therefore these classes did not need to diversify their livelihood options. Individuals and households across Africa have responded differently to these macro-economic processes based on the nature of their employment, skills, access to resources, socio-economic background and place of residence. Individuals employed in the public or private sectors as well as private entrepreneurs have all attempted to diversify their sources of income, although the motivations for doing so vary. Many formal sector employees supplement their incomes with part-time informal- sector jobs, such as cab driving or petty trading. Other members of their households may also supplement the family income by engaging in similar activities. For instance, many civil servants in Kampala engage in urban agriculture and poultry keeping, own taxis, or operate small kiosks and about two-thirds of households in Accra engaged in at least two income generating activities. Such multiple livelihood strategies have become "the way of doing things" in many African cities, and as a result, the traditional distinction between the formal sector and the informal sector has become more blurry and complex.

The proliferation of multiple livelihood strategies in Sub-Saharan Africa (SSA) cities has significant implications on urban economic studies and on urban planning in the region. First, it signifies the need to revise African city models to include urban cultivation as a legitimate urban activity. Urban agriculture poses a difficult challenge for planners and policy-makers because it is "a ubiquitous, complex and dynamic feature of the urban and socio-economic landscape in Africa." However, the practice cannot simply be wished away in African cities because of its widespread nature; rather it should be seen within the broader context of the urban economy, urban management, and urban development. This would require documenting the benefits and disadvantages of urban agriculture and finding ways of creatively integrating the practice into the urban fabric. Second, one effect of the participation in multiple economic activities is that the notion of the house or dwelling as a mono-functional (residential) unit is increasingly becoming out of sync with the reality in many African cities. Many urban residents of different socio-economic backgrounds have economic enterprises that are located in their homes—a space that conventional planning reserves for residential use only. Urban planners in the region need to introduce relevant changes in zoning regulations and housing design standards, because the multiple function of the house in African cities and the proliferation of home-based enterprises do not appear to be a stop-gap measure. Third, multiple livelihood strategies have also led to the emergence of non-traditional household living arrangements that challenge the conventional definition of households and the distinction between urban and rural residence. Historically, SSA households have used migration as a strategy for ensuring their survival, especially in Southern Africa, but involvement in multiple livelihood strategies requires different and more creative living arrangements. To overcome the limitations of a particular local economy or to expand their options, some households adopt flexible arrangements that allow members to participate in multiple urban and/or rural economies. The final issue relates implication of the increased involvement of public sector employees in multiple economic activities for public sector efficiency. While participation in multiple economic activities by public sector employees in the African contexts benefits those directly involved in the practice, the overall impact on society is often negative. As the involvement of civil servants in multiple income-generating activities became widespread, the moral authority of supervisors to reprimand moonlighting staff was compromised, especially when the officials themselves were guilty of the same. Thus the relationship between the proliferation of such practices among public sector employees and the effects of such practices upon the efficiency of public institutions deserves serious scrutiny.

Source: Francis Owusu (2007) "Conceptualizing Livelihood Strategies in African Cities: Planning and Development Implications of Multiple Livelihood Strategies," *Journal of Planning Education and Research* Vol 26 No. 4 pp 450–463.

The wide range of seemingly unsolvable problems has led some to conclude cities in Africa just "don't work." Others, like the urban scholar AbdouMaliq Simone, prefer to see them as "works in progress," driven forward by inventive ordinary people. In city after city, urban residents rely on their own ingenuity to stitch together their daily lives. SSA cities are often distressed places in need of good governance, management, or infrastructure, greater popular participation in decision making, sustainable livelihoods, and expanded socioeconomic opportunities. Yet they are much more than some form of failed urbanism. To see SSA cities more complexly, we must appreciate the historical specificity and heterogeneous cultural vibrancy of different cities in Africa.

HISTORICAL GEOGRAPHY OF URBAN DEVELOPMENT

Simply because SSA is often considered among the least urbanized world regions, outsiders assume that its cities must be recent. Because European colonialism was such a pervasive regional experience, it is also assumed that the urbanization of Africa ought to be attributed to colonialism. In fact, many SSA urban settlements are much older than the colonial era, and the relationships between formal colonialism and the urbanization process in Africa are more complicated than they first appear. Roughly speaking, we may divide contemporary African cities into categories, including urban areas with origins in the: (1) ancient or medieval pre-colonial period; (2) period of the Trans-Atlantic Slave Trade or European trade and exploration; (3) period of formal colonial rule; and (4) post-colonial period. However, it

rapidly becomes difficult to differentiate cities by these categories. For instance, take the case of Zanzibar, Tanzania, where an indigenous urban center with origins in the 1100s was refashioned under the domination of outsiders from Portugal in the 1500s and Oman in the 1690s, then became caught up in the slave trade and trade with Europe and the Americas in the 1700s and 1800s, then became a British colonial capital, and then the symbolic heart of a post-colonial socialist revolution. Like so many hybrid cities of contemporary Africa, Zanzibar has elements of its fabric that belong to all four of the categories above. Rather than making sharp breaks between city types based on their origins, it is more helpful to simply lay out some of these different types of origin stories and to appreciate that most contemporary African cities are woven together from threads of each origin.

Ancient and Medieval Pre-Colonial Urban Centers

Many urban centers that were prominent before 1500 CE—and in some cases, prominent before the Common Era even began— are ruins now. Other prominent centers of ancient and medieval times were bypassed by the new economic geographies that arose in Africa's relationships with Europe and the New World after 1500, which developed strong associations with coastal urbanisms.

There were at least five major centers of urbanism before 1500, with the oldest being the ancient Upper Nile/Ethiopian centers of Meroë, Axum, and Adulis (fig. 8.4). The medieval Sahelian (or Western Sudan) cities of West Africa's great trading empires, such as Kumbi Saleh, Timbuktu, Gao, or Jenne, arose as middle-agent ports of a network of

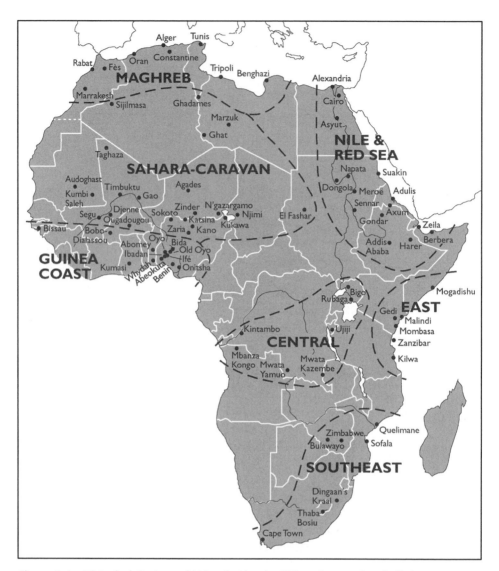

Figure 8.4   Historical Centers of Urbanization in Africa. *Source:* Assefa Mehretu

caravan routes that crisscrossed the Sahara. They achieved significance in the medieval world as nodes of empires, trading entrepots, or centers of learning. Timbuktu, Gao, and Jenne were widely regarded for their scholarship in medieval times, but they disappeared or stagnated after the 15th century. Timbuktu has about the same population as it had seven hundred years ago, and it is only now, in the early 21st century, attempting to restore its extraordinary libraries of medieval thirteenth and fourteenth century histories and religious texts, while Gao and Jenne no longer exist. Other early Western Sudan urbanisms survived the new circumstances of the post-1500 world and developed into important contemporary settlements, for instance the Hausa cities of today's northern Nigeria and southern

Niger, particularly Kano, because their nineteenth century rulers derived great strength from Islamic religious *jihad* movements.

This adaptation and growth after 1500 was even more common for many ancient and medieval cities of Nigeria, like Oyo, Ibadan, and Benin in the Benin-Yoruba area of early urbanism. The Yoruba cities of southwestern Nigeria had developed metalwork artistry and skill unsurpassed in the first millennium world. Benin-Yoruba cities, and neighboring urban areas further to the west, were well positioned to capitalize on the new trade with Europeans after 1500, as is seen below.

Some of the trading city-states along the Swahili coast and the East African coast more broadly, including Mogadishu and Mombasa, also grew after 1500; but many coastal settlements, like the settlements further to the southern interior (the Zimbabwean zone of urbanism, in particular) with whom they traded, largely disappeared. The ruins of the Great Zimbabwe in today's Zimbabwe still demonstrate the remarkable organizational and architectural features of the medieval empire whose central city was located there. The southern interior cities were connected by trade to those along the coast for many centuries before 1500.

Coastal trading centers on the Red Sea and Indian Ocean arose in ancient times, and an extensive trade linking the African interior from Zimbabwe north to Lake Victoria with the Arab and Persian peoples of Asia flourished for more than a thousand years. Beginning in the 9th century, the significance of the Swahili coast ratcheted upwards with increased trade with the Arabian peninsula and the Gulf area, based around the export of gold, ivory, and slaves from Africa in exchange for textiles, jewelry, and other commodities.

East African coastal centers such as Kilwa, Malindi, and Mombasa derived their growth, character, and political organization from the encounters and exchanges between the African mainlanders that founded them and small numbers of Arab, Persian, and even South Asian settlers who made permanent homes there. Their rise is considered part of the medieval golden age of Swahili civilization. The greatest of these, Kilwa, now in ruins in southern Tanzania, had diplomatic exchanges with China in the 15th century.

Urban Development after 1500

Nearly all SSA urban centers of the pre-1500 era were comparatively quite small, with less than 50,000 residents. Europe's impact on SSA changed both the locations and the sizes of major centers. European influence began with the Portuguese in the 15th century. For about two and a half centuries, most contact between European traders and Africans occurred in coastal installations, from which Europeans gradually developed trade networks for various tropical commodities. The slave trade arguably contributed the most to the development of many coastal trade centers between about 1500 and 1870 CE, but this impact was not an unambiguously positive one. During those years, more than 20 million Africans were forcibly relocated to the Americas or died en route; roughly an equal number died or were displaced within Africa. Nonetheless, it is remarkable how many of the major and secondary cities of coastal West and Central Africa in particular grew up in the midst of the trans-Atlantic slave trade.

The Portuguese established the first of these towns, St. Louis, in the 1440s at the mouth of the Senegal River, later creating

Figure 8.5   As an international hub of banking, transportation, and communication for much of South Africa, Cape Town's skyline testifies to the post-apartheid economic growth of the city and its attraction for international businesses and in-migrants. (Photo by Stanley Brunn)

centers at Bissau in today's Guinea-Bissau; Luanda and Benguela in Angola; and Lourenço Marques (now Maputo) and Mozambique in Mozambique. The Dutch, French, and British followed the Portuguese lead. The Dutch founded Cape Town in 1652 (fig. 8.5), and the French and British established West African coastal towns such as Conakry in Guinea or Calabar in Nigeria. Most towns were merely forts—Accra, for example, was originally the site of Fort Ussher, established in 1650 by the Dutch, and Fort James, founded in 1673 by the British.

During the 19th century, the trans-Atlantic slave trade declined, superseded by what was termed the "legitimate trade" in African raw materials. In combination with competition between European firms and states as the

century progressed, African urban areas that participated in this increasingly high-volume and high-value trade grew dramatically. Pre-colonial towns such as Ibadan in Nigeria witnessed considerable growth. The nineteenth century also saw the rise of new or rejuvenated urbanisms in Eastern and Southern Africa. The city-state of Zanzibar grew into the "island metropolis of Eastern Africa" as the center of a mercantile empire whose tentacles stretched to the Congo. Khartoum emerged in the Sudan. A number of South Africa's major cities, including Port Elizabeth, Durban, Bloemfontein, East London, and Pretoria, were founded via European settlement (box 8.2).

During the period of European contact before formal colonialism in the 1880s, SSA's

Box 8.2   South Africa's New Names for Cities

Symbolic renaming went along with the process of dismantling apartheid's geography—the two old provinces that remained largely unchanged had new names (the Orange Free State became simply Free State, distancing it from the Dutch ancestry of the white settlers that had established apartheid, while Natal became KwaZulu-Natal, an integration of the former British colony with the Zulu majority Bantustan). The wholly new ones also became African-ized in some cases. The former Cape Province was sub-divided into the Provinces of Western, Northern, and Eastern Cape, respectively, while the former Transvaal was re-configured into Limpopo (for the river at its northern edge), Northwest, Mpumalanga (meaning Eastern), and Gauteng (SeSotho for "at the gold," home to Johannesburg).

New African names have been created for many of South Africa's cities and towns that had names that symbolized apartheid or the colonial past. The approach often involved incorporat-ing formerly whites-only cities and the non-white townships that surrounded them, retaining the old name (for example, Pretoria) for the area it used to apply to, but governing it as a part of a new Metropolitan Municipality (in this example, Tshwane, "Place of the Black Cow" in Tswana). South Africa now has six metropolitan municipalities, including Tshwane. Port Elizabeth was reconfigured as a constituent part of Nelson Mandela Municipality in honor of the multiracial democracy's inspiring first President (1994–1999). Durban is now a segment of eThekwini ("By the Bay" in Zulu), and East Rand just a part of Ekurhuleni ("Place of Peace" in Tsonga). Johannesburg and Cape Town retain their names, albeit with many local nicknames taking their place in ordinary conversation or media (iGoli—"place of Gold" in Zulu—for Jo-hannesburg, for example). Although the earlier city names still remain in use informally, and even formally in the case of specific cities as areas within the metropolitan municipalities, some urban names have formally and informally been expunged. One dramatic example of this came just west of Johannesburg, in Sophiatown. The apartheid regime forcibly removed the multiracial and income-diverse cultural bastion of Sophiatown early in its reign, demolishing it in an effort to make a "black spot" into a middle-class white suburb like the neighborhoods around it. Apartheid's planners chose the Afrikaans name, Triomf (Triumph), for the new white area when their courts gave them the right to remove the black spot. The new South Africa's Board of Geographical Names promptly erased this example of arrogance and renamed the newly re-integrating settlement Sophiatown again. In keeping with the new openness and technical savvy of the multiracial rainbow nation, this board accepts applications for place name changes online on a continual basis, in any of the eleven national languages.

urban geography began to take form, but under constraints. First, most European contribu-tions in settlement development were coastal with minimal impacts in the interior. Sec-ond, many coastal settlements were intended as transshipment points for trade and lacked regular urban facilities, except those structures that served as European housing or as port and defense establishments. Third, there was a lack of diffusion of European technology

Figure 8.6   The old CBD of Mombasa still evokes the ambiance of the colonial era in this tropical city near the equator. (Photo by Jack Williams)

and culture to the interior's indigenous urban centers.

African Urbanization in the Era of Formal Colonial Rule

The European Scramble for Africa lasted from the 1880s through the 1914 outbreak of the First World War. By that point, virtually the entire continent had fallen under European domination. Ethiopia and Liberia remained independent states, and South Africa became an independent, white-minority-ruled state in 1910; but the British, French, German, Italian, Portuguese, Belgian, and Spanish colonial powers controlled the rest of SSA. Urbanization followed suit, since social and physical aspects of urban development followed the social and political objectives of these European powers. Colonial regimes moved aggressively into the interior of their colonies; and urban settlements sprang up or expanded from existing towns along infrastructure lines (roads or railroads), near mines or large-scale plantation

areas, or in regions requiring administrative centers. Virtually all coastal ports and railheads from Dakar to Luanda became the capitals and/or primate cities in their respective countries, with external trade as their major function. In East Africa, where the resource hinterlands are far in the interior, towns such as Kampala, Nairobi, and Salisbury (Harare) were linked by railways to ports in or near each country, such as Mombasa in Kenya (fig. 8.6), and Beira in Mozambique. Other East African centers, such as Dar es Salaam (fig. 8.7) and Maputo, became important ports.

In South Africa, the pattern was somewhat different. Major European settlement in the interior predated the formal colonial era of most of SSA (*i.e.*, the 1880s to the 1960s), and major mining and agro-industrial towns were well established by 1900. In South Africa, as a result, there are now numerous urban centers in the interior served by a number of ports all around the southern tip of the continent. The railway pattern is much more intensive, with a high degree of connectivity between urban

Figure 8.7   Along this commercial street in present-day Dar es Salaam, one can buy a hot dog, a pool table, a school uniform, jewelry, chocolate, and seeds, according to shop signs. (Photo by Assefa Mehretu)

centers in the plateau hinterland as well as between the interior settlements and the port cities.

In most of SSA, under European colonialism, little real industrialization occurred. Colonial regimes prioritized the export of minerals, metals, or primary goods to Europe, so that industrial development was most intensive in places like the Zambian Copperbelt (in cities such as Ndola or Kitwe) and the neighboring mining province of Shaba (Katanga) in Congo. In many colonies, and in white-ruled South Africa, severe limits were placed in African residency in urban areas. To support the scale of trade that flowed between Africa and Europe and to control the colonies, larger administrations emerged, leading to an outsized service sector for the comparatively shrunken state of secondary sector activities. Urban services were also generally quite warped by race and class (fig. 8.8).

As a result of the limited economic opportunities and restrictions on movement, many SSA urban centers remained relatively small until after World War II. The so-called second colonial occupation of the post-war era, when colonial regimes invested in African development largely in an effort to shape decolonization movements away from the influences of the Soviet Union, led to the growth of investments in many urban areas. Relaxation of migration and residency regulations with independence brought massive rural-to-urban migration in SSA.

There are differences between the respective colonial powers (particularly Britain, France, and Portugal) in terms of their legacies in urban areas, but there are also facets of their legacies held in common. British colonies with substantial white settlement developed more highly segregated urban settlement patterns regulated by more rigid building rules and land

Figure 8.8   These colorful apartment houses dot the landscape of Cape Town's Bo-Kapap area. It is known as the "Cape Malay" district. (Photo by Brennan Kraxberger)

laws than would be the case for colonial cities in the interior areas of many French West African colonies, for instance. Cities with significant white populations in the colonial era tended to have larger investments from colonial states for infrastructure and from the private sector for industrial development. Yet exceptions to these differentiations existed, and the distinctions between different colonial powers' strategies in urban areas are often overridden by the commonalties. One can still see some distinctly British features of eastern and southern African cities in architecture (many colonial government buildings still in use were designed by the British architect Herbert Baker and an army of his protégés) or urbanism more generally

Figure 8.9   A dramatic air photo of Lusaka, Zambia, today shows the formerly all-white township of Roma. *Source:* Garth Myers

(the many small urban parks just adjacent to Central Business District (CBDs) with the same strict use rules on signs at their entrance that one sees in London or Hong Kong). Many of the neighborhoods formerly segregated by race are now just as segregated, but by class, as illustrated by the dramatic air photo from the late 1990s in Lusaka, Zambia, of what was until 1964 the whites-only and separately-governed township of Roma and the informal settlement of Ng'ombe on its eastern edge (fig. 8.9). Today,

Roma is populated predominantly by the African professional class and the political elite of Lusaka, and their maids and gardeners—also African—still live in Ng'ombe. Distinctively French architectural or planning legacies will also be in evidence up until today in former French colonies. But over time, cities all across the region are becoming more and more alike in their hybrid form and function as the post-colonial era brings unprecedented urban growth to SSA.

Figure 8.10   Along Great East Road in Lusaka, Zambia, the informal economy punctuates the streets as vendors sharpen the pitches they need to clinch each sale. (Photo by Angela Gray Subulwa)

Post-Colonial Urbanization

From the 1960s through the 1980s, SSA contained the world's most rapidly urbanizing countries. Eastern and southern African countries have led the world in urbanization rates for nearly half a century. Some countries had annual rates of urban population growth near or above 10 percent from 1960–91. Even during the 1990s and 2000s, when many observers noted a slowdown in African urbanization, several countries had estimated urban growth rates near or above 5 percent. In less than 50 years, some eastern and southern African countries have gone from being largely rural societies to being places where almost half the people live in or around cities.

The rapid growth of many cities and the path of urbanization in most countries have been somewhat distinct from what has been seen in other regions, particularly in wealthy European or North American settings. With some exceptions, the extraordinary story of urbanization in Sub-Saharan Africa has not

Table 8.1 Female & Male, age 15–24, in Informal Employment (in percent)			
	Female	*Male*	*Year*
Benin	79.8	69.8	2006
Burkina Faso	82.8	20.7	2003
Cameroon	69.9	62	1998
CAR	96.8	75.2	1994
Chad	83.6	60.2	2004
Comoros	93.4	54.5	1996
Congo	92.5	52	2005
Cote d'Ivoire	77.6	35.3	1998
Ethiopia	69.9	16.8	2005
Gabon	75.6	71	2000
Ghana	85.2	30.1	2003
Guinea	98.6	-	2005
Kenya	63.8	5.3	2003
Madagascar	77.7	-	1997
Malawi	72.6	-	2000
Mali	91.2	53.3	2001
Mozambique	70.9	8.5	2003
Namibia	38	-	2000
Niger	92.1	57.2	1998
Nigeria	59	16.8	2003
Rwanda	60	23.2	2005
Senegal	84	23.9	2005
South Africa	39.3	-	1998
Togo	94.3	60.1	1998
Uganda	74.4	14.9	2001
Tanzania	70.6	4.7	2004
Zambia	68.7	11.4	2002
Zimbabwe	53.6	-	1999

From: Global Urban Indicators Database 2010

SSA, the ever-expanding numbers of urban residents have become increasingly dependent on what are termed informal activities—small scale, low-technology manufacturing, petty wholesale trading, and informal service provision—for basic needs and daily life (fig. 8.10). Many of these informal economic activities are highly gendered activities that are defined by—and often challenge—traditional understandings of gendered divisions of labor (table 8.1). Especially in the case of southern Africa, the informal sector is sometimes associated with another problem, HIV/AIDS (box 8.3)

Current Urbanization Trends

Compared to the other regions, SSA still has one of the lowest levels of urbanization. According to UN estimates, only 40 percent of the region's population lived in urban areas in 2010. There are however significant differences in the levels of urbanization within the region. Coastal western Africa and southern Africa have the most developed urban hierarchies. Eastern Africa is the least urbanized (table 8.2).

Unlike the other regions of the world that play dominant roles in the globalization process, SSA lacks any major "world city" given its marginality to the world economic system. Although Johannesburg plays a dominant regional role, most large SSA cities are centers

accompanied a substantial economic transformation of society towards industry and manufacturing. In some countries, notably South Africa, industrial development occurred with the urban-ward trends. But in much of

Table 8.2 Urban Population as Percentage of Total Population		
Regions	*1995*	*2010*
AFRICA	34.6	40.0
Eastern Africa	21.9	23.6
Middle Africa	32.8	43.1
Southern Africa	49.7	58.7
Western Africa	36.3	44.9

Box 8.3   HIV/AIDS and Urban Development

In the past 25 years, a new scourge has appeared in Sub-Saharan Africa (SSA) that is creating havoc among the already weak and struggling nations. HIV/AIDS is a worldwide threat, but is particularly concentrated in SSA. According to the 2010 Report on the Global AIDS Epidemic, a little more than one-tenth of the world's population lives in the region, but SSA is home to almost 68 percent of all people living with HIV—22.5 million. About 2.3 million of them are children younger than 15 years of age, with women comprising of about 60 percent of adults living with HIV in the region. The report also suggests that new HIV infections across the region peaked in the late 1990s, especially in Kenya, Zimbabwe, and in urban areas of Burkina Faso. Southern Africa appears to be the global epicenter of the epidemic—almost one in three people infected with the disease globally live in this sub-region. The hardest hit countries in 2010, in terms of the percentage of adults living with HIV were (in descending order): Swaziland (25.9 percent), Botswana (24.8 percent), Lesotho (23.6 percent), South Africa (17.8 percent), Zimbabwe (14.3 percent), Zambia (13.5 percent), Namibia (13.1 percent), Mozambique (11.5 percent) and Malawi (11 percent). These countries are concentrated in the southern part of the continent, but most countries in SSA are affected to some degree. In many of these countries, the rate of infection among pregnant women is even higher.

Why such a disproportionate share in this already beleaguered part of the world? The reasons are many:

- Governments in SSA have been in denial, refusing to recognize the existence or severity of the threat.
- The health care system, including the availability of effective drugs and blood-screening procedures, is woefully inadequate to meet the challenge.
- Widespread poverty, illiteracy, and ignorance make it extremely difficult to educate people about the dangers of the disease and how to prevent infection.
- There are large numbers of people in high-risk groups (sexually active workers, sex-industry workers, migrants, military personnel, truck drivers, an intravenous drug users who share needles).
- Male-dominated societies put women at particularly high risk.

HIV prevalence is higher in the growing cities of SSA. According to the South African Cities Network (SACN), rapid urbanization, often associated with the growth of informal settlements, appears to provide a favorable environment for the spread of diseases, including HIV/AIDS. Cities, by concentrating large numbers of people in small areas, increase the speed of transmission of HIV. Indeed, there appears to be a higher prevalence of HIV in urban areas as compared to non-urban areas worldwide. In an international survey of countries with the highest HIV prevalence rates, UNAIDS found a strong trend towards higher HIV prevalence rates in urban areas as compared to nonurban areas. The urban/nonurban differences in SSA are particularly significant in Botswana, Congo, Ethiopia, Lesotho, Mozambique, Namibia, Rwanda, Uganda, and Zambia. The only Sub-Saharan countries where HIV prevalence rates in non-urban areas were found to be higher than in urban areas were Liberia and the Democratic Republic of Congo.

The high prevalence rates in cities threaten to make even worse the many problems plaguing these countries and their cities. The disease mostly affects adults in their early productive years, thus directly impacting development of social capital and creating large numbers of orphans. This witches' brew obviously makes it even harder for SSA nations to attract foreign investment and turn the tide economically. There are glimmers of hope, nonetheless, in various parts of SSA, where a wide variety of initiatives are making some progress, often in localized areas. The rate of prevalence has dropped on the continent from 2001–2009, according to UNAIDS, from 5.9 percent to 5.0 percent, and the decline is very significant in some countries. Tanzania's rate, for instance, dropped from 7.1 percent to 5.6 percent in this decade. A good example of a local group that seeks to educate the public is a group in Cape Town called Wola Nani (wolanani@Africa.com). This group airs radio announcements in the various languages of South Africa and provides counseling, home care, workshops, training for those living with someone infected with HIV, community AIDS education, and other services.

From an urban development standpoint, high AIDS rates might mean that some cities in SSA will stop growing, or even decline, in population in coming years. Declining economies and political/social instability in the cities might well discourage people from migrating to the cities, thus slowing or even reversing the current trend of increased urbanization. SSA thus might be the one region of the world in the 21st century to deviate significantly from the trends observable in cities of most other developing realms.

of national economies. Despite its marginal position in the global economy, urbanization in the region has continued. The proportion of its population in urban areas was 15 percent in 1950; it then jumped to 25 percent in 1970; and it is projected to exceed 47 percent by 2025. Between 1950 and 1995, SSA's urban population increased by an average of 5 percent per annum—this represents about twice the average population growth rate of the region. Since then however, the growth rate has slowed and it is projected to be slightly below 3 percent by 2030. There are also significant variations in the urban population growth rates of the countries in the region.

Although the overall growth of urban population in SSA has slowed in recent years, most of the major cities continue to increase their populations. For instance, the 2010 population of Lagos, Nigeria's metropolitan area is estimated to be over 12 million; Kinshasa-Brazzaville is 10.5 million; Johannesburg 8.2 million, and Abidjan 4.3 million. Between 1996 and 2006, Conakry doubled its population to 2 million, and Luanda also more than doubled its population from 2.2 million in 1995 to 4.7 million in 2010. Also, some of the secondary urban centers have experience some growth since the 1960s due to deliberate government policies to slow the growth of the capital cities.

Most of the largest cities in SSA are the national capitals of their country. In most cases, the other urban centers are much smaller, except when such cities house major economic activity like a mine or a port. Much of development efforts and the concentration of a host of functions including administrative functions and as a port and center of communication, industry, commerce, education and culture, have focused on such national cities while ignoring the many smaller urban centers.

Many SSA cities derive their importance from the role they played during the colonial and/or post-colonial eras. An important post-colonial SSA urban development phenomenon is the creation of newly planned cities. The first of such cities was the port of city of Tema, Ghana, built in the early 1960s in anticipation of the country's industrial development. Several countries have since established new cities and their capitals, including Dodoma in Tanzania, Lilongwe in Malawi, Yamoussoukro in Côte d'Ivoire, and Abuja in Nigeria. These new capitals were meant to give the nation a "fresh start" and to direct growth away from existing cities. However, considering that none of the new capitals, other than Abuja, have grown to more than about half a million inhabitants, one can say that these new cities have not had any significant influence on the growth of the already established cities. And, in Abuja's case, the staggering rates of growth that the city has experienced (its population jumped from 1.3 million to 2 million between 2005 and 2010

alone) have far outstripped Nigeria's planning capacity for coping with it.

Another characteristic of SSA urban centers is the importance of port cities. Apart from the land-locked countries in the region and those that have created new capitals, many of the rest of the countries have port cities as their capital city. This is often a carryover from the colonial period when the main function of the capital city was to provide access to the metropolitan country. In addition, SSA's role during the colonial era as the producer of natural resources led to urban development that was based on resource exploitation. Zambia provides a good example of urban centers that grew out of the copper mining centers. For instance, Chingola grew up around the Nchanga copper mine; Kitwe is at the site of the Nkana mine and Luanshya stems from the Roan Antelope copper mine. Another important feature of SSA's urban evolution is increasing importance of tourism cities. Mombasa, the second largest city in Kenya and the center of the coastal tourism industry, continues to attract immigrants from the interior of Kenya because of the employment opportunities in the tourist industry. Gorée Island, located just off the Dakar Peninsula also attracts many tourists annually because of its slave history. Similarly, Cape Coast and Elmina in Ghana attract many tourists who are interested in the experiences of the transatlantic slave trade to the castles from where the slaves were shipped to the New World.

DISTINCTIVE CITIES

Kinshasa: The Invisible City

About 35 percent of the Democratic Republic of the Congo's (DRC) population lives in cities,

Box 8.4   Kinshasa's Imaginative and Generative Side

Kinshasa is often seen as one of the worst examples of what has gone wrong in Sub-Saharan Africa's (SSA) cities. It has grown very rapidly without corresponding industrial manufacturing growth, and has endured decades of mismanagement amidst severe governance crises in the Democratic Republic of Congo (DRC). Its vast sprawl and poor infrastructure are part of why it is often portrayed as an example of relocalization: where a city becomes a set of villages distinct and cut off from each other. And yet, at the same time, Kinshasa has endured as a major engine of creativity in music and the arts, and its people display tremendous ingenuity in manufacturing the means to survive the challenges of life in the city. In recent years, its residents have farmed its cemeteries and reclaimed stretches of the Malebo Pool in the Congo River as arable land. Filip de Boeck has estimated that Kinois (the people of Kinshasa) have now empoldered more than 800 hectares (1,977 acres) of the Malebo Pool. More than 80 farmers' associations govern this vast urban agricultural garden belt essentially outside of government control. What de Boeck calls an "organic approach to the production of the city" certainly does not occur without conflicts, but its freedoms and innovations need to be recognized in any future attempts to come to grips with this megacity. Unfortunately, the creativity of Kinois residents is more frequently subjected to harsh and capricious crackdowns, street-sweeps, or programs of demolition.

The new democratically elected regime of Joseph Kabila has, since its 2006 election, invested heavily in remaking Kinshasa's downtown, with Chinese, Indian, Pakistani, UAE, or Zambian engineers, contractors, or investors. Ubiquitous billboards advertise the global ambitions for Kinshasa among the DRC's new elites, including a proposed gated condominium community, *La Cité du Fleuve*, to be built on two artificial islands in the Congo River. Kinshasa's artists, such as Bodys Isek Kingelez, have re-imagined Kinshasa as well. Kingelez's most striking piece, *Projet pour le Kinshasa du troisième millénaire*, is a multimedia model imaginarium for the DRC's megalopolis in the future, the 'Third Millennium.' In all likelihood, the future of Kinshasa belongs to the farmers reclaiming Malebo Pool for their gardens more than it does to the dreamland of billboards or dioramas of its glory. But Kinshasa, far from being an "invisible" city, actually makes itself a visible symbol of all that is wrong, but it also is a symbol of all that is marvelous, about the DRC.

See: Filip de Boeck, 2010, "Spectral Kinshasa: Building the City through an Architecture of Words," paper presented to the workshop, "Beyond Dysfunctionality: ProSocial Writing on Africa's Cities," Nordic Africa Institute, Uppsala, Sweden.

with that percentage expected to top 50 percent by 2025. Kinshasa's population was conservatively estimated to be 10.5 million in 2009. This capital city has more than 14 percent of the DRC's population. Instability and warfare (especially from 1996–2002) have hindered Kinshasa's economic development, even while enhancing incentives for Congolese people to migrate to it. Its rapid growth in the last half-century has outstripped the government's political and economic capacity to provide for its needs (box 8.4).

For Sub-Saharan Africa's second-largest city, Kinshasa, has a relatively brief and notably turbulent history. The British-American explorer-agent for the Belgians, Henry Morton Stanley, built a new city just adjacent to a set of pre-existing settlements in 1881, naming it Leopoldville to honor Leopold II, the Belgian king. A railway connection with the coastal port of Matadi soon made Leopoldville a key town for linking the vast interior of the basin with the world economy, and in 1923 Leopoldville became the Belgian Congo's capital. Eventually, the Belgians extended the city's boundaries; at independence, this expanded entity became Kinshasa.

Kinshasa had only about 30,000 residents in the 1880s, but this had risen to 400,000 by independence in 1960. As was the case in many cities in SSA, colonialism held the population down by enforcing restrictions on urban residence. At independence, the new government ended these controls, opening the doors for massive rural-to-urban migration. For much of the last half-century, the annual growth rate of the city's population has been above 10 percent.

Until 1945, most of Leopoldville's Africans lived not in the city itself, but in adjacent riverine settlements. After the Second World War, new neighborhoods arose, some planned for African workers by the colonial regime. These planned neighborhoods were nearly the only serious investments in African areas of the city made during the Belgian era; Leopold II's Congo Free State and the Belgian Congo that replaced it in 1908 are considered by many scholars to demonstrate the worst case for negative impacts of colonial rule,

with extremely limited investments in human welfare or security in the colony's capital city. Thus Kinshasa's infrastructure woes are not entirely the result of warped post-colonial era's politics—the colonial regime failed to provide urban services to African areas even when investing heavily in European areas of the city.

The governments that have ruled Kinshasa since 1960 and the private sector entities that have invested in the Congo's vast resources have not improved matters. Both government and the formal private sector failed to keep pace with Kinshasa's housing, infrastructure or employment demands. Despite this, migration to Kinshasa continues to rise. Warfare, violence, hunger, insecurity, and the departure of industries from rural areas drive people to Kinshasa. Their perceptions do not match with realities—some 60 percent of Kinshasa's work force is estimated to be unemployed, housing and sanitation conditions remain poor, and environmental health problems are rampant.

The largest zones in the city population-wise are at the far eastern and far western edges of the ever-expanding urbanism. Growth in these and other areas is mostly unregulated and uncontrolled. Post-independence efforts to provide public housing, credit facilities, or transport have been marred by gross corruption, mismanagement, and negligence, particularly under the notorious dictatorship of Mobutu Sese Seko, who ruled the DRC (which he renamed Zaire) with brutal inefficiency from 1965–1997. As a result, Kinshasa's residents do as much as they can informally, outside of the state's purview or the formal private sector. So much of what comprises Kinshasa in both physical and economic terms

Figure 8.11   Kinshasa workers wait by the side of the road at an "artificial" taxi and bus stop. Note the poor road and heavy air pollution, plus the use of median space for roadside gardens. (Photo by Daniel Mukena)

is undocumented, giving rise to discussion of it as an Invisible City. Geographers Guillaume Iyenda and David Simon estimate that three-fourths of Kinshasa's houses are self-built by their owners, often in such close proximity to one another as to prohibit sufficient road construction. Roads, railroads, airports, port facilities, river transport, bridges, and public vehicles in Kinshasa have deteriorated steadily.

Industry in Kinshasa has been declining for thirty years or more. Rioting, looting, and urban violence in the 1990s and 2000s reduced the city's industrial capacity still

further. Kinshasa's manufacturing sector still produces soft drinks, beer, cigarettes, textiles, soap, matches, plastics, newsprint, and other lower order goods, but in declining volume. The service sector completely dominates Kinshasa's economy, accounting for three-fourths of all urban activities (fig. 8.11). Yet the extraordinary degree of urban primacy that Kinshasa still maintains means that it continues to dominate the DRC's economy, accounting for between 19 percent and 33 percent of all firms or establishments in each sector.

Despite the negativity that surrounds most descriptions of and scholarship about

Kinshasa, this megacity is a thriving center of the arts, particularly for popular music. Kinshasa's musicians have produced chart-toppers and dance-hall favorites across SSA and Europe for decades, inventing new styles of music and dance and pushing on through every new twist in the city's political and economic malaise. The 2006 democratic elections in the DRC appear to have marked a turning point toward peace and stability, and Kinshasa is—possibly, and arguably for the first time—poised to benefit from the DRC's tremendous and untapped base of natural resources. DRC was once among the top five producers of industrial diamonds in the world; and these are still estimated to account for more than half of the country's export earnings, alongside extensive copper, cobalt, coffee, palm oil, and rubber exports. Such riches might provide for a rebirth of possibilities for Kinshasa—although they might just signal a continuation of missed opportunities.

Accra: African Neo-Liberal City?

Ghana's urbanization rate for 2009 was 51 percent, and it is projected to increase to over 58 percent by 2020. The two most important cities in country are Accra, which is along the coast, and Kumasi, in the interior. Accra, however, with an estimated 2009 population of 4 million, is the undisputed primate city of Ghana. Accra's dominance also manifests itself in the political, administrative, economic, and cultural spheres. The country's open economic policies and its relative political stability in a region characterized by instability have also elevated Accra's influence internationally. Accra has experienced a surge in business and industry, becoming a destination

for many foreign visitors to West Africa. At the same time, a significant proportion of the city's residents have not benefited from the good economic fortunes.

Accra began as a coastal fishing settlement of the Ga-Adanbge people in the late 16th century. Although there were other trade and political centers in the inland of the country at the time, there is no evidence that they were connected in any way to Accra. During the 17th century, a number of forts were established in the area by the Europeans. The rise of Accra as an urban center began in 1877 when it replaced Cape Coast as the capital of the British Gold Coast colony. Unlike many SSA capital cities such as Dakar that were selected because of pre-existing economic advantages, the choice of Accra was influenced by the colonialists' desire to find a newer area that would protect Europeans from native-borne diseases. Accra's new status as the capital made it an attractive location for many merchants and investors, and by 1899, the city had been transformed into the busiest port on the Gold Coast with the largest number of warehouses. The colonial administration used legislation to limit the development of manufacturing in the city, so at independence in 1957, Accra had developed a reputation not as a factory city but a warehouse city. As the first city in Africa to become the capital of a new nation after World War II, Accra also became an important political center for the struggle for independence in Ghana and in Africa.

Post-independence governments in Ghana continued to promote the city's development by concentrating governmental functions and economic opportunities in the city and ignoring the other important cities in country, such as Kumasi. As a result, Accra expanded as the administrative functions for the entire country expanded. In addition, the development of Tema port led to the abandonment of Accra harbor as a commercial port. However, like many cities in Ghana, Accra's growth began to slow down significantly in the 1970s and 1980s because of the economic crisis that engulfed the country.

The Ghanaian government accepted a World Bank-supported economic reform package in 1983 and agreed to pursue liberal economic policies, including the privatization of state owned enterprises, deregulation of currency markets, promotion of the private sector and foreign direct investment, reduction in the public sector and trade liberalization. These free market policies are essential for understanding the contemporary urban economy of Accra. The policies helped to transform the state-controlled business environment in the country and encouraged the development of the private sector. It became easier to import many commodities, including building materials, leading to rapid residential development in the city and an expansion in the number of motor vehicles that has cluttered Accra's roads. In addition, the infrastructure-building program created visible signs of development in the city such as major road construction (new thoroughfares and flyovers), upgrading of the international airport in Accra and the Tema port, and the creation of export processing zones to attract foreign investors. As the hub of Ghana's economic activities, Accra has also become the host to a number of the national, regional, and multinational financial and business institutions that have flocked into the country. These economic activities have exceeded the ability of the old central business district to house them; and as a result, many of the headquarters are located around the outskirts of the city.

Yet not all residents in Accra have benefited from the free market policies—the negative effects of the policies are also visible on the urban landscape of Accra. Income levels of most residents have not kept up with the rising cost of living. For instance, it is estimated that the number of households in poverty in Accra more than doubled between the 1988 and 1992—from 9 percent to 23 percent. Lack of employment opportunities for the majority of the residents has widened the gap between the rising small middle and the poor majority. The unequal distribution of wealth can be seen from the proliferation of new housing development in the outskirts of the city, including development of gated communities, luxury apartment buildings, expensive urban shopping malls, and the increased securitization of architecture presumably for protection against crime. Although the crime rate in Accra is low (compared to other SSA cities such as Lagos), the growth of the private security industry in the city reflects the feeling of insecurity by the residents and emphasizes the need to address this emerging problem. The poverty in the city can also be seen from the increasing number of street traders on busy intersections and other hot spots. Many of the poor, including young children, make a living by hawking anything that they can find, especially the ubiquitous water in plastic bags. The effect of the liberalization policy has also been the flooding of the city with vehicles, which combined with lack of comprehensive transportation planning, has created insurmountable traffic problems in the city.

Ghana celebrated its 50 years of independence in 2007, with the country booming with many activities of varying proportions; and Accra played an important role, not the least as the nation's capital. In the same year, Ghana also discovered oil at the offshore Jubilee Field and began pumping the oil in 2010. It is hoped that the nationalistic overtones of the 50th independence anniversary and the county's oil resources will help address the challenges facing the majority of the city's residents who are have so far not benefited from the liberalization of the city but are bearing the brunt of the government's policies.

Lagos: Megacity of SSA

Nigeria's population is more than 49 percent urbanized. Lagos, with almost 11 million inhabitants in 2010 and over 16 million projected for 2025, qualifies as SSA's most populous city and one of the world's megacities. The development of the petroleum industry in Nigeria has given a boost to urban development, including that of Lagos. It is often said that Lagos owes its growth and dynamism to European influence. Yet, it is also true that Lagos, in its development dynamic, owes much to early African urban development. Lagos was established in the 17th century, when a group of Awari decided to cross over the lagoons and settle in a more secure setting on the island of Iddo. They later crossed over to Lagos Island in search of more farmland. In this manner, the three important parts of the city of Lagos were founded as fishing and farming villages by the indigenous population well before the impact of major external influences in the 18th century were felt.

Another important historical factor in the development of Lagos is its significance in the slave trade between 1786 and 1851, in which Africans, especially the Yoruba, played important facilitating role. Lagos was not a slave market until 1760, but it soon became one of the most important West African ports in the slave trade. Lagos Island became an important center where slaves were barricaded as they awaited their export along with primary commodities, particularly foodstuffs and Yoruba cloth, which reached markets as distant as Brazil. Although in 1807 the British passed an act to abolish the slave trade, Lagos, because of its locational advantage, continued the trade until it was halted by the British invasion of the city in 1851. British bombardment of the city also caused a temporary decline in the city's population. With the cession of Lagos to Britain as a colony in 1861, the colonial era for Lagos had begun.

People continued to move into the "free colony," leaving behind slavery, war, and instability in the interior. Freed slaves also returned from Brazil as well as Sierra Leone and made their homes in Lagos. Toward the end of the 19th century, Britain intervened to stop internal hostilities and established a protectorate over the whole of Nigeria. A railway from Lagos, begun in 1895, reached Kano in 1912. As its effective hinterland now expanded to the interior of Nigeria, Lagos became even more important as a trade and administrative center. By 1901, the city had a population of more than 40,000; and, by this time, the future prominence of the "modern metropolis" was pretty much established.

Lagos has experienced spectacular population growth and spatial expansion in the past four decades. The population grew at an estimated 14 percent per annum in the 1960s and early 1970s. This growth rate slowed down to an estimated 4.5 percent per annum in the late 1980s—a trend which has continued as the cost of living in the city and other problems continue to rise. Many of city's current problems are rooted in its rapid growth. It has been called the "biggest disaster area that ever passed for a city." That may be overstating it, but Lagos has acute, sometimes incomprehensible, problems of congested traffic, inadequate sanitation, housing, and social services, and urban decay.

Lagos is a primate city. The disparity in socioeconomic status between the elite and the mass of urbanites is very wide. That also means the city reflects two contradictory modes of living: one that is an extension of European style brought about by those who can afford the luxuries of a high level of technology and another that is an extension of the traditional mode of living, which has been distorted to fit an urban milieu. This curious amalgam, as it reflects itself in an African urban setting, loses the beauty, charm, and convenience of either of its parts and becomes a nuisance, as exemplified by the traffic congestion and slum dwellings of Lagos.

As a primate city, Lagos has the typical problems of rapid population growth and insufficient employment opportunities. The net effect of these problems is enormous. It depresses urban wages to almost marginal subsistence levels and adds to the pressure on urban amenities and housing, as well as to numerous other social problems, especially in the slums of Lagos and peripheral residential communities. In fact, an estimated 64 percent of the residents of Lagos reside in slum areas. Lagos is a good example of an African primate city whose growth rates and attendant problems in distorted consumption patterns have created a stultifying effect that a weak and often disorganized city government is incapable of handling.

There are, however, some positive developments underway. Abuja, in the vicinity of the confluence between the Niger and Benue rivers, was designated as the new capital city of the country and all government functions have steadily moved to that more central location. This has meant a major step toward decentralization, and it has reduced the concentration of functions in Lagos.

Lagos, even in the colonial period, had less than 5,000 expatriates. Hence, compared with Dakar, Nairobi, and Kinshasa, the character of the city and its spatial organization were considerably less a function of the impact of the Europeans. The process that Lagos is undergoing, if there is any recognizable process at all, may throw light on the problems of indigenization of African primate cities that have been, and in most cases still are, enclaves of European economic systems and often as alien to their people as cities in Europe. Lagos is a bona fide African city and, as disorganized as it is, it may offer a lesson on the transition from a colonial to an indigenous urban environment.

Nairobi: Urban Legacies of Colonialism

East Africa has commonly been taken to be SSA's least urbanized region, and until quite recently, it was estimated that only 20 percent of Kenya's population lived in cities. That percentage had risen to 43 percent by 2007, and it is expected to hit the 50 percent mark by 2015. Nairobi, the capital and primate city of Kenya, is currently estimated to have 2.9 million residents. It is a transport hub for much of East Africa, as well as a key site for international diplomacy, and it serves as the headquarters or conference space for many international organizations. Despite its short history, Nairobi has grown into the major industrial urbanism in its region, and the economic engine of Kenya. Nonetheless, it is a city with substantial rates of poverty, great disparities between rich and poor, and faltering urban services.

Nairobi is something of an accident of geography. Uninhabited forest and swampland in the 1890s, Nairobi was by 1906 the site of the new capital of British East Africa (renamed Kenya Colony in 1920). Nairobi became the headquarters of the BEA's Uganda Railway, conveniently positioned near the Nairobi River and the mid-point of the rail line. The site's drainage and health problems did not prevent the colonial rulers from seeing the new railway headquarters as an ideally situated forward capital for the colony.

By 1906, the new city contained more than 13,000 people. By 1931, this population had grown to 45,000, nearly 60 percent of whom were Africans. Nairobi became the most important colonial capital in the region, as the seat of Britain's High Commission for East Africa (including the colonies of Kenya, Uganda, Tanganyika, and Zanzibar). By 1948, the city had more than 100,000 people, and its growth continued steadily through to independence in 1963.

Nairobi's colonial legacies continue to haunt it. The first element of this legacy is its physical location. It might have been convenient as a site for railway administration and management, and its geographical centrality might have assisted the efficiency of colonial rule; but physically, colonial Nairobi was, as the early colonial administrator, Eric Dutton, once put it, "a slatternly creature, unfit to queen it over so lovely a country." Sanitation and urban services more generally lagged—and continue to do so—in part because much of the city lies in or near wetlands.

The second legacy scars Nairobi more heavily, and that is the legacy of colonial segregation. Nairobi was built to be the capital of what its small population of European settlers claimed as a "white man's country"; and, though Africans were a majority in the city by 1922, most were not legally given rights of residency in the city under colonial rule. When the colonial regime did begin to formally plan for the African areas in Nairobi in the 1920s, these were consistently laid out in the lowest-lying, least desirable eastern areas of the urban zone. Whites invariably were situated in the higher elevation areas west of Nairobi's downtown. Since the colonial regime also at times encouraged the immigration of Indians and Pakistanis to East Africa, Nairobi quickly developed a substantial Asian population that took up residence in the middle, literally and figuratively.

Under colonial rule, the Europeans—themselves divided between the administrative functionaries and an elite class of settlers—controlled the government, the resources, and the finances of Kenya despite the paltry percentage of the population they represented. More than half of the urbanized land of Nairobi in the 1960s still remained in white hands, when whites comprised less than 5 percent of the city population. The Asians of the colonial era were mostly shopkeepers and merchants or skilled artisans. Eventually, much of the land of the eastern, African-dominated areas

Figure 8.12   This modern office complex in Nairobi points to the city's role in building Africa's future as a command-and-control center for East Africa's economy. (Photo by Garth Myers)

of the city came to be Asian-owned. From the beginning, in Nairobi, Africans occupied the lowest ground and the lowest rungs of the economy. Most Africans lived in rented housing built by the city or their employers. The African residential zone of Eastlands, for example, was characterized by high turnover rates, high unemployment, and poor environmental conditions.

Although the tripartite racial geography of the city has faded somewhat in the forty-eight years since independence, Nairobi remains a heavily divided city, in class terms now as much or more so than in racial terms. The western and northwestern suburbs remain low density elite—albeit increasingly multiracial—areas. Some African elites have moved into traditionally European and Asian neighborhoods, but this is a minority in the upper echelons of political society. Upper Nairobi and the "Nairobi Hill" residential areas continue to be dominated by European single-

unit and fashionable homes complete with servants' quarters. Well-to-do Asians inhabit Parklands, adjacent to the historically European sector. Poorer Asians live in Eastleigh. Some of the Asian population has moved to a second Asian quarter in Nairobi South. The CBD and middle-class or working-class zones predominate in Nairobi's geographical center, and most of the city east of downtown is dominated by informal squatter settlements. The unregulated growth of the latter has been the major story of the post-independence landscape of Nairobi, and occasionally these interrupt the general geographical pattern.

Although both its growth rate and its economic health have declined in the past twenty-five years, Nairobi has become a major African metropolis. Its primate city role for Kenya is augmented by its role as an international center for East Africa and even SSA more generally (fig. 8.12). Nairobi is a prime African example of a *splintering urbanism*, where one portion of the city is highly integrated with the world economy while another larger portion is disintegrated, literally and figuratively. Compared to many SSA cities, Nairobi had a good record in industry in the 1950s and 1960s and an important financial services sector in the CBD. The average European, African elite, or Asian in the city lives in a comfortable home in Upper Nairobi or Parklands and works in the CBD or some other similar enclave. Many African residents are not integrated into the core functions of the city and find themselves locked out of the Nairobi economy that most Western visitors see—as in the development of gated communities for elites in the city. The informal economy provides the overwhelming majority of job opportunities and residences in Nairobi now.

The CBD of Nairobi represents one of the busiest spots in the continent. Its most prominent functions are commerce, retailing, tourism, banking, government, international institutions, and education. It gained unwanted international notoriety with the 1998 bombing of the US embassy and an adjacent office building that caused 284 deaths, all but ten being Kenyans. It returned to the news again with a wave of post-election violence in December, 2007–January, 2008 that left thousands of Nairobi residents—especially in the notorious Kibera slum not far from downtown—displaced for months afterward. Despite these tragedies, the Nairobi CBD still provides SSA with arguably its most picturesque and captivating skyline of multistory buildings. Its double-lane streets, with their busy streetlights, full buses, massive congestion, gigantic billboards, neon signs, and automobiles from all over the world

make Nairobi the ultramodern heartbeat of Kenya. The reconciliation and constitutional changes that have followed the cessation of post-election violence in early 2008 brought hope to many Nairobi residents that renewed government attention and reinvigorated foreign investments would reverse their city's steady decline. The government established a cabinet-level Ministry for Nairobi Metropolitan Development and released an ambitious plan for making Nairobi a world-class city by 2030, even as it re-invested attention into grassroots, decentralized and community-based development in the city. While no real reversal of Nairobi's declining fortunes is in evidence, the city is re-emerging as a lively and creative cultural center for Kenya and the region around it (box 8.5).

Dakar: Senegal's City of Contradictions

Senegal is about 43 percent urbanized. With 2.8 million inhabitants, Dakar is a principal primate city in West Africa. The city is known for its beauty, modernity, charm, and style, as well as its agreeable climate, excellent location, and urban morphology. But, of course, this image applies to only part of Dakar. As with Nairobi, Dakar is a city of phenomenal contradictions.

The city was founded in 1444, when Portuguese sailors made a small settlement on the tiny island of Gorée, located just off the Dakar Peninsula. In 1588, the Dutch also made the island of Gorée a resting point. Although the French came to the site in 1675, they did not move onto the mainland until 1857; they used Dakar as a refueling and coal bunkering point. A number of developments expanded Dakar's functions, leading it to be, in a short

Box 8.5   Political Geography and Popular Songs in Nairobi

Nairobi in the early 21st century is a highly cosmopolitan city with a great deal of racial and ethnic diversity. Yet its sociocultural map still reflects colonialism's racial division of space, albeit with Nairobi now much more a city divided by class-based zones. The worlds of Asian, white, and African elites overlap, but they are largely disconnected from those of the African poor and working-class majority.

The divides of the splintering urbanism extend into cultural life. Joyce Nyairo has shown how one of Nairobi's most famous nightclubs, Carnivore Simba Salon, segments its clientele by developing special nights that cater to followers of Indian bangra music, Kikuyu music, and Western pop. Meanwhile, on the other side of Nairobi, the new popular musical form of Kenyan rap has blossomed to play an increasingly significant role even in the country's social and political life. The rap duo of Gidi Gidi Maji Maji, for example, provided Kenyans with what in effect became the soaring theme song of their country's transition out of authoritarian rule with their hit, *Unbwogable*, inventing a new word in the title, as a Luo-English mishmash meant to convey the unstoppable cause of democratization. Other rappers like Ndarlin P, Googs and Vinnie, or Deux Vultures are speaking for the underclass of the rapidly growing and changing metropolis. In Ndarlin P's remarkable song, *4 in 1*, the rapper plays four different characters (a *matatu* (minibus) driver, an Asian, and two rapper characters, one of whom is Ndarlin P as himself) as they live lives that intersect in Nairobi's sprawling Eastlands. Nyairo shows how Ndarlin P maps onto his song the consistent pattern of deprivation Eastlands residents have faced, poking fun at elites, at government, and at himself in equal measure.

Nairobi's new music emphasizes the heterogenous character of the city, even while it celebrates margins of the city into which few outsiders or local elites will go. As Nyairo puts it, through these new songs' lyrics "one captures the spatial and the social geography of these areas, and gains a better understanding of the people who inhabit these spaces, and of the pleasures and dreams that daily lubricate their existence." In songs like *4-in-1*, Nyairo argues, we hear a mimicking voice that "parallels Nairobi's own mimicry of global concepts of town planning and urban design. It is as if to say that Nairobi only pretends to be a modern enclave whereas in fact its character is derived more from the ethnic ingenuity that has collapsed all plans of a green city and made illegal densification and informal settlements the order of the city's development."

Source: J. Nyairo, "(Re)Configuring the City: the Mapping of Places and People in Contemporary Kenyan Popular Song Texts," in M. Murray and G. Myers (eds.), *Cities in Contemporary Africa* (New York: Palgrave Macmillan, 2006), pp. 71–94.

time, the most important colonial port on the west coast of Africa. In 1885, Dakar was linked to St. Louis, the old Portuguese port, by rail; and this gave it an added importance as a trading center. Because of its situation and site advantages, Dakar soon became a focus for French colonial functions in the region. In 1898, Dakar became a naval base and in 1904 it became the capital of the Federation of French West Africa. Dakar's location on the westernmost part of the continent made it the most strategic point for ships moving between Europe and southern Africa and from Africa to the New World. As capital of French West Africa until 1956, it served a hinterland stretching from Senegal in the west to the easternmost part of Francophone West Africa, which included Mali, Burkina Faso, and Niger.

After the French moved from the island of Gorée to the peninsula, there was some uneasiness about living in quarters surrounded by African villages. Although a policy of racial segregation was not officially pursued, the French settlers had always wanted to keep the two communities separate. However, only because of a natural calamity that befell the Africans could the French finally accelerate the establishment of their exclusive holdings. Progressive displacement of African dwellings was underway before the outbreak of a yellow fever epidemic in 1900, but the Europeans, invoking sanitation requirements, displaced the Africans at a greater rate afterwards, pushing them northward. Between 1900 and 1902, numerous African homesteads were burned down as a "sanitary measure," and the occupants were relocated after receiving compensation for their holdings. Another epidemic in 1914 again brought destruction of African

homesteads in the south and more relocation of Africans to the north. On the eve of World War II, the French succeeded in almost completely dominating downtown Dakar, often called Le Plateau or Dakar Ville, concentrating the Africans in what became known as the African Medina, in the north-central part of the peninsula. The problem of "cohabitation," as the French called it, was at the root of the whole displacement campaign. Although the colonial authorities would never admit a policy of official segregation, many recommendations were made to openly enforce a system based on race. A commission charged with the study of Dakar in 1889 put forth a recommendation for separate residential quarters for European and African populations. In 1901, another report proposed relocating the Africans outside the confines of the city. A new plan, implemented in 1950–1951, gave further excuse to the colonial administrators to displace more Africans.

The present internal morphology of Dakar reflects this historical background. The city is composed of four main divisions. Although rigid, exclusive ethnic domains are no longer in evidence, Le Plateau still contains one of the most Westernized sectors in Africa. It compares easily with any European city—with high-rise buildings, expensive shops, exclusive restaurants, business offices, and many European residents. Characterized by its white-painted, tree-lined boulevards, Le Plateau is the most modern sector of the city, and contains upper-class residential quarters, commercial and retail functions, and government offices and institutions. The African Medina, by contrast, reflects its background as a concentration of Africans into high-density housing projects and bidonvilles (squatter settlements). It is the popular area of the city, is still densely populated, and houses many popular markets and clubs. Its functions are primarily residential, but it also contains shops, markets, and cultural features. It contains the industrial laborers and those employed in the informal sector, both outside and inside the Medina. The population of the Medina, and the adjacent bidonvilles of Ouakem and Grand Yof, is uniformly poor and live in poorly serviced parts of the city. Recent expansions of the city have also resulted in the development of a sector called Grand Dakar, which contains a variety of neighborhoods ranging from the well-to-do through middle income and poor sectors and includes a mixture of modern residential quarters, industries, and bidonvilles. There is also the Dakar industrial sector, which houses the bulk of the city's industrial activities.

Another important part of Dakar that deserves attention is Gorée Island. This island served for many centuries as one of the principal factories in the triangular trade between Africa, Europe, and the Americas. The popular *Maison des Esclaves* (Slave House) built by the Dutch in 1776 serves as a poignant reminder of Gorée's role as the center of the West African slave trade. The Slave House with its famous "Door of No Return" served as a place where Africans were brought to be loaded onto ships bound for the New World. The Slave House has been preserved in its original state and attracts thousands of tourists each year.

As with many African primate cities, Dakar faces the problem of rapid population growth. In 1914, the city had a total population of 18,000; by 1945 it had 132,000; and by 2010 it had almost 3 million. Clearly most of the growth is attributable to rural-urban migration, which is characteristic of all Sub-Saharan primate cities. The rate of natural increase of the city's population has also been much higher than the national average on account of better sanitation and medical services.

There is no doubt that Dakar is still an important center whose functions reach far beyond its national boundaries. Its ideal location still makes it a center of maritime as well as airline traffic. Many international organizations are located in Dakar because of its geographic situation and agreeable urban environment. Many international conferences and meetings are held there. Above all, it is one of the most favored vacation spots in West Africa for European tourists, especially those from the Mediterranean, who find a familiar climatic comfort in exotic surroundings.

The future of Dakar, nevertheless, depends largely on confronting two challenges. One is to stem the tide of rural-urban migration by a sound policy of regional and rural development, including development of satellite cities and subsequent decentralization. The other, as in other African primate cities, is to bridge the gap between the city's ultramodern sectors and the bidonvilles and make Dakar a true African city.

Johannesburg: A Multi-Centered City of Gold

South Africa has long been among the most urbanized countries in SSA. Some 58 percent of its population now lives in cities. With more than 3.6 million inhabitants, Johannesburg is the largest city in South Africa. Gauging its population is, however, both easier and more complicated at the same time, in comparison with most SSA cities. On the one hand, the relative wealth of the Republic of South Africa affords it the possibility of keeping more regular and reliable census figures than most SSA countries, and the South African Cities Network is one of the continent's best repositories of urban data. On the other hand, the reorganization of municipal and local government in South Africa (box 8.6) and the presence of Johannesburg in the geographical center of Africa's greatest example of a polynodal conurbation (many-centered urban area) make it more complicated to decide where Johannesburg begins and ends. The major cities of Tshwane (formerly Pretoria) and Ekurhuleni (formerly East Rand), both of which have more than 2 million residents, and Vereeniging (1.1 million residents) are within the metropolitan area of Johannesburg, and other big towns adjoin it as well. Thus the metropolitan conurbation is said to contain more than 8.2 million residents.

South Africa also has the most deeply developed urban hierarchy in SSA. The common issues surrounding primacy in African urban hierarchies are moot here. Johannesburg's population of over 3.6 million is nearly equaled by eThekwini (formerly Durban, 3.2 million) and Cape Town (3.4 million); and both Nelson Mandela Bay (formerly Port Elizabeth) and Vereeniging are over 1 million, giving South Africa seven municipalities (including Tshwane and Ekurhuleni) with more than 1 million residents. Five more cities have more than a half a million. This significantly dilutes any primacy Johannesburg might claim, although, when one considers the immediate proximity of Tshwane and Ekurhuleni it is still possible to recognize the greater Johannesburg area and Gauteng Province as the core of the South African economy. Johannesburg is, moreover, frequently the

only SSA city to be considered a world city. It holds the largest mining and industrial center on the African continent. It is home to Africa's largest stock exchange, busiest airport, most diverse manufacturing sector, and the ugliest urban racial history in Africa.

Johannesburg owes its establishment and phenomenal growth to the discovery of gold in 1886. The rush of settler communities from the south to share in the riches of the land caused the town's population to increase to 10,000 within a year of its birth. By 1895, hardly ten years after its establishment, the city had about 100,000 people, half of whom were European. Johannesburg was the creation of the mining companies, which until recently probably had more to do in determining the spatial organization of the city than did the civil authorities.

The City of Gold has the unfortunate distinction of having been at the heart of a notorious experiment in social engineering. This experiment was built around the notion of separate development for settlers and the indigenous African population. Johannesburg's separate development started with the assertion in 1886 that no native tribes could live within 70 mi (112 km) of the site of the new town. When the "native" problem first arose in 1903 and when it surfaced again in 1932, with the creation of the Native Economic Commission, the European settlers argued that Johannesburg had been built by the Europeans, for the Europeans, and belonged to them alone. They maintained that the "natives" were needed for unskilled labor and came to the city to work but not to live in it, mainly because of their inability to handle European civilization. In this manner, the largest of the "European" cities in Africa came into being. Africans were denied permanence of dwelling while they worked for

Box 8.6   Geographic Information Systems (GIS) in the New South Africa

The racist apartheid policies of South Africa's white minority regime from 1948 to 1994 built on the segregationist policies that preceded them. Both colonial urban policies and those of the independent white minority regime from 1910–1948 shared with apartheid a central role for geography in the creation and maintenance of separate and unequal socioeconomic development by racial group. The highly regulated authoritarian apparatus of the apartheid regime increasingly depended on high technology—particularly in the form of weapons technology and police procedures—to enforce its order, particularly in South African cities.

Beginning in the early 1990s, though, new geographic technologies came into use in a very new and different way—in the cause of dismantling apartheid and reorganizing society along equitable multiracial lines. GIS has been central to the recreation of South Africa as a rainbow nation, particularly within and around urban areas. This reorganization and the role of GIS within it have not been without contention, but the new South Africa does provide geographers with a vital example of the possibilities and limitations of GIS for the betterment of society.

Apartheid's Homelands Policy and its Group Areas Act provided the legal framework by which the geography of separate development took shape throughout much of the second half of the 20th century. Through the first, the minority regime sought to legislate the displacement of the country's black majority to poor, marginalized rural homelands, termed *bantustans* in Afrikaans (fig. 8.13). Through the latter, apartheid's architects sought to

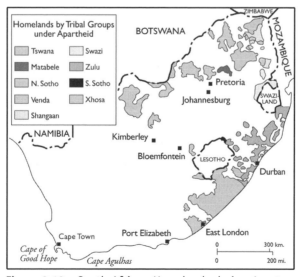

Figure 8.13   South African Homelands during Apartheid. *Source:* Compiled from various sources

exclude blacks from residing in the cities. Apartheid's urban system was built around the separation of cities, which were set aside as "white-by-night," from townships, which housed the black working classes that made the cities function (other than domestic workers in white homes). Colored (mixed-race) and Asian groups also each had their areas. Townships were typically located at a distance from cities, with physical features, railways, or highways used as dividing lines. When pre-existing black neighborhoods interrupted the racial map, the apartheid regime followed a program of forced removals for what it bluntly termed black spots.

The dismantling of apartheid literally required South Africa to remake the internal maps of its human geography. Two of the country's four former provinces were subdivided into seven new ones (giving the country nine provinces). This was part of an effort to integrate government services for all races in a decentralized democracy, including the dissolution of Bantustan governments. Remaking South Africa's urban geography has been a much more complicated task. New boundaries needed to be created for every level of local government. The new government's 1998 Municipal Structures Act led to the establishment of a Municipal Demarcation Board (MDB) in 1999 to restructure the urban system from the bottom up using Geographic Information Science.

In keeping with the new era, the South African government sought to make the demarcation process as democratic and decentralized as efficient management would allow. Using the Internet, the MDB and the Ministry for Provincial and Local Government produced more open access to spatial information than South Africa—or Sub-Saharan Africa—had ever seen before. The government seemed to genuinely want to use GIS data to provide for technocratic solutions to deeply entrenched problems. While generally applauded for this and for making so much information available transparently, questions have been raised about the process in a number of cities. The criticism has been most notable in Johannesburg, where the formerly white-by-night city became integrated into the massive townships that surrounded it. When presented with a number of different GIS-derived internal boundary scenarios for the new Greater Johannesburg Metropolitan Council, the government apparently bowed to pressure from the ruling African National Congress elites to ignore what the geographers and GIS experts involved considered the most appropriate sub-divisions for maximum equitability. Similarly, GIS applications have not prevented the skewing of policies like that of the provision of free basic water away from the interests of the poorest of South Africa's urban poor. GIS itself will not solve South Africa's laundry list of urban problems, but it is clear to most observers that GIS has been a useful tool in apartheid's dismantling and its replacement by a multiracial democratic order. In cities, in particular, GIS has enhanced urban management and the efforts of the new black majority regime to serve its urban constituents.

the city and were absolutely barred from living in the city, and were restricted to guarded compounds or distinct townships during the tenure of their urban employment. The "pass law" (requiring all Africans to carry passes, or internal passports) begun in 1890 and the "compounding system" (restricting Africans to certain residential areas) contributed to severe urban structural problems with which Johannesburg still has to cope.

Johannesburg became Africa's largest manufacturing center, and a principal center of culture and education. The prosperity of the city was derived from the labors of all the races, but was appropriated by the European minority, who enjoyed perhaps one of the highest living standards in the world. Today, the city is clean, with well-planned streets, skyscrapers, and extremely plush residential quarters. The downtown area is similar to that of any industrial city in Europe and North America, with high- rise development to house the offices of the numerous companies, trading firms, and government institutions. In the suburbs of Johannesburg, such as Sandton, are residential homes for the well-to-do Europeans, whose architecture and amenities more than match those of their European and North American counterparts.

Separate development became a formal national policy under the name *apartheid*, after the 1948 election of the white racist National Party into power in whites-only polls. The Nationalists' apartheid built on decades of gradual evolution and enforced unequal separation in an extremely geographical manner. Apartheid was a socially engineered, hegemonic tool to maintain a privileged status for the European settler population so that its small number could appropriate the vast amount of wealth that

was being generated in the country. The impact of this policy was perhaps felt more in places like Johannesburg than anywhere else in the country. Apartheid's application was severely tested in the dynamic environment of Johannesburg, which was attracting a great number of Africans to supply the labor requirements of a rapidly growing industrial conurbation. The authors of apartheid followed a myopic vision by engineering an unsustainable institution of separate development for Africans in their native homelands. They were later forced to tolerate settlements such as Soweto, growing by leaps and bounds in the shadows of Johannesburg (fig. 8.14). Apartheid was doomed to succumb to the social disorder its authors never anticipated. The 1976 race riot in Soweto was a watershed in the development of nonracist South Africa. The impatience of the world community with the brutal regime brought moral outrage from outside and increased violence from within. Under the weight of these two dynamics, and aided by a visionary leader, Nelson Mandela, South Africa emerged from its nightmare by the early 1990s, when apartheid came to a formal end, symbolized by Mandela's election as president in 1994.

The future of Johannesburg (and other South African cities) lies in how the root causes of urban instability created by apartheid are dismantled while maintaining the city's ability to continue as South Africa's most important industrial and business center. The challenges that Johannesburg faces are evident from what has been happening in attempts to resolve the severe socioeconomic disparities. With the enforcement mechanisms of the apartheid influx control laws gone, an orderly transition from a divided city into an

Figure 8.14   This is one of the famous shantytowns, or squatter settlements, near Johannesburg. It is part of Soweto (South Western Township). (Photo by Brennan Kraxberger)

integrated city has been a daunting task for policy makers and city planners. Johannesburg residents suffer from high rates of violent crime and continuing insecurity. Johannesburg is becoming a megalopolis, as a series of mining towns and industrial areas merge together. The previously marginalized townships such as Soweto and Alexandra are now firmly integrated into metropolitan life. Though still high, income inequality within Johannesburg has declined, as has the violent crime rate. Johannesburg (as home to two of the stadiums used to host of the final games) led the rest of South Africa in celebrating the success of the 2010 FIFA World Cup in soccer—the first time the world's biggest sporting event had ever been held in Africa. While con-

troversies persist over the massive investment in stadiums and infrastructure that hosting the World Cup necessitated (at the expense of potentially significant investments in urban developments more beneficial to the poor), Johannesburg's new Gautrain rapid transit railway, remodeled international airport, fully upgraded highway system, and many other manifestations of FIFA 2010 remain on the landscape, alongside a deep pride among urban residents across the country in South Africa's ability to produce a first-rate World Cup without the mishaps and fiascos many outsiders (almost gleefully) seemed to expect. With a nonracial central government at the helm and with ample human and physical resources for economic progress, all signs point to a more progressive trajectory for cities like Johannesburg. Success depends on whether people of all ethnic groups living in the city can deal responsibly with the history of their relations and choose to build a diverse society in which everyone has a stake in the new South Africa's development.

URBAN CHALLENGES

Urban Environmental Issues

Because most SSA cities have experienced more limited industrialization processes than similarly sized cities of Europe, Asia, or the Americas, specifically urban environmental problems are often thought to be of a smaller magnitude. Yet problems of solid waste management, air and water pollution, toxic waste disposal, or environmental health are profound issues in much of urban SSA (fig. 8.15). In part due to the generally smaller formal industrial sector and smaller manufacturing value-added base in urban Africa, revenues that accrue to urban local government are typically not significant enough to support the broad array of urban services expected of city governments. This array includes environmental management services such as solid waste management, water and sanitation supply, and any form of environmental monitoring or oversight.

As a consequence, formal and regulated supply of these services in SSA cities is often in very short supply. Solid waste services are one crucial example of the interlocking environmental problems that can result. Many cities with more than a million inhabitants in the region report that the proportions of the residential solid waste produced that actually makes it to a landfill range from 3 percent to 45 percent. This means that the majority

of solid waste remains in urban neighborhoods. In an earlier era of smaller settlements, the burial or burning of such waste was not taken to be a major problem, because its content was overwhelmingly organic and biodegradable. The increasing use of plastics and other inorganic materials, along with ordinary source items for toxic waste (such as batteries and household insecticides in aerosol canisters), and the staggering growth of unregulated settlements means that the lack of proper solid waste management has indeed become a severe crisis. Buried in congested neighborhoods, solid wastes can and do pollute the water supplies of untold millions of urban Africans. Pollutants sourced to uncontrolled landfills have been shown to enter into the fruits and vegetables urban farmers pluck from downstream gardens in Dar es Salaam, Lusaka, and elsewhere. For example, just along the Roma-Ng'ombe border (as evident on the air photo) urban women gardeners utilize a "rich" section of sewage-infested wetland for growing tomatoes, vegetables, and sugar cane for sale in downtown Lusaka, as well as in the surrounding compounds such as Ng'ombe. Burned on the surface, the waste causes serious damage to the air quality of such neighborhoods. Left in ditches, the waste can inhibit proper drainage, leading to flooding or the increased presence of standing water that becomes breeding space for malarial mosquitoes. Mounds of waste left for months on the surfaces of African urban neighborhoods provide habitat for vermin that carry serious public health risks.

The interconnected water, sanitation, waste, air quality, and environmental health problems of African cities, though they may seem at first to pale by comparison to the infamous environmental crises of Southeast

Figure 8.15   Erosion scars the suburban landscape on the outskirts of Zanzibar. (Photo by Garth Myers)

Asian or Central American megacities, may in fact be as bad or worse, proportionally. This is because the problems have great potential to magnify one another in the absence of much regulation or amelioration. Even where significant industrial development is associated with African urbanization, such as in Nigeria's oil-rich Niger Delta, Zambia's Copperbelt, or South Africa's Gauteng Province (i.e., greater Johannesburg), colonialism, transnational capitalism, or repressive governments (or, in the latter case under apartheid, all three at once) make for a heady combination of roadblocks to environmental control. The levels of heavy metals pollution downstream from the Copperbelt's largest copper smelter, for example, are mind-boggling, and yet even as the technology exists to prevent or significantly reduce the smelter's air and water pollution, successive colonial and post-colonial governments for many decades now have shown reluctance to force environmental controls onto an industry that provides more than 90 percent of all of Zambia's export earnings. Despite such limited or politically circumscribed capacity for urban environmental management, many African cities are witnessing substantial efforts to bring environmental crisis points under control. In line with the prevailing development models of the day, many cities are experimenting with private sector urban service provision, including in environment-related sectors. Dar es Salaam, as the pilot city for the United Nations Sustainable Cities Program, privatized solid waste management services and produced an increased rate of deposition from under 10 percent of residential waste to more than 40 percent in less than a decade. Other cities have privatized water supply or even sanitation services. Still more have attempted public-private partnerships where private sector companies have joined forces with government to provide services. Not all of the new innovations have been driven by the private sector, nor have they been automatically friendly to the environment. South Africa's post-apartheid regime has carried out a policy for free basic water provision, for instance, that did increase the supply of clean water for the poor; but many critics point to the limitations in that system that are causing poor urbanites to seek unclean water alternatives that exacerbated a cholera epidemic. In other cities, the driving forces for change in urban environmental management belong with grassroots community groups, such as

Nairobi's Mathare Sports Club, whose local environmental planning and consciousness-raising earned them global attention at the World Summit on the Environment in Rio de Janeiro in 1992 that in turn improved the club's soccer match gate revenues. Regardless of the paths taken, though, it is clear that much more needs to change for African cities to gain control over the daunting array of environmental problems confronting them.

Primate Cities

Urban primacy continues to dominate the African scene (fig. 8.16). Indeed, one of the more significant factors in urban transformation in Africa in the post–World War II and post-independence period has been the dramatic growth of the primate cities. Primate cities contain more than 25 percent of the total urban population in SSA. In places such as Lesotho, the Seychelles, and Djibouti, primate cities contain 100 percent of the urban population. Primate cities in SSA are also not limited to small countries such as Burkina Faso or Guinea Bissau; they can also be found in such large countries as Angola or Mozambique. Generally, in those countries where urbanization has had a relatively long history, the ratios are lower. But the degree of primacy will continue to be a significant pattern in Africa's urban development for quite a long time.

In the 1960s, most primate cities in Africa accounted for about 10 percent of the urban population. By the year 2000, many cities, such as Kinshasa, Lusaka, Accra, Nairobi, Addis Ababa, Luanda, Dakar, and Harare, increased their share of their respective nation's urban population to over 20 percent (fig. 8.17). Currently, many primate cities account for over 30 percent of the urban populations in their

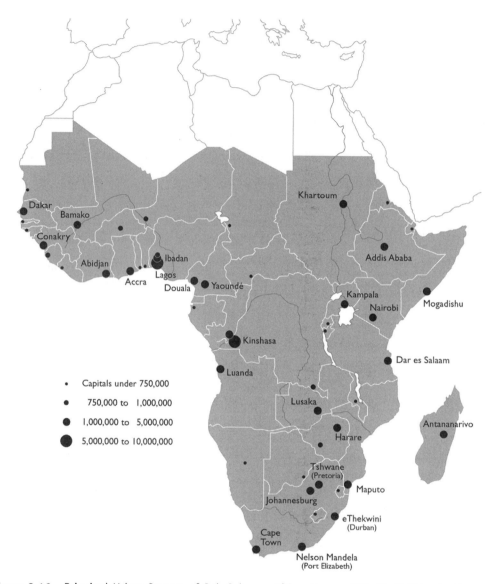

Figure 8.16   Principal Urban Centers of Sub-Saharan Africa. *Source:* UN, *World Urbanization Prospects: 2001 Revision,* http://www.un.org/esa/population/publications/wup2001/wup2001dh.pdf

respective countries. It is also important to note that since the second half of the 1980s, some encouraging signs of deconcentration around primate cities have been observed and the ratios of urban populations that reside in primate cities seem to be stabilizing. For instance, between 1990 and 2005, the share of

the urban population in the primate city in several countries has declined. The decreases in the percent of urban population in the largest city were highest in Angola, Burkina Faso, and Guinea.

While the dominance of primate cities in SSA has historical roots traceable to colonial

Figure 8.17   Getting your hair cut and styled is one of the basic services provided by every culture. Around Kaunda Square in Lusaka, entrepreneurs earn a bit more by adding telephone services to their business model. (Photo by Angela Gray Subulwa)

administration policies, post-colonial governments have perpetuated this pattern by making them the centers of modern development and governance. SSA's primate cities dominate the political processes, often reinforcing the status quo but sometimes creating avenues for change (fig. 8.18). For example, urban political processes in some SSA cities have created new spaces of engagement, particularly for professional women. For example, women comprise over 30 percent of parliament in Burundi, Mozambique, Rwanda, Tanzania, Angola, Uganda, and South Africa (in comparison, the female representation in the US Congress is less than 20 percent) and over 25 percent of min-

isterial positions in Niger, Burundi, Mozambique, Gambia, Uganda, Lesotho, South Africa, and Botswana. Formal and informal solidarity organizations and movements of women, operating in SSA's primate cities, have also emerged as significant agents of change. The Liberian peace movement, for example, was heavily influenced by such coalitions of women and ultimately ushered in SSA's first female head-of-state, Ellen Johnson Sirleaf.

Primate cities in SSA not only tend to be the capital city of the country; they often also have more disproportionate influence than is warranted by the magnitude of the populations living in them. They dominate the political,

Figure 8.18   By using billboards to help change human behavior, Lusaka, Zambia, tries to ceate a greener capital city as a role model for the nation. (Photo by Angela Gray Subulwa)

economic, infrastructural, and cultural scene of their countries. The strong influence they exercise also enables them to preempt a good portion of the national social and industrial investments. Concentration of power in these cities has produced large disparities in standards of living between those who live in them and the population in the rest of the country. Primate cities also have some of the most serious urban problems in the region. These include mounting unemployment and the resultant increased crime and youth unemployment; severe housing problems, reflected in overcrowding and the spread of slums and squatter settlements (over 80 percent of the urban populations reside in slums in Mozambique, Niger, Angola, Chad, Central African Republic, and Sierra Leone); and in the immense pressures on urban infra-structure and services such as water, sewage, and transportation.

Rural-to-Urban Migration

The rapid growth of rural-urban migration has been a common feature of SSA countries. Between 50 percent and 60 percent of the urban growth in SSA comes from rural-to-urban migration of young adults seeking jobs and other livelihood opportunities in urban areas. This migration is one of the most crucial problems facing African governments at present. A massive flow of people into the few urban centers that became the locus of power and investment, particularly in the period following decolonization of the continent, has strained the carrying capacity of most urban centers. Concern about the rate of SSA urbanization would hardly seem justifiable, considering the very high magnitude of its rural population. However, the urban growth pattern, often dominated by the primate city and the high rate of growth of such centers, is far beyond the capabilities of the urban socioeconomic system to generate the needed employment opportunities, housing, and social services. As a consequence, the primate city has become, in most instances, a liability to the overall development process.

The paradoxical fact that rural-urban migration continues to grow in spite of rising unemployment in urban centers of Lesser Developed Countries (LDCs) has given rise to a migration theory based on rural-urban income differentials. This theory assumes that migration is "primarily an economic phenomenon" and that the potential migrant makes a calculated move in order to realize much higher "expected" earnings with varying probabilities. According to this theory, the wide differences between urban and rural wages, coupled with the fact that the long-run probability that a migrant could secure wage employment in the urban area, explains the motives behind the increases in rural-urban migration. The structural adjustment programs implemented across the region were meant to bridge this gap by increasing the prices of agricultural exports and instituting payment for previously free or subsidized services consumed mostly by urban dwellers.

Rural-to-urban migration in SSA is also explained with reference to *push factors* that operate in rural areas and *pull factors* that attract migrants to urban areas. The push factors include the deteriorating socioeconomic conditions in rural areas, including lack of access to agricultural land, which literally forces people to leave rural areas. The pull factors emphasize the attractions and socioeconomic opportunities available in urban areas. The economic opportunities in urban areas, as well as social, cultural and psychological factors, including escaping social controls in rural areas also attract people to urban areas.

Whether rural-urban migration in SSA is caused primarily by the difference in expected wage from migration (urban wage) versus an agricultural wage or the balance between pull and push factors might be open to debate. It is important to note that more recent evidence seems to suggest that rural out-migration has not only abated in SSA, but that its counter-stream (i.e., urban out-migration) has progressed and even, in some cases, outnumbered the rural-to-urban flow of people. For instance, a number of major towns and cities in Ghana and Zambia experienced a negative migratory balance between the 1970s and 1980s. Zambian census data also indicated that the population of some urban areas decreased between 1990 and 2000, due in large part to growth in some secondary cities, but also due to opening up large farms blocks and resettlement schemes designed to attract some urban dwellers and retirees. Similar patterns have been observed in Francophone West African countries of Burkina Faso, Guinea, Côte d'Ivoire, Mali, Mauritania, Niger, and Senegal, where many secondary towns registered a negative net loss of migrants between 1988 and 1992. In some of these countries, the net migration rates of rural areas suggest that rural-to-urban migration may not be as important as expected or that the reverse movement has increased.

While the apparent slow down of rural-urban migration is important, the fact remains that the SSA primate city is a major factor in the contradictions between urban and rural life. Because of its monopoly of economic opportunities and political power, the city has created a perception on the part of rural-urban migrants of certain and immediate opportunities for socioeconomic improvement, a perception that is rarely realized. The phenomenal growth of shantytowns and squatter settlements and the proliferation of informal employment in the cities of SSA are the result of this miscalculation.

A HOPEFUL VIGNETTE

Namushi comes from family of seven—four boys and two girls living with their mother—and was born in Mabumbu village in western Zambia. Throughout her childhood, Namushi and her family survived on what they managed to grow from their fields, together with any small, temporary jobs to access limited cash. Often times, Namushi's family struggled to provide more than one meal a day. At the age of 20, Namushi, together with four other boys from her village, made the (mis)calculation to set out for the capital city of Lusaka in search of a better life. According to Namushi, she set out with an unquestioned belief in the socioeconomic possibilities (in fact, certainties) that life in Lusaka would offer.

Upon arrival, Namushi quickly realized that her beliefs and perceptions about life in Lusaka were indeed miscalculations. Her childhood in rural, western Zambia did not equip her with the tools (linguistic or otherwise) to easily navigate and negotiate the complexities of life in Lusaka. Although she struggled to learn Nyanja and English, she remained committed to the idea that succeeding here would translate

to success for her entire family back home in Mabumbu. For the first few days (before they could locate anyone who lived in the city from Mabumbu's nearby villages), Namushi and her four travel companions slept at the bus station and set out each morning in search of employment. Luckily, Namushi was able to secure employment as a housemaid in the suburbs of Lusaka by her third week in the city.

For the next one year, Namushi worked as a housemaid, while sending a significant portion of her earnings back home to her village. It was during this time that Namushi was able to improve her English, which later helped secure better employment at a newly opened gas station. While working at the gas station, Namushi realized that she would continue to struggle and fail to realize her goals if she continued to send all of her earnings directly home to Mabumbu. She decided to reduce her remittances to the village and direct some of her income into savings, with the goal of opening a small business in the nearby compound, Kaunda Square. Over the course of a year, Namushi saved enough cash to construct a small, mobile store (*katemba*) from which she sold vegetables, candy, candles, salt, and the daily essentials of life in the compound. As her business stabilized, Namushi continued to save cash, hoping one day to open a larger, permanent structure grocery in the compound market area. Within three years of saving and navigating the intricacies of grocery marketing in the compounds, Namushi finally succeeded as she opened her grocery in Kaunda Square in 2005.

When asked, Namushi characterizes her successful navigation of life in Lusaka as a story of constant negotiations and calculations of risk, coupled with the determination to survive that she learned from her grandmothers

Figure 8.19   Namushi and her grocery shop on Kaunda Square in Lusaka. (Photo by Angela Gray Subulwa)

back in Mabumbu. Namushi reflected on the overlapping challenges she faced—coming from a rural area, navigating the informal urban economy, lacking any support system or network, and most critically doing all of this as a woman. As a woman, Namushi felt compelled to carry herself a "bit rough" in order to guard against those who saw an opportunity to take advantage of a young, single, rural woman in the city. Another problem she faced as a woman entrepreneur in the informal economy was the issue of transportation (from the wholesales to the compound). Namushi found it difficult to secure transport without the fear of being taken advantage of by the overwhelmingly male drivers—fears ranging

from that of simply being overcharged to fears of thieves and even fears of being raped along the way. Another obstacle Namushi faced came from the wholesalers themselves (in Lusaka, the wholesale market is dominated by men of Indian descent). Namushi often found herself served last even if she arrived first. And while her male counterparts were able to negotiate for small credits and loans on their wholesale purchases, Namushi was unable to negotiate similar deals.

In the face of all of these obstacles, Namushi did succeed in opening her small, permanent grocery stop on a busy corner near the Kaunda Square vegetable market in 2005 (fig. 8.19).

By 2010, she had expanded her original grocery shop and had opened another six shops in Kaunda Square (three additional groceries and three cosmetics/pharmacies). Namushi returned to Mabumbu, collected four of her brothers, and returned with them to Lusaka. The boys now assist Namushi in maintaining and expanding her shops, often mediating some of the gender constraints that remain in Lusaka's informal economy. While her brothers help her with the shops, Namushi is sending her nieces to school, in the hopes that they will do greater things than she had accomplished herself.

SUGGESTED READINGS

Bryceson, Deborah F. and Deborah Potts, eds. 2006. *African Urban Economies: Viability, Vitality, or Vitiation?* London: Palgrave Macmillan. A wide range of scholars examine urban economic issues in select African cities.

Coquery-Vidrovitch, Catherine. 2005. *The History of African Cities South of the Sahara: from the Origins to Colonization.* Trans. by Mary Baker. Princeton, NJ : Markus Wiener Publishers. Discusses the evolution of pre-colonial urban centers in Africa.

De Boeck Filip, and M.-F. Plissart. 2004. *Kinshasa: Tales of the Invisible City.* Antwerp: Ludion. Brings cutting-edge theory to bear on this decrepit and despairing yet vibrant city.

Enwezor, Okwei, *et al.*,eds. 2002. *Under Siege: Four African Cities: Freetown, Johannesburg, Kinshasa, Lagos.* Ostfildern-Ruit, Germany: Hatje Cantz. Famous architects, art critics, and urban studies academics look at life in four very different African cities.

Hansen, Karen T. and Mariken Vaa, eds. 2004. *Reconsidering Informality: Perspectives from Urban Africa.* Uppsala, Sweden: Nordic Africa Institute. Contains insights from Nordic, European, and African experts.

Locatelli, Francesca, and Paul Nugent, eds. 2009. *African Cities: Competing Claims on Urban Spaces.* Leiden: Brill. Problem-oriented examination of the processes that are shaping the 21st century African city.

Murray, Martin J, and Garth A. Myers, eds. 2006. *Cities in Contemporary Africa.* New York: Palgrave Macmillan. Includes chapters on both the largest cities (Lagos, Kinshasa, Johannesburg) and those less frequently analyzed (Bulawayo, Kano, Luanda).

Myers, Garth A. 2011. *African Cities: Alternative Visions of Urban Theory and Practice.* London: Zed Books. A multi-disciplinary, cutting-edge take on the emerging field of African urban studies that builds from both original research and analysis of research from across the continent; includes case studies of more than a dozen cities

Simone, A. M. 2004. *For the City Yet to Come: Changing African Life in Four Cities.* Durham and London: Duke University Press. Combines policy-centered analysis with what is in effect storytelling about four different urban African experiences in Douala, Dakar, Johannesburg, and the African pilgrimage community within Saudi Arabia.

Simone, A. M., and A. Abouhani. 2005. *Urban Africa: Changing Contours of Survival in the City.* London: Zed Books. A fine example of the types of interdisciplinary and multi-national research efforts that the Council for the Development of Social Research in Africa has fostered.

SUGGESTED FILMS

The Constant Gardener (2005) Although not without flaws from an African perspective, this Hollywood film provides an unforgettable vision of Nairboi's genteel expatriate elite and its vast informal settlement of Kibera.

In a Time of Violence (1995) Explores the grand drama and the micro-politics of South Africa's rough transition out of apartheid with Johannesburg's Soweto and Hillbrow areas as the setting.

Tsotsi (2005) Chronicles the life of a small-time criminal and provides vivid imagery of life and survival in Johannesburg.

Pray the Devil Back to Hell (2008) Captures the story of brave Liberian women who came together to end a bloody civil war and usher peace into their country.

Figure 9.1   Major Cities of South Asia. *Source:* UN, *World Urbanization Prospects: 2009 Revision,* http://esa.un.org/unpd/wup/index.htm

9

Cities of South Asia
IPSITA CHATTERJEE, GEORGE POMEROY, AND ASHOK K. DUTT

KEY URBAN FACTS

Total Population	1.64 billion
Percent Urban Population	29.9%
Total Urban Population	492 million
Most Urbanized Countries	Pakistan (35.9%)
	Bhutan (34.7%)
	India (30.0%)
Least Urbanized Countries	Sri Lanka (14.3%)
	Nepal (18.6%)
Afghanistan (22.6%)	
Annual Urban Growth Rate	3.41%
Number of Megacities	5 cities
Number of Cities of More Than 1 Million	61 cities
Three Largest Cities	Mumbai, Delhi, Kolkata
World Cities	Mumbai

KEY CHAPTER THEMES

1. The duality of prosperity alongside poverty in the midst of a vibrant mosaic of language, ethnicity, and faiths make cities in South Asia unique.
2. There are three basic types of South Asian cities: bazaar-based, colonial, and planned.
3. There have been five major influences on the development of South Asian cities: the Indus Valley civilization, the Aryan Hindus, the Dravidians, the Muslims, and the Europeans.
4. The current urban system most distinctly reflects the dominance of Presidency Towns during the colonial era.

Box 9.1 Call Centers, SEZs, and Sweatshops

Have you done any of the following of late? Called to reserve a rental car? Telephoned technical support for help with your new computer? Spoken to someone via phone to straighten out a credit card issue?

If you have, then chances are very good that the person on the other end of the line was in India. If, indeed, this was the case, then it was your personal encounter with "outsourcing," a phenomenon that is transforming the way business is done across the globe. With the adoption of global free market policies, corporations now have the freedom to take their production activities outside their home country and situate it anywhere they find more advantageous. The result is a dismantling of the heavy, assembly line, factory-based manufacturing, and the beginning of a more flexible-style production—different parts of the production process can be geographically dispersed or *outsourced*. Outsourcing first gained attention as U.S. automakers began to subcontract the manufacturing of certain auto components to other firms in the U.S. For example, Ford would contract with a smaller, independent firm (perhaps in a nearby city) for wheel assemblies.

Offshore outsourcing is when a firm takes activities and moves these overseas. Giant corporations in the U.S., Western Europe, and Japan prefer outsourcing, because it allows them access to cheap labor, tax rebates, and relaxed environmental norms, hence higher profits. The rise of *business process outsourcing firms (BPOs)* shows that even service-sector employment can be outsourced. Business process outsourcing involves taking accounting functions, customer services, computer programming, and other activities outside the company and usually offshore.

India has several advantages, which make it an attractive destination for BPO opportunities. First, it has a well-developed system of universities and technical colleges that produce a large supply of technically qualified and well educated personnel. Second, English proficiency, a legacy of British colonial rule, has provided post-secondary graduates with the language skills needed to work in "call centers." Third, the cost differential between hiring U.S. workers and hiring those in India may be as high as 10 to 1. This represents a potential cost savings that is hard for any firm to ignore. Finally, with the rise of modern information technologies (telephone, internet), distance has collapsed, and the cost of doing business

5. The urban form of South Asian cities is reflected in two basic models: the colonial-based city model and the bazaar-based city model, with permutations of both.

6. India has a relatively well-balanced urban hierarchy; Pakistan has a dominant southern city and dominant northern one; all other South Asian countries are characterized by urban primacy.

7. Massive rural-to-urban migration in recent decades has led to exploding urban populations and has overwhelmed urban systems.

over great distances has in many ways vanished. Bengalūru, Mumbai, Pune, Hyderabad, and Chennai represent important BPO destinations; Accenture, Citibank, Dell, IBM, Infosys, Microsoft, Office Tiger, Verizon, and Wipro are some of the corporations setting up shop there. The call center workers function as customer service representatives answering 1–800 calls. They undergo accent training and are briefed about American sports and weather so that they can politely chat with a customer. They are often given more "relatable" names like Dave or Nancy so that customers are comfortable.

Outsourcing has become the subject for heated debates because of its controversial impacts on labor and the environment. A large portion of the outsourced jobs include flexible style manufacturing jobs—a shirt can be stitched in China, the label sown in Guatemala, and the buttons stitched in Mexico, before it comes back to the American consumer. On the one hand, this represents a job loss from the outsourcing nation and has therefore become the focus of political debates in outsourcing nations. On the other hand, corporations outsourcing jobs to cheaper locations are accused of exploiting labor and environment in the outsourced locations. The result is a world of sweatshops and Special Economic Zones (SEZs). Since India's economic integration, SEZs have expanded at a rapid rate. These SEZs require huge land areas, often a minimum of 1000 hectares and usually include appropriation of agricultural land adjoining major cities to be used for processing export goods. The SEZs are known to employ labor at exploitative rates—wages are 34 percent lower than non-SEZ jobs; and the workers are forced to work longer hours. Women and children bear the brunt of this exploitation, because SEZs prefer women and children as they are seen as "nimble" and "compliant."

City municipalities often encourage investment by corporations like Nike, Reebok, and Adidas by providing them land, giving them tax rebates, and relaxing environmental laws, because they look upon outsourced ventures as contributing to the export earnings. Manufacturing hand-stitched soccer balls in South Asia has become a controversial case: 5–14 year olds in Pakistani cities are employed by global corporations like Nike to work for twelve hours a day in near-slavery conditions. United Students Against Sweatshops (USASS) is an international grass roots organization of students trying to pressure their respective universities to ensure that university clothing is not produced in sweatshop conditions.

8. Civil wars and political instability have been major contributing factors to the destabilization of urban areas over the decades, most recently in Sri Lanka and Afghanistan.
9. Planned cities and new towns have played an important but subsidiary role in the region for a long time, with Islamabad (Pakistan) and Chandigarh (India) as notable recent examples.
10. Globalization, which began with the adoption of economic reforms, has resulted in growing affluence among a rising middle class; but it has also increased urban poverty, spatial exclusion of the poor, and urban violence.

In the cities of South Asia, a vibrant optimism and a newfound confidence abounds (fig. 9.1). Led by the 250 million strong Indian middle class, a rampant consumerism illustrates a giddy self-assurance and sense of hope. One merely needs to step into the bright, flashy, and glamorous automobile showrooms, where eager upper-middle income buyers may be seen purchasing not just a car, but in some cases a *fifth* family car and at prices over $23,000—*more than 46 times* the average per capita income in India. Glossy malls, housing dominant retail giants like Gucci and Prada, boutiques selling designer clothes, discotheques packed with hip youngsters, McDonalds and Pizza Huts filled to the brim with kids wanting the "American experience," amusement parks and movie theaters run by the latest digital technologies, and beauty parlors and spas define the landscape of fast-globalizing South Asian cities. Globalization is here to stay, and the growing South Asian middle class want more of it. Computer literacy, software proficiency, and business management skills define the youth component of the middle class, who are increasingly acquiring lucrative jobs in the local branch offices of global corporate giants. The grim determination of the post-independence era (post-1947) to produce scientists, engineers, and doctors who could build the nation, is slowly being replaced by a global dream to produce CEOs, accountants, software professionals, who could afford the consumptive lifestyle of the American middleclass. Mumbai, the dominant financial, commercial, and movie hotspot in India, is considered a world city of the first order because it sends and receives massive financial, commercial and cultural flows. Delhi, Bengalūru (Bangalore), Hyderabad, and Kolkata (Calcutta) in India, Dhaka in Bangladesh, Colombo in Sri Lanka,

and Karachi and Lahore are also world cities because they are economically and culturally integrated with global flows of goods, investments, images, and people. Beginning in the late 1980s, the countries in South Asia dissolved protectionist economic systems which shielded their domestic markets through tariff walls, licensing, and quotas. The "License Raj" (as it was known in India) was abolished, and through rounds of structural adjustments, a New Economic Policy of liberalization or free market globalization was adopted. The adoption of this policy opened up South Asian markets and also its people to global corporations and their investments. The economic reforms have produced tremendous urban impacts, many of which have been contradictory and controversial (box 9.1). While the middle class has become a global labor source, manning call centers, foreign banks, and corporate offices, the majority of urban poor have been relegated to sweatshop-like conditions in Special Economic Zones (SEZs). Many others have lost formal sector jobs as manufacturing units closed down unable to keep up with global competition in this free market regime. Still others have suffered evictions under city greening, beautification, and slum demolition policies

While annual rates of urban growth have slowed in the region, massive migrations of rural poor to the cities have put tremendous pressure on urban infrastructure. Increasing urban poverty has been coupled with an increase in low-paid informal jobs resulting in visible landscapes of poverty. Mumbai, the city which best demonstrates the affluence and consumption noted above, also presents urban poverty at its most daunting. Within the city proper 6.7 million slum dwellers comprise about 54 percent of the population.

Figure 9.2   As cities fill up with people, streets become more congested with not only cars, but bicycles and camels as well. (Photo by George Pomeroy)

Across the Mumbai urban agglomeration these numbers swell further. Dharavi, with over 600,000 people (and perhaps as many as 1 million!) crowded into an area of 432 acres (216 hectares), just under a full square mile, is Mumbai's largest slum and perhaps the best known (though, contrary to rumor, not the world's or even South Asia's largest). Pavement dwellers, those who have no shelter at all and sleep on sidewalks, doorsteps, and the like, number over 600,000. Across each of the region's megacities the story is the same—tremendous numbers of people living in conditions of poverty with inadequate shelter, lack of clean water, and filthy living conditions. Already racked with unemployment and underemployment, each swells with new migrants each and every day (fig. 9.2).

This duality of prosperity alongside poverty in the midst of the vibrant mosaic of language, ethnicity, and faiths of South Asia is what makes cities in this region unique. Throughout South Asia, cities remain receptacles of hope and serve as powerful engines of social and economic change.

URBAN PATTERNS AT THE REGIONAL SCALE

South Asia has 494 million urban dwellers, and this number will balloon to over 853 million in 2030. This means that South Asian cities will expand by more than the populations of the United States, Canada and Mexico combined! Most of the urban population is in India, Pakistan, and Bangladesh, which currently rank as the world's second, sixth, and seventh most populous countries. Together, the three comprise 95 percent of South Asia's total population; three out of four live in India alone.

With its immense population, India, at 30 percent urban, largely determines the overall regional average, which is also about 30 percent. Pakistan has just over one in three living in urban areas; Bangladesh just under one in four. The smaller countries of Afghanistan, Nepal, and Sri Lanka have smaller urban shares to match. Bhutan, although also small, has demonstrated a burgeoning urban expansion in the recent decades; its cites are growing at an annul rate of 4.8 percent. While most of South Asia has seen steady urbanization, Afghanistan, Sri Lanka, Nepal have suffered civil conflict, revolt, or insurgency, making city growth and development more difficult and erratic.

While cities of all sizes in South Asia have been growing, the five megacities have grown the most. Mumbai is already among the world's five largest urban agglomerations with over 20 million people. It is projected to grow to over 24 million over the next decade, at which time it will rank second only to Tokyo in population. Delhi and Kolkata, in India occupy sixth and eighth ranks among the world's largest metropolitan areas. Dhaka, Bangladesh's capital, and Karachi, Pakistan, are not far behind, ranking 9th and 11th, respectively. Altogether, over 61 cities in South Asia are currently "millionaire cities" having populations of over one million residents. Forty-eight of those millionaire cities are located in India, eight in Pakistan, three in Bangladesh, and one each in Nepal and Sri Lanka. Twelve cities have population over 5 million, eight of them in India, two in Pakistan, and two in Bangladesh. There are twelve urban agglomerations in the region, each accounting for five million people or more.

Table 9.1   South Asia's Twelve Largest Cities

City-Country	Population (2010)
Mumbai-India	20 million
Delhi-India	17 million
Kolkata-India	15.6 million
Dhaka-Bangladesh	14.7 million
Karachi-Pakistan	13.0 million
Chennai-India	7.56 million
Bengal ru-India	7.23 million
Lahore-Pakistan	7.0 million
Hyderabad-India	6.76 million
Ahmedabad-India	5.72 million
Pune-India	5.01 million
Chittagong-Bangladesh	5.01 million

If one takes into consideration all cities of 500,000 or more in population, the number is over 197.

Within India, six megacities dominate the urban hierarchy. Mumbai, Delhi, and Kolkata anchor points of a northern urban triangle. Chennai, Bengalūru, and Hyderabad form a southern urban triangle. Together, these six cities make up over one-fifth of India's urban population. Cities of the two triangles are being tied together by a great highway-building project known as the Golden Quadrilateral (fig. 9.3). The Golden Quadrilateral is a critical part of India's version of the U.S. Interstate Highway system. This project is a component of the country's most ambitious plan to improve transportation infrastructure since independence. The Quadrilateral will run through thirteen Indian states and connect the nation's four largest cities with 3,600 miles (5794 km) of four- and six-lane highways. Construction is now largely complete. The overall scheme is to widen and pave 40,000 miles (64,374 km) of highways over a fifteen year period.

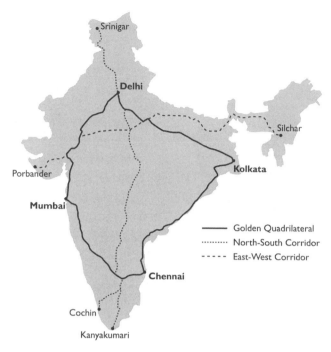

Figure 9.3   The Golden Quadrilateral of express highways is nearing completion. It links the anchor cities of India's urban hierarchy: Delhi, Mumbai, Kolkata, and Chennai. *Source:* Compiled from various sources

With the Golden Quadrilateral as its linch-pin, the revamped highway system is intended to facilitate regional trade and commerce, reduce transportation costs, and contribute to economic efficiency generally. Truck transportation comprises 4 percent of India's GDP and the potential for saving 40 percent on travel time translates into significant savings. "These micro gains make for macro benefit," noted the *New York Times* (12–4-05) as it profiled a milk seller who sliced one-third off his 90 minute commute to market, a child who took half as long to travel to school, and a truck driver who reduced his travel time by half.

The construction of the Golden Quadrilateral is remarkable not only for its scale,

but also for how it is being built. Prior to the 1990s there were limits on foreign investment in India. As trade liberalization and deregulation have occurred over the last two decades, the presence of foreign goods, services, and contractors has expanded dramatically. In constructing the Quadrilateral, 19 of the 35 component projects involve foreign contractors. For the highway system as a whole, construction companies from over 30 nations are involved.

Pakistan's urban hierarchy is dominated by a pair of cities with a combined population of over 20 million: Karachi dominates southern Pakistan and Lahore (the smaller of the two) dominates the north (fig. 9.4). Urban primacy characterizes the remaining countries of the

Figure 9.4   A picture of hope in this street scene from Lahore, Pakistan, in which a teacher uses the side of a boulevard as a makeshift outdoor classroom for his young charges, part of a broader UNICEF program to help the urban poor. (Photo courtesy of UNICEF—John Issac)

region. In Bangladesh, Dhaka is the center of gravity as it is over three times the size of Chittagong, the second largest city. Kabul, Afghanistan, and Kathmandu, Nepal, are both national capitals and are over six times the size of the next largest city. Most extreme is Sri Lanka's capital of Colombo, which is over 11 times the size of the next largest city.

South Asian cities may be divided into three basic types: *traditional cities, colonial cities*, and *planned cities*. Traditional cities are those which were part of a thriving urban system prior to Western colonialism, whether as centers of trade and commerce, centers of administration, or as religious pilgrimage destinations. Varanasi, for example, remains

the leading pilgrimage center in South Asia for Hindus and Jains; it attracts millions of pilgrims each year. Traditional cities could often be distinguished as being trade centers, such as Surat on India's west coast.

Several of the region's largest cities, most notably Mumbai, Kolkata, Chennai, were the products of British colonialism. In the post-independence period, basically since 1947, various traditional and colonial cities have evolved further into commercial/industrial cities to serve the expanding economies and national development plans of the region.

South Asia's third basic urban type is the new planned city, of which there are two kinds: (1) political and administrative centers such as Islamabad, Pakistan, and Chandigarh, India; (2) industrial centers for steel and other related heavy industrial activities, such as Durgapur, in the state of West Bengal and Jamshedpur in Bihar, both in India. Islamabad was built in the 1960s to take the place of Karachi as Pakistan's capital city. Chandigarh was also built after independence as the new state capital of India's Punjab. When a new state was carved out of Punjab in the 1960s, Chandigarh found itself on the border and has since served as capital of both Punjab and newly formed Haryana.

A recent and remarkable change in the urban fabric of South Asia is in India. With prosperity and growth has come the development of "urban corridors" broadly akin to Jean Gottmann's Megalopolis in the United States. Spurred by the construction of new express highways, city-regions are emerging along transportation corridors.

HISTORICAL PERSPECTIVES ON URBAN DEVELOPMENT

The cultural, linguistic, and religious diversity, and in turn the urban fabric, of South Asia are derived from five distinctive influences. Chronologically these are (1) the Indus Valley civilization, 3000 to 1500 B.C.; (2) the Aryan Hindus since 1500 B.C.; (3) the Dravidians since about 200 B.C.; (4) the Muslims since the 8th century; and (5) the Europeans since the 15th century.

Indus Valley Era

The Indus Valley is one of the "cradles of civilization" and among the world's oldest urban hearths. The preeminent cities of the Indus Valley civilization, Mohenjo Daro and Harappa, located in what is now Pakistan, were established as planned communities as early as 3000 B.C. They flourished for about 1500 years. As the largest urban centers of the region, they anchored an extensive settlement system that included at least three other large urban centers and perhaps over 900 smaller ones. First excavated in 1921, the ruins of Mohenjo Daro reveal a carefully constructed city reflective of a highly organized and complex society. No other city outside the Indus civilization possessed such an elaborate system of drainage and sanitation, signifying a generally high standard of living. This urban civilization came to an end around 1500 B.C., when the newly arrived Aryans—less civilized but more adept in warfare than the Indus people—overpowered the Indus civilization and turned the northern part of the South Asian realm to a mixture of pastoralism and sedentary agriculture.

Aryan Hindu Impact

Eventually the demands of trade, commerce, administration, and fortification gave rise to the establishment of sizeable urban centers, particularly in the middle Ganges Plains. Originating from a modest 5th century fort, Pataliputra developed into the capital of a notable Indian empire, the Maurya (321–181 B.C.). Its location coincides with that of present-day Patna. Pataliputra was organized to conform to the functional requirements of the capital of a Hindu kingdom: residential patterns based on the function-based, four-caste social system along with the requisite royal administrative features.

While caste is not as important in determining contemporary residential patterns in cities

Figure 9.5   The dhobi-wallahs, or "washer-men," make their living washing clothes, here drying in the open air. (Photo by George Pomeroy)

as it was in Pataliputra, it remains socially significant today. Caste, the designation of social status by birth through the caste of one's parents, places each person into one of four broad groups—*Brahmins*, *Kshatriyas*, *Vaisyas*, and *Sudras*. Within the *Sudra* designation are those of very 'low' caste and even those without caste or status. The people "without caste" are today more popularly referred to as *Dalits*. In the past, Dalits were often referred to as "untouchables" because the touch or even a shadow cast by one was thought to "pollute" someone of higher caste. Traditionally each caste and its subcastes had certain designated occupations; for example, "washer-men" or *dhobi-wallahs"* (fig. 9.5). *Brahmins* (priest caste) are at the top, followed by *Kshatriyas* (warrior caste), *Vai-*

syas (commercial and agricultural caste), and *Sudras* (manual labor caste). Caste still plays a major role in society. Marriages are generally within caste and there remain broad connections between caste status on one hand and income, quality of life, and social connections on the other. This remains true even after years of legal reforms such as a "reservation system" (similar to affirmative action in the U.S.) and broader social campaigns led by people such as Mahatma Gandhi, spiritual leader of India's independence movement. Even today, urban land use patterns and socioeconomic structures somewhat reflect the caste system (fig. 9.6).

In ancient Pataliputra one could clearly see the spatial distribution of castes. Near the center and a little toward the east were the temple

Figure 9.6   Low-caste Hindus wait to see a puppet show by proselytizing Muslims. (Photo by John Benhart, Sr.)

and residences of high-ranking *Brahmins* and ministers of the royal cabinet. Farther toward the east were the *Kshatriyas*, rich merchants, and expert artisans. To the south were government superintendents, prostitutes, musicians, and some other members of the *Vaisya*. To the west were the *Sudras*, including Untouchables, along with ordinary artisans and low-grade *Vaisyas*. Finally, to the north were artisans, *Brahmins*, and temples maintained for the titular deity of the city. A well-organized city government, hierarchical street network, and an elaborate drainage system accompanied this functional distribution of population in Pataliputra.

After the eclipse of the Maurya Empire in the 2nd century B.C., rulers of the Gupta Empire (320–467 A.D.) made it their capital, but the city lost its importance thereafter and was eventually buried under the sediments of the Ganges and Son Rivers. Only small parts of the old city have recently been excavated. Other Hindu capitals that developed in both North and South India changed and modified the many urban forms used in Pataliputra.

Dravidian Temple Cities

In contrast to the subcontinent's north, Hindu kingdoms in south India had relative control for most of the historical period; distinctly Hindu forms of city development evolved uninterruptedly. The rulers of south India constructed temples and water tanks as nuclei of habitation. Around the temples grew commercial bazaars and settlements of Brahmin priests and scholars. The ruler often built a palace near the temple, turning the temple-city into the capital of his kingdom; Madurai and Kancheepuram are examples of such lofty

Figure 9.7   The Taj Mahal has become the single most recognized icon of India. It was built in Agra as a tomb for Shah Jahan's wife and is now a UNESCO World Heritage Site. (Photo by George Pomeroy)

and grand temple cities. Such city forms were also exported to Southeast Asia; Angkor Wat (802–1432 A.D.) in Cambodia is one example.

Mentioned by Ptolemy, Madurai, the second capital of the south Indian Hindu kingdom of Pandyas, dates to about the beginning of the Christian era. Though Madurai is now several times the size of the old walled city, with imprints from a brief Islamic period and a longer British dominance, its religious importance is nearly comparable to Varanasi in north India, which is the foremost Hindu pilgrimage center.

Muslim Impact

The first permanent Muslim occupation that significantly influenced the subcontinent began in the 11th century A.D. and resulted in adding many Middle Eastern and central Asian Islamic qualities to the urban landscapes of South Asia. Shahjahanabad is a particularly good example of Muslim impact. The Moghul emperor Shah Jahan, who planned the Taj Mahal as a tomb for his wife in Agra (fig. 9.7), moved his capital from Agra to Delhi (about 125 mi or 200 km) and started the construction of a new city, Shahjahanabad, on the right bank of the Yamuna River. The city was built near the sites of several previous capital cities and took nearly a decade to complete (1638–1648). In its architecture was a fusion of Islamic and Hindu influences. Though the royal palace and mosques, with their arched vaults and domes, adhered to Muslim styles, Hindu styles were found in combination. The Muslim rulers were most concerned with the magnificence of their royal residences and courts, and massiveness of their fortresses. These features are represented most vividly in Shahjahanabad. Surrounded by brick walls without a moat, the city was completely fortified.

Situated at the east end of the city was the Red Fort (fig. 9.8). Planned as a parallelogram with massive red sandstone walls and ditches on all sides except by the river, the fortress

Figure 9.8   The Red Fort, in Old Delhi, remains a potent feature of nationalism. (Photo by Don Zeigler)

had an almost foolproof defense. Inside were a magnificent court, king's private palace, gardens, and a music pavilion. All were built of either red sandstone or white marble. The Red Fort remains a central feature of Delhi today, serving often as a platform for political proclamations and as a well-known tourist destination. A main thoroughfare, Chandni Chowk ("silver market") ran straight westward from the Red Fort toward the Lahore Gate of the city. The Chandni Chowk was one of the great bazaars of the Orient.

Shahjahanabad ceased to be the capital of India, more precisely north India, when British rule started in the 18th century, but it continued to be a functional city. Today, the area is known as "Old Delhi" and is part of the Delhi metropolis. Most of the city walls are gone, though all the basic structures of the Red Fort and Jama mosque remain intact. Chandni Chowk continues to be a busy traditional bazaar. It is very densely populated with a mixture of Hindus and Muslims (fig. 9.9). Large parts of Old Delhi are gradually being transformed into commercial and small workshop uses.

Colonial Period

After Vasco de Gama discovered the oceanic route via Africa's Cape of Good Hope and landed on the southwestern coast of India in 1498, the European powers of Portugal, Holland, France, and Britain became greatly interested in developing a firm trade connection with South Asia. Though initially all four powers obtained some kind of footing in India, the sagacious diplomacy and "divide–and–rule" policy of the British succeeded in ousting the other Europeans from most Indian soil. Eventually, the British established three significant centers of operation—Bombay, Madras, and Calcutta.

The British needed a firm footing in seaports for the convenience of trading and

Figure 9.9 To the left is a Muslim neighborhood and to the right a Hindu one in Old Delhi. (Photo by John Benhart, Sr.)

receiving military reinforcements from the parent country. Bombay, Madras, and Calcutta were the seaports. Hence, these three cities were designed as the headquarters for the three different Presidencies into which the British divided South Asia for administrative purposes. Consequently, the cities are referred to as the "Presidency Towns." In the 1990s they were renamed Mumbai, Chennai, and Kolkata, respectively, to reflect indigenous cultures and further downplay India's colonial heritage.

The Presidency Towns

When Mumbai, Chennai, Kolkata and Colombo were established as Presidency Towns, their nuclei were forts. Outside these forts were the cities. Inside the cities, two different standards of living were set for two different classes of residents: Europeans and "natives" each had their own parts of the city. The rich were composed of absentee landowners from rural areas, moneylenders, businesspeople, and the newly English-educated elite and clerks. The poor comprised servants, manual laborers, street cleaners, and porters. The rich needed the service of the poor, and therefore the houses of the native rich in many instances stood by the houses of their poor, native service providers.

As local industries grew in the 19th century, a new working class developed. In Kolkata, the industrial workers worked mainly for jute mills and local engineering factories; in Mumbai, for the expanding cotton and textile-related industries; and in Chennai, for tanning and cotton textiles. As these trades grew, the Presidency

Figure 9.10   The "Gateway of India," a distinctive symbol of India's largest city, was built in 1911 on Mumbai's harbor to commemorate the landing of King George V. (Photo by George Pomeroy)

towns turned from water to railways and roads for inland transportation. Train services were started in South Asia in 1852. The Presidency Towns also developed huge hinterlands that catered to the needs of the colonial economy. The hinterlands supplied the raw materials to the three seaports for export to the United Kingdom; in return, the British sent consumer-type manufactured goods through the same ports. Thus, the Presidency towns became the main focus of the colonial mercantile system.

Colonial architecture was designed in the image of Western Gothic and Victorian styles (fig. 9.10). Most public buildings of the Presidency towns were built in these designs. One of the best examples of the use of these architectural styles in combination with some Indian forms is the Victoria Memorial Building (built 1906–1921) in Kolkata. The building was done in white marble with European designs, and the architects originally intended it to surpass the Taj Mahal in massive grandeur.

MODELS OF URBAN STRUCTURE

Early on in the study of South Asian cities, some theorists would mechanically try to apply Western models, such as Burgess's Concentric Zones, irrespective of their applicability. No comprehensive model explaining the growth patterns of indigenous cities structures has been offered, though three basic models have been proposed to explain the distinctive form of South Asian cities: the *bazaar-based city model*, the *colonial-based city model*, and the *planned city model*. The basic characteristics of the three types of cities have been summarized in a matrix (table 9.2). Whichever model is used, however, it is important to note that

Table 9.2 Topological Characteristics of South Asian Cities

	Land Value	Population Density Gradient	Physical Aspect	Land Use Composition of the City Center	Historical Roots	Mixture of Three Forms
Bazaar City	Highest at the center; declines as one moves to the periphery	Highest at the center and declining inversely as one moves to the periphery	Narrow streets, commercial establishments at the center occupying the road; front, back, and second-/third-floor residential; generally congested and dirty	Retail and wholesale business mainly, with limited recreation and combined with high-density residential	May have origins from ancient medieval, or recent times and accordingly may have imprints from Dravidian, Hindu, Muslim or Western forms at the periphery	When a bazaar city was implanted with colonial aspects, gardenlike, semiplanned "civil lines" were added; similar addition of planned neighborhoods may also be at the periphery after independence
Colonial City	Highest at the center; generally declines as one moves to the periphery, but relatively higher in the European town compared to the "native" town	Center with minimum density; with highest densities around the CBD, creating a "carter effect" at the center; thereafter declines as one moves to the periphery	Wide streets at the center and the European town, with garden like, affluent appearance; the "native" town characterized by narrow, sinuous, streets and generally shabby condition	Offices, banks, main post office, transport headquarters, government buildings with large open space, hotels, retail and recreational activities and residential use	Origins no earlier than 16th century; Victorian, neo-Gothic, and other Western forms widely prevalent; native forms also implanted	Parts of colonial city—particularly adjacent to the CBD and some specific locations in the "native" town—evolving characteristics of the bazaar center; planned neighborhoods added, particularly adjacent to the periphery of the European town after independence
Planned City	May vary according to predetermined values for different locations	May vary at different locations of the city, but the initial plan is for low density	Organized and generally pleasing appearance	Combination of retail, office, and recreation in a systematic fashion	May have origins in any historical period, but over time bazaar aspects will begin to dominate the center if restrictions are not strictly adhered to	Planned towns with addition of "civil lines" at the periphery and bazaar central forms evolving at one or many locations

Source: A. K. Dutt and R. Amin, "Towards a Typology of South Asian Cities", *National Geographic Journal of India* 32 (1995): 30–39.

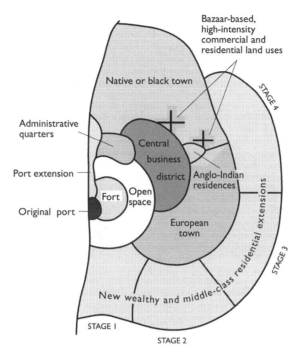

Figure 9.11   A Model of the Colonial-based City in
South Asia. *Source:* Ashok Dutt

two principal influential forces—colonial and traditional—have combined to create the existing forms of South Asian cities.

The Colonial-Based City Model

The need to perform colonial functions demanded a particular form of city growth, relevant to the subcontinent that produced the following characteristics (fig. 9.11).

1. The need for trade and military reinforcements required a waterfront location accessible to oceangoing ships because the colonial power operated from Europe. A minimal port facility was a prerequisite and was the starting point of the city.

2. A walled fort was constructed adjacent to the port with fortifications, white soldiers' and officers' barracks, a small church, and educational institutions. Sometimes inside the fort, factories processed agricultural raw materials to be shipped to the mother country. Thus, the fort became not only a military outpost but also the nucleus of the colonial exchange.

3. Beyond the fort and the open area, a "native town" or town for the native peoples eventually developed, characterized by overcrowding, unsanitary conditions, and unplanned settlements. It serviced the fort and the colonial administration, as the CBD and further administrative activities developed near the fort and the native town.

4. A Western-style CBD grew adjacent to the fort and native town, with a high concentration of mercantile office functions,

Figure 9.12   One of the first buildings constructed (1889) in Allahabad University is typical of British colonial structures, with a melding of Gothic and Indian architectural styles. (Photo by Ashok Dutt)

retail trade, and low density residential areas. The administrative quarters consisted of the governor's (or viceroy's) house, the main government office, the high court, and the general post office. In the CBD there also were Western-style hotels, churches, banks and museums, as well as occasional statues of British royals and dignitaries (fig. 9.12).

5. The European town grew in a different direction from the native town. It had spacious bungalows, elegant apartment houses, planned streets with trees on both sides, and a generally European look; clubs for afternoon and evening get-togethers and with European indoor and outdoor recreation facilities; churches of different denominations; and garden-like graveyards.

6. Between the fort and the European town (or at some appropriate nearby location), an extensive open space (*maidan*) was reserved for military parades and Western recreation facilities such as race and golf courses, soccer and cricket. On Saturdays, for instance, whites and a few moneyed native people frequented the horse races to gamble.

7. When a domestic water supply, electric connections, and sewage links were available or technically possible, the European town residents utilized them fully; whereas, their use was quite limited in the native town.

8. At an intermediate location between black towns and white towns developed the colonies of Anglo-Indians. They were the offspring of mixed marriages (European and Indian), and they were Christians. Never were they fully accepted by either the native or the European community.

9. Starting from the late 19th century, the colonial city became so large that new living space was necessary, especially for the native elite and rich people. Extensions to the city were made by reclaiming the

lowland or developing in a semiplanned manner the existing non-urban areas.

10. From the very inception of such colonial cities, population density was very low at the center, which housed Europeans, while a much larger group of "natives" lived outside the colonial center. When the European center was gradually replaced by a Western-style CBD during the second half of the 19th century, there was a further decline in the population of the center, giving rise to a density gradient with a "crater effect" at the center.

As the colonial system became deeply entrenched in the Indian subcontinent and an extensive railway network was made operational, waterfront locations accessible to oceangoing ships were no longer a prerequisite for a colonial headquarters. Kolkata, Mumbai, Chennai, and Colombo were not the only suitable locations on the subcontinent for high levels of administration.

Cantonments, railway colonies, and hill stations were three other lesser (but numerous) colonial urban forms that were introduced to the subcontinent to serve very specific purposes.

Cantonments (from the French word *canton*, meaning "district") were military encampments, some 114 in all by the mid-19th century, which housed a quarter of a million soldiers (both European and natives). Strict segregation by class and ethnicity was practiced in these camps.

Railway colonies surrounded a railroad station or a regional headquarters for railway operation and administration, also with strict segregation in their design. Often situated near urban centers, they eventually formed part of the greater urban area.

Hill stations, at altitudes between 3,500 and 8,000 feet (1,067 to 2,440 m), served as resort towns for Europeans to escape hot summers on the plains and spend time in the midst of a more exclusive European community. By the time of independence, there were 80 such stations, such as Simla and Darjeeling (fig. 9.13).

The Bazaar-Based City Model

The traditional bazaar city is widespread in South Asia and has certain features that date to pre-colonial times. Ordinarily, the city grows with a trade function originating from agricultural exchange, temple location, transport node, or various administrative activities (fig. 9.14). Usually, at the main crossroads a business concentration occurs where commodity sales dominate. In north India such an intersection is known as *Chowk*, around which cluster houses of the rich.

The bazaar, or the city center, consists of an amalgam of land uses that cater to the central-place functions of the city. The commercial land use, dominating the center, consists of both retail and wholesale activities. Perishable goods, such as vegetables, meat and fish—which are bought fresh daily because many homes lack refrigeration facilities—are sold in specific areas of the bazaar. These areas often lack enclosing walls and instead have a common roof. In the process of bazaar evolution, functional separation of retail business occurs: textile shops stay together, attracting tailors; grain shops cluster with each other near the perishable goods market; and pawnshops are adjacent to jewelry shops. Sidewalk vendors are present almost everywhere in the bazaar.

Wholesale business establishments also form part of the bazaar landscape (fig. 9.15).

Figure 9.13   Simla serves as the companion hill station for Delhi and remains a popular destination today. Sited at 7,500 ft (2,300 m), the elevation provides relief from the heat at lower elevations. (Photo by Ashok Dutt)

Situated near an accessible location, they tend to separate according to the commodities they deal in, such as vegetables, grains, and cloth, depending on the size of city. Traditionally, public or nonprofit inns provide modest overnight accommodation in the bazaar for a nominal fee. However, as a result of Western impact, some hotel accommodations are now available in the medium-sized and larger cities. Prostitutes or dancing girls, once a source of evening entertainment in the bazaar, have been supplanted by cinemas that in turn have declined due to the prevalence of television, VCRs, and DVD players. Traditionally, shops selling country-made liquor were never located in the bazaars, probably because drinking alcohol in public places was considered ill mannered by both Hindu and Muslim societies. Only in recent years have Western bars and liquor stores started to appear in city centers. Barbers, who used to work outdoors, now have regular shops like those in Western countries, but many still operate on the sidewalks. Long distance private telephone centers along with Internet access enterprises have become commonplace in city centers.

Beyond this inner core, in a second zone, rich people live in conjunction with poorer servants, but not in the same structure. The rich need the poor as domestic servants, cleaners, shop assistants, and porters. The residences of the poor surround this second zone in a third area, where the demand for land is less and its price low. Beyond the third zone, Civil Lines were established during British colonial rule. Here, particularly after independence, the native rich and middle class settled in neighborhoods and squatter settlements developed alongside.

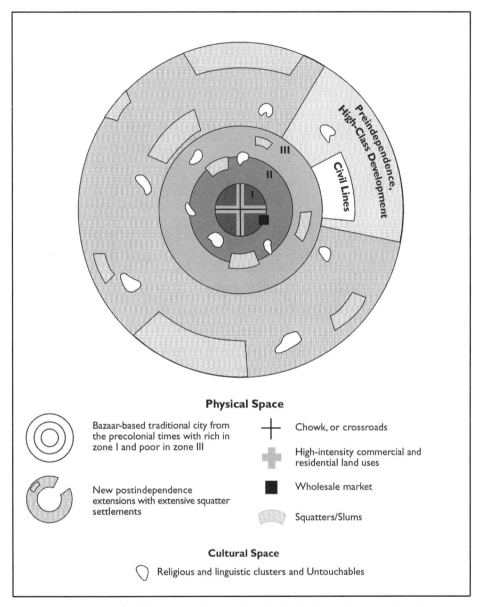

Physical Space

Bazaar-based traditional city from the precolonial times with rich in zone I and poor in zone III

New postindependence extensions with extensive squatter settlements

Chowk, or crossroads

High-intensity commercial and residential land uses

Wholesale market

Squatters/Slums

Cultural Space

Religious and linguistic clusters and Untouchables

Figure 9.14   A Model of the Bazaar-based City in South Asia. *Source:* Ashok Dutt

As bazaar cites grew, ethnic, religious, linguistic and caste neighborhoods were formed in specific areas in accord with the time of settlement and availability of developable land. The "untouchables" always occupied the periphery of the city, although sometimes other housing developed later beyond their neighborhoods. In Hindu-dominated areas of India, Muslims always formed separate neighborhoods. Similarly, Hindu minorities in Muslim-majority Srinagar and Dhaka lived in enclaves of the old cities. Often migrants from other linguistic areas formed specific neighborhoods of their own.

Figure 9.15   A produce vendor in Chennai typifies the bazaar-based city. (Photo by George Pomeroy)

Planned Cities

Though there were several planned historic cities in the subcontinent (Mohenjo Daro and Pataliputra, for instance), they did not survive. There were, however, others that were planned during pre-colonial, colonial, and independence periods that not only survived but also formed nuclei for major urban developments. Jaipur is an example from pre-colonial times, and Jamshedpur from the British Colonial period. Jaipur, India's twelfth largest city with an estimated 2.3 million people, was founded as a planned city in 1727. A hierarchy of streets divided it into sectors and neighborhoods. Though the planned city covered only 3 square miles (5 sq km) and is now surrounded by a built-up area of about 22 square miles (35 sq km), an urban morphology provided by 18th century planning has endured.

Similar is the case with Jamshedpur, planned as a company town for the first steel mill in the subcontinent by the Mumbai-based industrial family of the Tatas. Jamshedpur is about 150 miles (90 km) southeast of Kolkata. The raw materials for the steel making—iron ore, coal and limestone—are found nearby. The city underwent four different plans during colonial times and one after independence. As the city approaches its 100th anniversary, the population approaches 1.5 million residents. True to the characteristics of the planned-city model, Jamshedpur (like Jaipur and other such cities) has a planned central core that has undergone modifications over time and is surrounded by unplanned traditional developments and semi-planned post-independence extensions.

Mixtures of Colonial and Bazaar Models

The functional demands created by activities in colonial, bazaar-type, and planned cities generated interaction among city types. British administrative requirements in the traditional cities resulted in the establishment of *Civil Lines*, generally on the urban periphery. The Civil

Lines were composed of a courthouse, treasury, jail, hospital, public library, police facilities, club houses, and residential quarters for high administrative and judicial officials. The streets for the Civil Lines were well planned, paved, and had trees planted on both sides.

During the 20th century, both before and after independence, when the local rich needed to build houses of their own, several governmental, semi-governmental, cooperative, and private agencies planned developments surrounding the traditional city. Consequently, many traditional cities developed new extensions on their peripheries. For the most part, however, colonial cities never grew as conceived by their colonial masters. Traditional factors played an unavoidable role in altering colonial forms. The traditional bazaar, inherent to the indigenous cityscape, always interacted with other city forms. The bazaar thrived side by side with the CBD. As a result, to correctly model classic colonial cities, such as Kolkata, Mumbai, Chennai and Colombo, it is essential to consider the impact on them by the traditional bazaars. All colonial cities bear imprints from bazaar forms, just as the bazaar cities were impacted by colonial functional demands. Planned cities, too, as they expanded, often added colonial and bazaar city forms, and sometimes the two latter city types created new planned entities as their expanded appendages.

DISTINCTIVE CITIES

Mumbai: India's Cultural and Economic Capital

Mumbai is India's largest urban agglomeration and the nation's most cosmopolitan city. Two dimensions have been critical to its rise to preeminence, one commercial and one cultural. The skyscrapers in the Nariman Point area are the heart of the city's—and nation's—corporate and financial sectors and signify the city's role as the single most important command-and-control point in the national economy. The two largest stock exchanges of Mumbai handle an overwhelming majority of the country's stock transactions and figure among the world's largest in volume and value. In addition, 40 percent of the country's foreign trade is conducted through the city. Mumbai has also emerged as the nation's cultural capital through its prolific film industry, which is among the world's largest: films produced in "Bollywood" are eagerly consumed not only by the viewing public in India, but also in Bangladesh, the Middle East and Africa.

In 1672, Mumbai became the capital of all British possessions on the west coast of India. The 17th century British possession of the seven islands, which now form the oldest part of this city, initiated the construction of the fort. Beyond the fort grew a "native town," where sanitary conditions were miserably poor and drainage was a serious problem. The "European town" grew around the fortress on higher ground, and protective walls were erected around it. The native or Black town was separated from the European town by an Esplanade, which was kept free of permanent houses. A main spur to Mumbai's development occurred when Britain's supply of raw cotton temporarily diminished in the 1860s during the U.S. Civil War. India became an important supplier of cotton, most of which moved through Mumbai. This resulted in an amassing of huge reserves of capital by Mumbai-based businessmen, and the city became the main cotton textile center of the realm. In 1853, the opening of railways eventually

Figure 9.16   The recreational side of Mumbai may be seen on Chaupati along Marine Drive, and wherever you find people, you will find vendors. (Photo by Ipsita Chatterjee)

connected Mumbai with a hinterland covering almost all of west India. Mumbai further prospered with the opening of the Suez Canal in 1869, which enhanced the city's trade advantage by further cutting the distance to Europe. This closer proximity helped earn the city its nickname: "Gateway to India."

When the fort area developed into a Western-style CBD, the British, followed by rich Indians, moved to Malabar Hill, Cumballah Hill, and Mahalakshmi on the southwestern portion of the island. These remain exclusive neighborhoods today. In the 1940s, the rich settled in another attractive area of the island, Marine Drive (now renamed) which lay along the Back Bay (fig. 9.16). Because of the ever-increasing demand for commercial and residential land in the fort area and the lack of land on the narrow peninsula and the island city, dozens of skyscrapers have been erected since the 1970s at Nariman Point, generating a skyline resembling a miniature Manhattan. The most recent phenomenon of Mumbai's commercial land use change is the partial shifting of office- and financial-

related activities to the newly built high-rise buildings of Nariman Point, though the old Fort area is still considered as the main core of the CBD. Mumbai is an expensive city to conduct business in and ranks 6th in the world with respect to office occupancy costs. The poor, the middle class, and a few native businessmen settled mostly at the center and the north of the island. At present, the colonial influence created by European settlements can be observed in the southern part of the city proper. The more traditional influences have remained observable in the north. Unsanitary slums, built of flimsy materials and serving as habitat of the poverty-ridden, mushroom all over the city.

As a state capital and the largest metropolis in South Asia, Mumbai has also become the largest port of the entire subcontinent; not only does it handle the largest share of foreign trade, it collects 60 percent of India's duty revenues. Though employment in cotton textiles manufacturing remains important, other sectors including general engineering, silk, chemicals, dyeing and bleaching, and information technology (IT) are now emerging as important employment sources. Total industrial employment has declined and the service sector is increasingly prominent. Still, the Mumbai Metropolitan Region accounts for a disproportionately large share of India's industrial employment and fixed capital, and the Mumbai-Pune corridor is India's second most important center of employment for IT.

Mumbai attracts an enormous number of migrants from the western and central parts of India, thus giving it a religious and linguistic diversity that surpasses all other cities in South Asia. Yet, it also shares some religious characteristics with other South Asian cities. For example, the decline in the Muslim population resulted from the partitioning of British India into India and Pakistan in 1947, prompting the mass exodus of Hindus and Sikhs from West Pakistan (now Pakistan) and Muslims from India. The partition also led to the flight of many Hindus from East Pakistan (now Bangladesh) to India. Today, Mumbai's population is 69 percent Hindu, 14 percent Muslim; 7 percent Sikh; and smaller populations of Christians, Jains, and Buddhists. Two other minority religious groups play a socioeconomic role far beyond their numbers. First, the Zoroastrians, or Parsis as they are called in Mumbai, are a very significant minority group. Even though their numbers in Mumbai are small, more Parsis live here than anywhere else across the globe. Also significant are the Jains, mainly migrant businessmen from nearby Gujarat State, who were drawn by Mumbai's increasing commercial attraction. In terms of linguistic characteristics, no other metropolis of the subcontinent has Mumbai's uniqueness. The regional language, Marathi, is spoken by less than half the population. Mumbai is a linguistic microcosm of India itself.

Bengalūru and Hyderabad: India's Economic Frontier

When asked to identify economic success stories in South Asia, two cities immediately leap to mind: Bengalūru (Bangalore until 2006) and Hyderabad. Globalization is the vehicle that both cities have ridden to prosperity, as each has become a center for IT development and business process outsourcing. The success of each is built upon the country's supply of capable, technically-skilled, and English-proficient (but underemployed) college graduates, combined with the forces of technology and globalization that have reduced distances and hence costs. Other elements distinctive to these two cities are the presence of an entrepreneurial spirit, government flexibility, and critical investments in infrastructure. Both cities also serve as state capitals.

Bengalūru's association with IT dates to the arrival of Texas Instruments in the mid-1980s. Even before that, however, the city had become a center of India's aerospace and defense manufacturing industries. By the late 1990s, so many multinational firms had established operations here that the city had been christened "India's Silicon Valley," an appropriate nickname because it accounts for over 1/3 of the nation's software exports. The concomitant wealth and affluence has given the city a rather cosmopolitan and trendy reputation.

Hyderabad's emergence as an IT center came in part through the visionary efforts of the state's chief minister during the late 1990s. He pulled out all the stops in providing incentives and infrastructure for high technology-related development. Today, the city prides itself as being referred to as "Cyberabad." It hosts a Genome Valley and a Nanotechnology park, outgrowths of the city's leading role in the nation's pharmaceutical industry. Microsoft's largest development center outside Redmond, Washington, is located here, as are many other multinational firms.

Delhi: Who Controls Delhi Controls India

Delhi, the seat of India's capital, combines a deep-rooted historical heritage with colonial and modern forms (fig. 9.17). The attraction for Delhi as the capital site was rooted in South Asia's physiography, locations of advanced civilization centers, and migration/invasion routes. Delhi occupies a relatively flat drainage divide between the two most productive agricultural areas of the realm, the Indus and Ganges plains, where the most notable centers of civilization and power developed in the past. The control of Delhi was so vital to the rule of north India that a popular saying arose: "Who controls Delhi, controls India."

Old Delhi (the former Shahjahanabad) is a traditional bazaar-type Indian city, with the Chandni Chowk as the main commercial center (fig. 9.18). Here sanitation used to be one of the main problems, but after independence the area has been fully provided with underground sewers and piped water. Rich merchants and ordinary working people live close to each other. West of the Yamuna, connected by bridges with the main city, is a post-independence semi-planned area as well as squatter developments, where one-third of Delhi's population live—some in unhealthy slums with very little or no basic facilities of water, sewer, electricity, or paved roads.

New Delhi, situated south of Old Delhi, is a majestic colonial creation that emerged as a new city after the capital of British India

was moved from Kolkata. The capital was temporarily moved to Delhi's Civil Lines and Cantonment in 1911 before being installed in New Delhi (1931). New Delhi was planned by a British architect, Edwin Lutyens, in a geometric form that combined hexagons, circles, triangles, rectangles, and straight lines (fig. 9.19). Spacious roads, a magnificent viceroy's residence (now the President's Palace), a circular council chamber (which is now Parliament House), imposing secretariat buildings, a Western-style shopping center (Connaught Place) with a large open space in the middle, officers' residences in huge compounds, and a garden-like atmosphere formed the main elements of New Delhi. The new capital was separated from the congested, unsanitary, and generally poor conditions of Old Delhi by an open space.

The main economic base of New Delhi is government services. After India's independence, an increasing demand for housing was created by new government employees, which led to large-scale public housing developments around earlier settlements of New Delhi. The most noticeable feature of such developments was the segregation of larger neighborhoods according to the rank of the government employees and foreign residents. Class rather than caste determined the new neighborhood composition. Delhi has expanded in a planned manner as the Delhi Development Authority (DDA) has worked to coordinate the land development process with an innovative revolving funding scheme. The dark side of this strategy is that the lower and middle classes are left without affordable housing options (fig. 9.20). That one in five Delhi residents lives in a slum blurs any gains registered.

Delhi's problems are so intense it is numbered among the five worst cities in the world

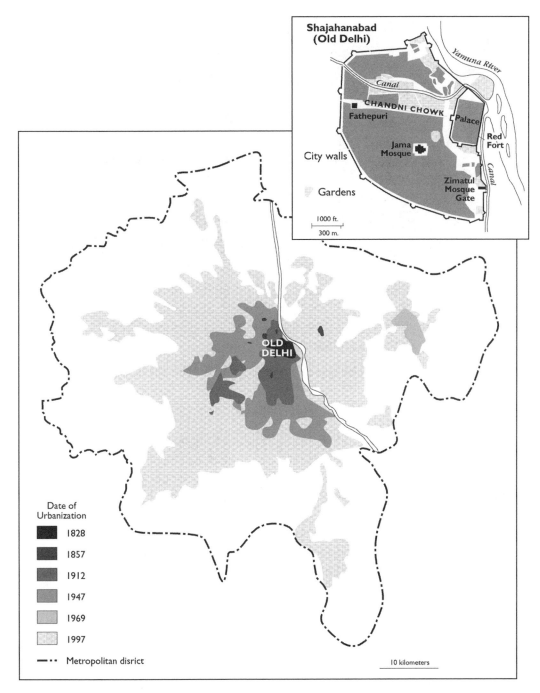

Figure 9.17   Delhi and Shajahanabad (Old Delhi). *Source:* Ashok Dutt and George Pomeroy

Figure 9.18   Sidewalk dwellers are a common scene in Delhi and across the cities of South Asia. (Photo by Ipsita Chatterjee)

with respect to air pollution. Motor vehicles are the leading source of air pollution, followed by industrial activities. Steps taken to ameliorate the problems over the last decade include banning of leaded gasoline, conversion to compressed natural gas (CNG) buses, mandated use of low-sulfur diesel fuel, and other tighter emission restrictions. With these steps, pollution levels have stabilized.

Kolkata: Premier Presidency Town

The popular image of Kolkata to foreign commentators through time has evoked a number of nicknames, including geographer Rhoads Murphey's "City in a Swamp," Dominique LaPierre's "City of Joy," Rudyard Kipling's "Cholera Capital of the World," and "City of Pavement Dwellers." As administrative capital for much of the colonial period and as an important commercial center since before then, Kolkata set the tone for urban imagery in South Asia.

"City in a Swamp" appropriately describes the city even today. Sited on the levees sloping east and west from the riverbank, the 40-mile (64 km) long metropolitan district is for the most part less than 22 feet (7 m) above sea level. This flood prone elevation is further aggravated by the monsoon, which brings most of the city's annual rainfall of 64 inches (1600 mm) between June and September, coinciding with the river's highest level. Waterlogged soils and extensive flooding substantially impacts a majority of slum dwellers. Despite these physical disadvantages, the city's location on the river Hugli (a distributary of the Ganges) provided advantages

Figure 9.19   This spacious road illustrates the harsh imperial geometry stamped upon Delhi by Edwin Lutyens. The domed building in the distance is Rashtrapat Bhavan, the President's residence. (Photo by John Benhart, Sr.)

for industrial growth. Its location 60 miles (97 km) upriver from the Bay of Bengal allowed access to 19th century ocean vessels provided a populous hinterland with rich mineral and agricultural resources. The city's seaport facilitated the import of wholesale machinery for the jute industry, which became most significant industrial activity of the metropolis. Finally, the establishment of the trading post and military garrison by the British in 1756 provided the mechanisms through which trade was conducted.

Spatial growth of the city, and the business district in particular, centered on the original fort and continued from that point even after its relocation several years later. A European component grew from the south end and a native town in the north. The "native town" included a wealthier and middle class component that resided at the northern edge of the

CBD. This area, the Barabazar, is reminiscent of the traditional bazaars (fig. 9.21). Immediately to the south is the Western-style CBD, mainly an office, administrative, and commercial district with a low density of residential population. As the city grew, its northern part reflected more traditional characteristics, while the southern section presented more of a European look. New areas were later reclaimed in the southern and eastern portions of the city to be inhabited mostly by wealthy Bengalis, the native inhabitants of the state.

Post-independence manufacturing activity was initially hurt by the partition of the subcontinent, which severed the jute mills of the city from their supply areas, contributing to the city's relative industrial decline. Today it is the engineering industry that is an important component of the local economy. Other

Figure 9.20   The use of makeshift materials, as seen here in Delhi, is an indicator of slum housing. (Photo by John Benhart, Sr.)

important industries are paper, pharmaceuticals, and synthetic fabrics. Information technology firms and related employment are growing, but Bengal ru and the Mumbai-Pune corridor remain ahead of Kolkata. Commercially, the city now ranks second to Mumbai. It has headquarters of native business firms, banks, and international corporations.

Before the partition, Kolkata attracted migrants from many different parts of northeastern India and East Pakistan (now Bangladesh). The city, therefore, demonstrates a multilingual demography. Two-thirds of the population speak Bengali, another fifth speak Hindi, and one in ten speaks Urdu. Hindus constitute 83 percent of the population; most of the remainder is Muslim.

Karachi: Port and Former Capital

Situated by the western edge of the Indus River delta and with approximately 13 million people, Karachi is Pakistan's largest city and former capital. It is highly industrialized and relies upon cotton textiles, steel, and engineering for its economic base. It remains by far the nation's leading port, and has become an important center for educational and medical services. Despite its leading role as a vibrant and cosmopolitan center of finance and trade, Karachi has gained a notorious international reputation for its lawlessness.

One block down and just around the corner from the U.S. Consulate, which was shaken by a car bomb, stands the restaurant where *Wall Street Journal* reporter Daniel Pearl disappeared. Three blocks farther along is the Sheraton Hotel, where another car bomb killed 11 French engineers. From there it's just a hundred yards to the bridge where a man with an AK-47 shot dead four Houston oil company auditors on their way to work. And this would be Karachi's very best neighborhood. (Source: *Washington Post*, June 24, 2002 "In a Dangerous City of Dreams, Survival Rules")

Figure 9.21 Fishmongers are widespread in Kolkata. Not only does the city have a huge consuming population, it is along the coast. (Photo by Ipsita Chatterjee)

Karachi's job opportunities have long provided an urban "pull" not only for Pakistanis, but also for large numbers of Muslim refugees leaving India, especially in the period immediately following partition. In the 1960s only 16 percent of Karachi's citizens were native born, while 18 percent were in-migrants from different parts of Pakistan, and 66 percent were Indian Muslim immigrants from the late 1940s.

While the city has become more religiously homogenous as a result of the in-migration and the effects of the 1947 partition, it has also become more linguistically and ethnically diverse. The partition led to the departure of all but a few thousand Sindhi Hindus, while Urdu-speaking Muslim immigrants came in large numbers from the north and central parts of India, and there was a preexistent base of Gujarati Muslim migrants from India since the pre-partition times. Thus, apart from English, three languages are prevalent in the city: Urdu, Sindhi, and Gujarati. A considerable influx of Pushtu-speaking Pathans from the Northwest Frontier Province (NWFP) and Afghanistan in the 1980s often clashed with the Urdu-speaking migrants from India. The combination of linguistic, religious, and ethnic differences, exacerbated by ineffective and corrupt law enforcement and a bureaucratic judicial system and availability of arms supplied by the US during the Afghan War in the 1980s have contributed to an alarming level of violence. This violence has escalated and continues today, and along with declining employment opportunities and crumbling infrastructure, has led to urban discontent and near anarchy.

The center of Karachi still represents a true bazaar model: high population density, high intensity of commercial and small-scale industrial activity, and a relatively higher concentration of rich people. Toward the east from the center of the city were the planned cantonment quarters that originated during the British occupation dating back to 1839. After independence, new suburban residential developments occurred surrounding the eastern two-thirds of the colonial city, while planned industrial estates were built mainly toward the northern and western fringes.

Dhaka: Capital, Port, and Primate City

Modern Dhaka is Bangladesh's largest urban center and capital. Situated in the center of the country, Dhaka is sited on the left bank of a tributary of the Meghna-Ganges system. Founded in the 7th century A.D., the city was subsequently governed by Hindu Sena kings (from the 9th century) and Muslim rulers (1203 to 1764), for which it served as an intermittent regional capital and attained great commercial importance. Toward the end of the 17th century, Dhaka had over 1 million inhabitants and stretched for 12 miles (19 km) in length and 8 miles (13 km) in breadth. The loss of regional capital status in 1706, however, precipitated a decline that continued well into the colonial period. Kolkata's status as a colonial capital and trade center eclipsed Dhaka's sphere of economic and political influence.

The early 20th century and the city's status as a temporary regional capital brought about important changes, resulting in a resurgence of growth that took the population from 104,000 in 1901 to 239,000 in 1941. Partition in 1947 prompted large-scale immigration of Muslims from India and stimulated industrial growth from an almost non-existent base. Spearheading the industrial growth was jute processing, which was now taking place in Bangladesh because of the new international boundary that provided an effective barrier to the jute mills of Kolkata. Subsequently, Dhaka's Urdu speakers swelled in numbers. Later, Dhaka became the national capital of Bangladesh, which separated from West Pakistan (now Pakistan) and became independent in 1971. The resultant multifarious administrative-commercial activities provided the necessary pull factors to trigger the most rapid growth in Dhaka's history. Today the metropolitan population of approximately 14 million ranks the city fourth largest in South Asia (fig 9.13). However, the contrast between rich and poor is glaring (box 9.2).

With over 90 percent of the city's population Muslim, mosques are so numerous in Dhaka that some refer to it as the "City of Mosques." Hindus are the most important minority group (about 8 percent); the rest are Christians and Buddhists. Linguistically, Dhaka is very homogenous, as over 95 percent of the population speaks Bengali, the native language.

Kathmandu, Colombo, and Kabul: Cities on the Edge

Colombo, Kabul, and Kathmandu are the premier cities and national capitals of Sri Lanka, Afghanistan, and Nepal, respectively. Officially, however, the capital of Sri Lanka has moved to Sri Jayawadenepura, located within the Colombo metropolitan area. As the former capital, the city of Colombo itself continues to be the seat of administration and decision-making. These cities on the edge of

Box 9.2   Micro-Credit Comes to Main Street

Few things better validate a "good idea" than a prize, and no prize is bigger than the No-bel Peace Prize. So, when Bangladeshi Mohammed Yunus and the Grameen Bank, which he founded, were awarded the Nobel Peace Prize in 2006, it was in recognition of his tremen-dous accomplishments. He and the bank pioneered the innovative idea of micro-credit. Once thought to be useful only in rural areas, this poverty-alleviation strategy is now successfully making its appearance in cities.

While the idea of micro-loans has been around for over a century, the current model was firmly established by Yunus and his bank. Yunus' first loan—$27 to 42 families for making bamboo furniture—was made in 1976, shortly after Bangladesh suffered a terrible series of natural disasters. Since then, his bank has made 97 percent of its loans to women, has established over 2,000 branches, and has a loan recovery rate of over 95 percent! Today, the model may be found across the developing world.

Micro-credit, or micro-lending, is the idea of making very small loans to people who, lacking collateral, regular employment, or credit history, cannot get credit from standard financial institutions such as banks. Upon receiving a micro-loan, these borrowers invest the money in their own informal or formal sector small business. The business savvy, which they may already possess, combined with the funding, allow many of these entrepreneurs to flourish. These individuals really are "bankable"; they simply did not possess the formal qualifications needed to obtain loans through standard financial channels. One reason micro-credit is seen as a success is that the rate of repayment is very high. The initial funds loaned by these "banks" may start with grants and loans from aid organizations, but oftentimes it consists in part of the pooled funds of the local poor. To ensure repayment, small informal support groups of borrowers, bank representatives, and others are required to attend weekly or bi-weekly meetings.

There are many advantages to micro-credit. Foremost of these is that micro-credit builds wealth so that the borrowers can exit extreme poverty. It helps those most who have had the least access to traditional financing—women, minority populations, and the poor. In South Asia, this would include those of lower caste, especially widowed women of lower caste. Dignity is enhanced, too, as the borrowers are not just receiving aid as a handout. Instead, they can rightfully see success as their own creation instead of someone else's. Finally, the socioeconomic status of the poor and other socially excluded groups is enhanced.

Micro-credit is seeing success in the cities of South Asia and beyond. In Bangladesh, for example, the Shakti Foundation for Disadvantaged Women is using the Grameen Bank model in Chittagong and Dhaka. In Dhaka, one-third of the 105,000 borrowers have now crossed the national poverty line. As South Asian cities swell and urban underemployment persists, it is clear that the micro-credit model will be of increasing relevance in urban settings.

South Asia differ greatly in size: Kabul with 3 million; Colombo with 660,000; and Kathmandu with one million.

Kabul is the oldest, even being referred to in the *Rig Veda*, a 3500-year-old Hindu scripture, and by Ptolemy in the 2nd century A.D. Kabul's strategic position by the side of the Kabul River and at the western entrance of the famous Khyber Pass, gave the city great political and trading significance. Sited at an elevation of 5900 feet (1800 m), it has served as a regional or national capital for numerous regimes over the centuries, most notably for the Moghul Empire (1504–1526) and after 1776 for an independent Afghanistan. Later attempts by both the Russians and the British to subjugate the country and its capital failed. After 1880, modern buildings and gardens were constructed but these have not diminished the identity of the old bazaar city.

The city has suffered greatly since the 1970s due to international conflicts being played out in its backyard. The overthrow of the king in 1974 initiated a series of events that included the establishment of a Soviet-backed government and subsequently, United States intervention to arm mujahideens (Afghan insurgents) against the pro-Soviet government (1970s), the establishment of a weak puppet state (1989), a period of warlord-dominated chaos (1992), and establishment of the Taliban regime (1996). Allegations that Afghanistan's difficult physical geography and a sympathetic Taliban regime provided havens for extremist elements involved in the World Trade Center attacks of September 11, 2001, invited retaliation by the United States leading to intense and destructive bombing of Kabul, as well as Kandahar and other cities. In the wake of almost continuous strife, it remains to be seen if the government will be able to

rebuild the city and provide a stable socioeconomic setting. The post-Taliban regime seems determined to establish stability and democracy, but it remains only nominally in control of the country beyond the vicinity of Kabul.

Kathmandu, in a valley of the same name, lies in the mid-mountain region of the Himalayas, at an elevation of about 4400 ft (1,350 m). Kathmandu occupies a central position for most of Nepal and is the most important commercial, business, and administrative center of the country. The Gurkha ethnic group, after conquering Nepal, made Kathmandu its capital in 1768. Reconstruction efforts after the devastating 1934 earthquake and post-World War I developments added more buildings, but the older section of the city has retained its bazaar characteristics. Though the overwhelming majority of Kathmandu residents are Hindus, there are also some Buddhists. Autocratic rule by the king and sporadic violence by Maoist guerillas made situations unstable in the past. In 2008, the monarchy was abolished and a Maoist led government was democratically elected. The city remains a major center of South Asian tourism and is a launching ground for treks in the Himalaya Mountains.

Situated on the west coast of the pearl-shaped island of Sri Lanka (formerly known as Ceylon to the West), Colombo has functioned as an important port city since at least the 5th century. Nonetheless, it was Western contact, beginning with the Portuguese settlement in 1517, which really started the growth of Colombo as the key port for Sri Lanka. The Dutch occupied the port in 1656, but the British replaced them in 1796 and turned the city of Colombo into their main administrative, military, and trading place in Sri Lanka. After independence in 1948, Colombo became the

the country's capital, but it has been challenged to build a single nationality in a county where Buddhist Sinhalese and Hindu Tamils have been at odds for most of Sri Lanka's history.

Colombo continues to be Sri Lanka's most important city in terms of business, administration, education, and culture. It is a colonial-based city. The CBD-like center with high-level government offices situated within the former fort area of colonial times lies by the side of the old section of the city, Pettah ("the town outside the fort" in the Tamil language). Pettah represents the characteristics of the Bazaar city enclave. Cinnamon Gardens, the former cinnamon-growing area of the Dutch period, has been turned into a high-class, low-density residential quarter. The city has expanded significantly since independence and has industries that mainly process the raw materials that are exported through the port of Colombo. In order to diversify the national economy through industrialization, an export-oriented Free Trade Zone has been established near the port of Colombo.

GLOBALIZATION, CITY MARKETING, AND URBAN VIOLENCE

The contemporary geography of South Asian cities is defined by globalization and its various economic, cultural, and political impacts. Urban landscapes of South Asia represent a complex mixture of global influences and local particularities. Globalization has been defined as the intensification of interaction between previously far-away places and people so that local happenings are now shaped by distant events. Economic liberalization, moves towards free market reforms, and the

information and technology revolution are considered to be the main forces propelling increased spatial interaction. Telemarketing, electronic banking, plastic money, high speed internet have made national borders porous. Economic liberalization has encouraged the erosion of tariff barriers and the free flow of investments across national borders. As a result, the cities in South Asia have increased abilities to tap global business in the form of corporate offices, export processing endeavors, hotel, hospitality, tourism (spiritual and medical), and retail industries. The global flows of investment and goods from headquarters of advanced nations search for emerging markets among the growing middle class in South Asia. Global corporations also look for cheap pools of unorganized labor in populous South Asian cities. This give-and-take between cities allows for what Erik Swyngedouw calls "glocalization"—local places like cities engage with global flows and ground them in locally specific ways so that both the cities and global flows are altered. For example, McDonalds, which is a global fast food chain, is globalized in Indian cities when it sells the 'no-beef' "Maharaja (King) burger" instead of the Mac burger, because Hindus do not eat beef.

South Asian cities now manifest a hybridization that symbolizes this tension between local imperatives and global impetuses. City municipalities are increasingly urged to "go global" in their search for funds. This results in a "new urban politics" of city-versus-city competition to acquire foreign business and investments. In India for example, the central government launched the Jawaharlal Nehru Urban Renewal Mission (JNNURM) in 2002. The purpose of the JNNURM was to select 63 cities, which would be given funds to "go global" by becoming entrepreneurial, profit-seeking, and "world class," like New York, Tokyo, and London. The JNNURM urged city governments to shed their pro-poor social agendas (e.g., providing cheap infrastructure and slum up-grading), and instead, focus on marketability. Redistributive measures like rent control, which were put in place to control the concentration of wealth, were abolished to allow entrepreneurialism and private investments to flourish. This push towards urban entrepreneurialism has been answered differently by different cities, but the dominant strategy has been 'place marketing'—repackaging urban areas to make them attractive to global capital. One place-marketing strategy includes green entrepreneurialism which involves greening, cleaning, and developing urban gardens and parks, introducing manicured traffic islands, and switching to Compressed Natural Gas (CNG) instead of gasoline. The idea is to present a global image of an environmentally friendly, sustainable, and smart city that will be attractive to foreign business and tourists. This green entrepreneurialism has become controversial, because greening is often accomplished by evicting the poor, and forcing them to adopt the more expensive CNG, while the urban rich are allowed to own multiple numbers of gasoline operated vehicles. Greening also concretizes uneven geographies within the city where affluent neighborhoods gain parks and open spaces, while poor neighborhoods continue to suffer from stagnation (box 9.3). These exclusions brought about by greening strategies have often been touted as "bourgeois environmentalism" or "elitist environmentalism." For example, in Ahmedabad, India, greening under the Green Partnership Program has benefited the more affluent west Ahmedabad, while the poorer parts of east Ahmedabad

lacks open spaces. The exclusionary politics of enforcing CNG in Delhi has also been well documented.

Another place marketing strategy includes city beautification through urban renewal. Beautification involves giving the city a facelift so that it can project the image of an efficient growth engine—a spruced-up city can potentially out-compete other cities in attracting investment. City governments therefore, go out of their way to sell communal/public land to private construction companies who are then supposed to 'up-scale' the city with promenades, boulevards, water parks, state-of-the-art offices, malls, parking lots, and high speed transit corridors. The cultural impact is often a homogenization of once unique landscapes. Mumbai, Delhi, Colombo, and Karachi demonstrate this growing loss of cultural diversity as small businesses, local food cultures, indigenous handicrafts, and local embroidery disappear to make way for the world's McDonalds and the Benettons. Delhi, for instance has seen a growing grass root movement of hawkers challenging the government's policy of evicting street vendors under a city 'cleaning project.' A more problematic dimension of urban renewal is that it is often achieved by 'liberating' spaces through demolition of slums with little compensation for the slum-dweller. Mumbai, Delhi, and Ahmedabad have become dominant sites for massive demolitions. In the post-liberalization era, all of these cities have developed their individual Master Plans to become 'world class,' and in a desperate effort to out-compete each other, have produced various forms of public-private partnerships that have engineered violent evictions. The municipalities in these cities represent the public entity that maintains a rhetoric of "liberalization with a human face,"

Box 9.3   Greening the City

Waquar Ahmed

Increase in environmental pollution has proved intrinsic to industrial and economic growth. As India has embarked on a path of rapid economic growth, its cities have become more polluted. According to the "white paper on pollution in Delhi," published by the government's Ministry of Environment and Forests, concentrations of sulphur dioxide in Delhi's air, between 1989 and 1996, increased by 118 percent, and that of nitrogen dioxide by 82 percent. In a joint study conducted by the World Bank and the Asian Development Bank on air pollution, New Delhi was classified as the most polluted among 20 major Asian cities (between the years 2000 and 2003). Kolkata and Mumbai were also among the top 10 polluters.

Even as air pollution in Delhi has increased, the source of pollution has undergone change. In 1970, domestic sources, industries, and vehicles contributed 21percent, 56 percent, and 23 percent, respectively, to Delhi's air pollution. By 2000, the corresponding figures for domestic sources, industries, and vehicles changed to 8 percent, 20 percent, and 72 percent, respectively. Proliferation of two- and four-wheeled motor vehicles, thus, has become a bane for the city's respiratory health.

Figure 9.22   Land is being clearned for the Sabarmati Riverfront Development in Ahmedabad.

Concern for Delhi's declining air quality has engendered environmental activism, supported particularly by the judiciary. Given Delhi's poor air quality, environmental activism and its fallout, ideally, should have been a welcome development. This, however, has not been the case. Environmental activism and judicial intervention, in fact, has been elitist in its manifestation, in other words, detrimental to the interest of Delhi's poor. Responding to the public interest litigation filed by environmental groups in the city, and confirming a judicial ruling that had been made two years ago, India's Supreme Court, in 2001, ruled that Delhi must replace its entire fleet of buses with compressed natural gas (CNG) powered "pollution-free" vehicles immediately. This Supreme Court decision plunged the city into chaos as thousands of buses were taken off the road, depriving the poorer population of transport, and in turn, of their livelihood. Public buses, just like other motor vehicles in Delhi, have certainly contributed to air quality decline. But given that public buses are used by one-third of Delhi's total population—especially those with low incomes—to commute, Delhi's environmental activism and the role of the judiciary in this episode assumed an elitist character. Another point to be noted here is that by the year 1999, 14 percent of commuters in Delhi—especially those with higher incomes—used scooters and motorcycles (two-wheelers) for their daily commute. These two-wheelers made up for two-thirds of Delhi's total vehicles and accounted for more than 70 percent of hydrocarbon and 50 percent of carbon monoxide emissions. Even as public transport was singled out and pushed into chaos, private transport, including two-wheelers were left untouched.

while an entourage of private construction companies are given the go-ahead to do the "needful." In Ahmedabad a gigantic project called the Sabarmati River Front Project was launched to develop the riverfront into a fast track "world class" corridor (fig. 9.22). NGOs claim that over 6000 families will be evicted by this project (box 9.4).

City marketing also manifests as gated communities replete with gyms, sports facilities, swimming pools, and shopping complexes mimicking the "good life" of the American middle class. In Bengalūru, where a burgeoning group of software professionals have quickly become rich in India's own Silicon Valley, the gated communities represent spaces of social mobility. Professional elites with sizable disposable incomes are increasingly seduced by the globalization of home-garage-pool life styles. These spaces of affluence have been touted as spaces of exception—they geographically materialize the growing gap between the urban rich and urban poor. The gated communities also cripple informal economies of vendors and hawkers who are no longer allowed to enter the "sanitized spaces" of the rich.

The global-local tensions of glocalization are manifested not only in the uneven geographies of affluence and deprivation, but also through morphologies of violence. Cities in South Asia, therefore, not only internalize the daily violence of exclusion, but are often also the sites of inter-community riots and global terrorism. In India, Hindus account for the majority religious group and

Box 9.4 Urban Renewal in Ahmedabad City

In keeping with the "new urban politics" and economic liberalization, India's Ahmedabad city, located in the western state of Gujarat, has launched a vociferous place marketing policy. The purpose is to outcompete any other city in India, including Mumbai and Delhi, and become the region's premier globalizing city. The Ahmedabad Municipal Corporation (AMC) is projecting an urban vision of a clean and beautiful business destination. A Global Investor's Summit has been held every year since 2007 to wine and dine foreign investors. Sabarmati River Front Development Project has been launched to renew the city. The river Sabarmati divides the city into east and west Ahmedabad, and 16,000 poor Hindus and Muslims live in slums along its banks. These families are employed in low-paid informal work, like selling used clothes, transportation, as domestic helpers, and as venders—6,000 families are slated for eviction. The AMC plans to reclaim both banks of the river, fill it with soil, and sell it to private construction companies. The aim is to convert this stretch of the river into a commercial corridor for big business and urban elites. Hotels, amusement parks, water parks, gardens, boulevards, and promenades are to be constructed to give the city a face-lift. The poor on the banks of the Sabarmati, however, argue that this vision of urban renewal is elitist and exclusionary. The urban poor claim that the budget allocation for rehabilitation is low, when compared with expenditure on "earth-filling" and "garden development." Residents claim that the original plan had set aside a stretch of the bank to develop subsidized housing for those slated for eviction. However, contemporary urban politics indicate otherwise—the Hindu poor fear that they will be relocated outside the city, and the Muslim poor fear that (given the far rightwing Hindu fundamentalist bias of the AMC) Muslims will not be rehabilitated at all. When endless marches, demonstrations, sit-ins, and protests did not yield results, a representative group from the banks of the Sabarmati filed public interest litigation against the AMC. A social movement has emerged among the poor. However, this movement is far from consolidated—ethnic, caste, and linguistic differences often fracture its unity. While Ahmedabad globalizes into a "world class" city, a battle is being fought on the banks of the river Sabarmati—a battle to reclaim the urban future of Ahmedabad.

Muslims account for the largest minority community (13 percent approximately). Hindus and Muslims share a contentious history because of the colonial policy of "divide and rule," the partition of India and the creation of Pakistan, and the horrific violence that resulted from it. In the post-independence context, Hindu-Muslim violence has been concentrated in cities. The Mumbai riots of 1992–93 and the Ahmedabad riots of 2002 were the deadliest in post-partition India. The Mumbai riots claimed 2000 lives—a regional far right political party is said to have engineered the riots and was allegedly responsible for horrific atrocities against Muslims. Many Muslims were killed along with Hindus and many others were displaced. The 2002 riots in Ahmedabad lasted for two-and-a-half months, involved systematic destruction of Muslim homes, property, and businesses. Mobs led by cultural and political affiliates of another far right party engineered the riots. Victims claim that the rioters came with lists of addresses of Muslim homes—2000 Muslims were killed, 100,000 displaced, mosques were demolished and replaced with temples and roads, the displaced now live in all-Muslim ghettoes outside the city. The urban riot machinery in India is fuelled by a political ideology, which believes that Indianness equals Hinduness, and therefore other religious minorities are considered invaders/foreigners, hence their patriotism is always suspect. In the post-September 11 context, the global narratives of "war on terror" and "terrorism and Islamophobia" are also adopted and localized. This "golden age of Hindu India" is not envisioned as isolated from global economic integration and global discourse of terrorism. A local politics is creatively juxtaposed with economic reforms, where non-Hindu foreign capital and foreign corporations are welcome. Local "Islamophobia" is also juxtaposed with global narratives of terrorism—the global-local tensions of glocalization are creatively imprinted on urban space.

In the Sri Lankan context on the other hand, the majority Sinhala Buddhist community benefiting from small scale self-enterprises under a pre-economic liberalization regime, faced increased hardship due to the crowding out of indigenous firms in the post economic liberalization context. Increased urban hardship led to increased ethnic polarization in the 1980s when the minority Hindu business community was targeted by the majority Buddhist community in the 1983 ethnic violence in Colombo and elsewhere. Apart from intra-national conflicts, South Asian cities have also become a hotspot for international terror. The Taj Mahal hotel shootings in Mumbai in 2008 was touted by the media as "India's September 11." (box 9.5)

URBAN CHALLENGES

The euphoria of globalization and economic growth poorly masks the enormity of the challenges facing South Asian cities. Most countries inherited a colonial legacy of poverty and extreme rural-urban dichotomy. In the contemporary context, urban realties depict a simultaneous juxtaposition of affluence and poverty. An expanding middle class, proliferation of malls, extreme poverty, inadequate housing, lack of public services, unemployment, and environmental degradation inscribe the South Asian urban landscape.

In the post-colonial period, most South Asian nations pushed for food self-sufficiency and basic industrial development subsidized by the government. The idea was to develop the villages, with towns and cities acting as complementary industrial hubs. However, rural land distribution was extremely unequal because of a feudal land tenure system, which was never rectified in the colonial era. In the post-independence period, inability to launch cohesive land reforms exaggerated rural poverty. In that context, the primate cities, which had already experienced infrastructural development in the colonial period, continued to attract masses of rural poor in the post-independence period. Although, manufacturing received a major boost in most countries through government initiated import substitution industrialization, the rate of growth of manufacturing jobs could not keep pace with rural-to-urban migration. Moreover,

Box 9.5   Cities and Terror

In November 2008, terrorists attacked Mumbai's Taj Mahal hotel. At the same time, another team of terrorists also attacked another hotel, a Jewish center, a railway station, a hospital, a restaurant, and a movie theatre complex. Many people, especially British and American guests, were taken hostage in the Taj Mahal hotel. Combat, which lasted all night, ensued between Indian commandoes and the attackers. Over a hundred people were killed and nine gunmen died. The media proclaimed the Taj Mahal hotel incident as India's September 11. A lot of intelligence effort went into understanding why Mumbai was chosen or why the Taj was the targeted location. It is generally believed that Mumbai's status as a "world city" gave it a certain symbolic value and global exposure which the terrorists wanted to exploit. The Taj Mahal hotel has been a site often favored by the Indian and international glitterati alike. Foreign businessmen and investors are wined and dined there, and foreign tourists sample "world class" Indian hospitality at the Taj. It was claimed that the terrorists wanted to send a message to globalizing India and to the world that all was not well. Mumbai's symbolic importance as the financial capital of South Asia, the center of a giant movie industry, the center of tourism and economic investment, and a past history of anti-Muslim riots (1992–1993), were important urban variables that defined that geography of terrorism. The choice of New York City as a target on September 11, 2001, can similarly be argued as strategic. The World Trade Center represented the seat of American capitalism, the hub from which global flows of finance and investment emanated. The Pentagon on the other hand, represented the center of American military might. Commuter trains in Madrid in 2004 and underground trains and buses in London in 2005 were similarly important from a symbolic point of view. A densely packed urban agglomeration like Mumbai provides the potential for spectacular destruction in terms of lives, infrastructure, and property, and hence, provides the terrorists with the ability to capture global attention. Big cities like Mumbai are also excellent camouflage: they afford densely packed environments, relaxed security, and too many distractions. Finally, big cities have ample resources for shadowy networks of cell phone and internet communication and plenty of black money to generate an appropriate infrastructure for terror. What is important however is a deeper understanding of why a certain group of people feels compelled to adopt terrorism as their chosen vehicle of expression to the extent that they are willing to give up their own lives. Economic reforms and globalization, it is argued, have disproportionate impacts on ordinary people. Some are able to "upscale" because of skills like computer literacy and English proficiency, while others face unemployment, eviction, displacement, and cultural negation. The contestations become most entrenched in cities of the world, a phenomenon Mike Davis calls, "Urbanization of the Empire." The old hardships of colonization become the contemporary reality for many. The "New Empire" is consolidated by corporations and urban elites. Inter-class, inter-community, inter-region, inter-city, and intra-city gaps widen. In the face of intense economic hardships and cultural annihilation, many poor people in cities may form non-violent social movements, justice movements, environmental movements, and right-to-the-city movements, while many others, may resort to extremism. Cities in South Asia are faced with increasing contradictions caused by economic and cultural impacts of globalization—urban violence may become a new modality for confronting such contradictions.

most rural migrants were unskilled and hence incapable of employment in modern industries. The result was a swelling informal sector consisting of low-paid jobs that offered no security (e.g., porters, rickshaw pullers, domestic servants, construction workers and other manual laborers). These informal workers often had to live on pavements, in railway and bus stations, and in other instial spaces. Others were "lucky" enough to find a one-room home in already overflowing slums.

Estimates of the number of pavement dwellers vary widely. For Mumbai alone, estimates vary from 250,000 to 2 million. In most places, they are concentrated in the central areas of the city and irregularly employed in low-skill occupations that pay the least. For Kolkata, about one-half of the pavement dwellers are employed in transport. They are predominantly males aged 18 to 57 years, though nearly 1/3 are children, most of whom add to the family income by working as child laborers, or by begging and scavenging.

Slums, or *bustees*, have developed in almost all the major cities of South Asia. The name *bustee* is used in Kolkata and Dhaka, *jhuggi* is used in Delhi; *chawl* in Mumbai. The bustee has been defined by the Indian government's Slum Areas Act of 1954 as a predominantly residential area where dwellings (by reason of dilapidation, overcrowding, faulty arrangement, and lack of ventilation, light, or sanitary facilities—or any combination of these factors) are detrimental to safety, health, and morals. Moreover, the slums mainly consist of temporary or semi-permanent huts with minimal sanitary and water supply facilities and are usually located in unhealthy water-logged areas. They develop ubiquitously in the metropolises, though there is always a greater concentration of large slum areas away from the central business area. They begin with temporary settlements, sometimes started by landlords, but oftentimes by illegal squatting on public lands, sides of railroad lines or canals, unclaimed swamp like lands, public parks, and any other vacant land. In Mumbai it is estimated that nearly 60 percent of the population lives in slums; this proportion is representative of most large cities in South Asia.

The South Asian nations were forced to embrace structural adjustment programs at the behest of the World Bank and the International Monetary Fund. These programs forced open South Asian economies under the policies of free market liberalization. Governments were supposed to roll back their control over the economy, stop subsidizing industries and other sectors like health and education—the task of development was to be left in the hands of the market. Opening up markets brought foreign corporations, their Toyotas and Macs, their call center jobs, and dreams of a consumptive lifestyle. The English-educated, computer-literate middle class took advantage of the economic reforms; many acquired jobs with salaries equal to their First World counterparts. These inflated salaries in poor countries afforded conspicuous consumption and life in bungalows and gated enclaves. Their consumption drew more global business. Construction companies, encouraged by the boom in foreign investment and domestic consumption, pushed for renewal of the cities. The old, ugly, and poor gave way to gloss and glitter. City municipalities were often incentivized by the central government to decentralize and become market oriented. Place marketing and urban renewal were adopted to achieve world class urban status. This trend has been described as "Manhattanization" or "Shanghaization"; others have called it "bourgeois urbanism." Place marketing calls for greening, cleaning, and beautifying the city to create affluent spaces so that the rising middle class and global business and service industries can find their niche. Culturally, this means that cities lose their personality and uniqueness, homogenized through the impact of "McDonaldization." Politically, this means that South Asian cities are increasingly acquiring symbolic capital and are more integrated into the global geopolitics of violence, often becoming prominent targets for extremist groups. Socially and economically it means an increased gap between the rich and poor and the places they occupy. The poor find themselves increasingly pushed out through urban renewal, greening, and beautification schemes that demolish their already flimsy *bustees, jhuggis* and *chawls* without any promise of relief or rehabilitation. Right-to-the-city struggles of the poor are rising in many cities of South Asia. These struggles aim to reclaim the city, alter the vision of urban development, and push for a more inclusive urbanism. South Asian cities therefore embody the tensions between local imperatives and global push.

SUGGESTED READINGS

Ahmed, Waquar. Amitabh Kundu, and Richard Peet. 2010. *India's New Economic Policy: A Critical Analysis.* New York and London: Routledge. A critique of the economic forces that are reshaping India's cities.

Das, Gucharan. 2002. *India Unbound: The Social and Economic Revolution from Independence to the Global Information Age.* New York: Anchor. Well-received book that provides insight into India's rapid development.

Chapman, Graham P. 2003. *The Politics of South Asia: From Early Empires to the Nuclear Age,* 2nd ed. Aldershot, Eng.: Ashgate. General in scope but a very useful introduction to South Asian history, culture, and development.

Chapman, Graham P., Ashok K. Dutt, and Robert W. Bradnock. 1999. *Urban Growth and Development in Asia: Making the Cities,* 2 volumes. Aldershot, Eng.: Ashgate. Companion volumes contain a number of chapters devoted to cities, urbanization, development, and planning in South Asia.

Luce, Edward. 2007. *In Spite of the Gods: The Strange Rise of Modern India.* New York: Doubleday. A newspaper correspondent's perspective on India's rapid economic development.

King, Anthony D. 1976. *Colonial Urban Development: Culture, Social Power, and Environment.* London: Routledge and Kegan Paul. A comprehensive analysis of colonial urban forms of New Delhi, the hill station of Simla, and cantonment towns.

Nair, Janaki. 2005. *The Promise of the Metropolis: Bangalore's Twentieth Century.* New York: Oxford University Press. A timely, well-written, informed and empirically rich case study of the South Asian city most closely associated with globalization.

Noble, Allen G., and Ashok K. Dutt, eds. 1977. *Indian Urbanization and Planning: Vehicles of Modernization.* New Delhi: Tata McGraw-Hill. A classic work containing more than 20 chapters contributed by leading geographers and planners.

Ramachandran, R. 1989. *Urbanization and Urban Systems in India.* New Delhi: Oxford University Press, 1989. Straight-forward and comprehensive treatment of urban India.

Turner, Roy, ed. 1962. *India's Urban Future.* Berkeley and Los Angeles: University of California Press. Of particular interest in this classic collection of articles is the contribution by John E. Brush, "The Morphology of Indian Cities."

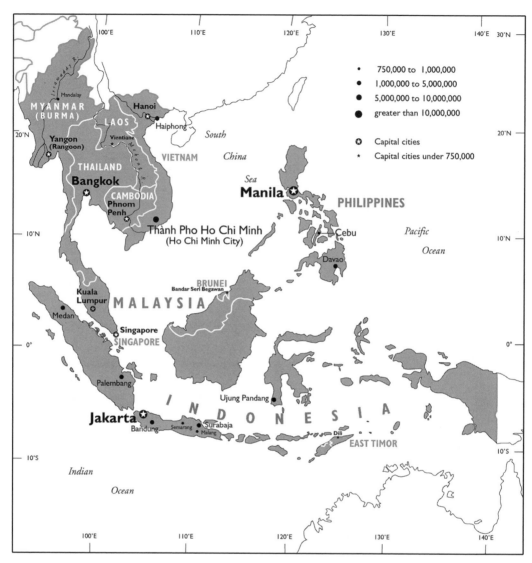

Figure 10.1   Major Cities of Southeast Asia. *Source:* UN, *World Urbanization Prospects: 2009 Revision,* http://esa.un.org/unpd/wup/index.htm

10

Cities of Southeast Asia

JAMES TYNER AND ARNISSON ANDRE ORTEGA

KEY URBAN FACTS

Total Population	590 million
Percent Urban Population	41.9%
Total Urban Population	247 million
Most Urbanized Countries	Singapore (100%)
	Brunei (75.7%)
Malaysia (72.2%)	
Least Urbanized Countries	Cambodia (20.1%)
	Timor-Leste (28.1%)
Vietnam (30.4%)	
Annual Urban Growth Rate	2.88%
Number of Megacities	2 cities
Number of Cities of More Than 1 Million	18 cities
Three Largest Cities	Jakarta, Manila, Bangkok
World Cities	Singapore

KEY CHAPTER THEMES

1. Urban landscapes of Southeast Asia have been shaped by Chinese, Indian, Malay, and international influences, especially colonialism and more recently globalization.
2. All of the world's major religions are represented in the landscapes of Southeast Asia's cities.
3. All of the major cities of the region have experienced rapid population growth since independence.
4. Primate cities (notably Jakarta, Manila, and Bangkok) dominate the region, but the key urban center of Southeast Asia is the city-state of Singapore.
5. Foreign influences, especially through foreign direct investment, play a critical role today.

6. Land reclamation is increasingly used in port areas to provide space for urban expansion.
7. Land-use patterns in the cities are very similar throughout the region.
8. Many cities are restructuring their economies to become IT ("information technology") cities.
9. Some of the world's largest cargo ports—notably Singapore—are located in this region.
10. Transnational cities, which reach across international boundaries in their influence, are becoming more important.

Towering glass-encased skyscrapers, flashing Coca-Cola signs, McDonald's restaurants——the increasingly universal symbols of central cities around the world—are very evident in the cities of Southeast Asia, especially the larger ones, giving the cities a deceiving sense of familiarity. Closer examination, however, reveals many subtle, and sometimes not-so-subtle, differences. Southeast Asia as a whole is a cornucopia of cultures, with hundreds of different languages and many distinct religions. Nestled between two dominant cultural hearths, China and India, and exhibiting a storied colonial past, Southeast Asia is a blend of indigenous and foreign elements. This diversity, not surprisingly, has been and continues to be inscribed on the region's urban landscape, from the lotus-blossom-shaped stupas of Buddhist temples in Bangkok, to the brightly colored Hindu temples in Singapore; and from the golden-domed Muslim mosques of Kuala Lumpur to the Roman Catholic cathedrals of Manila and Ho Chi Minh City.

Yet for many travelers to the region, the extent of Southeast Asia's urban regions comes as a surprise. The typical image of the region is agrarian: thatched huts perched atop stilts and brilliant green rice paddies with water buffalo. The reality is very different. Flying over Manila, Bangkok, or Ho Chi Minh City is like flying over Los Angeles, New York, or Tokyo. The landscape reveals not a dense green three-tiered canopied jungle but instead a dense concrete jungle of apartment complexes, shopping malls, financial districts, and amusement parks.

Southeast Asia's major cities are focal points of political and cultural activity and centers of commercial circulation and exchange (fig. 10.1). Some, such as the "post-socialist" cities of Ho Chi Minh City, Hanoi, and Phnom Penh are undergoing phenomenal political and economic changes; others, such as Rangoon (Yangon) remain aloof from broader global trends. The cities of Southeast Asia are also sites of vast inequalities between the rich and the poor as well as the healthy and the malnourished. In Bangkok, Manila, and Jakarta, Toyota Land Cruisers and Louis Vuitton designer stores are as much a part of the urban landscape as are shantytowns and raw sewage.

Beyond the limits of Southeast Asia's primate cities are a host of medium, or intermediate, cities. Many, such as the Philippines' Cebu and Thailand's Chiang Mai and Chiang Rai, are fast becoming important regional urban centers in their own right. And still other, predominantly rural areas, such as the Central Highlands of Vietnam, are sites of contestation and conflict resulting from indigenous land-use practices and national urban policies.

Figure 10.2   The Central Market in downtown Phnom Penh was built in 1937 in art deco style. It is the soul of the city, a place where you can purchase just about anything. (Photo by James Tyner)

The urban and urbanizing areas of Southeast Asia are more than just containers of people and commodities. They are agents in their own right and, in the coming years, will continue to influence, and be influenced by, local, national, and global affairs.

URBAN PATTERNS AT THE REGIONAL SCALE

Downtown Phnom Penh, the capital of Cambodia, is dominated by the mustard-colored Central Market (fig. 10.2). Built in the Art Deco style of the 1930s, the market is cruciform in design, with four halls radiating out from a central, cavernous dome. Inside, hundreds of venders ply their wares. Care to buy hand-woven silks or a traditional Khmer scarf (known

as *krama*)? Perhaps you're in the mood for some fresh vegetables or pork? No matter your taste, whatever you seek can probably be found at Phnom Penh's Central Market. And if not, it is but a short journey by motorbike to visit Phnom Penh's Russian Market. On the surface, the two markets are very much alike. Many of the same fruits, vegetables, and souvenirs, for example, may be found. However, the sweeping arches and vaulted ceilings of the Central Market give way to the Russian Market's dimly lit and claustrophobic feel. The Russian Market is a rabbits-den of activity, as shoppers jostle elbow-to-elbow with merchants and tourists. Inside, the air is stifling, a sweltering mix of too many people and too many cooking pots bubbling stews of fish and vegetables.

Stepping inside any of Southeast Asia's historic (and even many of the region's new)

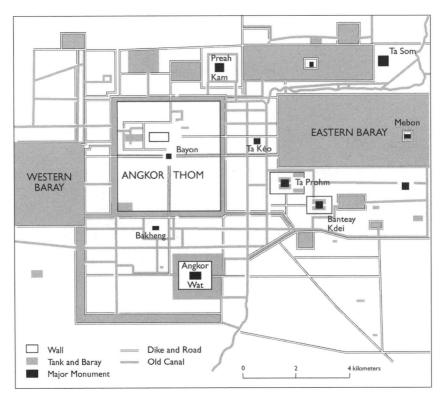

Figure 10.3   "Plan of the Angkor Complex, ca. A.D. 1200." *Source:* T. G. McGee, *The Southeast Asian City* (New York: Praeger, 1967), 38

markets is like Alice stepping into Wonderland. Whether you find yourself wandering the stuffy aisles of Binh Tay Market in Ho Chi Minh City, window-shopping in Bangkok's upscale River City shopping complex, or sipping an iced coffee while shopping along Orchard Road in Singapore, you are guaranteed to be dazzled with new sights, sounds, and smells. The intoxicating aroma of sandalwood incense combines with the smells of fresh fruits, vegetables, and spices to provide an aromatic bouquet that is found nowhere else. And unlike the sedate, antiseptic shopping malls of North America (which are, incidentally, increasingly popping up in Southeast Asia), the labyrinthine markets of Cambodia, Thailand, Vietnam, and elsewhere

seem to embody much of the region's urban geography.

As a whole, Southeast Asia remains one of the least-urbanized regions of the world. Only four countries—Singapore, Brunei, Malaysia, and the Philippines—are more than 50 percent urban. Other states are considerably more rural in character: Cambodia, Laos, Burma, and Vietnam, for example, are all less than 30 percent urbanized. However, recent years have seen massive growth. Many countries have registered startling urban population growth rates in excess of 3 percent per year; and Cambodia and Laos stand at more than 6 percent.

Such urban growth is not new to Southeast Asia (fig. 10.3). Before the Portuguese ever

Figure 10.4   Angkor Wat, built between 1113 and 1150 by Suryavarman II, is but one of hundreds of wats spread throughout Cambodia. Because it symbolizes the golden age of Cambodian history, its image can also be found on the nation's flag. (Photo by James Tyner)

arrived in Malacca, or the Spanish landed in the Philippines, Southeast Asia was home to some of the world's most impressive cities: Angkor in Cambodia, Ayutthaya in Siam (now Thailand), and Luang Prabang in Laos. Their names continue to evoke rich histories of commerce and conquest. For it was through Southeast Asia that the fabled spice trade coursed. And it was through Southeast Asia that ships laden with goods from China, India, and beyond sailed. Then and now, the cities of Southeast Asia were centers of economic, religious, and cultural exchange.

Historically, one or two urban areas would dominate the region. The Kingdom of Angkor, for example, exerted its influence between the 9th and 14th centuries over much of present-day Cambodia, Laos, and Thailand (fig. 10.4). Centered on the city-state of Malacca, the Srivijayan empire ruled much of

insular Southeast Asia from the late 14th century to the early 16th century. And today, most countries in Southeast Asia continue to exhibit phenomenally high levels of urban primacy. In Thailand, the capital city of Bangkok stands as second to none. Not to be outdone, both Jakarta and Manila exert their dominance over Indonesia and the Philippines, respectively.

But Southeast Asia's urban patterns of the 21st century reveal many remarkable differences from previous eras. Whereas many earlier cities were densely populated and compact, the cities of Southeast Asia today are densely populated and sprawling. Rapid urban growth is occurring on the peripheries of Manila, Bangkok, Phnom Penh, Jakarta, and Ho Chi Minh City. This growth, some planned, some not, has led to conflicts over land use; it has threatened once-prime agricultural lands; and

Figure 10.5   New residential, leisure, and commercial developments rise on the outskirts of Manila, taking the place of former sugar cane plantations. (Photo by Arnisson Andre Ortega)

it has spurred attendant economic problems of land speculation and landlessness (fig. 10.5).

In light of persistent problems of over-urbanization—traffic congestion, pollution, unemployment—local government officials throughout Southeast Asia have initiated regional economic development projects. Often these projects are multipurpose in scope: promoting economic growth and development in more peripheral regions (*i.e.*, northeast Thailand) and lessening the burdens of primate cities. Still other governments have relocated entire cities for unknown reasons. The always-secretive leaders of Burma, for example, have recently relocated its capital from Rangoon (Yangon) to the interior, semi-rural town of Pyinmana. The few western journalists who have been fortunate enough to visit the new Burmese capital—named Naypyidaw ("Seat of Kings")—write of expansive residential areas, reminiscent of any suburban development found in North America, and impressive office buildings. But no one lives there! Streets are empty; most Burmese are denied access to the new city.

Similar to Phnom Penh, the cities of Southeast Asia reflect the past and the future. They are centers of intense commercial activity and cultural exchange (fig. 10.6); and, while shar-ing many commonalities, they—like Phnom Penh's Central and Russian Markets—exhibit remarkable differences.

HISTORICAL PERSPECTIVES ON URBAN DEVELOPMENT

Precolonial Patterns of Urbanization

Southeast Asia is characterized by more coastline than perhaps any other major world region, and much of this coast is accessible to sea traffic. It is understandable, therefore, that maritime influences have contributed significantly to the Southeast Asian urbanization process.

The region, but most especially mainland Southeast Asia, also contains many fertile river valleys, which gave rise to densely populated settlements. These include the Chao Praya, Irrawaddy, Mekong, and Red Rivers, along with their tributaries. Bangkok, Phnom Penh, Hanoi, and Ho Chi Minh City all continue to reflect the importance of these highways of water.

Southeast Asia's physical geography, its complex environment of river systems and coastlines, contributed to the region's importance as a crucial crossroads of commerce

Figure 10.6   In Pleiku, Vietnam, a woman makes a living by selling fresh fruits and vegetables—proudly displayed as in an American supermarket—to shoppers in the early morning hours. (Photo by James Tyner)

between China, India, and beyond. And it was this factor that precipitated the urbanization process of Southeast Asia. Although it is commonplace to speak of the global economy as beginning in the 16th century, it is important to recognize that international trade existed long before European states like England and Spain began colonizing the Americas, Africa, and Asia. Indeed, long-distance trade existed between China and India, and linked these areas with places as far afield as Africa, as far back as the early centuries of the first millennium A.D. The importance of long-distance trade in the eastern Indian Ocean and South China Sea regions, in fact, led to the appearance of a series of cities and towns along the coast of the Malay Peninsula and on the islands of Sumatra and Java. In time, Southeast Asia would be home to some of the largest urban centers in the world. Indeed, prior to the era of European colonialism (from the early 16th to the mid-20th century) Southeast Asia was one of the world's most urbanized regions. As late

as the 15th century, for example, the population of Angkor (in present-day Cambodia) had a population in excess of 180,000; Paris, in contrast, had a population of only 125,000.

Southeast Asia's geographical location made it a natural crossroads and meeting point for world trade, migration, and cultural exchange. A century before the Christian era began, seafarers, merchants, and priests traversed the region, contributing to the urbanization process. In turn, the nascent towns and cities of Southeast Asia became centers of learning through the diffusion of new religious, cultural, political, and economic ideas. Most Southeast Asian societies (with the exception of Vietnam and the Philippines) were influenced primarily by India, and this is most pronounced in the religious and administrative systems of the region. The process of "Indianization," however, was not marked by a mass influx of population like the movement of Europeans into North America. Neither was this a process of replacing indigenous

Southeast Asian culture with Indian elements. Rather, the influence of India on Southeast Asia represented a more gradual and uneven process of exposure and adaptation. China provided the other major cultural impetus, although this impact was greatest in Vietnam and through tributary arrangements with various maritime Southeast Asian kingdoms bordering the South China Sea.

Two principal urban forms emerged in precolonial Southeast Asia: the *sacred city* and the *market city*. Although both types of cities performed religious as well as economic functions, the two exhibited many differences. First, sacred cities were often more populous; wealth was gained from appropriating agricultural surpluses and labor from the rural hinterlands. Market cities, in contrast, were supported through the conduct of long-distance maritime trade. Through the market cities passed the riches of Asia, including pearls, silks, tin, porcelain, and spices. Second, sacred cities were sprawling administrative, military, and cultural centers, whereas market cities were mostly centers of economic activity. In physical layout, sacred cities were planned and developed to mirror symbolic links between human societies on earth and the forces of heaven. Monumental stone or brick temples commonly occupied the city center. Market cities, in contrast, tended to occupy more restricted coastal locations, and thus had more limited hinterlands. These cities were more compact in their spatial layout, with much activity associated with the port areas. Lastly, compared to sacred cities, market cities were ethnically more diverse, populated by traders, merchants, and other travelers from all parts of the earth.

The earliest city to emerge in Southeast Asia was apparently Oc Eo, located along the lower reaches of the Mekong Delta, in present-day Vietnam. Flourishing between the 1st and

5th centuries A.D., Oc Eo was an important center for the exchange of cargo, ideas, and innovations. It served as an important city for both Chinese and Indian traders, as well as other seafarers from as far away as Africa, the Mediterranean, and the Middle East.

After the decline of Oc Eo, Srivijaya emerged as an important maritime empire, flourishing between the 7th and 14th centuries. It depended on international maritime trade and China's sponsorship through a tributary system, which meant paying tribute (goods and money) to the Chinese emperor in exchange for independence. Located on the straits of Malacca on the island of Sumatra, Srivijaya controlled many important sea lanes, including the Sunda Strait. Evidence suggests that the Srivijayan Kingdom had numerous capitals, one of which was Palembang, located on the southern end of Sumatra. Palembang provided an excellent, sheltered harbor and served as an important Buddhist pilgrimage site. To this day, Palembang remains an important port city and marketplace in Indonesia.

Another example of a market city is Malacca (fig. 10.7). Founded around 1400 on the western side of the Malay Peninsula, Malacca was a counterpart to Palembang and emerged as an important entrepôt and a key node in the spice trade. Although Malacca never had a permanent population of more than a few thousand, it was an extremely vibrant city, inhabited by many foreigners as well as indigenous Malays. In recognition of its multicultural heritage and its role in blending cultures from east and west, Malacca was recently named a UNESCO World Heritage Site. Other important market cities located throughout Southeast Asia included Ternate, Makasar, Bantam, and Aceh.

Sacred cities often occupied more inland locations. One of the earliest was Borobu-

Figure 10.7   Malacca is one of the oldest towns in Southeast Asia and has changed hands many times as successive rulers and colonial powers came and went. Today, it is an important commercial and tourist city in Malaysia. (Photo courtesy Malacca government)

dur, situated on the island of Java. It is at Borobudur that the world's largest Buddhist temple is located. Built between 778 and 856 A.D., the ten-level Borobudur temple corresponds to the divisions within the Mahayana Buddhist universe and is one of the great cultural treasures of Southeast Asia. A U.N.-sponsored program rebuilt the complex several decades ago to preserve the site for future generations.

Arguably, the best known and most famous of all inland sacred cities is Angkor. Centered at the northern end of the Tonle Sap basin, the Angkorian Empire, at its peak, included present-day Cambodia and parts of Laos, Thailand, and Vietnam. The Angkor Kingdom was founded in 802 A.D. and by the 12th century contained a population of several hundred thousand. According to some historians, it may even have exceeded one million. The temples at Angkor—there are more than

70 recognized sites—were designed to mirror the complex Hindu and, later, Buddhist cosmologies.

By the 16th century many of the once-prosperous inland sacred cities were in decline. In part, internal factions, economic collapse, and foreign intervention hastened the collapse of Angkor and other empires. Coastal market cities, however, continued to thrive on maritime trade. For the region as a whole, the coming years of European colonial dominance would irrevocably alter the course of urbanization in Southeast Asia.

Urbanization in Colonial Southeast Asia

Nutmeg and cloves, cinnamon and sandalwood: these were the prized commodities that drove the world's economy for hundreds of years. And these were the goods that spurred European colonial activity in Southeast Asia.

Five hundred years of colonial and post-colonial influence dramatically affected cities in Southeast Asia. Compared to other world regions, Southeast Asia was relatively urbanized by the time of European colonialism. By the 16th century, there were at least six trade-dependent cities that had populations of more than 100,000: Malacca, Thang-long (Vietnam), Ayutthaya (Siam, present-day Thailand), Aceh (Sumatra), and Bantam and Mataram (both on Java). Another half-dozen cities had at least 50,000 inhabitants. Only the Philippines, because of its more peripheral location vis-à-vis the major sea-lanes, lacked an urban tradition. But even there, by the early 16th century, the seeds of urbanization had been planted. The sultanate of Brunei had extended his authority into the Philippine archipelago and with this came the spread of Islam.

In 1511, however, a Portuguese fleet captured the port city of Malacca, thus ushering in nearly 500 years of European colonialism. The Portuguese came primarily to gain access to and control of the lucrative spice trade. They were soon followed by the Spanish (1521), the British (1579), the Dutch (1595), and the French (mid-17th century). Other colonial activities, such as religious conversion, were present but less important at this time.

The early years of European colonialism in Southeast Asia were similar to the colonial practices found in Africa and the Americas. Europeans captured or built garrisons in coastal cities, established treaties with local rulers, and thus brought about a transformation of the urbanization process in Southeast Asia. Many former empires and kingdoms, as well as their cities, suffered tremendous population declines. Malacca, for example, once the premier entrepôt on the strait that shares its name, declined in size and importance after its capture by the Portuguese. From a peak of over 100,000 inhabitants, its size dwindled to 30,000 inhabitants in a very short time.

During the first three centuries of colonialism, European influence was most pronounced in two regions: in Manila (the Philippines) under the Spanish and in Jakarta (Indonesia) under the Dutch. The first permanent Spanish settlement, Santisimo Nombre de Jesus (Holy Name of Jesus), was established in 1565 on the Philippine island of Cebu. Five years later the Spanish occupied a site on the northern island of Luzon, situated on the Pasig River and proximate to Manila Bay (fig. 10.8). Two existing fishing villages known as Maynilad (from which the present city takes its name) and Tondo were occupied and expanded. Apart from accessibility, defense was often an important consideration in early city planning. In the Philippines, for example, the Spanish had to contend with European rivals, namely the Dutch and Portuguese, as well as Chinese pirates. Consequently, after 1576 construction began on a fortified structure known as the *Intramuros* (walled city). In time, Manila would become the commercial hub of the Philippines and a key node in the Spanish galleon trade that stretched from India to Mexico.

The early European presence was also pronounced in on the island of Java in Indonesia. The Dutch East India Company during the 17th century established a few permanent settlements, one of which, Batavia, would become the largest city in the region. It is now known as Jakarta. From its inception Batavia exhibited numerous situational advantages. Geographically, it was located near both the Sunda Strait and the Strait of Malacca, thus

Figure 10.8   A statue in Manila honors Raja Solayman, the city's Muslim prince, who defended the town against the Spaniards in the 1500s. (Photo by Arnisson Andre Ortega)

allowing easy access to maritime trade. The first Dutch building, a combination warehouse and residence, was built in 1611, and by 1619 a plan was laid out for the city. Much of early Batavia was modeled after the cities of Holland; canals were dug; and the narrow, multi-storied Dutch residences were also copied. However, the architecture found in Europe was not functional in hot, humid locations such as Java, so building styles were altered to better fit the tropical environment. Batavia would emerge as the preeminent city of Java and serve as the key node of the Netherland's Southeast Asian empire.

Many of the great cities of Southeast Asia today trace their roots to European colonialism (fig. 10.9). Singapore (from the Malay words *singa*, lion, and *pura*, city), for example, began as a small trading post located on an

island south of the Malay peninsula at the southern entrance to the Strait of Malacca. The town was known as Temasek (Sea Town) before 1819, when Sir Stamford Raffles of the British East India Company signed an agreement with the Sultan of Johor allowing the British to establish a trading post at the site. Benefiting from its strategic location and deep natural harbor, the incipient Singapore began to attract a large number of immigrants, merchants, and traders. As a British colony, Singapore emerged as one of the paramount trading cities of the world—a role it has maintained to this day.

Saigon (now Ho Chi Minh City), which became the capital of French Indochina, likewise began as a small settlement, one that included a citadel and fortress, surrounded by a Vietnamese village (fig. 10.10). Saigon's

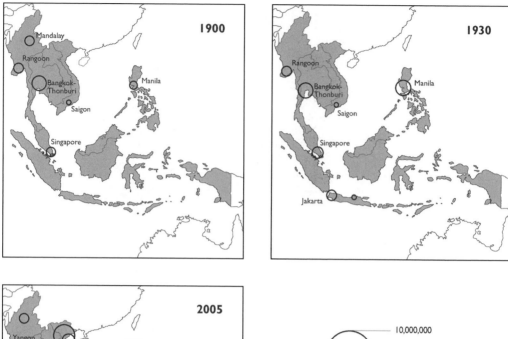

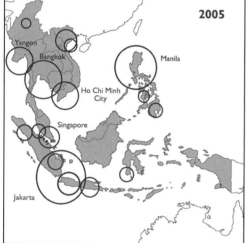

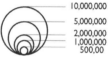

Figure 10.9   Urban Growth in Southeast Asia, 1900–2005. *Source:* compiled by the author from various sources

location in the Mekong delta region—one of the world's great rice granaries—provided the city an important function as an agricultural collection, processing, and distribution center. The French, also, were intent on refashioning their colonial capital as a microcosm of French civilization. In 1880 the French erected Notre Dame Cathedral in the heart of Saigon.

Stones were imported. Ten years later saw the completion of Saigon's French-built Central Post Office. This latter structure, complete with an iron and glass ceiling, was designed by Gustave Eiffel.

Bangkok, the current capital of Thailand, was never colonized by the European powers, yet it still reflects considerable Western

Figure 10.10   Fast food—or "good food fast"—is widely available on the streets of Southeast Asian cities. Here, early morning breakfast is served in Ho Chi Minh City (Saigon). (Photo by James Tyner)

influence. Bangkok proper is a relatively new city; it was not founded until 1782. Prior to this date the capital of Siam, as Thailand was then known, was located at Thon Buri along the Chao Phraya River. Beginning in the 1780s, however, construction began on an easily defensible but swampy site located opposite Thon Buri, on the east bank of the Chao Phraya. Later, major public works were initiated, often with Western advice and assistance. This is particularly evident in the expansion of rail and road networks, port facilities, and telegraph services.

The most significant impact of colonialism was the establishment of urban nodes, such as Bangkok, Manila, Batavia, Saigon, and Rangoon (Yangon), that would grow into primate cities. Geographically, these cities were located at sites that provided access to seas or rivers. These locations afforded the European colonizers easy access, so that their ships could export primary products from the region and import secondary products from Europe and elsewhere. Such a dependence on maritime trade, and the subsequent concentration of political and economic functions in these selected cities, contributed to the decline of other, more inland cities. In this manner, the urban system of Southeast Asia was turned inside out, as development was encouraged on the coast and suppressed inland.

The primate cities also served multiple functions. Thus, political, commercial, financial, and even religious activities were concentrated in these urban areas. Saigon, for example, was

Figure 10.11   The Overseas Chinese are a pervasive presence in Southeast Asia, especially in cities. This Chinese temple is in Georgetown, Penang, one of the original Straits Settlements that began British colonial rule in Malaya, and which brought the Chinese here. (Photo by Jack Williams)

an administrative and manufacturing center, as well as the dominant trading port of Indochina. Primate cities also became exceptionally large. Stemming from the increased concentration of economic and political functions, these cities served as magnets for both internal and international migration. Moreover, their populations were characteristically diverse. In many colonies, the colonizers encouraged contract labor. The British, for example, actively encouraged the importation of labor from China and the Indian subcontinent (fig. 10.11). By the 19th century, immigration facilitated the establishment of Chinese communities in cities such as Singapore, Manila, and Bangkok, and an Indian community in Singapore. As a result, segregated foreign quarters and Chinatowns emerged, for example, Cholon in Saigon and Binondo in Manila. It was not uncommon for these segregated areas to arise as a result of European force and prejudice. In Manila,

for example, the Spanish government issued a series of decrees that required, with few exceptions, all Chinese, Japanese, and even Filipinos to leave the Intramuros section of Manila before the closing of the city gates at nightfall. Additionally, Spanish authorities enacted regulations enforcing ethnic segregation and commercial activity.

Apart from the establishment of primate cities, a second impact European colonialism had on Southeast Asian cities was the establishment or transformation of smaller cities. This included the establishment of mining towns, such as Ipoh in Malaysia, or regional administrative centers such as Medan on the island of Sumatra and Georgetown on the Malay Peninsula. Also notable was the emergence of upland resort centers or hill stations. To escape the oppressive heat and humidity of the lowlands, colonial powers would erect cities high in the mountainous regions.

Bandung, for example, a small city located in a deep mountain valley on the island of Java, was established by the Dutch. The cool climate of Bandung served as a welcome relief from the tropical climate and also facilitated the cultivation of coffee (still known as "Java" worldwide), cinchona (for quinine), and tea. Other examples of hill stations include Dalat, a French-built city located 4,800 feet (1,463 m) above sea level in the Central Highlands of Vietnam; Baguio City, a mountain resort atop a 4,900-foot (1,493 m) plateau in the Philippines, developed by the Americans in the early 20th century; and the Cameron Highlands of peninsular Malaysia.

Lastly, European colonialism significantly affected the development of regional transportation and urban systems within Southeast Asia. Along the Malay Peninsula, for example, the British-built railways ran from the Perak tin-mining areas to the coast and were later expanded along a north-south axis to provide access to the ports and tin-smelting facilities in Penang and Singapore. Consequently, the major urban areas along the western coast of the Malay Peninsula—Kuala Lumpur, Ipoh, Seremban, and Singapore—formed an interconnected urban system that remains evident today. Similar patterns are also visible in Indonesia, where road and rail networks reflect access between sites of resource extraction (plantations and mines) and ports. To a lesser extent a similar colonial-derived infrastructure remains in the Philippines and the former French Indochina. Colonial powers also exploited river systems (such as the Irrawaddy in Burma, now Myanmar) for internal transportation systems, to exert political and economic control, to link administration functions, and to provide access to sites of resource extraction.

Recent Urbanization Trends in Southeast Asia

In 1940 no city in Southeast Asia registered a population of more than a million; by 1950, just two cities had surpassed this mark. In 2000, however, 13 cities exceeded the one million threshold. Three megacities—Jakarta (13.7 million), Manila (11.8 million), and Bangkok (10.3 million)—have also emerged. These massive urban agglomerations contain a disproportionate share of the region's urban population and stand as exemplary primate cities. Bangkok, for example, contains more than 54 percent of Thailand's urban population, while Manila accounts for nearly a third of the Philippines' urban population. Such growth is indicative of the pattern of urbanization that has characterized Southeast Asia throughout the 20th century.

Recent urban growth in Southeast Asia is the result of three basic demographic processes. First, urban areas in Southeast Asia have increased in population size resulting from the excess of births over deaths. In general, this natural increase accounts for about one-half of urban population growth in Southeast Asian countries. It is important to remember, however, that while overall natural increase may contribute to *urban growth*, it may not contribute significantly to *urbanization*, that is, the increasing proportion of people living in urban areas relative to rural areas.

Second, cities in Southeast Asia have increased through a net redistribution of people from rural to urban areas though migration. Studies reflect that, overall, rural-to-urban migration has resulted from larger regional and global economic transformation, and subsequently has played a major role in both the rapid urbanization and urban growth in Southeast Asia over the past two decades. Internal migration is extremely important to urban growth in Thailand, for example. In 1990, the Thai census recorded more than 1.5 million rural-to-urban migrants, although there were more than twice as many rural-to-rural migrants. Currently, approximately one out of every seven urban residents in Thailand is classified as a recent migrant. In Malaysia, with economic growth concentrated in Kuala Lumpur and Johor Bahru, the city closest to Singapore, these cities have attracted sizeable numbers of migrants in recent years. The Central Highlands of Vietnam have also experienced rapid urban growth through both government-sponsored and spontaneous migration. At the beginning of the 20th century, for example, the four provinces that comprise the Central Highlands— Kontum, Gia Lai, Dak Lak, and Lam Dong— had a population of approximately 240,000; today, the region's population exceeds four million residents. The majority these inhabitants, approximately 75 percent of the total, are lowland migrants, their children, or refugees. Unfortunately, such remarkable urban growth has resulted in an escalation of land conflicts and political violence (box 10.1).

Third, immigration has contributed to the growth of urban areas in Southeast Asia. This occurs because, in general, overseas migrants tend to move to urban areas as opposed to rural areas. Singapore and, to a lesser extent, Kuala Lumpur, are two of the major immigrant receivers in Southeast Asia. That said, many governments (including Singapore's) often try to prevent immigrants from settling permanently. Overall, however, urban growth in Southeast Asia through international migration remains a relatively insignificant component. Emigration also is inscribed on the urban landscape (box 10.2).

A final component of urban *change* that must be introduced is "reclassification." As a result of bureaucratic decisions, urban populations may change simply through administrative acts. In Malaysia, for example, the number of people needed for an area to be classified as urban changed from 1,000 to 10,000 in 1970. Such statistical changes reflect the changing perception of urbanization by people and their governments.

Within Southeast Asia over the past two decades the combined components of migration and reclassification account for the largest portion of urban growth (table 10.1). There was also considerable variation among countries. Urban growth in Burma, the Philippines, and Vietnam, for example, has occurred primarily through natural increase; whereas, natural increase has assumed a lesser role in Cambodia, Thailand, Malaysia, and Indonesia. The importance of internal migration to the urbanization process of Cambodia, of course, is a consequence of the forced displacement of urban-based people during the murderous Khmer Rouge regime (box 10.3). At 100 percent urban already, the Republic of Singapore's population growth consists mostly of natural increase.

Aggregate numbers such as these mask significant social and economic changes that are occurring. Internal migration in Southeast Asia, for example, is increasingly dominated by female migrants, a process that mirrors that occurring elsewhere in the world, such as China. In Thailand, for example, the share of female migrants increased to more than 62 percent of all Bangkok-bound migrants in the 1980s. The increased feminization of internal migration in Thailand is related to structural changes occurring in both rural and urban areas. The majority of these women—most of whom are in their early twenties—originate in northeastern Thailand, one of the most impoverished regions in the country. Faced

Box 10.1   Conflict in the Central Highlands

Located between the plains of eastern Cambodia, the Annamite mountains of northern Vietnam, and the coastal lowlands of southern and central Vietnam lie the Central Highlands of Vietnam. Although not as daunting as the Himalayas or the Alps, the Central Highlands--with peak elevations over 5,000 ft (1,524 m)--the Central Highlands are impressive nonetheless. In the Central Highlands, dense jungle-covered forests give way to dazzling rice paddies; rain-fed waterfalls cascade down to white-capped rushing rivers. Clusters of stilt-houses occupy the cool valley bottoms.

The Central Highlands are home to approximately one million indigenous peoples. Known generically by the French term *Montagnard*, meaning "mountain dwellers," these peoples are actually composed of 54 officially recognized groups, with the Jarai, Rhade, Bahnar, and Koho among the most numerous. For centuries these people have lived in the region, practicing swidden agriculture and fishing. By custom and tradition, agricultural lands were defined by family use rights. Collective lands, such as streams, grazing pastures, and drinking water sources, were managed by village elders. Owing to a French colonial presence, an estimated 230,000 to 400,000 highlanders are followers of evangelical Protestantism.

History, however, has not been kind to the indigenous peoples of the Central Highlands. During the long Indochina War (known as the Vietnam War in the United States), many areas of the Central Highlands were targeted for aerial bombing raids and chemical defoliants were also widely used (fig. 10.12). Both the United States and the North Vietnamese enlisted

Figure 10.12   Pleiku, in the Central Highlands of Vietnam, was important during the Vietnam War. It has been the focus of considerable in-migration from northern Vietnam and the surrounding area has been beset by conflict between the Vietnamese and the Montagnards. (Photo by James Tyner)

indigenous peoples to serve as scouts, spies, or soldiers. After the war, the victorious Democratic Republic of Vietnam continued to view the Central Highlands as an area of potential ethnic and religious strife.

Vietnamese urban policies, though, have fueled rising tensions in the region. Supposedly in response to increasing population densities and urban development in the lowlands of northern and southern Vietnam, many ethnic Vietnamese—known as *Kinh*—have been resettled into the Central Highlands. These government-sponsored migrations have been accompanied with sizeable flows of spontaneous migration. State agencies maintain that resettlement contributes to the overall economic development of the country. However, these resettlements carry political and economic undertones. Land is at the heart of these conflicts.

As the Kinh continue to move into the Central Highlands, traditional lands used by the highlanders are being converted to sprawling coffee and rubber plantations. In fact, Vietnam has emerged as the world's second largest exporter of coffee. The economic benefits of this trend, though, have not been felt by the indigenous peoples. Indeed, a lack of land security, an escalation in the state confiscation of land, and, consequently, rising levels of landlessness have led to considerable unrest by the highlanders. The indigenous peoples, likewise, charge that they have been neglected by the government in terms of education and employment opportunities. In light of rising protests, the Vietnamese government has responded aggressively. Many highlander organizers have been arrested and detained; reports circulate of torture and other forms of government reprisals.

Here, on the slopes and in the valleys of the Central Highlands, the conversion of pristine agricultural lands to row upon row of coffee plants provides a tangible reminder that the confluence of religious freedom, ethnic autonomy, economic development, and urban growth does not occur without struggle.

Table 10.1 Components of Urban Growth in Southeast Asia (percentage of urban growth)

	1980–1985		1990–1995		2000–2005	
	Natural Increase	*Migration and Reclassification*	*Natural Increase*	*Migration and Reclassification*	*Natural Increase*	*Migration and Reclassification*
Southeast Asia	49.1	50.9	44.9	55.1	41.7	58.3
Cambodia	70.9	29.1	49.5	50.5	30.6	69.4
Indonesia	35.2	64.8	37.0	63.0	36.7	63.3
Laos	43.8	56.2	44.7	55.3	43.8	56.2
Malaysia	22.0	78.0	38.0	62.0	40.0	60.0
Myanmar	110.0	−10.0	63.2	36.8	44.5	55.5
Philippines	66.0	34.0	62.4	37.6	57.0	43.0
Singapore	100.1	−0.1	100.1	−0.1	98.9	1.1
Thailand	39.6	60.4	31.4	68.6	31.2	68.8
Vietnam	71.7	28.3	50.5	49.5	38.1	61.9

Source: Graeme Hugo, "Demographic and Social Patterns," in *Southeast Asia: Diversity and Development*, edited by Thomas R. Leinbach and Richard Ulack, 74–109 (Upper Saddle River, NJ: Prentice Hall, 2000), table 4.17.

Box 10.2   A Geography of Everyday Life

Mong Bora is a twelve-year-old boy. He lives in stilt-house with his mother, father, and four sisters. The village in which he lives is approximately one hour north of Phnom Penh. During the rainy season much of the area is inundated with water, hence the necessity to live in houses perched on stilts. In the surrounding vicinity of Bora's village are acres of rice fields and fish ponds. His diet is typical for many Khmers: a staple of rice and fish, coupled with fresh fruits and vegetables. Bora is particularly fond of ripe mangoes, watermelon, and papayas.

Increasingly, Bora's village is being encroached upon by urban sprawl from Phnom Penh. This is considered both a blessing and a curse. On the one hand, villagers are concerned about maintaining their way of life. On the other hand, they recognize that urban growth may translate into better economic opportunities. Many residents of Phnom Penh, for example, visit the area for Sunday picnics, seeking a respite from the chaotic hustle-and-bustle of the capital. And many villagers are able to supplement their incomes from these weekly picnickers. Villagers, like Bora's mother, sell lotus-blossom seeds as snacks, or bottled water. Others rent "cabanas" or tent awnings, under which the visitors can escape the intense heat.

Bora's house sits near the base of two hills, the larger of which is called Phnom Reach Throp, or "hill of the royal treasury." In part, it is because of this hill that people, both locals and international travelers, come to the area. Between 1618 and 1866 Cambodia's capital was located on this site. Known as Udong (meaning "victorious"), the former capital once dominated the landscape. Today, however, little remains of Udong's former glory. The passage of time, but mostly the effects of war and genocide, has devastated much of Udong's former architectural greatness. Several *stupas* remain at the site, as does a colossal Buddha figure. Blown up the by Khmer Rouge, the Buddha and many of the *stupas* are actively being restored.

On any given morning Bora travels 2 km (1.2 mi) by foot to the other side of Phnom Reach Throp to attend school. Among his favorite classes is English. He also enjoys learning about Cambodia's ancient history and geography. It is Bora's dream that these subjects will help him in his chosen career. You see, when Bora is not attending school, or playing soccer with his friends, he is informally working as a tour guide for the many visitors who travel to Udong. Walking step-by-step with tourists, both local and international, Bora happily details the specifics of the former capital: how many stairs from one *stupa* to the next, the height of the Buddha, the dates of former kings. Bora he gets to practice English and earn some extra money to help his family.

Perhaps, if your future travels include the ancient site of Udong, you may be approached by a youthful man by the name of Bora who will offer to guide your tour. Be sure to accept, for Udong is the young man's home.

Box 10.3   Devastated by Genocide

On April 17, 1975, the Khmer Rouge—the Communist Party of Kampuchea (CPK)—marched through the hot and dusty streets of Cambodia's capital city, Phnom Penh. Their arrival marked the beginning of a four-year period of unimaginable horrors, and a genocide that would claim nearly 3 million people, or approximately one-third of Cambodia's total population.

The Khmer Rouge, led by a mysterious figure who called himself Pol Pot, sought to transform Cambodia into a communist utopia. The revolution of the Khmer Rouge was envisioned to be complete. A massive program of social and spatial engineering was set in motion, designed to eradicate all previous social, political, and economic relations.

Flush with victory following years of civil war, the Khmer Rouge began to immediately evacuate Phnom Penh and other major cities throughout Cambodia. Cities, according to the ideologues of the Khmer Rouge, were viewed as bastions of immorality, vice, and corruption. These were also, according to the Khmer Rouge, sites of foreign dominance. Phnom Penh, for example, contained a sizeable minority population of Vietnamese-Khmer and Chinese-Khmer people. The Khmer Rouge also viewed cities as home to those people who resisted, or opposed, their revolution: members of the monarchy, foreigners, and merchants. A forcible evacuation was deemed the most efficient means to de-mobilize any potential opposition, and to ensure security for the newly imposed government. Lastly, the de-population of urban areas would provide a surplus of labor that could be exploited on the agricultural communes and collectives that were initiated under the Khmer Rouge regime.

In 1979 the Khmer Rouge were defeated by invading Vietnamese forces. Decades of painful rebuilding ensued. This resurgence, however, has come with a price. Following the genocide, thousands of survivors have moved back into the cities, especially Phnom Penh. Many of these refugees had nothing. They established shantytowns; and many of these makeshift encampments, located along the muddy banks of the Bassac, Mekong, and Tonle Rivers that converge at the capital, have lasted for nearly three decades. But now the government is attempting to rebuild and to redevelop the city. As a result, local governments have been evicting residents of the shantytowns. In their place are new luxury housing projects; but the vast majority of Phnom Penh's residents will never be able to afford these new homes and apartments. Other lands, formerly occupied by those less well off, are being cleared for the rapidly proliferating five-star hotels in hopes of capitalizing on rising flows of international tourism.

The future of Cambodia, as well as its cities, is unclear. Decades of civil war, genocide, occupation, and now urban land-use conflicts have taken a toll on the Khmer people. One thing is certain: the cities of Cambodia, particularly Phnom Penh, will remain sites of struggle.

with minimal prospects in the rural areas, these women are increasingly moving to Bangkok to obtain employment in factories, the service sector, or the informal sector. Also, a certain number of these migrants find employment in the sex trade and end up working in brothels, massage parlors, or strip clubs. Likewise, internal migration to Jakarta, Manila, and Phnom Penh has also become more feminized in response to the structural transformations occurring in these cities.

Not all internal migration is permanent. Indeed, many cities in Southeast Asia, but especially Bangkok and Ho Chi Minh City, are impacted by daily or seasonal circular migration. Three factors are readily identifiable. First, circulation is highly compatible with work participation in the urban informal sector. Migrant laborers are able to circulate between rural villages and urban sites depending on the season. Circulation thus offers a flexible solution to the seasonality of labor demands; laborers are able to work on the nearby farms during the peak agricultural period, while during downtimes these same workers are able to participate in the informal economy in the city. Second, circular migration diversifies families' income-generating activities. Depending on the relative economic strength of urban and rural areas, workers may alternate their activities accordingly. Third, circular migration has, with advances in transportation systems, become a more viable option. Improvements in mass transportation systems, such as paved roads and mass-transit bus lines, have permitted people to move with greater ease, thus contributing to the growth of suburban residential areas. The spectacular urban growth to the south and west of Ho Chi Minh City is indicative of this process.

Globalization, Urbanization, and the Middle Class

The most significant development in the world economy during the past few decades has been the increased globalization of economic activities. The transnational operations of multinational firms have given rise to a new international division of labor, one that has witnessed a shifting of manufacturing sector enterprises from developed to developing economies, and the emergence of new corporate headquarter activities, producer services, and research/development sites. Final assembly and testing of audio-video equipment are located in Singapore and Penang, Malaysia; the assembly and packaging, low-skilled and labor-intensive, in Bangkok, Jakarta, and Manila; and marketing and sales functions, mid- to high-end manufacturing, in Singapore. In Southeast Asia, these far-reaching changes are evidenced by the spectacular growth of assembly plants in Phnom Penh as well as the emergence of Cyberjaya, Malaysia's high-technology city (the "Silicon Valley of the East") that forms the hub of that country's Multimedia Super Corridor.

Shifts in the structure of Southeast Asia's economies have led to remarkable societal and occupational changes. Declines in agricultural workers are matched by increases in the number of workers employed in the service and manufacturing centers. Consequently, the increasing portion of clerical, sales, and service workers, in particular, has translated into redefined social categories. One change that is especially salient is the emergence of a new middle class. And given that many of the economic transformations have occurred disproportionately within the urban areas of Southeast Asia, it should come as no surprise that the emergent middle class in Southeast Asia is likewise urban-based.

The rise of Southeast Asia's new urban middle class has drastically altered the urban landscape. Demographically, Southeast Asia's middle class tends to have smaller families; economically, it tends to have high and rising levels of consumption and to spend money on non-essential items, such as luxury cars. Many members of this emergent class express "Western" middle-class fantasies to materially project their newfound social status. In terms of housing, Southeast Asia's middle class demands more space and more privacy (hence leading to demand for western-style housing in the form of detached and semi-detached single-family dwellings). Having the ability to purchase an automobile, the middle class is able and willing to commute longer distances to work, thus fueling the sprawl of cities into traditional agricultural hinterlands. Others prefer to live closer to the traditional downtown districts, fueling the proliferation of condominiums and apartment complexes. The emergence of the middle class, and its growing spending power, is likewise reflected in the mushrooming of shopping malls and country clubs and in the proliferation of leisure activities and nightclubs. It is seen in the growth of gourmet restaurants, coffee bars (Starbucks are becoming all-too pervasive), theaters, galleries, and boutiques.

Cities in Southeast Asia have historically been segregated. During the colonial years, British, French, Spanish, and America authorities often restricted residential and commercial activities by ethnic classification in their respective colonies. The Spanish, for example, disallowed Filipinos from living in *Intramuros* Manila; the French restricted Vietnamese settlements in Saigon. Today, segregated areas remain, but these often reflect class differ-

ences as much as anything. This segregation is epitomized by the rise of gated communities.

The desire among the new middle class for gated communities results from demands for privacy, security, and prestige. Many of these new housing developments are equipped with strictly controlled gates that are fully secured by armed private guards and monitored by CCTV. In exclusive villages, entrance is permitted only to residents with proper photo ID or to their friends and acquaintances. Also, many of these villages carry Western-themed names and architecture and are fully equipped with top-notch amenities such as tennis courts, club houses, golf courses, swimming pools, and spacious houses; some even provide heliports for their residents. Everyday life in these communities is heavily controlled by home-owner association rules from the kinds of designs permitted for houses to curfew hours.

The growth of Southeast Asia's middle class is a major contributor to the sprawl of cities. A key prerequisite for the construction of gated communities, for example, is land. In Jakarta, Ho Chi Minh City, and Manila, local governments have allowed the spread of new middle-class enclaves on their peripheries and the conversion of old land uses for middle class condominiums in the cities. Many lower cost housing units in Phnom Penh, for example, have been razed in order to erect higher cost apartments and condominiums for the middle class. In Manila, informal settlement communities have been demolished or were dubiously destroyed by fire to make way for new mixed-use commercial business districts. Such changes have resulted in conflict.

Often, these new middle-class enclaves sit side-by-side with ever-rising numbers of the urban poor, who continue to migrate toward cities in search of jobs. Displaced by land

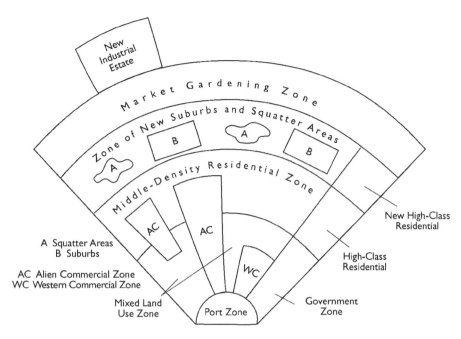

Figure 10.13   A Generalized Model of Major Land Use in the Large Southeast Asian City. *Source:* T. G. McGee, *The Southeast Asian City* (New York: Praeger, 1967), 128

scarcity and the mechanization of agriculture, these rural-to-urban migrants must compete with the wealthy, and now with the middle class, for limited resources in the cities. The poor provide their own housing, usually makeshift structures of corrugated tin, cardboard, or plywood. Often, these structures sit in sharp contract next to the golf courses and gated, elitist communities of the better off. The urban landscape of Southeast Asia thus reflects the social and economic transformations of a region enmeshed in much broader changes; the growing disparity between the "haves" and the "have-nots" is all too apparent.

MODELS OF URBAN STRUCTURE

For more than three decades, thinking about Southeast Asian urbanization has developed out of T. G. McGee's model of city structure (fig. 10.13). Building on a long tradition of urban modeling in North America, McGee's model presumed, first, that no clear zoning characterizes land use of the large cities of Southeast Asia. Instead, he proposed that only two zones of land use remained relatively constant, these being the port district and, on the periphery of the city, a zone of intensive market gardening. In between were areas of mixed economic activity and land use with other areas of dominant land use, such as spines of high-class residential areas and clusters of squatter settlements.

As cities in Southeast Asia have experienced rapid urban growth, in part associated with more intense integration into the global economy, our understanding of these places has changed considerably. While remnants of McGee's initial model are still visible, *e.g.,* a central port zone and mixed land use patterns, other aspects of Southeast Asian cities have been radically transformed. Urban scholars now speak of *extended metropolitan regions* (EMRs). This distinctive form of Asian urbanization has its origins in the way in which Asian cities (not just limited to Southeast Asia) have been incorporated into the global economy. The colonial-influenced primate cities are increasingly penetrating their surrounding hinterlands, urbanizing the countryside, and drawing rural populations deeper and deeper into the urban economy. In certain respects, EMRs are similar to metropolitan regions in the United States. However, EMRs in Southeast Asia differ from their North American counterparts in that the former exhibit a greater population density in both the urban cores and surrounding rural periphery.

EMRs may be differentiated into three basic forms. The first is the *expanding city-state*. Singapore provides the only example. In recent decades, Singapore has extended its political and economic influence into the territory of its neighbors, Indonesia and Malaysia. A second type is the *low-density* EMR, exemplified by Kuala Lumpur. These EMRs have been able to maintain relatively low population densities through the successful development of satellite cities that form a fringe around the dominant urban area. In other words, low-density EMRs reflect controlled and managed growth, as opposed to the more rapid and unplanned growth of other cities. Ho Chi Minh City likewise reflects aspects of a low-density EMR. The third and most prevalent type is the *high-density* EMR. This form, epitomized by the massive cities of Jakarta, Manila, and Bangkok, exhibits a chaotic spillover of urban economic functions into the rural hinterland, with an accompanying conversion of agricultural land to residential and industrial development.

Related to EMRs is a new urban form, identified by McGee as a *desakota*. The term itself is derived from the Indonesian words for village (*desa*) and town (*kota*) and is meant to capture the process whereby urbanization overtakes its surrounding hinterland. A number of elements have been associated with *desakotas*. First, these cities exhibit considerable diversity in their land use. Characteristically, a *desakota* region encompasses cities with mixed residential and industrial land uses, as well as a densely populated wet-rice agricultural area. Within a *desakota*, moreover, there is significant interaction between village and town. This is made possible by an integrated transportation system that permits high levels of population mobility. Indeed, the increased daily commuting patterns seen in Bangkok testify to this increased circulation within that *desakota*. Second, these regions also are strongly integrated into the global economy. Foreign investment is generally important in these areas, as multinational corporations tap into large and readily available labor surpluses. Lastly, it is the *process* of formation of the *desakota* that is perhaps more important than the resulting pattern itself, because surrounding rural areas become urbanized without the transfer of population that occurs in, say, rural-to-urban migration. On many occasions, urban transformation of rural areas involves dispossession and displacement of farmers and other rural residents. For example, there has been a large-scale conversion of agricultural lands in provinces adjacent to Metro Manila into industrial parks, eco-tourism developments, and gated communities. These developments have effectively displaced many farming communities and often involve violence between resisting farmers, the military, and landlords' security forces (box 10.4).

Box 10.4   From Hacienda to Mixed-Use Suburbia

"You don't have the right!" ("*Wala kayong karapatan!*"), shouted the lead architect of a development firm to a large group of farmers on the picket line. On May 21, 2010, violence erupted between a group of pro-development supporters (including the architect, surveyors, and members of both the military and the local police) and resisting farmers in Buntog, an upland community in Canlubang, Laguna, Philippines. During the incident, around a hundred farmers and activists were hurt. Eleven people, including a pregnant woman and a 70-year old grandmother, were illegally jailed, two were badly beaten by the police, and one elderly man suffered a heart attack. This is just one of the many cases of violence faced by farmers resisting urban development in the past decades.

Buntog is part of the former Canlubang Sugar Estate or Hacienda Yulo. The 7,200-hectare (4,201-acre) estate is owned by one of the Philippines' most powerful elite families, but the farmers of Buntog have been living in the area since the early 1900s when the region was still an unoccupied forest. The farmers cleared the forest and planted coconut trees and other crops. In the ensuing years, the vast tract of land near Buntog was purchased by an American conglomerate, Ehrman-Switzer, that established the Calamba Sugar Estate. Nationalist calls for independence and World War II provided an effective context that transferred the estate to the hands of Filipino elite families. Thus it was that Jose Yulo purchased the estate using a war reparation loan and renamed it the "Canlubang Sugar Estate." After the War, Buntog farmers filed a formal petition to acquire the title of their land; but, surprisingly, found out that Yulo's sugar estate had extended into their community. In the process, farmers were able to find employment as sugarcane plantation workers.

In 1988 the Philippines enacted the Comprehensive Agrarian Reform Program that sought to redistribute large landholdings to tenant farmers. But the Yulos of Canlubang had different plans. In fact, they had been converting parcels of the whole estate into industrial sites and golf courses a decade prior to land reform. They successfully classified the area as industrial and therefore received an exemption to redistribute their lands to farmers in the area. In 1996, the sugar estate closed and farming was prohibited in large sections of the estate. Through their many corporations, the Yulo family entered into joint venture agreements with top real estate firms to develop mixed-use urban projects that combined gated communities, golf courses, and commercial districts. The most ambitious of these development projects is Nuvali, a 1,700-hectare (4,201-acre) joint-venture master-planned township project with Ayala Land Corporation. This "city of the future" is being promoted as the Philippines' first "eco-community" and boasts a blend of "environmentally sustainable" residential, commercial, and recreational developments, such as mixed-use business and retail hubs, schools and universities, a human-made lake, a bird sanctuary, and four residential communities that cater to upper middle class residents. Adjacent to Nuvali are the Canlubang Golf Course, Carmelray Industrial Park, and other smaller gated communities.

Interestingly, these sustainable urban developments are built on lands previously occupied by farming communities. To build these projects and compel the farmers to leave their communities, the landowners either gave small amounts of money to farmers to "self-demolish" their houses and move or used physical force through the military and local police to force them off the land. Resisting farmers who remained faced increased intimidation and harassment that even led to the deaths of several farmer activists in the area.

Buntog is one of the few remaining parcels of land in Canlubang that are awaiting development. There are plans to expand a golf course and to build a mix of eco-tourism components and residential developments that will link the area to other tourist attractions. Because of this, surveyors and architects, accompanied by the military and security forces, have been intensively inspecting and mapping the area in recent years. Numerous cases of military intimidation and illegal cutting of coconut trees have been reported. The whole estate has been littered with security posts that serve to control the flow of residents in Buntog. Farmers have been prohibited from taking their produce to nearby markets and from bringing in much-needed materials for house repair. In addition, a water tank facility project in the community was halted by the landlords arguing that the whole area is their "private" property. In response to these abuses, Buntog farmers have joined the national Peasant Movement of the Philippines (*Kilusang Mangbubukid ng Pilipinas*) and have set-up a community organization called SAMANA BUNTOG (*Samahan ng mga taga-Buntog*) that regularly meets to plan ways to confront purges by the military and landlords' security groups and join protest marches and demonstrations along with other farming groups in the country. Many Buntog farmers are determined to continue their fight for land until death.

The case of Buntog is not isolated. For many former haciendas in the Philippines, like Canlubang of the Yulo family and Luisita of the Cojuangcos, and other large landholdings, land-use conversion and urbanization have become effective means by landlords to circumvent land reform and to squeeze land for profit. Around Manila and the surrounding region, many new mixed-use urban developments are built on lands with painful histories of dispossession and violence.

Moreover, it is the process of urban change that captures the broader political, economic, and social transformation that typifies Southeast Asian cities within a globalized world.

DISTINCTIVE CITIES

Singapore: World City of Southeast Asia

There is truly no other city in the world quite like Singapore. Its striking modernity, orderliness, and Disneyland-like cleanliness seem unreal, especially if one has just arrived from the sprawling, chaotic, and noisome cities of Bangkok, Jakarta, or Manila. To some, Singapore is a model of efficiency, a mosaic of well-manicured lawns, efficient transportation, and planned development. To others, Singapore represents a draconian police state masquerading as utopia. The reality, of course, lies somewhere in between, and depends on one's personal tastes and values.

Singapore is unique within Southeast Asia, and in the world, in that it is both a city as well as a sovereign state. And it is small. At just 246 sq mi (640 sq km), it is one-fifth the area of Rhode Island. But despite its size, Singapore is also very prosperous. In per capita income, the city-state is second only to Japan in all of Asia. In part, Singapore's economic success has been tied to its geography. Both historically and now, the city benefits from its strategic location and superb natural harbor. Furthermore, effective government policies and dynamic leadership have propelled Singapore into its role as Southeast Asia's leading port and industrial nexus, as well as a leading banking and commercial center. Indeed, Singapore ranks along with Tokyo and Hong Kong as one of Asia's three key urban centers in today's global economic system.

Singapore's economy is based on a strong manufacturing sector (initially processing raw materials such as rubber, but more recently electronics and electrical products), oil refining, financial and business services, and tourism. The economic success of Singapore has translated into a very favorable quality of life. Singapore registers the region's lowest infant mortality rate and the lowest rate of population increase, in addition to the highest per capita income. This quality of life is facilitated through subsidized medical care and compulsory retirement programs.

Singapore is a truly cosmopolitan world city. The affluence of Singapore is vividly seen along the city-state's major shopping and tourist corridor, Orchard Road. In the 1830s this area was home to fruit orchards, nutmeg plantations, and pepper farms. Now, the mangoes, nutmeg and peppers are gone, replaced by jeweled necklaces, designer clothes, and perfumes. Orchard Road currently extends 1.5 mi (2.4 km) and is lined with major shopping centers, upscale boutiques, luxury hotels, and entertainment centers such as the Raffles

Figure 10.14   This gaudy Indian temple in Singapore is one of the best-known cultural landmarks of the city and a center of the Hindu minority population. (Photo by James Tyner)

Village (built around the carefully restored classic Raffles Hotel, one of the great hotels of the past).

True to its place at the center of Southeast Asia, the urban landscape of Singapore reflects its rich ethnic heritage. Approximately 75 percent of Singapore's population is ethnic Chinese; Malays and Indians constitute 15 percent and 7 percent of the population, respectively. Consequently, Singapore has four official languages: English, Mandarin Chinese, Malay, and Tamil, a language of southern India and Sri Lanka. The immigrant history of Singapore, moreover, is well preserved in the city-state's architecture (fig. 10.14). The

Figure 10.15   Sir Stamford Raffles, founder of Singapore, stands along Collyer Quay with restored houses on the opposite bank as reminders of Singapore's colorful past. (Photo by Jack Williams)

Chinatown area of Singapore, for example, located at the mouth of the Singapore River, began in 1821, when the first junk load of Chinese immigrants arrived from Xiamen, in Fujian province. In 1842 these Chinese immigrants completed the Thian Hock Keng Temple and dedicated it to Ma-Chu-Po (Matsu), the goddess of the sea. Nearby is Nagore Durgha Shrine, built by Muslim immigrants from south India, and further down the road is the Al-Abrar Mosque, also known as Indian Mosque, which was built between 1850 and 1855. Because of land reclamation projects, the ultramodern skyscrapers of Singapore's financial district now eclipse Chinatown's oceanfront view.

Given the large number of historical temples and monuments, urban preservation is important in Singapore, although this has not always been the case. During the 1960s and 1970s, for example, many older buildings were demolished to make room for a more modern infrastructure. However, a movement to preserve Singapore's urban history was initiated, both for reasons of national prestige and tourism. Today, many areas, such as the waterfronts along Collyer Quay and Boat Quay, have been renovated (fig. 10.15).

Singapore is also instructive for its public housing programs. Beginning in the 1960s, the Singaporean government, primarily through the efforts of the Housing Development Board (HDB), moved to ensure adequate housing for its population. The result was the establishment of numerous new towns and housing estates. The Queenstown housing estate, for example, located in the central region of Singapore, was one of the earliest estates developed by the HDB. Tao Payoh New Town and Ang Mo Kio New Town, both located within the Northeastern Region, were initiated in 1965 and 1973, respectively. A more recent development is Woodlands New Town, located on the northern coast of Singapore. These new towns are designed to be self-sufficient communities of many thousands of residents. Moreover, in recognition of Singapore's ethnic diversity, the government has mandated ethnic-based occupancy rates to offset the emergence of hyper-segregated enclaves. By 2000, nearly 90 percent of Singaporeans lived in one of the high-density housing estates built by the HDB.

Despite these successes, urbanization in Singapore remains, and will continue to remain, hindered by two physical obstacles. First, Singapore cannot readily—or cheaply—expand its area because it occupies only a small island. In response, the Singaporean government has utilized land reclamation schemes and has also been working to expand its economic growth beyond its own political boundaries into neighboring Malaysia and Indonesia. In this manner, the government hopes to exploit the comparative advantages of Singapore and neighboring countries. A second, and perhaps more immediate, obstacle confronting Singapore is that of water. Water is a serious problem because the island contains no significant rivers or lakes to collect freshwater and so must rely on reservoirs and storm-water collection ponds to provide freshwater for its nearly 4 million residents. Most water is supplied by Malaysia. In response to its limited water resources, Singapore has pursued to two-fold strategy. First, it has initiated a series of conservation measures, including the installation of water-saving devices, water-recycling programs, and water-consumption taxes. Second, Singapore has pursued other means of obtaining water, such as the installation of desalinization plants, the construction of more reservoirs, and the possibility of acquiring water from Indonesia.

Kuala Lumpur: Twin Towers and Cyberspace

The skyline of Kuala Lumpur is one of the most recognizable sights in all of Southeast Asia. While Paris has its Eiffel Tower and Shanghai its futuristic TV tower, Kuala Lumpur has the 88-story Petronas Twin Towers. Standing like a giant double-barreled beehive, the Petronas Towers dominate the capital city of Malaysia (fig. 10.16). They are Kuala Lumpur's signature landscape and are symbolic of the lofty goals set by the Malaysian government.

Kuala Lumpur is a relatively young city, having been founded only in 1857 by Chinese tin miners at the swampy confluence of the Klang and Gombak rivers. In fact, the name "Kuala Lumpur" translates as "muddy confluence." The settlement grew rapidly however, and by 1880 had become the capital of the state of Selangor on the Malay Peninsula.

Despite its growing political importance, Kuala Lumpur throughout much of the early 20th century was still overshadowed in population by other cities along the Malay Peninsula, including Georgetown to the north and Singapore to the south. Although it was designated capital of the Federated States of Malaysia in 1963, Kuala Lumpur trailed both Singapore and Georgetown as preeminent commercial centers on the peninsula. In 1972, however, Kuala Lumpur gained city status and was declared a Federal Territory (similar to Washington's District of Columbia).

Kuala Lumpur experienced tremendous population growth throughout the late 20th century. Currently the population is approximately two million. However, unlike Bangkok and Manila, Kuala Lumpur has made a more concerted effort to manage urban growth. Planned satellite cities, for example, were designed to the urban congestion of the capital.

Figure 10.16   When Kuala Lumpur's Petronas Towers opened in 1999, they became the world's tallest, a title they held until 2004. (Photo by Richard Ulack)

In the 1950s, the satellite city of Petaling Jaya was established; it is now home to more than 500,000 residents and a major industrial center. Nearby, the satellite town of Shah Alam, initially planned to be half residential and half industrial, was built in the 1970s. Although shantytowns are visible in parts of Kuala Lumpur, as a whole the city exhibits a sedate orderliness more reminiscent of Singapore than of other Southeast Asian cities.

The economy of Kuala Lumpur is an exceptionally diverse mix of manufacturing and service activities. Many of these industries are clustered within the Klang Valley conurbation, an urbanized corridor stretching from Kuala Lumpur westward through Petaling Jaya and Shah Alam to the port city of Klang. Also, indicative of the information economies emerging in Southeast Asia, Kuala Lumpur is a key anchor in Malaysia's "Multimedia Super Corridor" (MSC). Planned to be a setting for multimedia and information-technology companies, the MSC is seen as the catalyst in propelling Malaysia's economy into the

Figure 10.17   The sprawl of Jakarta is visible in this bird's-eye view that simultaneously shows the old (the largest mosque in Southeast Asia) and the new (a line of the rapid-transit system designed to bring order to the city's traffic chaos). (Photo by Jack Williams)

global information age. In related developments, two new cities have been constructed to the south of Kuala Lumpur: Putrajaya, the "new" administrative capital of Malaysia, and Cyberjaya, billed as an "intelligent city," complete with a state-of-the-art, integrated infrastructure that attracts multimedia and information-technology companies.

Jakarta: Megacity of Indonesia

Most visitors to Indonesia arrive first in Jakarta, a sprawling metropolis situated on the north coast of Java. At two and a half times the size of Singapore, Jakarta is the largest city in population and land area in Southeast Asia (fig. 10.17). From its origins as a small port town called Sunda Kelapa, the Special Capital Region of Jakarta (Daerah Khusus Ibukota, or DKI Jakarta) has experienced phenomenal growth over the past five decades. From fewer than 2 million in 1950, Jakarta's population had swollen to more than 13 million by 2000. As with many primate cities in Southeast Asia, urban growth has sprawled into the hinterland. In recognition of this sprawl, in the mid-1970s officials began to refer to the entire region as Jabotabek, an acronym derived from the combination of *Ja*karta and the adjacent districts of *Bo*gor, *Ta*ngerang, and *Bek*asi. When the entire Jabotebek region is considered, the population of the Jakarta metropolitan area includes a mind-boggling 20 million people. Similar to other major cities in the region, Jakarta's population is also impacted by seasonal and daily commuting.

Figure 10.18   Only a minority of Jakarta's residents has access to piped water, so water vendors such as this wend their way through the streets selling this precious commodity. (Photo by Jack Williams)

Hundreds of thousands of workers, the majority of whom live in new residential communities in the Jabotabek region, commute daily to Jakarta.

Jakarta is Indonesia's largest and most important metropolitan area. It is the national capital and the principal administrative and commercial center of the archipelago. The city also plays a vital role in Indonesia's international and domestic trade and receives a disproportionate share of foreign direct investment. This investment, focused primarily on manufacturing but also on the construction and service sectors, operates as a multiplier effect for Jakarta's economy. It also accounts for a rapid rise in the middle class, with a corresponding impact on the urban landscape. In some respects, though, Jakarta has undergone a period of deindustrialization similar to that of other cities in the world, such as London. Thus, although the city remains an important manufacturing center, economic growth has been accounted for largely by increases in both tertiary and quaternary sectors (especially financial services, communications, and transportation).

Economic and social changes have also contributed to changing land-use patterns. The central core of Jakarta has experienced significant changes over the past decade, such as a conversion from residential to higher-intensity commercial and office land use, as well as the emergence of luxury high-rise apartments. In the Jabotabek region, the development of new towns (e.g., Lippo City, Cikarang New Town, and Pondok Gede New Town) and corresponding large-scale residential subdivisions has also transformed previously agricultural land into urban spaces. Indeed, upwards of eighty thousand new housing units are added each year to the Jabotabek region. Other changes include the emergence of larger industrial estates as well as leisure-related land uses (e.g., golf courses).

Figure 10.19   Unique to Manila, the jeepney is a form of urban transport that got started after World War II using old U.S. jeeps. Decked out with frills and gaudy decoration, the jeepneys play an important role and have become a symbol of Manila. (Photo by Jack Williams)

Jakarta also reflects the urban woes characteristic of primate cities, including a lack of adequate public housing, traffic congestion, air and water pollution, sewage disposal, and the provision of health services, education, and utilities (fig. 10.18). Because of the megacity's sheer size, not to mention the heightened political, social, and economic instability of Indonesia, Jakarta's problems are magnified to dangerous levels.

Manila: Primate City of the Philippines

Unlike walking the regimented, disciplined streets of Singapore and Kuala Lumpur, traveling within Manila is an experience unto itself. Indeed, with the possible exception of Bangkok, no other city in Southeast Asia is as famous—or infamous—as Manila for its traffic. Throughout the day, and frequently well into the night, traffic grinds slowly through the rabbit-den of highways and alleyways that constitute Manila's overburdened road network. Diesel-spewing *jeepneys*, rickety buses, and luxury sports utility vehicles all compete in bumper-car-like fashion (fig. 10.19). Turn lanes and stoplights are largely ignored. Yet beyond the chaos and congestion that puts the freeways of Los Angeles to shame, Manila exhibits its own charm and appeal. Indeed, much of Manila's charm stems precisely from its outwardly confusing appearance.

Politically, socially, economically, and in terms of total population, Manila far surpasses all other cities in the Philippines (fig. 10.20). Indeed, with the exception of Thailand, no other major country in the region has a higher primacy rate than the Philippines. Currently, Manila is approximately nine times as large as the Philippines' second-largest metropolitan area, Cebu. And similar to Jakarta's Jabotabek, Metro Manila is composed of many different

Figure 10.20   Traditional Manila contrasts with modern Manila as the city attempts to accommodate the rapidly expanding population by going up and spilling out onto the city's streets. (Photo by Arnisson Andre Ortega)

political units. In 1975, Metro Manila was formed through the integration of the four pre-existing, politically separate cities of Manila, Quezon City, Kaloocan, and Pasay, plus thirteen municipalities. The nature of governance and administration of the metropolitan region has changed over the years since its original inception, from a more centralized Metro Manila Commission (MMC) headed by Imelda Marcos as governor to its current monitoring and coordinating form as the Metro Manila Development Authority (MMDA).

The Metro Manila region, reminiscent of Harris and Ullman's multiple nuclei model, is a poly-nucleated area with many distinct personalities. Binondo, for example, located next to the Pasig River, was originally a Christian Chinese commercial district during the Spanish colonial period and to this day remains the heart of Manila's Chinatown. Nearby is Tondo, today an impoverished, densely populated district of rental blocks; prior to the arrival of the Spanish, it was a collection of Muslim villages. To the south, abutting Manila Bay, is Ermita. Once a small fishing village, the area developed into a prime tourist destination, packed with bars, nightclubs, strip shows, and massage parlors. In recent years, however, these establishments were closed down in Ermita and many have since relocated to other areas of Manila. As a final example, Makati, originally a small market village, is now Manila's major financial center, occupied by banks and multinational and national corporations. Makati also contains some of Manila's most expensive housing subdivisions, sprawling box-like shopping centers, and five-star hotels.

Within the greater Manila Metropolitan region is Quezon City. Named after Manuel Quezon (president of the Commonwealth of the Philippines from 1934–1946), Quezon City was the Philippines' national capital from 1948 to 1976. Now it is home to many important government buildings, medical centers, and universities, including the main campuses of the University of the Philippines and Ateneo de Manila University. Quezon City also consists of upscale, gated residential communities patrolled by armed guards. These estates, home to middle- and upper-class residents, are equipped with luxurious air-conditioned homes, tennis and basketball courts, golf courses, and swimming pools. However, reflective of Manila's complex land usage, as well as the highly polarized nature of Philippine society, just outside of these gated communities are numerous squatter settlements.

Characteristic of large cities in Southeast Asia, Manila exhibits an increasing number of consumer spaces. In recent years large shopping malls have been built, catering to a rising middle class. For many visitors, these malls are remarkably similar to those found in the United States and Europe. Major department stores anchor the malls, while in-between are dozens of specialty stores, food courts, and entertainment. The SM Megamall, for example, in addition to its numerous stores and restaurants, contains an ice skating rink, bowling lanes, a twelve-screen cinema, and

an arcade room. Meanwhile, multiple high-end condominium projects have been built in many parts of Metro Manila that cater to returning Overseas Filipinos and urban professionals. In many cases, these condominium projects were developed beside malls and shopping complexes. For example, the proposed Entertainment City in Manila, a mixed-use leisure complex of casinos, condominium units, and shopping malls, is planned for development on a previously reclaimed area near Manila Bay.

Similar to both Jakarta and Bangkok, the increased concentration of foreign direct investment into the Philippines has translated into rapid changes in the economy of Manila, as well as in land use. During the 1990s, the greater Metro Manila region and the surrounding provinces experienced remarkable industrial and manufacturing growth. This expansion, however, occurred at the expense of Manila's rural and agrarian hinterland. Metro Manila is, in fact, located toward the center of the Philippines' major rice-producing region, and continued urban sprawl is rapidly encroaching on these agricultural areas.

Urban poverty and landlessness continue to be major problems in Manila. The extent of these problems, however, remains a contested issue. Estimates of the number of poor vary widely, ranging from 1.6 million to more than 4.5 million. What is certain, however, is that landownership in Manila is decidedly uneven, with the majority of its population being landless. High urban land values mean that the majority of residents are unable to obtain legal housing, a situation exacerbated by continued high rates of in-migration. Their recourse is to resort to illegal housing and to settle in urban fringe areas, such as along railroad tracks and in vacant lots. Residents of squatter settlements are subject to deplorable health conditions and pollution problems, stemming from inadequate access to sanitary and plumbing facilities. They often must purchase fresh water from itinerant water vendors. Historically, squatter settlements have been demolished and their residents evicted. More recently, the Philippine government has attempted to provide low-cost housing for its urban poor through joint-venture agreements with private developers. But with relocation sites that are faraway from the city and insufficient facilities in many of the housing projects, many settlers end up returning to the metropolis.

Aside from poverty, Manila faces other serious problems. Accessibility to water, for example, looms large. Manila is also confronted with serious air and water pollution problems, as well as an inadequate sewerage system. Indeed, during the rainy season many streets throughout Manila, such as those in the port district and Tondo, become impassable due to flooding.

Bangkok: The Los Angeles of the Tropics

Bangkok, at 34 times the size of Thailand's second-largest city, is the textbook example of urban primacy. While only one-fifth of the country is urbanized, fully two-thirds of this urban population is concentrated in the Bangkok Metropolitan Region (BMR). Currently, the core of Bangkok has a population of about 6 million; when the entire BMR is considered, the region's population is more than 10 million. Furthermore, like the ocean's tides, Bangkok's population ebbs and flows, both daily and seasonally. An estimated one million people commute daily into Bangkok, while hundreds of thousands of other workers

seasonally circulate throughout the city in search of temporary jobs in the informal sector. This seasonal migration is particularly acute in the hot, dry months of February and March, a slack agricultural period.

The official name of Bangkok is Krung Thep, which translates as "The City of Angels" (the same meaning as Los Angeles). And in many respects, notably traffic, pollution, and urban sprawl, Bangkok might be considered the Los Angeles of Southeast Asia. For that matter, Los Angeles may be considered the Bangkok of the United States of America.

Currently, Bangkok remains poised to become an international communications and financial center, as well as a major transportation hub in Asia. Initially, much of Bangkok's growth was tied to massive amounts of investment brought about by the United States' involvement in the Vietnam War. During the 1960s, in particular, Bangkok served as a major military supply base. In subsequent decades, it continued to attract large sums of foreign investment. Between 1979 and 1990, nearly 70 percent of all foreign investment projects in Thailand were concentrated in the BMR. Economically, Bangkok has capitalized on its reserves of cheap labor, favorable tax incentives, and (until recently) political stability.

Similar to Jakarta and Manila, Bangkok is a multinucleated city. And while the "old city" remains the principal administrative and religious core of Bangkok, considerable expansion has occurred into surrounding districts. Bangkok has also experienced a rapid conversion of land use, with many residential areas in the city being converted to commercial use and former small shop houses being transformed into high-rise office buildings and large shopping complexes.

Figure 10.21   Bangkok, where private automobile ownership is widespread, has some of the worst traffic jams in the world despite the construction of urban expressways. (Photo courtesy Bangkok government)

The over-urbanization of Bangkok has resulted in serious environmental problems. Air and water quality have deteriorated in recent years, while the disposal of solid waste is an on-going problem. Also, Bangkok is sinking. Due to the overdrawing of well water, the city suffers from land subsidence, as its elevation drops at a rate of about 10 cm (3.9 in) per year. Indeed, some areas have subsided by more than 3 ft (0.91 m) since the 1950s. Global warming should be firmly on the minds of Bangkok's urban planners!

Bangkok is also plagued by severe transportation problems. The number of motor vehicles (excluding the ubiquitous motorbikes) increased from 243,000 in 1972 to more than a million in 1990; concurrently, only about 50 miles of primary roads were added. As a result, the average speed on most roads in Bangkok is less than six miles per hour (fig. 10.21). Numerous proposals and strategies have been advanced to rectify traffic congestion, including increased road capacity measures, improvements in public mass-transit systems, improvements in the traffic control system, and strategies to control the volume of traffic (e.g., staggered employment hours to reduce peak commuting traffic). The government is also encouraging the growth of satellite cities as a means of promoting regional economic growth and to the congestion of Bangkok.

Phnom Penh, Ho Chi Minh City, Hanoi: Socialist Cities in Transition

In early morning, the dusty streets of Phnom Penh are alive with swarms of noisy motorbikes that surge like schools of fish. Luxury cars and sports utility vehicles compete for limited space with the motorbikes. Plodding along the roadsides, in a vain attempt to escape the mechanized frenzy of Phnom Penh's traffic, are converted tractors with wooden trailers that ferry scores of young women—all dressed in identical green-and-white uniforms—to the foreign-owned assembly plants that ring the periphery of the city. Such is Phnom Penh in the 21st century: a frantic, disorderly city that is rebuilding after decades of tumultuous revolutions and genocide. The experience (and landscape) of Phnom Penh, combined with those of Ho Chi Minh City and Hanoi, provide vivid proof that urbanization is intimately associated with broader social movements, including revolutions.

Socialist cities in Southeast Asia have experienced, and reflect, a different pattern of urbanization than is the case with capitalist-based cities. The urbanization process in Southeast Asia's socialist countries (i.e., Burma, Cambodia, Laos, Vietnam) has been analyzed in a three-stage model. The first stage consists of a process of de-urbanization whereby major cities were depopulated. For example, as the Khmer Rouge assumed control of Cambodia in 1975, the socialist government undertook a forced evacuation of the capital city, Phnom Penh. Prior to this time, the population of Phnom Penh had swollen in size from around 700,000 in 1970 to approximately 2.5 million by 1975. This population increase resulted mostly from an inflow of refugees from the countryside escaping war. In 1975, however, the Khmer Rouge forcibly emptied Phnom Penh and other cities and villages throughout Cambodia. This process of de-urbanization was but one act of the Khmer Rouge's genocidal reign that took the lives of nearly 3 million people (box 10.3).

Today, Phnom Penh still bears the scars of its genocidal past, although these are slowing being repaired. The streets of Phnom Penh, unpaved and pockmarked with potholes in 2001, now shimmer darkly with fresh asphalt. Where once stood hollowed-out buildings, destroyed by war and neglect, now stand freshly painted apartment buildings and shopping complexes. Phnom Penh is indeed rebuilding, though not without difficulties; and this urban growth reflects a new orientation toward the global economy. For example, along the major road linking Phnom Penh and Cambodia's Pochentong International Airport, multinational corporations have established a visible presence, in the form of assembly plants and factories.

In Vietnam, a similar, though less brutal, process of de-urbanization occurred. The population of Saigon (now Ho Chi Minh City), like that of Phnom Penh, had also increased dramatically through in-migration and refugee flows. By 1975, Saigon had an estimated 4.5 million people. Following the communist victory, the new government of the Democratic Republic of Vietnam relocated about one million people.

The process of de-urbanization in these socialist countries was often accompanied by a re-fashioning of the cities. Initially, Western-style establishments and customs were replaced with a Spartan milieu. Cities, most notably Hanoi, were drab and monotonous, composed of row upon row of uniform, box-like buildings. Conforming to socialist ideology, the new governments attempted to eliminate the private sector; and shops, restaurants, hotels, and services were generally run by government enterprises or cooperatives. Consequently, cities were typically devoid of the mass advertising and consumer spectacles that are commonplace in capitalist cities. The new governments fostered symbolic changes as well. The renaming of Saigon to honor Ho Chi Minh, who led the fight against the French and established communism in

Figure 10.22   As a result of reforms known as doi moi, initiated in 1986, Vietnam at first tolerated and then encouraged private enterprise, as evidenced by these shops and vendors in Ho Chi Minh City. (Photo by Jack Williams)

Vietnam, provides the clearest illustration. Another visible difference between socialist and capitalist cities was the traffic. In Hanoi, the streets were practically empty of motor vehicles, save for an occasional Soviet-era limousine and a few battered and decrepit buses. Instead, bicycles thronged the streets, especially during peak hours, when residents cycled to and from work and school.

Following this initial stage, socialist governments entered into a second, bureaucratic stage wherein longer-term strategies of socialist urbanization were implemented. Especially in Vietnam, the socialist government developed spatial strategies to ameliorate the problems of large cities, including the provision of adequate food, employment, and housing. Policies were enacted to restrict population mobility, thereby affording relief to the infra-

structure of large urban areas, such as Ho Chi Minh City and Da Nang.

Socialist reform constitutes the third stage of the model. Although economic reforms were first introduced in Vietnam in 1979, it was not until 1986, with the initiation of *doi moi* (renovation), the slogan for the government's new development strategy, that substantial improvement occurred. *Doi moi* entails the gradual introduction of capitalist elements, including private ownership, foreign investment, and market competition. Vietnam remains politically committed to socialism, but economically the country is exhibiting a shift toward capitalism and a greater level of integration into the global economy (fig. 10.22).

Geographically, economic reforms initially focused on the southern region of Vietnam, and especially Ho Chi Minh City, because

Box 10.5   HIV/AIDS

HIV, the virus that causes AIDs, currently infects 40 million people worldwide. As of 2006 an estimated 7.2 million people were living with HIV/AIDs in Southeast Asia. This is the second-highest number of cases in the world outside of Africa. In Thailand alone there are over 580,000 HIV/AIDs cases, with an adult prevalence rate of 1.4. Statistics for other countries are equally grim: Burma (360,000 cases), Cambodia (130,000), Laos (3,700), and Vietnam (260,000).

HIV/AIDs is spread by many pathways. It can be transmitted through both heterosexual and homosexual intercourse or the sharing of tainted needles (*i.e.,* while injecting drugs such as heroin). HIV/AIDs may be transmitted from mothers to children in utero and during birth and breastfeeding. And HIV/AIDs may be passed on through the transfusion of infected blood.

Within Southeast Asia, key social, economic, and political factors have contributed to the spread of HIV/AIDs. A major contributing factor is the mobility of migrant workers, both internally and internationally. In Southeast Asia, all countries are engaged in sizeable and complex transnational migratory flows, as sending and/or receiving nations. Over one million workers from both Burma and Laos, for example, find employment in Thailand; thousands of migrants leave Cambodia to work in Thailand and Malaysia; while thousands of other workers leave China and Vietnam to work in Laos and Cambodia. Millions of other workers, from the Philippines, Indonesia, and Thailand, find employment throughout Europe and North America.

These networks of migrant labor frequently intersect with other transnational networks, namely the trafficking of sex workers. Throughout Southeast Asia, Europe, and beyond, many

of that city's much longer tradition with free-market economics and linkages with the outside world. Approximately 80 percent of all foreign investment flowing into Vietnam was directed toward the south. Investments in tourism, assembly, and manufacturing were concentrated in the larger urban areas. Ho Chi Minh City (still called Saigon by many residents) soon returned to its prewar capitalist character, with luxury hotels—Hyatt, Ramada, and Hilton—competing side-by-side with government-run hotels.

Hanoi, once the sedate, regimented, and subdued political capital of Vietnam, has itself undergone significant economic transformations. By the twenty-first century, Hanoi was increasingly showing the effects of globaliza-tion and now exhibits much of the color and dynamism of its rival to the south. Hanoi continues to showcase the symbols of Viet-namese nationalism, such as the Ho Chi Minh Mausoleum and the Ho Chi Minh Museum. But it also has become a bustling metropolis, an urban forest of hotels, restaurants, bars, nightclubs, and discotheques. Tourism, with thousands of global visitors interested in see-ing the heart of the Democratic Republic of Vietnam, is leading the change.

URBAN CHALLENGES

Cities in Southeast Asia are not immune to serious problems. Challenges run the gamut

women (and men) are forced or coerced into sex work. Some work in brothels, catering to migrant workers. In 2002, a survey revealed that 28 percent of all HIV-infected Filipinos were returning overseas contract workers. When these migrant laborers and sex workers return to their home villages, they may carry the disease with them.

Another form of population mobility includes regional and international tourism. And, within many countries of Southeast Asia, including the Philippines and Thailand, sex tourism is big business. As a result, the heterosexual and homosexual transmission of HIV/AIDs has increased precipitously.

The diffusion of HIV/AIDs is also associated with the prevalence of drug use. Southeast Asia is a key node in the global distribution of heroin, and the sharing of infected needles has been identified as a crucial factor in the rise of HIV/AIDs in the region.

It is not uncommon for these various pathways to come together in the major cities, such as Manila, Bangkok, Chiang Mai (Thailand), and Phnom Penh. Consequently, as people continue to circulate, the cities of Southeast Asia serve as "infection pumps"—centers of HIV/AIDs diffusion. It is for this reason that the governments of Vietnam, Cambodia, and Thailand have concentrated many of their prevention strategies in cities. Such efforts include educational campaigns, provision of blood screening test-kits, and the distribution of condoms and clean needles. Tragically, however, the diffusion of HIV/AIDs will be slowed only if many other contingent problems of Southeast Asia—civil strife, repression, censorship, political corruption, government neglect, and poverty—are addressed. As such, the cities of Southeast Asia will remain battlegrounds in the war on AIDs.

from health issues (box 10.5) to environmental concerns. Arguably, though, the inability of urban residents to obtain adequate employment and housing looms among the most serious issues faced by Southeast Asia's cities. For example, in the early 2000s in Manila, approximately 40 percent of the urban population was estimated to be living in squatter settlements, with another 45 percent of the population living in slum conditions. In Bangkok, 23 percent of the population was estimated to be living in slum and squatter settlements; and in both Kuala Lumpur and Jakarta, squatters constituted approximately 25 percent of the population. It should be noted, however, that most estimates on slum residents are measured at the national level and that there are, not surprisingly, conflicting numbers of squatter/slum residents.

The rise of squatter settlements is explained by factors other than population increase. Escalating land prices, compounded by real estate speculation, for example, exacerbate the problem of housing. So, too, does the creation of artificial land scarcity. In Metro Manila, for example, large tracts of land, even within the central business district of Makati, remain vacant. And lastly, the demolition of low-cost housing units, replaced by more affluent condominiums and gated communities, results in the rise of squatter settlements.

Sadly, many governments continue to view eviction and demolition as the most effective means of confronting squatter settlements.

In the Philippines, for example, more than 100,000 people were evicted from Manila each year between 1986 and 1992. Not surprisingly, a policy of relocating squatters to sites 20–50 mi (32–80 km) outside the city and placing them in high-density residential apartments proved ineffective. Only Singapore has achieved substantial results in the provision of public housing. Other cities, especially Manila, Bangkok, and Phnom Penh, trail woefully behind. While many of these governments have agencies charged with developing public housing, most lack the required economic resources and political resolve to be effective.

Many Southeast Asian governments also are unable to provide adequate services, such as clean water, sewerage, and other utilities. Only about 7 percent of Burma's urban population, for example, has access to piped water. In Jakarta, only a quarter of the population has solid waste collection; in the remainder of the city, it is collected by scavengers. One effective strategy has been Indonesia's Kampung Improvement Program (KIP). This program is a far-reaching initiative that concentrates primarily on the improvement of infrastructure and public facilities. Specific projects include footpaths, secondary roads, drainage ditches, schools, communal bathing and shower facilities, and health clinics. Since its inception, the KIP has been expanded to more than two hundred cities throughout Indonesia and has benefited more than 3.5 million people.

Both air and water pollution pose serious health hazards to residents and visitors alike in Southeast Asian cities. Jakarta, for example, exceeds the health standards for ambient levels of airborne particulate matter on more than 170 days of the year. Moreover, topographic features may augment pollution problems. The surrounding hills of the Klang Valley around Kuala Lumpur, and the mountains ringing Manila, for example, confine pollutants and thus exacerbate air quality problems.

Water pollution, likewise, remains a major obstacle to the quality of life in Southeast Asian cities. Many rivers, including the Pasig in Manila, the Chao Phraya in Bangkok, and the Ciliwung in Jakarta, are considered biological hazards. The canals and waterways in Bangkok, especially, are highly polluted from a combination of industrial and household discharge. Only 2 percent of the city's population is connected to Bangkok's limited sewerage system. Consequently, most solid waste is discharged into waterways. Compounding the problem is the fact that more than 15 percent of the garbage disposed of daily is left uncollected.

An additional problem is traffic congestion. The traffic problems of Bangkok and Manila were discussed previously. In Jakarta, likewise, private car ownership has outpaced road construction. Similar problems of congestion and pollution are being felt in Ho Chi Minh City, which is currently home to more than two and one-half million motorbikes, and, increasingly, in Phnom Penh as well. Some governments, including those of Malaysia and Indonesia, have utilized toll roads to reduce traffic congestion. Other efforts concentrate on the development of mass-transit systems, such as the construction of light-rail transit systems in both Manila and Kuala Lumpur. These projects, however, are extremely expensive and many have been temporarily halted. To this day, half-completed overpasses and bridges in Bangkok, and incomplete rail systems in Manila, stand as silent reminders of continued underdevelopment.

Amidst all these problems, there are efforts by governments, along with local businesses, to promote sustainable urban development. For instance, several environmental laws have been passed in different countries seeking the reduction of pollution or to respond to issues of climate change. In some cities, city planning programs pertaining to urban renewal and the promotion of sustainable urban life have recently been put in place. In Singapore, for example, the Ministry of National Development has actively directed the planning and implementation of policies and infrastructure projects that aim to create a sustainable city of knowledge, culture, and excellence. In other cities, new development projects are advertised as being constructed to satisfy global environmental standards. The real challenge, however, lies in the enforcement of these programs and policies so that they may truly contribute to a sustainable urban life. In fact, many residents are actually displaced in the name of urban renewal projects.

Despite the emergence of new urban developments that are intended to promote "urban sustainability," many inhabitants of these cities remain mired in poverty. As of the early 1990s, for example, more than 30 percent of Metro Manila's population lived below the poverty line, as did nearly 30 percent of Indonesia's urban population. Indeed, the figure for Indonesia escalated sharply in 2000–2001, as the country struggled with instability of all kinds.

Rampant poverty contributes to other serious problems, including political unrest, violence, and terrorist activity. In recent years, Southeast Asian cities have witnessed class-based political tensions and demonstrations. In Bangkok, "red shirt" protestors, mainly composed of rural working class residents and supporting ousted Prime Minister Thaksin Shinawatra, clashed with "yellow shirt" demonstrators composed of people from the urban middle class. Meanwhile, in Manila, two "revolutions took place in early 2001, reflecting the country's socioeconomic inequalities and class-based political tensions. Protest actions organized by civil society groups and the middle class, for example, ousted then-Philippine President Joseph Estrada over charges of corruption; and four months later, urban poor demonstrators and activists staged a similar demonstration that culminated in an attempt to storm the presidential palace.

Potential violence in cities of Southeast Asia comes in other forms. According to some scholars and regional experts, the main "terrorist" threat to urban life in many Southeast Asian cities is linked to radical Islamist groups, such as Jemaah Islamiah (JI), a group that has links to Al-Qaeda and Abu-Sayyaf. Over the past decade, JI has been linked to bomb attacks in many major Southeast Asian cities, such as the Bali bombing of 2002 that killed 202 people and the Jakarta bombing in 2004 that killed 11. In Manila, the Abu-Sayyaf has also been linked to several bombing incidents. In response to such terrorist threats, many cities in Southeast Asia have established security programs and policies. In Singapore, a large-scale preparedness exercise, named Exercise Northstar V, was conducted. The exercise involved simulated terrorist bomb attacks in multiple locations and included the participation of thousands of government personnel and civilians. In Manila, malls, MRT, and other establishments are guarded by armed security personnel who control the entrance of people.

AN EYE TO THE FUTURE

Like Gregor Samsa in Franz Kafka's novella *The Metamorphosis*, the cities of Southeast Asia have awoken from unsettled dreams to find themselves changed into something potentially monstrous: agglomerations of skyscrapers and street vendors, palatial residential neighborhoods and impoverished squatter settlements, overburdened utilities, and underdeveloped transit systems.

What does the future hold for the cities of Southeast Asia? Three themes come to mind. First, continued population pressures and environmental degradation will most likely accelerate rural-to-urban migration, thereby exacerbating over-urbanization problems. Consequently, the cities of Southeast Asia will continue to expand geographically. How this development occurs, however, and how governments respond or manage this growth, will greatly affect the livability of these cities. Will growth continue unabated, in an unplanned, haphazard manner, or will decentralization strategies and growth-diversion measures effect desired changes? In economically poor countries, and those saddled with massive foreign debts, fiscal capacity, management, and political motivation may hinder these attempts.

A second theme is that these cities will continue to be incorporated into the global economy. This holds true especially for the socialist cities of Phnom Penh, Ho Chi Minh City, and Hanoi. Consequently, manifestations of globalization processes at the local scale will become more apparent. For example, the mushrooming of McDonald's, Starbucks, and Kentucky Fried Chicken franchises will continue. But apart from these superficial changes lie deeper, structural transformations resulting from the infusion of foreign capital. Just as political revolutions had impacts on urban areas in the socialist countries, social changes, such as the emergence of a new urban middle class, are likely to stem from and reflect back on urban transformations.

Southeast Asia, because of its strategic location and long-standing ties to the global economy, is destined to grow ever more important in world affairs. The cities of Southeast Asia will continue to transform, and be transformed by, broader global changes.

SUGGESTED READINGS

Chulalongkorn University. *A look at various facets of Thailand's primate capital.*

Berner, Erhard. 1997. *Defending a Place in the City: Localities and the Struggle for Urban Land in Metro Manila.* Quezon City, Philippines: Ateneo de Manila University Press. An examination of the complex issues of land rights and squatters in overburdened Manila.

Bishop, Ryan, John Phillips, and Wei Wei Yeo. 2003. *Postcolonial Urbanism: Southeast Asian Cities and Global Processes.* New York: Routledge. A collection of essays that explores topics such as sexuality, architecture, cinema, and terrorism within the context of global urbanism.

Dale, Ole Johan. 1999. *Urban Planning in Singapore: The Transformation of a City.* Oxford: Oxford University Press. A study of the process of urban planning in Singapore from its early growth on the banks of the Singapore River to the present.

Evers, Hans-Dieter and Rüdiger Korff. 2000. *Southeast Asian Urbanism: The Meaning and Power of*

Social Space. Singapore: Institute of Southeast Asian Studies. Examines a variety of topics, such as the cultural creativity found in slum areas, related to the urbanization process.

Ginsburg, Norton, Bruce Koppel, and T. G. McGee, eds. 1991. *The Extended Metropolis: Settlement Transition in Asia.* Honolulu: University of Hawaii Press. A variety of authors look at various aspects of some of the key cities of Asia.

Logan, William S. 2000. *Hanoi: Biography of a City.* Seattle: University of Washington Press. An exploration of Hanoi's built environment and how the shape of the city reflects changing political, cultural, and economic conditions.

McGee, T. G. 1967. *The Southeast Asian City: A Social Geography of the Primate Cities of Southeast Asia.* New York: Praeger. An urban geography classic.

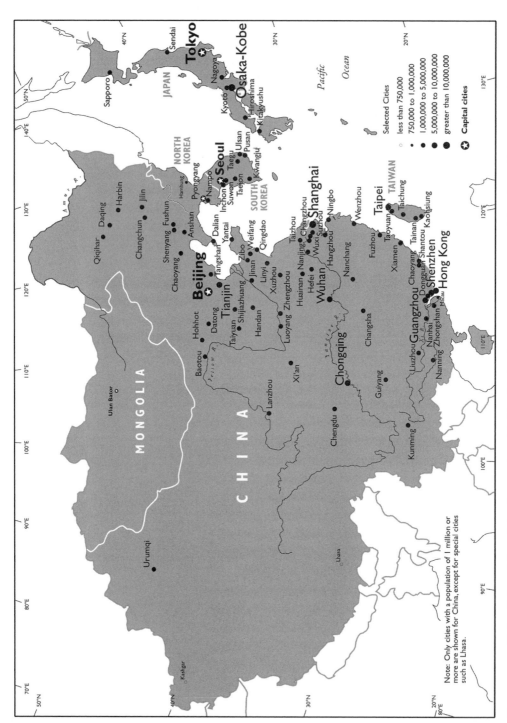

Figure 11.1 Major Cities of East Asia. *Source:* UN, *World Urbanization Prospects: 2005 Revision,* http://www.un.org/esa/
population/publications/WUP2005/2005wup.htm

11

Cities of East Asia
KAM WING CHAN AND ALANA BOLAND

KEY URBAN FACTS

Total Population	1.56 billion
Percent Urban Population	50.2%
Total Urban Population	785 million
Most Urbanized Country	South Korea (83.0%)
Least Urbanized Country	China (47.0%)
Annual Urban Growth Rate	2.18% (China is 2.62%; Remainder is 0.88%)
Number of Megacities	4 cities
Number of Cities of More Than 1 Million	106 cities (China has 89; Remainder have 17)
Three Largest Cities	Tokyo, Shanghai, Osaka-Kobe
World Cities	Tokyo, Osaka, Beijing, Shanghai, Hong Kong, Seoul
Global Cities	Tokyo

KEY CHAPTER THEMES

1. China is one of the original centers of urban development in history and has some of the oldest continuously occupied cities in the world.
2. Colonialism had a less important role in urban development in East Asia compared with other realms of the world, even though many large Chinese cities were treaty ports under colonialism. Hong Kong and Macau were entirely creations of colonialism, which formally ended in this region on the eve of the 21st century.
3. Japan, South Korea, Hong Kong/Macau, and Taiwan are already highly urbanized and deeply involved in the global economy, a status reflected in cities that are already in a "post-industrial" phase, with predominantly service-based economies and major high-tech sectors.
4. Since the late 1970s, China has been rapidly industrializing and urbanizing and is now the "world's factory" and a major player in the global economy.

5. China is unique in being the country with the largest total population on Earth and this population is about half urban in its distribution. It has the largest urban population and the greatest number of million-plus cities of any country in the world and has instituted a two-tier system of rural and urban citizenships.

6. Some of the world's most important world cities are in East Asia, especially Tokyo and Hong Kong. Cities such as Shanghai, Beijing, Seoul, and Taipei are in the second tier but many of them are quickly catching up. The larger cities especially reflect the wealth of the region.

7. North Korea is the lone holdout in East Asia in clinging to a rigid, isolationist, socialist system in its cities and in its economic development, in contrast with China and Mongolia, which are leaving that era to history.

8. Urban development in the region was heavily influenced by the Cold War, which lingers on in the Korean peninsula and in the ongoing tensions between Taiwan and mainland China. International trade has become another major driver of development of large cities in the last two decades.

9. Most major cities of the region show evidence of the concentric zone and multi-nucleic models of urban land use.

10. Most cities of East Asia have experienced the usual urban problems: environmental pollution, income polarization, and, especially in China, migration, both internal and international.

East Asia exudes power and success. Emerging in the past half century to rival the old power centers of the world in North America and Europe, East Asia's cities have been the command centers for the prodigious economic advances in much of the region. The region houses two of the world's largest economies, China and Japan, and is also the world's major exporter. Nowhere is this more evident than in East Asia's great cities, such as Tokyo, Beijing, Shanghai, Seoul, Hong Kong, and Taipei (fig. 11.1). They are also among the largest cities in the world; indeed, Tokyo has been widely recognized as the world's largest metropolis for the last three decades, and Shanghai is now the world's largest cargo port. Compared with the often-struggling urban agglomerations found in other realms of the world, especially in developing countries, East Asia's cities have been relatively more successful in coping with rapid growth and large size. Wealth does make a difference.

The region is still fairly sharply split between China, which is under a one-party system and which is rapidly urbanizing but it remains only about half urban, and most of the rest of the region, which is already more than 70 percent urban. This dichotomy is reflected in many ways, including the character of the cities and their policies and processes, both past and present, which have shaped them. In recent years, the region has witnessed a rapid rise in China's economic power and in South Korea's technological prowess, while

Japan has suffered from a series of fiscal and financial problems and has been devastated in many ways by the recent earthquake-tsunami in 2011.

THE EVOLUTION OF CITIES

The Traditional or Preindustrial City

East Asia, especially China, is one of the original centers of urbanism in world history. Many cities here can trace their origins directly back two millennia or more. One can see interesting parallels with the earliest cities in other cultural realms, with their focus on ceremonial and administrative centers planned in highly formal style to symbolize the beliefs and traditions of the cultures involved.

In its idealized form, the traditional city reflected the ancient Chinese conception of the universe and the role of the emperor as intermediary between heaven and earth. This idealized conception was most apparent in the national capitals, but many elements of this conception (grid layout, highly formalized design, a surrounding wall with strategically placed gates, etc.) could be seen in lesser cities at lower administrative levels. The Tang Dynasty (618–906 A.D.) capital of Changan (present-day Xi'an) was one of the best expressions of the classic Chinese capital city. Inevitably, the demands of modern urban development have necessitated, in the eyes of planners at least, the destruction of most city walls, and thus the removal of a colorful legacy of the past. The sites of the old walls commonly become the routes of new, broad boulevards. One of the few cities whose original wall has been retained almost in its entirety is Xi'an, because of the historic role it plays.

Of all the historic, traditional cities, none is more famous than Beijing (Peking), the present national capital of China. Although a city had existed on the site for centuries, Beijing became significant when it was rebuilt in 1260 by Kublai Khan as his winter capital. It was this Beijing that Marco Polo saw. The city was destroyed with the fall of the Mongols and the establishment of the Ming dynasty in 1368. Nanjing served as national capital briefly after that, but in 1421 the capital was moved back to the rebuilt city (now named Beijing, or "Northern Capital," for the first time), where it has remained with few interruptions since. The Ming capital was composed of four parts: the Imperial Palace (or Forbidden City), the imperial city, the inner city, and the outer city, like a set of nested boxes. It is the former Forbidden City that can still be partially seen within the walls of what is today called the Palace Museum.

The Chinese City as Model: Japan and Korea

Changan was the Chinese national capital at a time when Japan was a newly emerging civilization adopting and adapting many features of China, including city planning. As a result, the Japanese capital cities of the period were modeled after Changan. Indeed, the city as a distinct form first appeared in Japan at this time, beginning with the completion of Keijokyo (now called Nara) in 710. Although Nara today is a rather small prefectural capital, it once represented the grandeur of the Nara period (710–784). Keiankyo (modern-day Kyoto) was to survive as the best example of early Japanese city planning. Serving as national capital from 794 to 1868, when the capital was formally shifted to Edo (now Tokyo), Kyoto still exhibits the original rectangular form, grid pattern, and other features copied from Changan. However, modern urban/industrial growth has greatly increased the size of the city and obscured much of the original form. Moreover, the Chinese city morphology, with its rigid symmetry and formalized symbolism, was alien to the Japanese culture. Even the shortage of level land in Japan tended to work against the full expression of the Chinese city model.

Korea also experienced the importation of Chinese city planning concepts. The Chinese city model was most evident in the national capital of Seoul, which became the premier city of Korea in 1394. The city has never really lost its dominance since. Early maps of Seoul reveal the imprint of Chinese city forms. Those forms were not completely achieved, however, in part because of the rugged landscape around Seoul, which was located in a confining basin just north of the lower Han River. Succeeding centuries of development and rebuilding, especially in the 20th century during the Japanese occupation (1910–1945) and after the Korean War (1950–1953), obliterated most of the original form and architecture of the historic city. A modern commercial and industrial city, one of the largest in the world, has arisen on the ashes of the old city (reinforcing the popular name for Seoul, the "Phoenix City," after the mythological bird that symbolizes immortality). A few relics of the past, such as some of the palaces and a few of the main gates, stand today as a result of restoration efforts.

Colonial Cities

The colonial impact on East Asia was relatively less intrusive than what occurred in Southeast and South Asia, but was notable nonetheless.

First Footholds: The Portuguese and the Dutch

The Portuguese and the Dutch were the first European colonists to arrive in East Asia; the Portuguese were much more important in their impact in this region, the Dutch largely confining themselves to Southeast Asia. Seeking trade and the opportunity to spread Christianity, the Portuguese made some penetration of southern Japan via the port of Nagasaki in the latter 16th century. Their greatest influence was actually an indirect one, through the introduction of firearms and military technology into Japan. This led to the development of stronger private armies among the *daimyo* (feudal rulers) of Japan, which in turn led to the building of large castles in the center of each *daimyo's* domain. These castles, modeled after fortresses in medieval Europe, were commonly located on strategic high points, surrounded by the *daimyo's* retainers and the commercial town. These centers eventually served as the nuclei for many of the cities of modern Japan.

The Portuguese also tried to penetrate China. Reaching Guangzhou (Canton) in 1517, they attempted to establish themselves there for trade purposes, but were forced by the Chinese authorities to accept the small peninsula of Macau, near the mouth of the Pearl River, south of Guangzhou. Chinese authorities walled off the peninsula and rent was paid for the territory until the Portuguese declared it independent from China in 1849. With only 10 sq miles (26 sq km) of land (land reclamation in recent years has added a little to the total), Macau remained the only Portuguese toehold in East Asia, especially after the eclipse of their operations in southern Japan in the 17th century. Macau was most important as a trading center and haven for refugees. The establishment of Hong Kong in the 19th century on the opposite side of the Pearl River estuary signaled the beginning of Macau's slow decline, from which it has never fully recovered (fig 11.3).

In the post-1950 era, Macau survived largely on tourism and gambling (a downscale Asian version of Las Vegas, gangsters and all) In the 1990s, Macau attempted some modest industrialization, as it integrated economically with the Zhuhai Special Economic Zone just across the border. Since reversion to the People's Republic of China (PRC) in 1999, the emphasis has been on both gambling and tourism, with additional investments being made by Nevada gambling interests and by the construction of a number of new, gaudy casinos around the reconstructed harbor front that are starting to pull in large numbers of gamblers, especially *nouveau riche* from a booming China. Right next to the emerging casino quarter lies the historic heart of old Macau that has seen its colonial-era architecture restored and turned into a pedestrian-only area for the tourism business.

The Treaty Ports of China

It was the other Western colonial powers, arriving in the 18th and 19th centuries, that had the greatest impact on the growth of cities in modern China. The most important were the British and Americans; but the French, Germans, Belgians, Russians, and others were also involved, as were the Japanese, who joined the action toward the close of the 19th century.

It all began officially with the Treaty of Nanjing in 1842, which ceded to Britain the island of Hong Kong and the right to reside in five ports—Guangzhou, Xiamen, Fuzhou,

Ningbo, and Shanghai. Further refinements of this treaty in succeeding years gave to the other powers the same rights as the British. A second set of wars and treaties (1856–1860) led to the opening of additional ports. By 1911, approximately 90 cities of China—along the entire coast, up the Changjiang (Yangtze) River valley, in North China, and in Manchuria—with a third of a million foreign residents, were opened as treaty ports or open ports (fig. 11.2).

The treaty ports introduced a new order into traditional Chinese society. The Westerners were there to make money, but they also had the right of extraterritoriality, which guaranteed them protection by Western legal procedures. Gradually taxation, police forces, and other features of municipal government were developed by the colonial countries controlling the treaty ports. China's sovereignty thus was supplanted in the concession areas of treaty ports, as these areas were leased in perpetuity by the foreigners for modest rents paid to the Chinese government.

Shanghai

The most important treaty port was Shanghai ("On the Sea"), which had existed as a small settlement for two millennia. By the 18th century, the city was a medium-sized county seat with a population of about 200,000 and built in traditional city style, with a wall. The deposition of silt by the Changjiang river over the centuries, however, had made Shanghai no longer a port directly fronting the sea. The town was now located about 15 miles (24 km) up the Huangpu River, a minor tributary of the Changjiang.

Western control of Shanghai began with the British concession in 1846 and expanded over

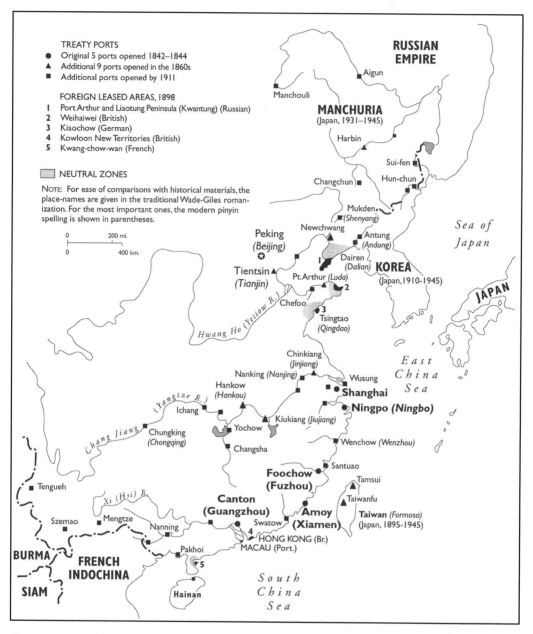

TREATY PORTS
● Original 5 ports opened 1842–1844
▲ Additional 9 ports opened in the 1860s
■ Additional ports opened by 1911

FOREIGN LEASED AREAS, 1898
1 Port Arthur and Liaotung Peninsula (Kwantung) (Russian)
2 Weihaiwei (British)
3 Kiaochow (German)
4 Kowloon New Territories (British)
5 Kwang-chow-wan (French)

▨ NEUTRAL ZONES

NOTE: For ease of comparisons with historical materials, the place-names are given in the traditional Wade-Giles romanization. For the most important ones, the modern pinyin spelling is shown in parentheses.

0   200 mi.
0   400 km.

RUSSIAN EMPIRE

Aigun
Manchouli
MANCHURIA
(Japan, 1931–1945)
Harbin
Sui-fen
Changchun   Hun-chun
Mukden
(Shenyang)
Newchwang   Antung
(Andong)
Peking   Sea of
(Beijing)   Japan
⊛
Tientsin ▲   Dairen
(Tianjin)   Pt. Arthur (Luda)   (Dalian) KOREA
Chefoo   (Japan, 1910-1945)
3
Tsingtao
(Qingdao)
Hwang Ho (Yellow R.)
Chinkiang
(Jinjiang)
Nanking (Nanjing)   Wusung   East
Hankow   China
(Hankou)   Shanghai   Sea
Ichang   Kiukiang (Jiujiang)   Ningpo (Ningbo)
Yangtze R.   Yochow
Chungking   Wenchow (Wenzhou)
(Chongqing)   Changsha
Chang Jiang   Santuao
Tengueh   Foochow   Tamsui
(Fuzhou)   Taiwanfu
Xi (Hsi) R.   Canton   Taiwan (Formosa)
Szemao   Mengtze   (Guangzhou)   Amoy   (Japan, 1895-1945)
Nanning   Swatow   (Xiamen)
BURMA   Pakhoi   4   HONG KONG (Br.)
FRENCH   5   MACAU (Port.)
INDOCHINA   South
SIAM   China
Hainan   Sea

Figure 11.2   Foreign Penetration of China in the 19th and Early 20th Centuries. *Source:* Adapted from J. Fairbank et al., *East Asia: Tradition and Transformation* (Boston: Houghton Mifflin, 1973), 577.

the years to cover most of the city. In 1863, the British and American concession areas were joined to form the Shanghai International Settlement, which in its heyday in the 1920s contained some 60,000 foreigners, the largest concentration of foreigners in China. Shanghai, as well as the other treaty ports, served as magnets for wealthy Chinese entrepreneurs and for millions of impoverished peasants seeking a haven in a disintegrating China. The wealthy Chinese invested in manufacturing and other aspects of the commercial economy; the peasants provided abundant cheap labor. By the end of its first century under Western control, just before World War II, Shanghai handled half of China's foreign trade and had half the country's mechanized factories. The city's population of four million made it one of the largest cities in the world; it was more than twice the size of its nearest rivals, Beijing and Tianjin.

Shanghai profited from its natural locational advantage near the mouth of the Changjiang delta for handling the trade of the largest and most populous river basin in China. During the 20th century, Shanghai's manufacturing was able to compete successfully with that of other manufacturing centers emerging in China, despite the absence of local supplies of raw materials, because of the ease and cheapness of water transport. This was achieved also in spite of a relatively poor site—an area of deep silt deposits, a high water table, poor natural drainage, an insufficient water supply, poor foundations for modern buildings of any great height, and a harbor on a narrow river that required regular dredging to accommodate oceangoing ships. Shanghai thus became one of the best examples in the world of how a superb relative location can trump a poor physical site to create a great city.

There was a colonial impact on other Chinese cities, too, of course. Particularly significant was the Japanese impact. In Manchuria (Northeast China), which the Japanese took over in the 1930s, many of the major cities were modernized and developed along the Western lines that the Japanese had adopted in the development of their own cities after

1868. Japan first expanded into this region to take advantage of the rich resource base of the region, particularly the region's iron, coal, timber, and agricultural products. Industry was concentrated in a string of major cities, particularly Harbin, Changchun, and Shenyang, connected by the railway network that the Japanese built. As with Shanghai, in these cities there arose a new Western-type commercial/industrial city alongside the old traditional Chinese city, which was eventually engulfed and left behind as a remnant of the past.

The Japanese Impact

The Japanese also greatly influenced the urban landscape in their two other colonies in East Asia. During their rule of Taiwan (1895–1945) and Korea (1910–1945), the Japanese introduced essentially the same Western-style urban planning practices, filtered through Japanese eyes, that they later brought to Manchuria's cities. Taipei (renamed Taihoku) was made the colonial capital of Taiwan and transformed from an obscure Chinese provincial capital into a relatively modern city. The city wall was razed, roads and infrastructure were improved, and many colonial government buildings were constructed. The most prominent was the former governor's palace, with its tall, red-brick tower, which still stands in the heart of old Taipei, and is now used as the Presidential Office and executive branch headquarters for Taiwan's democratic government. Like Taipei, Seoul was also transformed to serve the needs of the Japanese colonial rule of the Korean peninsula. In Seoul's case, however, this meant deliberately tearing down traditional palaces and other structures to be replaced by Japanese colonial buildings as part

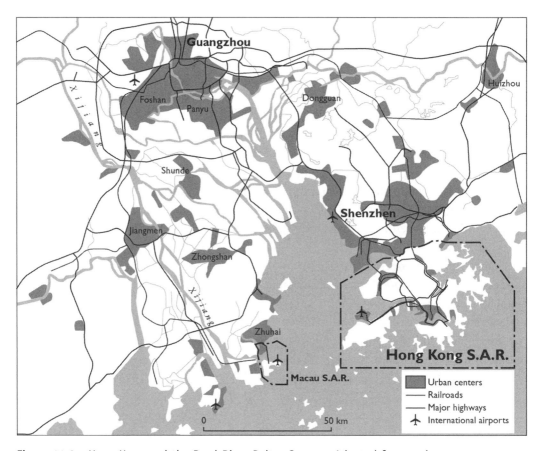

Figure 11.3   Hong Kong and the Pearl River Delta. *Sources:* Adapted from various sources

of a brutal effort to stamp out Korean resistance to Japanese rule.

Hong Kong

Hong Kong ("Fragrant Harbor") differed from other treaty ports in that there was little pretense of Chinese sovereignty there (though the Chinese government insisted after 1949 that Hong Kong was part of China). Hong Kong was ceded to Britain at the same time Shanghai was opened up in the early 1840s. Hong Kong became second only to Shanghai as the most important entrepôt on the China coast during the following century of colonialism.

The importance of Hong Kong was not difficult to find. In 1842, the city began with acquisition of Hong Kong Island (fig. 11.3), a sparsely populated rocky island some 70 miles (113 km) downstream from Guangzhou (Canton). The Kowloon peninsula across the harbor was obtained in a separate treaty in 1858. Then, in 1898, the New Territories— an expanse of islands and land on the large peninsula north of Kowloon—were leased from China for 99 years (hence, reversion to China took place in 1997), creating a total area of about 400 sq miles (1040 sq km) for the entire colony. The site factor that so strongly favored its growth was one of the world's great natural harbors (Victoria Harbor), between Hong Kong Island and Kowloon. Indeed, the advantages of the harbor outweighed the site disadvantages—limited level land for urban expansion, inadequate water supply, and insufficient adjacent farmland to feed the population. The city's location at the mouth of southern China's major drainage basin gave Hong Kong a large hinterland, which greatly expanded when the north-south railway from Beijing was pushed through to Guangzhou in the 1920s. Thus, for about a century, Shanghai and Hong Kong, two great colonial creations, largely dominated China's foreign trade and links with the outside world.

Japan: The Asian Exception

Following the classic capitals of Nara and Kyoto around the 8th century, other cities followed in Japan, principally the centers of feudal clans. Most of these were transitory, but a sizeable number survived into the modern era. One of the best-preserved historic towns today is Kanazawa, on the Sea of Japan in the Hokuriku region. The city was left behind by Japan's modernization after 1868 and escaped the devastation of World War II, since it had no industrial or military importance. Historic preservation since the 1960s has kept much of the lovely 19th century architecture and character of the old city, a rare exception to the urban development patterns found throughout most of Asia.

Japan is referred to as the "Asian exception" because it had only a minor colonial experience internally. Indeed, Japan was itself a major colonial power in Asia. Hence, the urban history of Japan involved an evolution almost directly from the pre-modern, or traditional, city to the modern commercial/industrial city. Japan did have treaty ports and extraterritoriality imposed on it by the Treaty of 1858 with the United States, which led to foreigners residing in Japan as they did in China. However, this colonial phase was short-lived. Japan was able to change its system and reestablish its territorial integrity by emulating rather than resisting the West. Extraterritoriality came formally to an end in 1899, as Japan emerged an equal partner among the Western imperial powers. Gradual political unification during the Tokugawa period (1603–1868) led to the establishment of a permanent network of cities in Japan. The castle town served as the chief catalyst for urban growth. One of the most important of these new castle towns to emerge at this time was Osaka. In 1583, a grand castle was built that served as the nucleus for the city to come. Various policies stimulated the growth of Osaka and other cities, including prohibitions on foreign trade after the mid-1630s, the destruction of minor feudal castles, and prohibitions on building more than one castle in each province. These policies had the effect of consolidating settlements and encouraging civilians to migrate to the more important castle communities.

The new castle towns, such as Osaka, were ideally located (fig. 11.4). Because of their economic and administrative functions, they generally were located on level land near important landscape features that gave the castle towns an advantage for future urban growth. Thus, Osaka emerged as the principal business, financial, and manufacturing center in Tokugawa Japan. The Japanese cities of that period were tied together by a network of highways that stimulated trade and city growth. The most famous of these early roads was the Tokaido Highway, running from Osaka

Figure 11.4   The beautifully restored castle of Hikone, in Shiga Prefecture, central Honshu, is a classic example of an old castle town from Japan's feudal past. (Photo by Jack Williams)

eastward through Nagoya (which emerged as another major commercial and textile manufacturing center) to the most important city of this period and after—Edo (Tokyo).

Among the major cities of Asia, Tokyo was a relative latecomer. It was founded in the 15th century, when a minor feudal lord built a rudimentary castle on a bluff near the sea, about where the Imperial Palace stands today. The site was a good one, however, for a major city—it had a natural harbor, hills that could easily be fortified, and plenty of room on the Kanto Plain behind the city for expansion. Tokyo really got its start, though, a century later, when Ieyasu, the Tokugawa ruler at that time, decided to make Edo his capital. Part of Tokyo still bears the imprint of the grand design that Ieyasu and his descendants laid out. They planned the Imperial enclosure, a vast area of palaces, parks, and moats in the very heart of the city. Much of the land on which central Tokyo stands today was

reclaimed from the bay, a method of urban expansion that was to typify Japanese city building from then on, reflecting the shortage of level land and the need for good port facilities. By the early 17th century, Edo already had a population of 150,000 surrounding the most magnificent castle in Japan. By the 18th century, the population was well over one million, making Edo one of the largest cities in the world.

Edo's growth was based initially on its role as a political center, tied to the other cities by an expanding network of roads. An early dichotomy was established between Osaka, as the business center, and Tokyo. With the restoration of Emperor Meiji in 1868, Japan's modern era began. The emperor's court was moved from Kyoto to Edo, which was renamed Tokyo ("Eastern Capital") to signify its additional role as national political capital. This transfer of political functions, plus the great industrialization and modernization program that was undertaken from the 1870s, gave Tokyo a boost that started it on its astounding growth during the 20th century.

INTERNAL STRUCTURE OF EAST ASIAN CITIES

It is not easy to generalize about the internal structure of cities in East Asia. This is partly because of the basic division between socialist and non-socialist urban systems that characterized the region for so long. It also is because of the lack of fit of Western urban models to even the non-socialist cities of the region. In most of East Asia, and now increasingly also in China after 1979, the forces that have produced and continue to shape cities are similar to those in the Western world, but with modifying local conditions. These forces include: (1) rapid industrialization focused in cities, combined with increasing inequalities between urban and rural residents, leading to high rates of rural-to-urban migration, rates which have now tapered off in the more developed economies (Japan, South Korea, Taiwan), but that are escalating in China (and Mongolia); (2) in the non-socialist world, private ownership of property and dominance of private investment decisions affecting land use; (3) high standards of living and consumption, and increasing reliance on the private automobile (or motorbike) for transportation, in spite of often very good public transport systems; and (4) a relatively high degree of racial (ethnic) homogeneity, but sometimes significant stratification into socioeconomic classes. These and other factors have had varying degrees of impact on the growth of cities, and on how space is used in cities, and hence on the types and severity of problems.

REPRESENTATIVE CITIES

With the exception of (British) Hong Kong and (Portuguese) Macau, the colonial era in East Asia ended with the defeat of Japan in 1945. The emergence of communist governments in the late 1940s in China and North Korea, joining the already communist government of Mongolia (established in the 1920s), split the region into two distinctly different paths of urban (and national) development: the path of the socialist cities of China, North Korea, and Mongolia, versus that of the market-economy cities of Japan, South Korea, Taiwan, Hong Kong, and Macau. This became the basic classification of cities of the region until the late 1970s at least. It also reflected the alignment of the Cold War era in this part of the world. In the late 1970s, China entered the post-Mao or Reform Era, in which market forces began to play a significant role in the economy and urban development. Only North Korea remained basically wedded to a rigid, orthodox "socialist" path, one of the very few in the world at present.

One can also classify the major cities of the region on the basis of function and size. From this perspective, several cities illustrate distinctive types: megalopolises or super-conurbations (Tokyo); recently decolonized cities (Hong Kong); primate cities (Seoul); regional centers (Taipei); and socialist cities undergoing transformation (Beijing and Shanghai).

Tokyo and the Tokaido Megalopolis: Unipolar Concentration

Japan illustrates especially well the phenomenon of super-conurbations or megalopolises. A distinctive feature of Japan's urban pattern is the concentration of its major cities into a

Figure 11.5   The A-Bomb Dome, officially the Hiroshima Peace Memorial, is now on the list of World Heritage Sites. It survived the nuclear explosion that ravaged the city on August 6, 1945, and now stands as a symbol of the need to eliminate nuclear weapons. The 2011 nuclear disaster at Fukujima in 2011 has added new meaning to this memorial. (Photo by George Pomeroy)

relatively small portion of an already small country. In spite of more than a century of industrialization, Japan did not pass the 50 percent urban population figure until after World War II. Between 1950 and 1970, the percentage of people living in cities with a population of 50,000 or more rose from 33 percent to 64 percent, while the total urban population reached 72 percent, a figure comparable to that of the United States. in the same year. In other words, Japan went through a process in 25 years that took many decades to accomplish in the United States. Since 1970, the proportion of urban population has continued to increase, but more slowly, reaching 78 percent by the late 1990s. As the urban population grew dramatically, so did the number and size of Japan's cities. Small towns and villages (those with fewer than 10,000 people) declined sharply in num-

bers and population, while medium and large cities grew rapidly, all the outcome of Japan's phenomenal economic growth after the war.

Almost all major cities are found in the core region. This region consists of a narrow band that begins with the urban node of the tri-cities of Fukuoka, Kitakyushu, and Shimonoseki at the western end of the Great Inland Sea, which separates the major islands of Japan and stretches eastward along both shores of Honshu and Shikoku to the Tokyo region. In between, especially along the southern coast of Honshu, are strings of industrial cities, such as Hiroshima, which grew to importance in the last century (fig. 11.5).

Within this core is an inner core, containing more than 44 percent of Japan's total population of 128 million, known as the Tokaido Megalopolis and consisting of the three urban/industrial nodes: (1) *Keihin*

(Tokyo-Yokohama), with more than 30 million people; (2) *Hanshin* (Osaka-Kobe-Kyoto), with more than 16 million; and (3) *Chukyo* (Nagoya), at nearly 9 million. There really are two distinct parts of Japan, in fact, in a classic core-periphery imbalance—the "developed" capital region centered on Tokyo, and the "underdeveloped" regions (in a relative sense) elsewhere in Japan (fig. 11.6). Rapid growth from the late 1950s through the early 1970s saw a shift from rural areas to big cities. Since then, the migration and growth have been increasingly toward Tokyo at the expense of the rest of the country, including Tokyo's long-standing rival, Osaka, a phenomenon dubbed *unipolar concentration* (*i.e.* urban primacy). Tokyo continues to expand, draining people and capital investment from the other regions, many of which are stagnating. The Osaka region has not seen much growth of new industries to replace the smokestack industries, such as steel and shipbuilding, while Osaka businesses continue to relocate to Tokyo. This combination encourages out-migration and depresses personal consumption. Nagoya has fared somewhat better than Osaka by managing to maintain employment and central city vitality. For people eager to be in the mainstream of modern Japan, living in or near Tokyo is essential. Of course, all these trends may have already been changing in the wake of a cataclysmic 9.0 earthquake and tsunami that took place in Fukujima, about 200 km (124 mi) north of Tokyo, on March 11, 2011 (fig. 11.7). The catastrophe caused about 24,000 deaths (including the missing) and a nuclear meltdown, with serious immediate and still not fully known long-term consequences.

In Japan, Tokyo is the primate city; its dominance is undisputable. Tokyo's roughly one-quarter share of Japan's total population is concentrated in barely 4 percent of the nation's land area. Whatever quantitative measures one employs, Tokyo has a disproportionate share of workers, factories, headquarters of major corporations and financial institutions, institutions of higher education, industrial production, exports, or college students. As the national capital, Tokyo has all the major governmental functions. All 47 prefectural governments have branch offices in Tokyo, in order to maintain effective liaison with the national government. One observer likened the situation to that of the Tokugawa era of the 18th century, when the provincial feudal lords were required to maintain a second household in what was then Edo (as hostages, in effect), as a means of maintaining the power of the Tokugawa Shogunate. The obeisance to Tokyo remains, albeit in a new form.

Tokyo City itself has increased in population only slightly, while the 23 central city wards have actually declined. By contrast, the three key surrounding prefectures (Saitama, Chiba, Kanagawa) have gained significantly, as evidenced by the sprawl into satellite towns and cities. Because land has become so scarce and expensive in Tokyo, virtually the entire perimeter of Tokyo Bay now consists of reclaimed land.

Western influences played some role in the prewar development of Tokyo. Unfortunately, the devastation of the 1923 earthquake and the urgent need for quick rebuilding precluded widespread adoption of Western urban planning ideas. World War II bombing had the same effect. In spite of ambitious plans drawn up immediately after the war, few were implemented. The result was a tendency for the city to grow haphazardly in a manner that

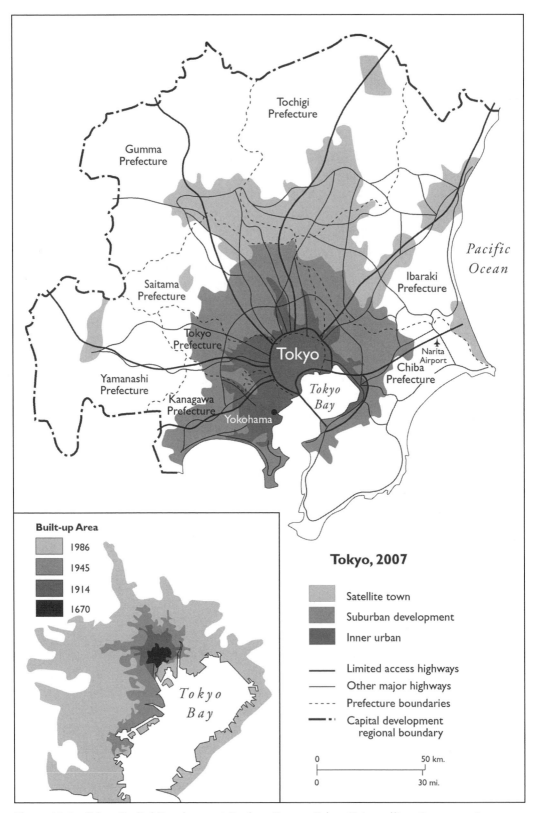

Figure 11.6   Tokyo Capital Development Region. *Source:* Tokyo Metropolitan Government

Figure 11.7   The headlines of the front page of an American newspaper convey the seriousness of the catastrophe at Fukujima—a major earthquake, followed by a tsunami, on March 11, 2011, which devastated a large stretch of Japan's northeast coastal region. *Source: The Seattle Times*, March 12, 2011

resulted in congestion and a disorganized city layout and lack of a clear Central Business District (CBD). In many ways, Tokyo came to resemble Los Angeles, except with a higher population density and vastly superior public transport. Growth has been concentrated around key sub-centers, such as Shinjuku (now the city government headquarters) and

Figure 11.8   The Ginza is the swankiest and most expensive shopping area in Tokyo. Major department stores and hundreds of boutiques compete for attention and wallets in this affluent Japanese society, the richest in Asia. (Photo by Kam Wing Chan)

Shibuya, and along the key transport arteries (rail and expressway) radiating outward from the old historic core of the Imperial Palace (Chiyoda District). The Ginza is the swankiest shopping area in Tokyo (fig. 11.8). The city has a distinct concentric ring pattern intermixed with elements of the multinucleic model. The city also has elements of the American-style "doughnut" model, because of spiraling land costs in the 1980s and desertion by middle-class people seeking affordable suburban housing. They commute to the central city to work in the daytime, but return to the suburbs in the evening. Unlike in U.S. cities, however, there are no serious racial residential patterns, except relating to the minority *burakumin* (untouchables, a social legacy of Japan's feudal past that stubbornly lingers on), and the Korean minority, who tend to live in their own ghettoes. Japan's "bubble" economy burst starting in the early 1990s, when both residential and commercial land prices peaked at 200–300 percent of what real estate prices were in the early 1980s. Land prices slid precipitously downward, returning to early 1980s levels or lower by the early 2000s. The recent catastrophe has added more problems to the already rather down economy. Will Tokyo, and broadly, Japan, bounce back after the Fukujima tsunami, just like they did in 1923 and post-WWII? (box 11.1).

Box 11.1   How Japan Will Reawaken

I thought of my grandmother as I walked the apocalyptic wastelands that had been tidy seaports just days before. Wheelchairs were some of the few recognizable jumbles of metal in the miles upon miles of detritus. Japan is the most rapidly aging society on earth. Because of a low fertility rate, the country's population is expected to shrink one-quarter by 2050. Many of those who perished in the earthquake and tsunami were simply too old to escape. Nursing homes are among the places that most urgently require aid. Elderly Japanese who evacuated to emergency shelters relied on the younger generation for help. This is a nation where Confucian respect for the aged holds. "If it wasn't for the young people in our family, we wouldn't have known anything," says 84-year-old Kimi Sakawaki, whose son surfed the Internet at home to find the evacuation center at Yonezawa gymnasium.

Still, the elderly who survived the March 11 catastrophe know better than any other Japanese how quickly their homeland can revive itself. My grandmother used to recall the U.S. firebombing of Tokyo during World War II, which reduced half the capital to rubble. The pictures of that era bear a haunting resemblance to the images coming out of northeastern Japan today. Yet within two generations, Japan had transformed itself from a defeated land into the world's second largest economy. Incomes were spread relatively equally, with little poverty to speak of. Japan took on a contented, comfortable air.

Source: Hannah Beech/Akaushi, *Time*, March 28, 2011, p. 46

Beijing: The Less Forbidden City

Beijing, the great "Northern Capital" for centuries, was a horizontal, compact city of magnificent architecture and artistic treasures of China's past grandeur when the "New China" began in 1949, although the magnificence of the old city had suffered greatly from general neglect during the century of foreign intrusion and civil war since the 1840s, and from the "revolutionary reconstruction" and "modernization" in the last 60 years. Centered on the former Forbidden City (Imperial Palace), Beijing was renowned for its sophisticated culture and refined society, a status linked to the city's function as the political center of a vast nation. Illustrating the city's influence,

the Beijing dialect (Mandarin) became the national spoken language (*putonghua*) after the collapse of the last dynasty in 1911. However, despite its political and cultural influence, there was little industry and a relatively small population.

In 1949 the city was chosen as the national capital of the new Communist government (Nanjing was the national capital during the Republican era, from the late 1920s to 1949). Since the communist takeover, the city has undergone several waves of demolition and construction, and expansion. Today's Beijing, as an administrative unit, covers a large territory of 6,500 sq miles (16,800 sq km), encompassing an urbanized core (high-density built-up area), surrounded by numerous scattered towns and large stretches of rural area, with a total population of 19.6 million (2010). But this *shi* (municipality or city) is a large administrative region and is not a "metropolitan area" as it is often mistakenly conceived. A rough delineation of the commuting zones (the suburbs) and urbanized area would suggest a much smaller "metropolitan Beijing," in the range of about 2,500–3,000 sq km (965–1,158 sq mi) in area and a population of about 12–13 million. Natural population growth, and, more importantly, net migration to metro Beijing and suburbanization have pushed the metro boundary further outward. During the 1960s and 1970s, migration control to Beijing was among the strictest in the country. Only the well educated and those needed by the central government could move to Beijing; for the rest it remained a "forbidden city."

Functionally, Beijing was also transformed into a *producer* city, as it became one of China's key industrial centers, while retaining its ongoing function as the center of government, culture, and education. Other functions such as commerce and services were greatly curtailed in this command-type economy which was geared to central planning and five-year plans, much like those of the then Soviet Union. Beijing became even more so as the power center of China, analogous to Moscow's role in the former Soviet Union.

The changes to Beijing's traditional urban landscape were enormous in the Maoist era (1949–1976), but continued to be significant in the last two decades. In the pursuit of "destroying the old and building the new" during the Maoist "revolutionary" era, many parts of the old city and the city walls were knocked down to make way for a new socialist capital city. These changes totally shattered the original form of Beijing and forever altered its architectural character. Beijing under Mao was a city with arrow-straight, wide boulevards and huge Stalinist-style state buildings, punctuated by seemingly endless rows of unadorned low-rise apartment blocks for the masses, with the emphasis on uniformity, minimal frills, and lowest possible construction costs. The city center lacked a human scale and was deliberately designed to emphasize the power of the party-state. A huge area in front of the *Tiananmen* (Gate of Heavenly Peace) was cleared to create the largest open square of any city in the world. Tiananmen Square became the staging ground for vast spectacles, parades, and rallies organized by the government. Mao and other party leaders would orchestrate the scene from on top of the gate like a latter-day imperial court. After Mao died in 1976 his body was embalmed and put in a crystal display case inside a huge mausoleum on the south end of Tiananmen Square, exactly along the north-south axis running through the Palace Museum. The parallel with the display of Lenin's body in Red Square in Moscow was intentional, as was the attempt to link Mao with the imperial tradition and the role of Beijing as the center of China. Though the square was designed and used mostly by those in power, it also became the staging ground of mass protests organized by students, intellectuals, and workers, from the famous May Fourth Movement in 1919 to the failed Pro-Democracy Movement in 1989 (fig. 11.9). Elsewhere in the city, the charm of many traditional middle-class courtyard houses in *hutong*, or narrow alleys, in the old city was totally lost in the need to subdivide the housing space for multiple families, but often without the necessary updates and maintenance (fig. 11.10).

Figure 11.9   The Monument of People's Heroes in Tiananmen Square in Beijing looms behind a parading group of PLA soldiers on a chilly morning in December 1989, five months after the crackdown of the student democracy movement that convulsed the country. The square is in the heart of Beijing and symbolic of so much that is modern China. (Photo by Kam Wing Chan)

To be fair, the government was faced with enormous problems, especially in providing housing and meeting basic human needs. Historic preservation tends to take a backseat to more urgent practical needs in almost every country. Moreover, the government kept the Forbidden City (plus some other national treasures, such as the magnificent Temple of Heaven and the fascinating Summer Palace in northwest Beijing) and worked toward restoration of its former grandeur, transforming the huge enclosure into the Palace Museum (*Gugong*)—a collection of former palaces, temples, and other structures, many housing collections of art from the imperial past. The motive was largely political, but the net result was indeed historic preservation. The Palace Museum remains one of the world's top cultural treasures.

China's large cities in the Maoist era were both production (manufacturing) centers and administrative nodes of an economic planning system that focused on national, regional, and local self-reliance. The functions of business and commerce were curtailed. Most cities tried to build relatively comprehensive industrial structures, resulting in much less division of labor and exchanges than would be found in a market economy. The huge surrounding rural areas (often confusingly called "suburban counties") included in the municipalities provided food, mainly vegetables, for the cities. Some satellite towns accommodated the spillover of industries. Without a land market, many self-contained work-unit neighborhoods dominated the landscape of large cities, which expanded in

Figure 11.10   Traditional, single-family courtyard houses in a hutong, or alley, of old Beijing. Many of these today house multiple families. These houses are disappearing rapidly to make room for high-rise apartments and offices. (Photo by Kam Wing Chan)

concentric zones. Beijing was no exception (fig. 11.11).

The new policies after Mao were meant to address some of the above weaknesses and transform Chinese cities through a series of market reforms. Those reforms have brought rising affluence, especially to the coastal provinces and cities, reflected in the urban consumption boom of the 1980s and after. In Beijing, this has resulted in proliferation of new stores and restaurants, including mammoth malls such as the one shown in Figure 11.12. Beijing now also has major commercial/financial districts, such as Xidan, a busy shopping area with modern architecture and expensive shops, and Wangfujing, an old retail strip that went through a major facelift in 1999.

Growing income of the residents, the pent-up demand left from the Mao's era combined with recent frenzied speculation, have continued to generate demand for housing in the last 30 years. The city has pushed outward and it has started to develop a noticeable daily commuting zone consisting of high-rise apartments, luxury detached houses, and often dilapidated "migrant villages" (fig. 11.13).

To attract industry, Beijing's government has established more than a dozen development zones. For example, Zhongguancun, set up in the northwestern part of the city close to top universities, is China's "Silicon Valley." Urban expansion also parallels a noticeable increase in income disparities and social differentiation. In the outskirts at the north of

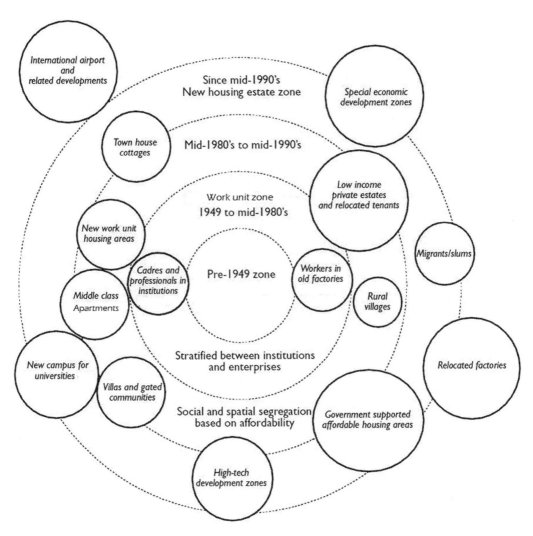

Figure 11.11   Model of the City in the PRC. *Source:* Adapted from Ya Ping Wang, *Urban Poverty, Housing and Social Change in China* (New York: Routledge, 2004)

Beijing, expensive, detached, Western-style bungalow houses have also appeared, catering to expatriates and the new rich. At the same time, with the relaxation in migration controls since the mid-1980s, Beijing now has a large migrant population of several million people. These mostly rural migrants fill many low-level jobs shunned by the locals. However, these migrants are not given legal residency status (*hukou*) of the city and are often denied access to many urban services (box 11.2). In addition, young college graduates who do not have Beijing *hukou* (because they come from other cities) also make their homes here. Several migrant communities have sprung up in Beijing's suburbs, such as "Zhejiang Village" (fig. 11.13) and "Xinjiang Village." These communities are named after the provinces from which most of their residents come from, creating regionally based enclaves within the

Figure 11.12   Golden Resources Shopping Mall in Beijing, one of the largest shopping malls in the world. (Photo by Kam Wing Chan)

city. Living conditions in these migrant villages provide a stark contrast with those of wealthier neighborhoods. In the inner city, local Beijing residents who are laid-off workers from bankrupt state enterprises are gradually forming the new urban poor.

In the last twenty years, the government has many massive-scale programs to "beautify" the city. They range from relocating steel plants from the city and implementing draconian measures to limit the use of automobiles to bring down air and water pollution levels, demolishing large numbers of old *hutong* houses and displacing tens of thousands of the city's poor, to what critics call "image projects" of building numerous very expensive ultra-modern architectural works. The 2008 Olympic Games provided the greatest stimulus for much beautification effort and for improvements in urban infrastructure

(such as newer expressways and subway lines, and a new airport terminal) and construction of several world-class sports stadia.

Shanghai: "New York" of China?

Shanghai is considered by many to be China's most interesting and vibrant city. This is because of its unique colonial heritage and because in many ways it is the center of change and new frontiers in social and economic behavior. Shanghai still is the largest and perhaps the most cosmopolitan city in China and has one of the highest standards of living. Shanghai city itself is part of the Shanghai administrative region, comprising an urbanized core, suburbs, and outlying rural areas, covers a total area of 2,400 sq miles (6,300 sq km) with a population of 23 million (in 2010), including several million

Figure 11.13   Lunchtime in the "Zhejiang Village" on the outskirts of Beijing, a community made up largely of migrants from Zhejiang Province. It is a major garment wholesale center in North China. (Photo by Kam Wing Chan)

inhabitants belonging to the "migrant population" category. According to one estimate, the population in the metropolitan area was about 16.7 million in 2003.

Shanghai came the closest to a true "producer" city in Mao's era. In that era, government revenues relied heavily on taxes on state-owned enterprises (SOEs); Shanghai, being the prime center of SOEs, was a major generator of such revenues. This revenue cash cow was naturally heavily favored by the central government and protected. China's Stalinist-type economic growth strategy gave priority to industry over agriculture, and as such, the strategy was greatly beneficial to Shanghai, which maintained its lead position in the national economy throughout Mao's years. As with many cities in that time, investment poured into industry in Shanghai but little in "non-productive" facilities such as housing and infrastructure. The downtown area, particularly around the Bund, or riverfront district, where the major Western colonial settlers built trading houses, banks, consulates, and hotels, had the look of a 1930s Hollywood movie set (fig. 11.15). In 1934, in the center of the city, the 22-story Park Hotel was built. It was then the tallest building in Asia and remained the city's tallest for almost another half century until 1983, when high rises were again constructed. It was the relative neglect of many cities, including Shanghai, that contributed to the impression that the Maoist policy was "antiurban," although the reality was quite the opposite.

With the reopening of China in the late 1970s, under the open policy, foreign investors returned to China, particularly the coastal zone in Guangdong in the south, this time at the invitation of the Chinese government.

Box 11.2   "Cities with Invisible Walls": The *Hukou* System in China

After the Communist Revolution in 1949, China opted for the Stalinist growth strategy of rapid industrialization based on the extraction of agriculture. This industrialization strategy led China to create, in effect, a dual structure: on the one hand the urban class, whose members worked in the priority and protected industrial sector and who had access to basic social welfare and full citizenship; and on the other hand the peasants, who were tied to the land to produce an agricultural surplus for industrialization and who had to fend for themselves. This in turn required strong mechanisms to prevent a rural exodus. In 1958, a comprehensive *hukou* (household registration) system to control population mobility was codified. Each person has a *hukou* (registration status), classified as "rural" or "urban," and tied to the locale where he or she stayed. The regulation decreed that all internal migration be subject to approval by the relevant local government, but approval was rarely granted. In essence, the *hukou* system functioned as an internal passport system, similar to the *propiska* system used in the former Soviet Union and the *ho khau* system in Vietnam. While old city walls in China had largely been demolished by the late 1950s, the power of this newly erected migration barrier functioned as invisible but effective city walls.

Since the late 1970s, development of markets and the demand for cheap labor for sweatshop productions for the global market have led to easing of some migratory controls. Rural migrants are now allowed to work in cities in low-end jobs shunned by urban residents, but they are still not eligible for basic urban social services and education programs. By the mid-1990s, rural-*hukou* migrant labor had become the backbone of China's export industry and, more generally, of the manufacturing sector. It is estimated that in 2010 about 150 million people were in this category of "rural migrant labor." This two-tier system of urban citizenship and the unequal treatment of the migrant population have drawn much concern from inside and outside China. Because migrant workers do not have local *hukou* in cities, they are often excluded from the population and employment counts, thereby causing serious inaccuracies or errors in many city statistics.

Although Shanghai was designated as one of the 14 "open cities" (for foreign investment) in 1984, Guangdong was really the initial region developed in cooperation with foreign (including Hong Kong) capital. Shanghai received a major impetus for development in 1990 in the aftermath of the 1989 Tiananmen crackdown, as the government struggled to regain foreign investors' confidence. China decided in 1990 to open up Pudong ("East of the Pu," *i.e.*, the Huangpu River, which bisects Shanghai), an essentially farming region on the east side of the old city core (fig. 11.14). The aggressively promoted World Expo 2010 was held in Pudong with a record number of 73 million visitors.

The Pudong development project was as one of China's most ambitious undertakings. It included massive investment in infrastructure (including a new airport and a 30-km

Figure 11.14   Since the early 1990s, Shanghai's new CBD has arisen across the river in Pudong, centered on the futuristic TV observation tower around which ultra-modern skyscrapers have sprung up. Pudong CBD is one of China's two major financial centers. (Photo by Alana Boland)

(18.6 mi) Maglev rail line) and a package of preferential policies, similar to those in China's special economic zones (SEZs), to woo foreign capital. These measures included lower taxes, lease rights on land, and retention of revenues. In Pudong, emphasis was given to high-tech industries and financial services rather than simply export processing. Among the foreign investors, Taiwanese businessmen have several thousand companies in the greater Shanghai region (including nearby cities like Kunshan and Suzhou), with an estimated one quarter million Taiwanese residing and working nearby. A "Little Taipei" has emerged in the Zhangjiang High-Tech Park in Pudong.

With the full backing of the central government, Shanghai has regained its role as China's prime economic (and financial) center in the new era (figs. 11.16–11.18). The Shanghai Stock Exchange, opened in 1990, is not only China's largest stock market (measured by market capitalization) but also the world's fifth in 2010. The skyline of Pudong is intentionally futuristic, with flickering neon-lit glass and steel skyscrapers, including a TV-observation tower that has become an icon for Pudong and the New China (fig. 11.14). It is quite a contrast to the neoclassical Bund on the other side of the river. Shanghai definitely has regained some of its pre-revolutionary glamour. Shops and architecture in some sections have a very cosmopolitan feel and again there is a sizeable expatriate community. There is, of course, a presence of the poor, if they are not the most obvious (box 11.3).

Many problems continue to plague the city and the region: serious inter-jurisdictional rivalries among local governments, inadequate port facilities, an unwise location of the new international airport in Pudong, an overheated real estate development, serious traffic congestion, and severe air and water pollution. Perhaps most important, Shanghai still lacks a well-established legal system that can truly protect citizens' rights and rein in officials from abuses of their powers. These are criticisms that could be directed at all of China today, for that matter.

Figure 11.15   Pre-WWII architectures in the old retail core of Shanghai give a glimpse of the city's past and glamour. (Photo by Kam Wing Chan)

Figure 11.16 and Figure 11.17   These two photos dramatically illustrate the transformation of Shanghai since the end of the Maoist era. In the mid-1970s, Nanjing Road, a key commercial artery of the city, was rather drab, punctuated with only revolutionary slogans. By the end of the 1990s, much of the road had been turned into a glittering mall full of billboards, stores, vendors, and shoppers. (Photos by Jack Williams)

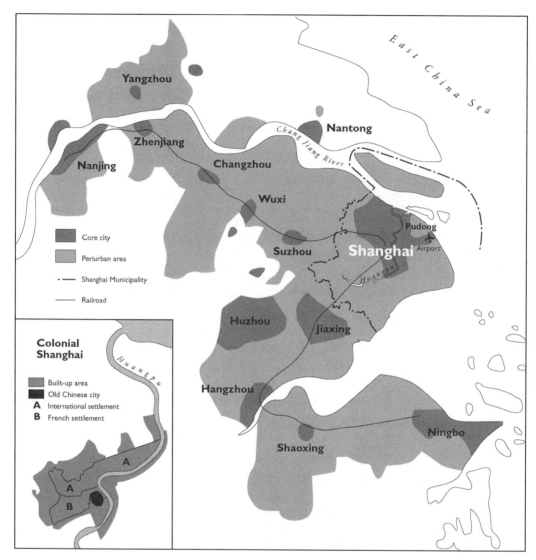

Figure 11.18   The Shanghai Region. *Source:* Adapted from Gu Chaolin, Yu Taofang, and Kam Wing Chan, "Extended Metropolitan Regions: New Feature of Chinese Metropolitan Development in the Age of Globalization," *Planner* 18, no. 2 (2000): 16–20

Hong Kong: Business as Usual?

At the stroke of midnight on June 30, 1997, Hong Kong was officially handed over to China and became the Hong Kong Special Administrative Region (HKSAR). This was an extraordinary historic event, marking the end of the colonial era in Asia and the rise of China's power. Hong Kong was one of the last two colonial enclaves left in all of Asia by the late 20th century. The other colony, Macau, likewise was returned to China by Portugal in December 1999, and became the Macau SAR. Hence, as China entered the new century, its

---

Box 11.3   Shanghai Impressions

Kam Wing Chan, December 17, 2005

Shanghai is enchanting, though quite cold (2°C). This morning I woke up in a warm big bed in the classy Park Hotel to seven windows of glistening modern and ultramodern skyscrapers with a rising sun behind them. What a different Shanghai, from what I knew nine years ago!

It was also intense. I got into Shanghai late last night. After checking into the hotel, I decided to get some food. In a nearby basement food court, I easily found my favorite beef noodle soup and dainty Shanghai wontons. This was a type of food court I had never seen before—with only major chain restaurants. The McDonald's, Burger King, Pizza Hut, and Yonghe (a noodle chain from Taiwan) all were fighting fiercely face-to-face for a share of the pie in a tiny basement in this global city. Behind all this (figuratively) and me (literally), there was a bunch of 6–7 year olds (where are they from?) running around. On closer look, they were snatching and gobbling up unfinished noodles and food left on the tables. One even approached me for the beef in the soup while I was still working on it! They were obviously hungry, but they also seemed to be enjoying themselves doing this, almost like playing a game. They ran and giggled—is that happy or sad?

When I got out of the food court, I realized that I was already on China's busiest pedestrian shopping street, Nanjing Road. Even though it was close to 10 pm, this place was still full of people; and many shops were still open. I decided to take a walk and get a feel for the town. There were many slim women, even in their winter coats, scarves, hats, and boots (they must be in vogue), quite chic, of course. I took some pictures of a few illuminated old Western, mostly neoclassical buildings, but soon I found myself caught in a struggle, having to continuously fight off solicitations by different kinds of strangers—three flirting women "wanting to make friends," four rather persistent "tour guides," and three beggars all within a span of only twenty minutes. What a busy place! With so many years of China experience, I thought I would mingle in a Chinese street crowd; apparently, I couldn't. I probably had chosen the color right, black, for my jacket, but I looked (I was?) a bit older than the average person on this street—I had left my Kangol hat in the hotel room. Or, perhaps any unattached man strolling alone on a Friday night was an obvious target.

---

humiliating experience with foreign colonialism finally ended after almost 160 years.

The 1997 event committed China to guarantee Hong Kong 50 years of complete autonomy in its internal affairs and capitalist system, under a model known as "one country, two systems." The latter refers to the "socialist" system in the PRC (which could likely mean the one-party system and the party control of the economy) and the "*laissez faire*" capitalist system in Hong Kong. China would have control only over Hong Kong's defense and foreign relations, leaving Hong Kong to be "ruled by Hong Kong people." In the years before the handover, the biggest concern was what would happen after 1997. Indeed, in the decade preceding the handover, these concerns triggered an exodus of about half a million Hong Kongers, mostly wealthy professionals, to Canada (especially Vancouver and Toronto), Australia, and the United States.

Back in 1949, when Shanghai and the rest of China fell to the communists, few thought that Hong Kong could long survive under British rule. The UN embargo on China during the Korean War effectively cut off most of Hong Kong's entrepôt trade with China. The population soared from half a million in 1946 to more than 2 million by 1950, with refugees fleeing the raging civil war and communist takeover in China. Huge squatter settlements appeared and the economy was in a shambles. The British, in collaboration with Chinese entrepreneurs, including many wealthy industrialists who had fled Shanghai and other parts of China, began to turn Hong Kong's economy around. They did it by developing products, "Made in Hong Kong," for export. It was a spectacularly successful transformation, with investment pouring in from Japan, the United States, Europe, and the overseas Chinese. Cheap, hardworking labor was available. Site limitations were overcome by massive landfill projects, and fresh water and food were purchased from adjacent Guangdong Province.

During the Cold War period of the 1950s and 1960s, Hong Kong commanded a unique geopolitical position. One of the paradoxes of Hong Kong was that China continued to permit this arch symbol of unrepentant Western capitalism and colonialism to exist and thrive on what was rightly Chinese territory. The Chinese did this partly because Hong Kong made lots of money for them,

too—several billion dollars a year in foreign exchange earned from the PRC's exports to Hong Kong and from investments in banking and commerce. Moreover, a struggling, isolationist, socialist China saw practical advantage in keeping the door open a crack to the outside world, and also in not being responsible for solving Hong Kong's then staggering problems. Banker's Row in Central District came to symbolize the financial powerhouse that Hong Kong had become, with the Bank of China, the Hong Kong Shanghai Banking Corporation (HSBC), and the Chartered Bank of Great Britain lined up side-by-side. The first two are regarded today as among the most important architectural structures of the 20th century and in some ways symbols of Hong Kong's emergence as a true world city.

As one of the top tourist meccas in the world, Hong Kong is, by any standards, a stunning sight whether one is arriving for the first time or the hundredth (fig. 11.19). The skyline is spectacular, especially at night, with its glittering, ultramodern high-rise buildings packed side-by-side along the shoreline and up the hillsides. There is so much money to be made in Hong Kong that every inch of space is extremely valuable and must be used to maximum advantage. The authorities have used remarkable ingenuity in designing the road system and other features of the built environment, especially on crowded Hong Kong Island. Kowloon, on the mainland side, has relatively more land, but even there the intricacy of the urban design is impressive. After the International Airport moved to Chek Lap Kok on the north shore of Lantau Island in 1997 (now widely regarded as one of the finest airports in the world), building height limitations in Kowloon were ended. The Kowloon side is now taking on the Manhattan-like

Figure 11.19   This view of Hong Kong Island, taken from Kowloon across the harbor, dramatically conveys the modernity and wealth of today's Hong Kong. The Central Plaza building towers over the wave-like profile of the Convention Center, where the handover to China took place in 1997. (Photo by Kam Wing Chan)

profile of Hong Kong Island. There seems no limit to the construction boom and demand for new buildings and other structures. Hong Kong remains incredibly dynamic.

Less eye-catching to the average tourist, but themselves impressive social accomplishments, are Hong Kong's public housing and new town programs. The two were begun simultaneously in the 1950s as measures to cope with the large influx of refugees from China. The programs gradually expanded into some of the world's largest. Today, about half of Hong Kong's population of seven million lives in public housing. Indeed, because of much lower rents, public housing has been a major mechanism for decentralizing the population outside of the main urban area. Reclamation has been a main strategy for creating new land for the city. Many of the large

new towns, such as Shatin and Tuen Mun, were built almost totally from scratch.

Until the late 1980s, the economy that had fueled the creation of this machine—where making money, as much and as fast as possible—was heavily based on consumer goods manufacturing and exports, especially textiles, electronics, and toys. The top markets became the United States, Europe, and Japan. But with China's opening in the late 1970s, integration with China escalated rapidly. Hong Kong quickly took advantage of the cheap land and labor in the Pearl River Delta and has steadily outsourced its manufacturing to the delta (while company headquarters remain in Hong Kong). Currently, well over 100,000 Hong Kong-invested enterprises operate in the delta region and employ several million workers. In economic terms, the delta and Hong Kong are now a highly integrated region, one of the world's major global export centers, with Hong Kong serving as the "shop front" and the delta as the "factory" (fig. 11.3). Tens of thousands of people cross the land border between Hong Kong and Shenzhen daily for work or for shopping.

Hong Kong has been the banking and investment center for the China trade, as well as regional headquarters for many international corporations since the late 1970s. Hong Kong has played a crucial role as the intermediary between China and the world, including serving as middleman for Taiwan's huge economic dealings with the PRC (though this role has been weakened in recent years as Taiwan moved to forge closer economic links with the PRC). Tourism remains vital, with the bulk of tourists now coming from the mainland, for whom Hong Kong is often their first taste of the real Western capitalist world, though increasingly, large cities like Shanghai and Beijing in many aspects (shops, skyscrapers, malls, etc.) look very much like Hong Kong, or even fancier.

Immediately after the handover, the Asian financial crisis of 1997–1998 pushed Hong Kong into a recession that continued past the events of 9/11. A series of mishaps further undermined peoples' confidence in the new SAR government. Rising economic competition, including from other Chinese cities (especially Shanghai, Shenzhen, and Guangzhou) made many wonder if Hong Kong could maintain its strength and standard of living. It has since pulled out of that recession, but a host of serious problems continue: (1) There is a widespread perception that Hong Kong lacks strong local political leaders in the current SAR structure, and this weakness has shown in several administrative and policy blunders in such incidents as the public health

scare brought on by the bird flu and outbreak of the serious respiratory disease known by the acronym S.A.R.S., and the issues of right of abode and enactment of an anti-subversion law. (2) Environmental problems, especially air and water pollution, have significantly worsened since 2000. A significant part of pollution is actually related to the recent rapid expansion of manufacturing in the Pearl River Delta. (3) Guangdong and its cities grow ever stronger, and there is already competition of the port and other functions currently dominated by the SAR. Hong Kong still has to define its niche in the evolving global economy of which China is a huge and growing part. (4) The new towns have lost some of their vitality because of change of the economy to a service-based economy. Allegedly patterned after the British "garden city" concept, as self-contained centers where people live and work without need to go to the central city each day, the new towns increasingly have reverted to nothing more than well-planned bedroom suburbs, with workers having sometimes long commutes to jobs. Once thriving factories in the new towns are now largely used as warehouses, or left vacant. (5) The incomes gaps between the rich and the poor have risen to an alarmingly high level—the highest in the developed world, according to a UN 2009 study. Observers have linked the several mass protests—some quite disruptive—in the last two years to this widening wealth gap and to the pervasive perception among the grassroots population that the government was too pro-business.

Perhaps the most critical issue for Hong Kong is maintaining a balance in the relations with mainland China, and it has been a very delicate one. Many people (in and out of Hong Kong) have concerns over the SAR's political and legal autonomy, and hence its ability to protect the city's cherished political and press freedoms, which are not enjoyed by compatriots in the cities of the mainland, and which are often viewed with skepticism by the PRC rulers. Whatever happens, Hong Kong's future is irrevocably tied to that of China.

Taipei: In Search of an Identity?

Although regional centers are found throughout East Asia, a particularly good example is the city of Taipei, which has been emerging from its provincial cocoon in recent years and acquiring some of the aura of a world-class city, following in the footsteps of Hong Kong. There is some ambiguity about how to classify Taipei, because after 1950 it became the "temporary" capital of the Republic of China (ROC) government-in-exile and, as such, experienced phenomenal growth beyond what it might have undergone if it had remained solely the provincial capital of an island province of China. For certain, if the communists had succeeded in capturing Taiwan in 1950, as they had hoped, Taipei would be a vastly different place today, probably something akin to present-day Xiamen across the Taiwan Strait. Instead, Taipei skyrocketed from the modest Japanese colonial capital city of a quarter million in 1945 to the present metropolis of more than 6 million that completely fills the Taipei basin and spills northeast to the port of Keelung, northwest to the coastal town of Tanshui (now a high-rise suburban satellite), and southwest toward Taoyuan and the International Airport. Functionally, the city shifted gears from being a colonial administrative and commercial center to becoming the control center for one of the most dynamic economies in the postwar world.

When the Republic of China's government retreated from mainland China to Taiwan in 1950, the provincial capital was shifted to a new town built expressly for this purpose in central Taiwan, not far from Taichung (fig. 11.20). Taipei was theoretically concerned with "national" affairs and hence had all the national government offices re-created there (transplanted with administrators and legislators from Nanjing). The provincial capital dealt with agriculture and similar island (local) affairs. This artificial dichotomy, designed to preserve the fiction that the ROC government was the legal government of all of China, held until the early 1990s, when the government finally publicly admitted it had no jurisdiction over the mainland. The impact on Taipei over the decades was great, resulting in a large bureaucracy and the construction of national capital-level buildings in the city. Huge tracts of land formerly occupied by the Japanese were taken over by the government after 1945 and the single-party authoritarian political system under the Kuomintang (KMT) allowed the government to develop the city in whatever way it desired, largely free of open public debate. After President Chiang Kai-shek died in 1975, a huge tract of military land in central Taipei was transformed into a gigantic memorial to Chiang, one of the largest public structures in Taiwan. As Taiwan's political system was democratized in the last two decades, this type of memorial and other commemorative fixtures of the KMT rule under Chiang have been under attack, especially during the period between 2000 and 2008 when the island was governed by a pro-Taiwan independence party, the Democratic Progressive Party, which remains a strong opposition party in the island's politics.

Taipei's metropolitan area, with an estimated population of about 6.5 million, has

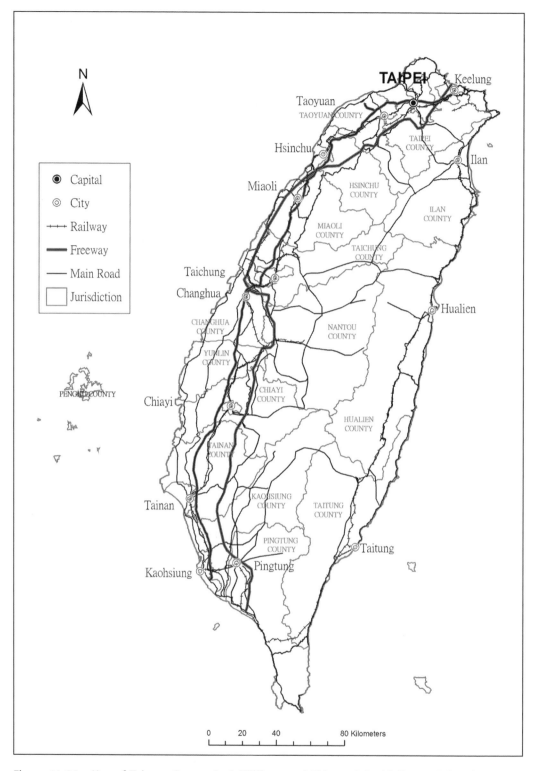

Figure 11.20   Map of Taiwan. *Source:* Jack Williams and Ch'ang-yi David Chang, *Taiwan's Environmental Struggle: Toward a Green Silicon Island* (New York: Routledge, 2008)

come to assume many primate-city functions, although it is statistically only thrice the size of Kaohsiung metropolitan area, the second largest in the island and the main heavy industrial center in the south. No longer touted as the "national" capital, with the gradual dissolution of the provincial government in central Taiwan in the 1990s, Taipei remains overwhelmingly the center of international trade and investment and includes a large expatriate community. Culture, entertainment, and tourism are all focused on Taipei. The Japanese especially like to visit Taiwan, because of its colonial heritage; the city's culture has a distinctly Japanese flavor to it. The metropolitan region also is one of the island's key industrial areas; most of the manufacturing is now concentrated in a number of satellite cities, to the west and south of the capital, but a noticeable portion has now been relocated to mainland China. The old port of Keelung, once the key link with Japan, serves as the port outlet for the north. As with Seoul, most of the city's huge population increase over five decades was the result of in-migration from the densely populated countryside, a migration that in recent years has been primarily toward the suburban satellite cities. Eastern Taipei ("New Taipei"), focused symbolically around the World Trade Center, has seen astounding growth, with hundreds of new high-rise luxury apartment buildings and office towers. Taipei in 2004 also became the site (temporarily) of the world's tallest building, with the opening of the 101-storey *Taipei 101*. Large-scale suburbanization has also taken place, as affluent yuppies have moved to the northern suburbs, or southward. Taipei, to some extent, reflects elements of the concentric zone model and the multi-nucleic model. In some respects, Taipei looks like Seoul on a

smaller scale, with modern buildings, broad, tree-lined boulevards, and a high standard of living. Substantial cleanup and improvements came with the 1990s as the political system was democratized, the environment became an important concern, and urban development became an open topic for public input. An excellent mass rapid transit system helps to ease the transportation crush, composed of hordes of motorbikes and increasing numbers of private automobiles.

Seoul: The "Phoenix" of Primate Cities

Seoul exhibits urban primacy in an especially acute form. The metropolitan area of Seoul is home to about 23 million people, slightly half of South Korea's total population of 48 million, putting it in the top ranks of the world's megacities. Seoul metropolitan region includes the smaller neighboring city of Incheon and the surrounding province of Gyonggi, which has a network of high-rise residential and commercial centers. This is remarkable given that in 1950, Seoul had barely more than one million people, just slightly more than second-ranked Pusan (Busan), the main port on the southeast coast. In 2010, Pusan had only 3.5 million people.

Seoul is the political, cultural, educational, and economic heart of modern South Korea, the nerve center for the powerful state that South Korea has become. As the capital city, Seoul has a large tertiary sector devoted to the national government and to the large military forces that South Korea must maintain. Manufacturing is another major employer, especially in electronics, machinery, and automobiles. The city is also the headquarters for many of the world's leading corporations such as Samsung, LG Group, and Hyundai Motors.

Though not yet on par with Tokyo or New York, Seoul is still regularly included in the ranks of major global cities. Over the past two decades, it has become increasingly cosmopolitan and globally connected via flows of capital and people, changes that are closely related to broader processes of democratization and globalization in South Korea.

The rise of Seoul to become one of the largest cities in the world is surprising, if only from a locational viewpoint. The city's site, midway along the west coast plain of the Korean peninsula, where most people live, was originally a logical place for the national capital of a unified Korea. However, since the division of the peninsula in the late 1940s and the bitter stalemate between North and South Korea since 1953, Seoul's location just 20 miles (32 km) from the demilitarized zone (DMZ) makes the city highly vulnerable. The city was in fact nearly leveled during the savage seesaw fighting in the Korean War, when the North occupied the city twice. In the 1970s, a greenbelt was established that encircled the city (approximately 15 km (9 mi) from the core) and limited its spatial expansion. This was done to contain urban sprawl as well as to protect Seoul from North Korean artillery attacks. Urban planning policies of the 1980s directed developments south, across the Han River, which was a strategy similarly informed by national defense concerns. Suggestions for moving the capital functions to a more southerly and theoretically more defensible site have regularly met with indifference or outright opposition. Instead, despite concerns about the safety of the country's political and economic power center, Seoul has experienced an increasing and seemingly unstoppable concentration of people and economic activity over the past four decades.

Growth since the 1960s has been primarily the result of massive rural-to-urban migration, encouraged by Korea's transformation into an urban/industrial society. The economic takeoff started in the 1960s, as the South embarked on an export-oriented industrialization strategy, much of it concentrated in the Seoul area. Urbanization thus accelerated and the South's population passed the 50 percent urban mark in 1977. Seoul has spilled over into other areas, including south of the Han River and completely filling the basin of the river, and covering an area equal to the island of Singapore. The southerly expansion was made possible by increasing the number of bridges across the Han from the original two to more than 20. The resulting urban landscape appears somewhat "centerless," as dense developments and high-rise buildings are found throughout the entire metropolitan area to create a multinucleic pattern (fig. 11.21).

The expansion away from the old city (north of the Han River) was driven in large part by a series of policy interventions intended to decentralize the economic functions by establishing new areas for residential and industrial development. The first wave of policy-led expansion occurred in the 1970s. It produced a form of suburbanization that involved massive new subdivisions dominated by high-rise apartment complexes, dense retail, new corporate headquarters, and the relocation of many public facilities (belonging both to the central and city governments). There was a second wave of development in the 1990s in response to demands for more affordable housing. Five large-scale new towns were established further from the city center. Their locations, 20–25 km away, were dictated in large part by the rather strict greenbelt policy that had created a barrier to continued

Figure 11.21   In Seoul, the urban landscape includes a dizzying mix of high-rise apartment blocks, mid-level residential and commercial buildings, with pockets of older 1–2 story buildings. This mixed pattern repeats itself throughout the city, accentuating the somewhat centerless feel of the city. (Photo by Lawrence Boland)

growth of Seoul's existing surburban areas. Approximately 20 percent, or about 2 million people, moved out from Seoul's central areas during the period of 1992–1999. With little industry of their own, these new towns functioned at first much like bedroom communities, linked by highways, and later by transit lines, to the central areas of Seoul. The combination of these two waves of expansion led to a "hollowing out" of the older parts of Seoul, north of the Han River, laying the foundation for an urban renewal program that began in the late 1990s. The older parts of the city have thus been transformed through a series of large-scale redevelopment projects intended to improve the quality of housing with construction of new high-rise apartments that in turn allow for more open spaces within the city center. Emblematic of this new emphasis on inner-city revitaliza-tion and improved pedestrian amenities is the dramatic restoration of the Cheonggyecheon urban steam through a freeway-to-greenway conversion project completed in 2005, after just over two years of work.

Changes in the built form and social landscape of Seoul have occurred in parallel with changes to its economy. While its early "take-off" occurred through heavy industrialization based on the availability of cheap labor, Seoul today is now better described as a post-industrial metropolis whose economic development is centered more on financial and corporate services, real estate, and in recent years, high tech and creative industries. With this shift, the city leaders have more explicitly sought to rebrand Seoul as a leading edge high-tech and sustainable city. The green belt, which was criticized in the past for failing to curb urban growth and leading to a greatly extended suburban sprawl, is now seen as an important green space winding through a dense metropolitan area. The satellite towns, once largely dependent on Seoul, have become more independent commercial centers, which has helped to relieve commuter traffic congestion. With improved mass transit, auto-dependency for commuting from suburban areas has been decreasing. Initially built to meet the needs of the 1988 Seoul Olympics, the impressive transit system is now much more extensive, combining subways, trains, and buses throughout the metropolitan area. Mirroring these improvements in transportation networks, Seoul has also transformed itself to be one of most wired cities in the world [see box 11.4]. The city's embrace of digital technologies is evident on the street-level, with the proliferation of neighborhood cyber cafes that have created new social spaces for the city's young people. And with Seoul's promotion of innovation and creative industries, its economic and cultural dynamism in the years to come will likely be linked closely to its digital development.

URBAN PROBLEMS AND THEIR SOLUTIONS

The relatively clear-cut dichotomy between the socialist path of China, North Korea, and Mongolia, and the non-socialist path of the rest of East Asia that characterized the region through the 1970s is no longer valid. China has abandoned orthodox socialism, at least on the surface, since the late 1970s though one-party rule remains. North Korea occasionally hints that it might be tempted to do so also, but then slips back into its Stalinist suspicion of the outside world (box 11.5).

Mongolia, like Russia, abandoned not only a socialist system but also single-party rule, and is now struggling to join the world, too. The colonial era is now completely over in the region. As a result of all these changes, urban problems and solutions take on new guises and, except for North Korea, look somewhat similar across the region.

The Chinese Way

Before 1979, China under communist rule pursued a Stalinist-type industrialization program, suppressing personal consumption (the "nonproductive" side of cities) and squeezing agriculture to help finance rapid industrial growth. To maintain the huge imbalance between city and the countryside, strict controls over migration to the city through the *hukou* (household registration) system was maintained. People in the city had some basic welfare and guaranteed jobs but their lives were also closely monitored through various policing measures. Such an approach kept Chinese people, even in the city, at the bottom rank of the living standards among East Asia's countries and has resulted in mass poverty in the countryside. When Mao died in 1976, the system had to change.

Starting in the late 1970s, China's leaders began to make significant changes in policy across the board, changing some of the key economic policies of the Maoist era though the one-party state remains intact, along with its authoritarian political system and the control of the economy by the state. A major policy change was in the *kaifang* ("opening") policy by opening China to foreigners for investment, trade, tourism, technical assistance, and other economic contacts. The policy of self-reliance was set aside. Rapid

Box 11.4   Digital Seoul

While Seoul's built environment might resemble other sprawling metropolises of Asia, where Seoul stands out as unique is in terms of its virtual environment. Considered the most wired city in the world, Seoul offers residents and visitors unprecedented levels of connectivity via high-speed cable, and increasingly, through various wireless technologies. This development began after the 1997 Asian Financial Crisis, when the government began to emphasize high-tech industry and the digital economy. South Korea is now home to some of the largest tele-communication and technology companies such as LG, Samsung, and KT Corporation. The high-speed Internet access rate in homes is impressive (over 90 percent), and an expansive wireless network has made it possible to get online from almost anywhere in the city for a small fee.

This coverage increases availability of information and entertainment, but has also be-come a key element in the city's development projects. Electricity grids are becoming "smart grids" that send information between supply companies and consumers, and allow people to control home systems while away to help save energy. The city is also setting up a system called "Ubiquitous Seoul," which allows residents to check air quality and traffic conditions from their own mobile devices or public access touch screens through the city. A new Bus Information Management System tracks the movement of every bus and provides arrival and waiting time estimates directly to commuters' mobile devices. Similar information is also available throughout Seoul's massive subway system, along with standard mobile voice and data services. These initiatives support the city's broader efforts to make mass transit a more convenient and attractive alternative to private cars.

The pervasiveness of the Internet in Seoul is also visible on city streets, where one can find cybercafés nestled in low rent spaces, such as the upper floors or basements of small commercial buildings. There are over 26,000 online game rooms, or "PC bangs" (PC rooms), throughout the city, where young adults hang out to play online games, video-chat, and just socialize. These PC bangs are the most recent variant on a much longer tradition in Korean communities, where a broad variety of bangs have offered residents an alternative social space outside of their typically small homes. The compact and dense nature of housing devel-opment in Seoul has also helped to accelerate broadband deployment by making it cheaper to connect residents to switching systems. The fact that much of the city's housing stock is relatively new also bodes well for an expansion of broadband infrastructure, as any necessary upgrades are easier to do in these more modern building structures.

Not to be outdone, many other cities in East Asia are also pursuing policies to help expand their digital networks. In Taipei, for example, much of the city is blanketed by one of the world's largest wi-fi grids. Beijing recently launched a similar network throughout one of its business districts. Like many other cities in the region, Beijing's recently constructed subway lines are design to allow passengers access to high quality 3G data and voice transmission through most its expanding underground transportation network. While not yet as developed as Seoul, other cities of the region are also developing state-of-the-art telecommunications systems to help with urban management and support growth in the new creative and digital economies.

Box 11.5   Isolation: Peripheral Cities

Isolation can be a huge handicap for cities, but isolation is a relative concept, in that it can be caused by both natural and man-made factors. Four cities in East Asia—Pyongyang, Ulan Bator, Urumqi, and Lhasa—play important roles in their respective regions, yet really are isolated—*i.e.,* by their peripheral geography and in terms of their linkages with the rest of the world.

Pyongyang is perhaps the biggest anomaly of the four. The government of North Korea rules this austere, reclusive nation of some 22 million from the capital city of Pyongyang. At an estimated 3.5 million in the metro region, Pyongyang is three times larger, in classic primacy fashion, than the next two largest cities, Nampo and Hamhung. This is hardly a surprise, given that the centrally planned, Stalinist system still hangs on, long after the Soviet Union, Maoist China, and communist Mongolia saw the light. Leveled to the ground during the Korean War (1950–1953), Pyongyang was totally rebuilt in the true socialist city model, with broad boulevards and massive government buildings, a superficially modern showcase of socialist dogma, but a city that gets terrible reviews from the few foreigners who have managed to visit. Pyongyang is little more than a grandiose monument to the whims of North Korea's autocratic rulers. The city may be geographically sited in the heart of East Asia, but it might as well be in the middle of Siberia.

By contrast, Mongolia's capital city of Ulan Bator (Ulaanbataar), although much smaller at about 800,000 people, is the center of a country now doing everything possible to integrate with the outside world. The main problems are Mongolia's tiny population (2.9 million), sprawling land area, and geographical isolation. Ulan Bator is also a primate city, many times larger than number two, Darkhan (about 71,000). As Mongolia sheds its socialist past and democratizes, the country is rapidly urbanizing and trying to find alternatives to the processing of animal products for its small economy. Tourism is growing, but industry is never likely to be a significant one here. It will be difficult to overcome the country's geographical limitations, and hence Ulan Bator will likely remain largely a minor regional center.

Urumqi is also a regional capital, for the Xinjiang Autonomous Region in China. An ancient city, Urumqi has become a booming metropolis of over 2 million, with a largely Han Chinese population, as the center of China's administration and development of Xinjiang. As such, Urumqi in recent decades has increasingly taken on the character and physical appearance of a Chinese city, very similar to those found throughout the eastern, more populous part of the country. Although geographically the most isolated of our four peripheral cities, Urumqi is actually very much in touch with the outside world, largely because of China's prodigious economic growth in recent decades. The city is the focal point of large-scale tourism, industrialization, and development of the region's oil and other resources. Urumqi is also the center of efforts by the Chinese government to contain separatist tendencies among Xinjiang's largely Muslim population (especially among the Uigur). Hence, the city's geopolitical importance may well exceed its economic role.

Figure 11.22   Lhasa, the capital of Tibet, is dominated by the Potala Palace, which used to be the home of Tibet's traditional ruler, the Dalai Lama. (Photo by George Pomeroy)

Lhasa, the capital city of Tibet, is similar in many ways to Urumqi, although much smaller (under 300,000 in the urban area). If not for Chinese rule, Lhasa would be even more geographically isolated as one of the world's highest cities (nearly 12,000 ft. elevation; 3,658 m). Also an ancient city and center of Tibet's unique Buddhist culture under the Dalai Lama (in exile in India), Lhasa was thrust into the modern world with China's takeover in the 1950s and became the focal point of China's efforts to contain Tibetan separatism, which drew much international attention in the process. Like Urumqi, Lhasa is rapidly becoming essentially a Chinese city, with the Han Chinese population steadily increasing, and Chinese urban forms displacing much that was traditional and Tibetan. Tibet remains one of China's poorest regions, and Lhasa's economy is largely dependent on tourism and services, subsidized by the Beijing government in its determination to ensure that peripheral regions like Xinjiang and Tibet (and their key cities) remain firmly within the People's Republic of China (PRC). One powerful demonstration of this effort was the opening in 2006 of the first railway linking Tibet with the rest of China (via Qinghai province to the north). Tibetan nationalists view the railway as one more tentacle of Beijing's grip. Beijing, in turn, sees the railway as an essential tool to further bring Tibet into the modern world and irrevocably into the PRC.

In sum, these four cities, in their historical context as well as recent development, illustrate that isolation can be imposed by either nature or by government, but overcoming isolation is no easy task.

growth in links with the outside world had a profound impact, but especially on cities and urban development in the coastal zone, which was earmarked for preferential treatment. The establishment of export-processing zones with concessionary tax policies to attract foreign investment included the designation of four special economic zones (SEZs) (Shenzhen, Zhuhai, Xiamen, and Shantou) in Guangdong and Fujian in 1979 and of 14 "coastal open cities" in 1984. At the end of the 1980s, Hainan Island became both a new province and the fifth SEZ, while Shanghai's Pudong district joined the category of "open" zones in 1990. By the mid-1990s, practically the entire coastal region contained thousands of "open zones," vying for foreign investments.

Another major change is the de-collectivization of agriculture and the return to private smallholdings (under the Household Responsibility System) in the early 1980s. This shift helped raise labor productivity quite drastically and brought a better quality of life for hundreds of millions of peasants. As many laborers were no longer needed on the farm, pressure was put on the government to relax restrictions on internal migration. Out of this easing of retrictions has emerged the "floating population" of migrant workers, estimated at 160 million in 2010, who provide the plentiful low-cost labor to make China the "world's factory." Migrants fill many industrial and service jobs shunned by urban workers, but under the *hukou* policy these migrants do not have the same citizens' rights and social benefits as ordinary urban residents do. This two-tier system of urban citizenship and unequal treatment of the migrant population has become a major concern and the source of many problems in the cities (box 11.2).

One of the most notable consequences of "opening" and rural de-collectivization was the creation of the new city of Shenzhen, a major export-processing center just across the border from Hong Kong. A small village at the border-crossing point became, within two decades, a city of several millions, with a large proportion being young migrants (without the local *hukou* status). Today its population is over 10 million, much larger than its neighbor, Hong Kong. Shenzhen today makes a striking sight when viewed from one of the vantage points in Hong Kong's New Territories. Most of the land along the Hong Kong side is undeveloped farmland or nature preserves, while immediately on the other side rises a sprawling Manhattan-like urban landscape.

The negative side of China's new policy, a combination of capitalist approach and one-party rule, has generated new imbalances between rural and urban areas, between provinces and different regions, and between socioeconomic classes. To some urbanites, life has unquestionably improved and appears increasingly similar to that of the rest of East Asia. Many big cities now offer a greater variety and quality of goods and services, including many luxury ones, a huge contrast from the Maoist years. In fact, some sections of many large cities today have the look of wealthier cities like Hong Kong or Taipei (fig. 11.26). To many others, urban life has also become a hectic struggle to make ends meet, especially in the face of escalating housing prices in recent years. Economic and social polarization is definitely on the rise. There is an expanding class of urban poor in China, consisting of migrants and laid off SOE, mostly older, workers. Unemployment remains a serious issue even in the face of a

Figure 11.23   Migrant workers shine shoes on a street in Wuhan, the largest city in central China. "Rural migrant workers," numbering about 150 million in 2010, are now everywhere in China's major cities, performing all kinds of low-skilled labor. The large pool of cheap labor is crucial to China's success in being the "world's factory." (Photo by Kam Wing Chan)

rapidly aging population: China simply still has too many more people than its economy can provide jobs for (fig. 11.23). Surplus labor in the countryside, especially in the older age groups remains serious. Moreover, impacting virtually everyone, rich or poor, is the critical state of the environment. Reputedly, a majority of the 10 most polluted major cities in the world are found in China today. This fact has not gone unnoticed as government at all levels has invested heavily in improving urban environmental conditions (see discussion later).

Other Paths in East Asia

As with big cities around the world, the industrial cities of East Asia are experiencing profound problems of overcrowding, pollution, traffic congestion, crime, and shortages of affordable housing and other amenities. This has been a region of impressive economic growth and advancement in recent decades, so that many residents of these cities have standards of living among the highest in the world (except for housing). Retail stores of all types provide every conceivable consumer good for affluent residents. At night, the cities glitter with eye-popping displays of neon lights, nowhere more dazzling than in Japan. Behind all this, in recent years there has been a serious concern of the widening wealth inequality. The poor are the elderly, the immigrants, or rural migrants. This is especially serious in China, where internal migrant workers constitute a whopping population size of about 150 million in 2010. Many of them work and live in the city in at a very minimal standard of living.

Expensive land is a major constraint to the development of these cities. Thus, the major cities are increasingly following in the footsteps of most other large cities with high-rise syndrome. There is even a growing competition among the cities of the region to see which can build the tallest skyscraper, as if having the tallest building somehow conveys status and superiority. Shanghai, Hong Kong, Taipei, Seoul, and others are involved in this one-upmanship. Even Japan's cities, long characterized by relatively low skylines (because of earthquake hazards), have succumbed to the trend toward high-rise construction, such as in the cluster of 50-plus-story buildings centered around the city government complex in Tokyo's Shinjuku District, or the new high-rise profile in the port of Yokohama. Japanese cities, and now increasingly other cities in the region, also make maximum use of underground space, with enormous, complex underground malls interconnected by subway systems.

Movement outward from the central city (suburbanization) is the only other alternative to going upward or downward. New communities have sprung up, including bedroom towns where people, for less money, can obtain better housing with cleaner air and less noise, even though doing so often means longer commutes to work. Fortunately, most of the large cities have developed relatively good public-transport systems. Nonetheless, the automobile culture is spreading rapidly, with the private automobile purchased as much for status as for convenience. The automobile culture first took hold in Japan in the 1960s, but other countries have followed suit and even China is now firmly on the private automobile bandwagon and has now replaced the United States as the world's largest market for automobiles in 2010.

Closing the Gap: Decentralization in Japan

The Japanese have been struggling for several decades to decentralize their urban system and reduce the relative dominance of Tokyo, by and large without success. The task really has two dimensions: improving living conditions in Tokyo, and physically decentralizing the urban system. Within the Tokyo area, the solution to overcrowding and the still high cost of land lies in finding new land, such as through continued landfill projects to expand the shoreline and make greater use of Tokyo Bay, as well as moving further outward toward less-developed areas in the Tokyo region. A prime example of the latter is the Tokyo Bay Aqua Line, an expressway across the middle of Tokyo Bay connecting Tokyo (in the north and west) with Chiba Prefecture that is designed to enhance development along the eastern and southern shore. The hope is that Chiba Prefecture, occupying the whole peninsula opposite Tokyo City, will be the new high-growth area for the Tokyo region, anchored around three new core cities—a high-tech Kazusa Akademia Park, Makuhari New City, and Narita Airport.

An alternative supported by many is to decentralize; but how to decentralize is a tough question over which there is anything but consensus. Over the decades, Japan has had a succession of National Development Plans, all of which have addressed in some way the need for more balanced regional development. Proposals for relocating the national capital out of Tokyo have gone nowhere. Various kinds of sub-center ideas have been tried; however, none has been substantially successful in reducing the drawing power of Tokyo. Neither has the decline of regions such as the Kansai been stopped, in spite of efforts such as the opening of the new Kansai International Airport on an artificial island off the coast of Osaka in 1990, in an effort to give Tokyo's Narita some competition. In 2001 the government launched a major effort to revitalize the Kansai, with limited results so far. Of course, as stated earlier, the recent Fukujima earthquake and the ensuing damage and nuclear contamination will lead Japan to rethink these efforts.

Seoul: The Problems of Primacy

Somewhat like its larger cousin, Tokyo, Seoul has suffered from the typical problems of urban primacy such as traffic congestion and housing shortages. Its pattern of dispersed development over the past few decades was meant to address these and related prob-lems. Much of the expansion was channeled into master-planned new towns outside of central Seoul in the 1990s, though the areas surrounding these new towns also saw haphazard and unplanned development due to land speculation. In the first years, the new towns were criticized for their lack of self-sufficiency, which meant traffic congestion became more severe as residents still regularly commuted to central Seoul for work, shopping, services, and entertainment. Over time however, these new towns have developed their own commercial, business, and educational facilities, meaning residents are making fewer trips into the central areas. This, in combination with ongoing expansion of mass transit networks outside of the city's core and the establishment of dedicated bus lanes, has seen improvements in the traffic conditions throughout the Seoul metropolitan region. Interestingly, while the planned new towns are dense enough to support mass transit, this is not the case with the smaller, more sporadic suburban developments that have sprung up around the new towns. It is these elements of Seoul's suburban landscapes that are a source of many of the development-related problems that plague the city.

While the city and national governments have sought to counterbalance the centralizing tendency of the Seoul's development, since the early 2000's, the city government has also been trying to avoid a decline in the core areas. However, in its efforts to revitalize older areas of the city through densification (by replacing two- to five-story buildings with high rise apartments) and provision of more outdoor open spaces, many lower income residents have been displaced from their old neighborhoods. Some areas have seen massive displacement of lower income tenants and

Figure 11.24   Cheonggyecheon Stream Restoration project in Seoul downtown. It took only three years to replace the road with this 5.8 km long urban
greenway that provides a unique open space cutting through the center of the
city. (Photo by Lawrence Boland)

small family-run businesses, unable to purchase or rent newly constructed units. Large
scale renewal projects such as the Cheonggyecheon Stream Restoration project arguably
further accelerate the gentrification process in
central Seoul by improving amenity spaces,
which in turn increases property values and
rent (fig. 11.24). There have been disputes surrounding the gentrification process and how
compensations packages are awarded. A 2009
protest against forced evictions in the Yongsan
area of Seoul turned violent, leaving six people
dead. As with other cities undergoing rapid
and dramatic redevelopment, it is not clear
whether Seoul will be able retain the social mix
that had traditionally defined life in its smaller,
more traditional neighborhoods as these areas
are replaced by the high-rise apartments and
broad boulevards that have become the trademark of the new East Asian city.

Taipei: Toward Balanced
Regional Development

Taipei has made dramatic progress in recent
years in solving some of its urban problems.
Completion of the Mass Rapid Transit system
and stepped-up enforcement of traffic rules
have brought order to what was once one
of the worst cities in Asia in terms of traffic
chaos. Pollution (especially air pollution) has
been drastically cut through various programs. While housing is still expensive, it is
becoming relatively more affordable. Overall,
the city is cleaner and decidedly a better place
in which to live or raise a family. Although
many people have moved to the suburbs, a
large residential population still lives within
the central city, so that the "doughnut" model
does not fit Taipei. The multiple nuclei model
is perhaps more applicable.

In part to solve the problems of overcrowded Taipei, the national government embarked on island-wide regional planning in the early 1970s. The end result was a development plan that divided the island into four planning regions, each focused around a key city. The Northern Region, centered on Taipei, has about 40 percent of the island's total population. Through a variety of policies, including rural industrialization, massive investment in infrastructure, and programs to enhance the quality of life and the economic base of other cities and towns, Taiwan has managed to slow the growth of Taipei (never as severe as the dominance of Tokyo or Seoul in their countries) and diffuse some of the urbanization. Taiwan, for instance, has developed its own sleek version of Japan's "bullet" train; it started operation in late 2006 and has cut the travel time between Taipei and Kaohsiung to about 90 minutes. The rivalry between Taipei and Kaohsiung is often fierce. Historically, these two cities have been controlled by the two main competing parties, the KMT and the DPP, which have their power bases in the North and the South, respectively. The current central government policy under the KMT has facilitated significant outsourcing of Taiwan's industry to mainland China to take advantage of China's cheap labor and land costs. As one can see from the experience of other East Asian economies, this has accelerated the economic structuring in the island; and it will have a noticeable impact on Taiwan's urban and economic structures.

The Greening of East Asian Cities

Cities of East Asia have faced many environmental challenges as they have carved out their different paths of development. Growing populations, rapid industrialization, and in recent decades, a general shift towards high consumption lifestyles, have escalated strains on air, water and land resources. It is only in past few years that cities in the region have begun to pay greater attention to quality of life concerns of residents. This change is linked to increasing awareness of the health and social costs of environmental degradation. It is also linked to broader transformations in the urban economies, dominated now by the service and the high tech industries. A similar development pattern has occurred in cities around the world, though for East Asia, it is the speed of change that is most remarkable. And perhaps, nowhere is this more true than in the cities of China.

In various listings of most polluted cities, China is often over represented. A commonly cited example comes from a 2006 World Bank announcement that China is home to 16 of the 20 cities with the most polluted air. The seriousness of the problem is recognized by all levels government, as demonstrated by the implementation in recent years various urban sustainability policies. One common strategy has been to close or move polluting factories currently located in urban areas and, in the case of northern cities, to replace the coal-burning boilers used for winter heating. Other programs, often with international funding, involve water improvement initiatives such as construction of new wastewater treatment plants and cleanup urban waterways through dredging of sludge and improved controls on industrial and agricultural activities. Cities have increased space for parks and local greenery as another high-profile strategy for improving the environmental quality of cities, while new subway construction has increased public transit capacity in many major cities.

Shanghai has been the most ambitious city, having in just fifteen years become one of the world's largest subway networks.

Clean-up efforts have paid off in many cities, particularly at the street level, which is where residents experience the direct impacts of environmental quality on their everyday lives. However, in the switch from "productive to "consumptive" cities, some problems have proved more intractable. Most notable in this regard have been the incredible rise in automobile use and the seemingly endless proliferation of solid waste on the outskirts of major cities. In Beijing alone, an average of 2,000 new cars were registered every day; even second-tier cities, such as Chengdu in western China, have seen a similar rush to buy private cars, with over 1000 new vehicles hitting the streets each day. There are also problems with once agricultural land in areas around cities being pulled out of food production to make way for new housing developments, golf courses, and large-scale factories. Other urban problems have their origins beyond the city boundaries. Emblematic here are the dusts storms that plague Beijing and other cities of northern China and the Korean Peninsula. These notorious dust storms have on occasion also swept through to reach more southern cities, including Shanghai. However, while perhaps most visible, dust from construction and sand storms, is arguably less worrisome than the more dangerous small sized pollutants emitted by the rising number of private automobiles and factories still operating within or near urban areas.

Similar trends characterize the mixed environmental records in other cities of East Asia. Cities such as Hong Kong, Tokyo, Taipei, and Seoul have all undergone similar transformations, namely a move towards less polluting forms of economic activity matched by increasing investment in and demand for cleaner air and water, and more environmental amenities such as public green spaces. In the case of Seoul, the plan to restore Cheonggyecheon was part of the then new mayor's strategy to "rebrand" the city by adopting a sustainability paradigm for urban management. It was this project that also fit well with urban revitalization efforts that were seeking to bring new life to what had become the somewhat barren streetscapes of the city's older core. While the project was initially criticized, after just over two years, 6 km (3.7 mi) of freeway cutting through the city center were replaced by a restored stream, flanked by a linear park and walkway for pedestrians and cyclists, with car traffic limited to roads on the sides of the greenway. To compensate for the loss of the freeway, the city simultaneously expanded its dedicated bus lanes serving the same area and improved links to Seoul's existing subway system. Both in the speed of its implementation and its generally positive effects on the downtown neighborhoods, the restoration of Cheonggyecheon has thus far earned praise for its direct environmental effects and the indirect benefits associated with increased flows of people back into the old city core, which had lost much of its street level life due to earlier waves of suburbanization.

While the restoration of Seoul's urban stream represents an ambitious multipurpose urban redevelopment scheme with obvious environmental benefits, East Asia is becoming known for even more ambitious urban environmental initiatives that are based on the creation of "eco-cities," some of which are to be built from scratch. These master planned cities include the New Songdo City (near Seoul), Dongtan (near Shanghai), and the Tianjin Eco-City (near Beijing). In most cases, the plans call for integration of residential, commercial, and industrial developments, incorporating cutting-edge green technologies with an emphasis on high-tech research and development. These eco-city projects are meant to be large and spectacular, with residential population targets in the hundreds of thousands. Such large numbers are consistent with the high-rise and high-density development pattern that is characteristic of urbanization in the region more generally. Among these planned eco-cities, Dongtan was the most high-profile project, though it appears to have been stalled, if not cancelled. The others have moved forward with construction, working their way through planning and consultations, often in collaboration with international design and engineering consortiums. Proponents of these projects argue that even if they are slow to be realized, they have positive impacts on the "urban vision" in other cities and help drive development of green urban technologies, such as low-carbon and resource-efficient heating. Some critics worry however that these utopian green communities will not achieve their ambitious goals, because they are more about improving the image of government officials and the design firms than promoting the projects. Other critics worry that even if built as planned, the new eco-cities will be accessible only to highly educated and wealthier residents, providing limited opportunity for the poor to enjoy the benefits of East Asia's sustainable urban futures.

PROSPECTS

Urban dwellers in many poorer cities around the world must look with some envy at the more prosperous cities of East Asia. To citizens of the region, however, and especially to urban planners, the overall problem of most cities of East Asia is how far short the cities still fall from expectations. As leading Japanese observers put it:

- "Japan's foremost urban centers lack anything resembling the character and depth of their European counterparts. Instead, they seem to be forever under construction."
- "There is huge potential demand for urban redevelopment."
- "Japanese city planning shows little vision regarding a living environment."

These may be excessively harsh criticisms from idealistic planners (perhaps biased because of their Western training). Much the same could be said of the rest of the region. The continuous demolition and construction that leaves little history and character are the price of rapid growth and economic success. But it is fair to say that cities of East Asia are also in many respects ingenious designs by generations of people trying to create hospitable urban habitats in the relatively unfavorable environment of high population pressure and scarce land.

So where do these countries and cities go from here? East Asia is likely going to be a region of continuing high economic growth with most countries pursuing a pro-business strategy. Urbanization will continue in China. There are vast amounts of capital in the region for urban development, for building urban infrastructure, and for expanding the use of information technology. But how to marshal that capital to maximize the quality of urban life and promote more equitable growth will remain major challenges facing all of these states. Undoubtedly, state-of-the-art technology, such as the bullet-trains, first pioneered in Japan, are now spreading quickly in China (figure 11.25) and other parts of the region, but will it benefit the masses or just the more affluent?

An analysis of China's current mammoth high-speed train project, funded almost totally

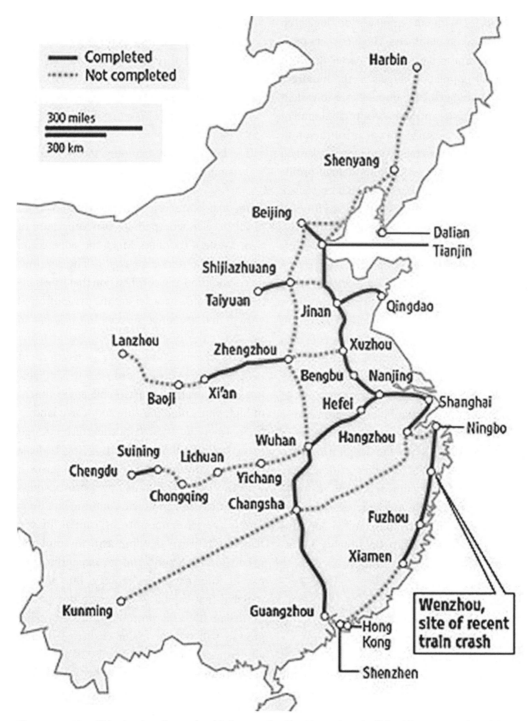

Figure 11.25   China's plan for major high-speed rails. More than half had been completed by mid-2011. *Source:* World Bank

by the government, shows that the project has served to benefit enormously the rich and the middle class (the top 25 percent of the income earners), who can afford the far higher fares. The project has also negatively impacted the lower income groups because many low-cost "slow trains" have been taken off the rails as a result of the arrival of the sleek bullet trains. The extra troubles and problems many migrants had to face in China's 2011 *chunyun*—the "spring movement" of an estimated 110 million people by train and bus home and back during the Chinese New Year break—clearly demonstrates the regressive nature of some of the "modern" projects.

There are increasingly louder voices from the grassroots population in many cities of East Asia, especially in Hong Kong and mainland China, and they cannot be always ignored as in the past. The cities of East Asia may be destined to play leading roles in world affairs in the 21st century, belonging as they do to one of the three power centers of the global economy. The top player of all may well be China—if things are done right—with its great cities superseding those of Japan, which dominated the region in the 20th century.

SUGGESTED READINGS

Brandt, Loren, and Thomas G. Rawski, eds. 2008. *China's Great Economic Transformation.* Cambridge and New York: Cambridge University Press. A comprehensive examination of various aspects of the Chinese economy, including the spatial dimensions of China's development.

Chan, Kam Wing. 1994. *Cities with Invisible Walls: Reinterpreting Urbanization in Post-1949 China.* Hong Kong: Oxford University Press. An examination of the major features of socialist urbanization in China, especially in relation to its industrialization strategy and the need for a migration control system.

Golonym, G. S., Keisuke Hanaki, and Osamu Koide, eds. 1998. *Japanese Urban Environment.* New York: Pergamon. An attempt to explain the success of Japanese urban design and planning, based primarily on Japanese scholarship.

Karan, P. P. and Krtistin Stapleton, eds. 1997. *The Japanese City.* Lexington: University of Kentucky Press. Readable review of various aspects of urban development, historical and contemporary.

Kim, W. B. *et al.*, eds. 1997. *Culture and the City in East Asia.* New York: Clarendon Press. A collection of chapters, most by geographers and urban planners, on the history and transformation East Asian cities, including Hong Kong and Taipei.

Solinger, Dorothy. 1999. *Contesting Citizenship in Urban China*, Berkeley: University of California Press. An insightful study of problems faced by millions of migrant workers in Chinese cities.

Sorensen, Andre. 2002. *The Making of Urban Japan: Cities and Planning from Edo to the Twenty-First Century.* London, New York: Routledge. An examination of Japan's urban development from earliest times to the present.

Wu, Fulong, JiangXu, and Anthony Gar-On Yeh. 2007. *Urban Development in Post-Reform China: State, Market, and Space.* London, New York: Routledge. A detailed look at various aspects of urban planning and development issues in China.

Wang, Ya Ping, 2004. *Urban Poverty, Housing and Social Change in China*, New York: Routledge. A comprehensive treatment of urban issues in China in the reform era with focus on urban poverty and housing.

Yusuf, Shahid, and Kaoru Nabeshime. 2006. *Post-Industrial East Asia Cities.* Stanford: Stanford University Press and World Bank. A book on technologies and innovations in several major cities in East Asia as they move away from manufacturing.

Zhang, Li, 2010. *In Search of Paradise: Middle-Class Living in a Chinese Metropolis.* Ithaca, NY: Cornell University Press. An ethnography of the different ways in which China's new middle-class is affected by the experience of home ownership.

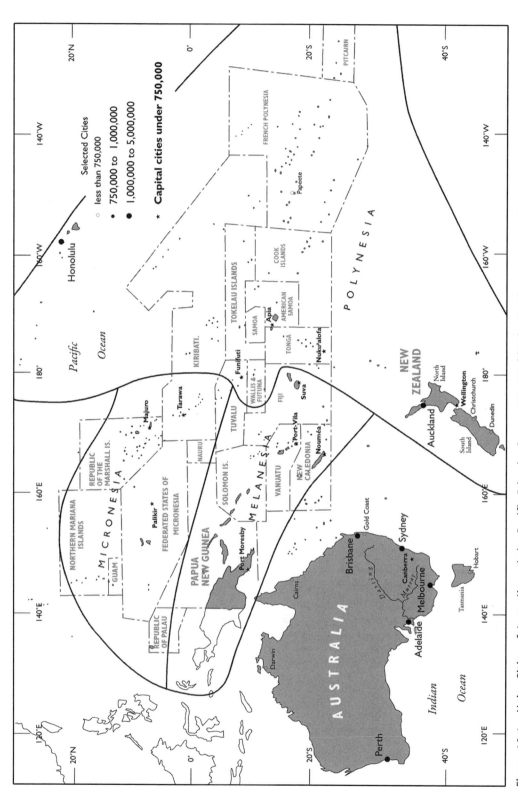

Figure 12.1 Major Cities of Australia and the Pacific Islands. *Source:* UN, *World Urbanization Prospects: 2009 Revision*, http://esa.un.org/unpd/wup/index.htm

12

Cities of Australia and the Pacific Islands
ROBYN DOWLING AND PAULINE McGUIRK

KEY URBAN FACTS

Total Population	35 million
Percent Urban Population	70.2%
Total Urban Population	25 million
Most Urbanized Countries	Nauru (100%)
	Guam (93.2%)
	American Samoa (93.0%)
Least Urbanized Countries	Papua New Guinea (12.5%)
	Solomon Islands (18.6%)
	Samoa (20.2%)
Annual Urban Growth Rate	1.57%
Number of Megacities	None
Number of Cities of More Than 1 Million	6 cities
Three Largest Cities	Sydney, Melbourne, Brisbane
World Cities	Sydney

KEY CHAPTER THEMES

1. Cities in this region may be understood as forming two groups—those of Australia and Aotearoa/New Zealand and those of the Pacific Islands—each with distinct characteristics.
2. All countries in this region are dominated by primate cities, but in the case of Australia primate cities are the capitals of states in federal union.
3. Australia and Aotearoa/New Zealand exhibit many of the urban characteristics of other developed countries, such as the United States.
4. The urban character of Pacific Island cities is similar to that of less developed countries though they are smaller and have considerably lower rates of population growth.

5. Sydney is by far the most globally linked city and the key economic center in this vast realm, though the global economic, cultural, and social connections of all cities have increased dramatically.

6. Many of the cities in the region were established as colonial or national capitals, and urban patterns and character are tied to this political influence.

7. In Australia, a popularly documented "sea change" phenomenon is drawing people away from the big cities toward small coastal towns.

8. Suburbanization and gentrification remain key residential forces in Australian and Aotearoa/New Zealand cities, and globalization is a central driver of urban economies.

9. A multicultural population is increasingly the norm in most cities in the region, especially in Australia and Aotearoa/New Zealand.

10. Awareness of the environmental impacts of urbanization is rising, with attempts to adapt planning frameworks and everyday life to sustainable outcomes.

11. Environmental vulnerability, especially to the direct and indirect consequences of climate change, is a key issue confronting the future of cities in the Pacific Islands.

The Pacific region is a constellation of islands of varying sizes (fig. 12.1). Australia (the island continent) and Aotearoa/New Zealand (now carrying both Maori and Pakeha, or settler, names) dominate the region geographically and economically. However, many smaller islands are to be found in those vast realms of the Pacific Ocean known as Melanesia, Micronesia, and Polynesia. Socially, politically, economically, and biophysically, this is a diverse region with diverse cities.

In this part of the world, it is easiest to understand cities as forming two main groups: those of Australia and Aotearoa/New Zealand, and those of the Pacific Islands. The former have cities with characteristics of more developed countries: industrialized, with a generally high level of affluence, and connected into global flows of people, money, information, and services. There are two key urban characteristics shared by both these nations. First, they are urban. Currently, over 88 percent of Australia's and 86 percent of Aotearoa/New Zealand's population live in urban areas. Second, they are, and long have been, nations of urban primacy: their urban pattern is dominated by a small number of large cities. Approximately one-fourth of all Aotearoa/New Zealanders live in just one city—Auckland—and Australia's two largest cities—Melbourne and Sydney—are home to more than 38 percent of the nation's population (table 12.1).

The islands within Micronesia, Polynesia, and Melanesia have starkly different urban characteristics. They have highly non-urban populations. Although reliable statistics are difficult to obtain, it is estimated that 35 percent of the population lives in urban areas, with a projected increase to over 50 percent by 2025. There are 35 towns and cities with a population greater than five thousand. Two-thirds of the southwest Pacific realm's urban dwellers are to be found in Papua New Guinea (PNG) and Fiji, the most populous nations

Table 12.1   **Australia and Aotearoa/New Zealand: Changes in Distribution of National Population**

Nation/Cities	% National Population 1981	% National Population 2006
	Australia	
Sydney	21.8	20.7
Melbourne	18.6	18.1
Brisbane	7.2	8.9
Perth	6.2	7.3
Adelaide	6.3	5.6
Hobart	1.1	1.0
Darwin	0.4	0.5
Canberra	1.6	1.6
	Aotearoa/New Zealand	
Auckland	26.1	29.2
Christchurch	10.1	8.7
Wellington	10.8	9.6
Dunedin	3.6	2.7

Sources: New Zealand Official Yearbook, 88th ed.; Year Book Australia; New Zealand Census of Population and Dwellings 2006; Australian Census of Population and Housing 2006

in the region (table 12.2). The region's largest cities—Port Moresby (PNG), Nouméa (New Caledonia), and Suva (Fiji)—are tiny by world standards. Negligible population growth is occurring in these cities, where economic opportunities remain limited. In Pacific Island nations, prestige and status are still very much tied to the land and the rural, rather than to cities and the urban.

HISTORICAL FOUNDATIONS OF URBANISM

Australia, Aotearoa/New Zealand, and the Pacific Islands have indigenous peoples with long histories of settlement, up to 40,000 years in the case of Australian Aboriginals. Cities in this part of the world are, however, very young.

Urban settlement began with the arrival of numerous colonizers in the 18th and 19th centuries. Australia became a penal colony of the British in 1788, with the arrival of convicts to Sydney and Port Arthur (near Hobart, Tasmania) and later to Brisbane. The continued arrival of convicts to these coastal towns and the establishment of additional settlements like Melbourne and Adelaide, for purposes of colonial administration, commerce, and trade cemented metropolitan primacy. The political independence of each of the British colonies (later to become states) also meant that the capital cities operated independently of each other throughout the 19th century, providing services to their rural hinterlands, acting as ports for the import and export of commodities to and from Europe, and functioning as centers of colonial administration.

Table 12.2   Population of Pacific Island Cities

ISLAND NATION/City	Population (2010)	% of Country's Population
FIJI	841,387	
Nasinu	88,566	10.5
Suva	85,754	10.2
Nausori	46,811	5.6
KIRIBATI	100,062	
Bairiki	47,946	47.9
Taburao	4,321	4.3
Bonriki	4,005	4.0
MARSHALL ISLANDS	54,185	
Rita	20,144	37.2
Ebeye	9,581	17.7
Laura	2,905	5.4
VANUATU	248,935	
Vila	47,510	19.1
Luganville	13,799	5.5
Port Olry	2,897	1.2
TONGA	102,368	
Nukualofa	24,310	23.7
Mua	5,190	5.1
Naiafa	3,965	3.9
SOLOMON ISLANDS	530,735	
Honiara	63,343	11.9
Auki	6,811	1.3
Munda	4,850	0.9
SAMOA	181,718	
Apia	36,440	20.1
Vaitele	7,333	4.0
Faleasiu	3,858	2.1
PAPUA NEW GUINEA	6,740,586	
Port Moresby	307,103	4.6
Lae	96,242	1.4
Mendi	43,005	0.6

Source: Country Watch 2010 Country Profiles, http://www.countrywatch.com/country_profile.aspx

Indeed, competition between the capitals further worked to bolster primacy. With each capital focused on ensuring continued economic growth, backed by political force within their respective territories, the establishment of alternative, prosperous, and comparable urban centers was made more difficult.

Two major events of the mid- to late-nineteenth century further enhanced the size, functions, and importance of Australia's six colonial capitals. Railroads were focused on the capitals, facilitating more efficient connections between the cities and their hinterlands. Industrialization similarly occurred within (rather than beyond) these coastal centers of colonial administration, though there were to be later exceptions like Wollongong and Newcastle in New South Wales, and Whyalla in South Australia. By the end of the 19th century, Australia had a total population a little less than four million. Sydney and Melbourne each had populations of approximately half a million, Adelaide, Brisbane, and Perth more than 100,000 each, and Hobart remained small at 35,000 people. Colonialism had hence been responsible for this uniquely Australian urban primacy and settlement pattern in at least two ways. First, the sites of European settlements (either convict or free), with their coastal locations and trading functions, formed the foundations of the colony and its growth. Second, the functions of colonial administration, and competition among the capitals fueled the growth of existing rather than new urban centers.

The first half of the 20th century saw urban Australia grow in the spatial pattern established by British colonialism. A manufacturing boom that began in the 1920s reinforced the primacy of each state capital. This era also saw the beginning of the systemic suburbanization of Australian cities. The establishment of middle-class suburbs in attractive surroundings away from the central city was facilitated by the development of public transport lines radiating out from the city center, as well as the activities of land developers and house builders. With the absence of inner-city slums on the scale of those in Britain, the social differentiation of Australian cities took on characteristics of the sector model related to transport links and features of the natural landscape.

The turn of the 20th century did see one challenge to the existing capital cities with the planning of the new city of Canberra. The federation of Australia's colonial territories in 1901 was designed to both create and unite a nation. The colonial capitals became capitals of states in the newly formed Commonwealth of Australia, and a new national capital—Canberra—was established between the two cities that dominated the national urban hierarchy—Melbourne and Sydney. Canberra's location between these two urban leaders was a compromise. The Australian Parliament did not formally relocate to Canberra until 1927, and the city remains comparatively small, with fewer than 400,000 inhabitants (fig. 12.2). Its dominating characteristic is the prominent role played by formal urban planning. A master plan developed by an American—Walter Burley Griffin—guided its development as a "garden city" built around a large lake, with a central focus on a "parliamentary triangle" and satellite suburbs with town centers of their own. Canberra's expansion was slow—only 16,000 people lived there in 1947—and its early economy was reliant on public service and diplomatic functions. Today, its economy is supplemented by a large student population which attends the relatively large number of public and private institutions of higher

Figure 12.2   Canberra's distinctive but controversial Parliament House is difficult to appreciate from the outside because much of the structure is underground. The inside is breathtaking, filled with beautiful art and materials native to Australia. (Photo by Donald Zeigler)

learning, including the Australian National University.

In Aotearoa/New Zealand, European settlement and modern urbanization began in 1840 with the signing of the Treaty of Waitangi between the British and the Maoris. Unlike the convict bases of Australia's settlements, free setters in Aotearoa/New Zealand were encouraged to migrate and invest, with the resultant economy largely dependent on pastoral activities like grazing sheep and cattle. Unlike Australia, urban primacy was not a 19th century phenomenon here, due to the originally more dispersed settlement pattern and more diverse reasons for urban settlement. For example, early towns like Wellington and Christchurch were established by trading and/ or religious interests; Auckland's natural harbor made it an ideal port (fig. 12.3); and gold

rushes underpinned the growth of Dunedin. Thus, by 1911, Auckland had a population of 100,000, Christchurch 80,000, Wellington 70,000, and Dunedin 65,000. Over half of the non-Maori population lived in urban areas. In contrast, throughout the 19th century and the first half of the 20th century, Maori settlement was predominantly rural.

Like Australia and Aotearoa/New Zealand, Oceania has had a long established indigenous population, and similarly it was the colonial context that underpinned the urban system of the region. Oceania was one of the last regions of the world to be colonized, with British, French, American, and Dutch powers establishing presences in countries like Fiji, Samoa, Tonga, and Vanuatu at various times across the 19th century. Towns first developed as trading ports, usually close to existing villages,

Figure 12.3   A city built on an isthmus and connected to a rich hinterland, Auckland now hosts many activities found in major world cities, including the famous Sky Tower and casino that dominates the skyline. (Photo by Donald Zeigler)

good harbors, and viable anchorages. These towns grew slowly, and some, like Levuka in Fiji, declined over time because of relative inaccessibility. They were never large: in 1911 Suva had a population of only 6,000 people, about 5 percent of Fiji's population.

The first half of the 20th century saw a diversification of urban functions and sporadic urban growth. Although widespread industrialization did not occur, the processing of agricultural commodities like sugar, and the extraction of resources through mining, diversified the economic base and saw the growth of cities in Fiji and New Guinea, where the mining towns were nearly as large as the colonial capital of Port Moresby. In Micro-

nesia, intense Japanese colonialism saw cities like Koror, on the island of Palau, grow substantially; other administrative capitals grew slowly. By the middle of the 20th century, urbanization remained limited.

CONTEMPORARY URBAN PATTERNS AND PROCESSES

The contemporary urban systems of Australia, Aotearoa/New Zealand, and the Pacific Islands are based upon the patterns established in previous decades. Economic, social, and political influences across the region have consolidated urban primacy. Urbanization processes, the overall urban pattern of the region, and the characteristics of cities within it, are far from uniform. For cities of the Pacific Islands, tourism, political independence, and instabilities, migration, and environmental hazards play significant roles. In Australia and Aotearoa/New Zealand, in contrast, industrialization followed by deindustrialization, globalization, international immigration, urban governance, and rural/urban population dynamics are the primary influences.

The Pacific Islands

The historical pattern of urban primacy in a largely non-urban region remains a hallmark of the Pacific's urban geography (table 12.2). By 1960, only Suva (Fiji) and Noumea (French Caledonia) had populations greater than 25,000, and even today the size of cities remains small. Political independence from colonial powers began in the 1970s. Only a few territories, such as French Caledonia, remain in colonial hands. Independence had a number of significant impacts on the region's urban system. Colonial administration was no longer the primary purpose of the largest cities in the region, but processes associated with independence cemented the primacy of these towns. In some, like Port Moresby, independence fostered urban growth because of new investment in urban housing and services. Across the region accelerated urban growth followed independence because of, for example, the removal of negative perceptions of urban living, or the establishment of some countries as tax havens (*e.g.,* Port Vila, Vanuatu). Independence also required bureaucracies in national capitals, and encouraged education and urban living in general.

Land and land tenure systems are a defining characteristic of Pacific cities. In Melanesia, Polynesia, and Micronesia, customary land tenures pose significant challenges for urban growth, housing, and infrastructure provision as well as the quality of urban life. In Port Moresby, for example, one-third of the city's total area is held by traditional owners, and land is seen as a communal resource. However, customary land tenure places limits on the land available to house urban residents and is associated with higher housing costs. It also provides a disincentive to invest in land development and urban infrastructure. A number of possible solutions to the limitations customary land tenure places on capitalist urban growth have been proposed. These include proposals to lease customary allotments, or the ability to use land to generate income through means other than compensation. Such proposals have been severely hindered by the limited capacity of urban governance across the islands.

Connected to issues of land tenure are the general housing characteristics of the urban Pacific. Palatial houses exist, but they are often built by expatriates and in gated communities. Formal housing of the type commonly found in Australian and Aotearoa/New Zealand cities exists as well. Far more common, however, are informal settlements. The great demand for housing, in the context of substantial urban poverty and limited employment opportunities, means that informal housing is common. Public housing is available, though waiting lists are extremely lengthy.

Finally, the present and future of the cities of the island Pacific cannot be understood without reference to environmental contexts and threats (box 12.1). Urban settlement has involved degradation of islands' fragile coastal environments. The waste and water requirements of growing urban populations threaten to overwhelm already stressed ecosystems.

Box 12.1 Urbanization and Human Security

"Existing insecurities (*e.g.,* income inequalities, environmental degradation, lack of services etc.) are catalysts in the process of urbanization (fig. 12.4). Rural to urban population movements, however, give rise to vulnerabilities within urban places, including pollution, exposure to hazardous substances, resource scarcities, and inequalities. Vulnerability is defined as having biophysical and social components. Biophysical vulnerability refers to the potential for loss from environmental threats. . . . Social vulnerability refers to the social and institutional capacity that defines both the susceptibility and the ability to cope with environmental threats . . . . [I]t is the interaction of social and biophysical vulnerability that contributes to the vulnerability of specific places. At the center of the Figure is the definition of human security, especially in the context of urbanization. The definition suggests that the state of human security will be determined by environmental threats, referred to in the

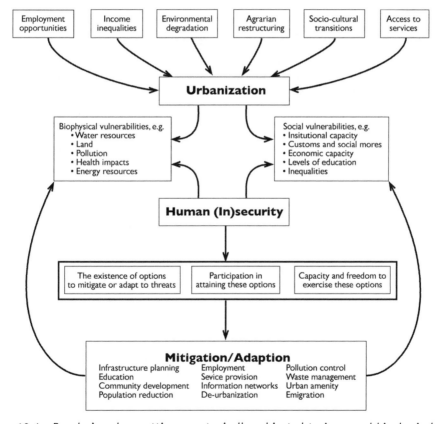

Figure 12.4   People in urban settings are typically subjected to increased biophysical and sociocultural threats to their security. *Source:* Redrafted from Chris Cocklin and Meg Keen, "Urbanization in the Pacific," *Environmental Conservation,* 27 (2000), 395

figure as biophysical vulnerabilities, which in an urban setting would include scarcities of basic resources, such as water, land, energy, and degradation of environmental quality. These threats have potential to undermine human security in myriad ways, but include most notably their implications for human health and physical well being, economic welfare, nutrition levels, and access to adequate housing."

Source: Abridged from Chris Cocklin and Meg Keen, "Urbanization in the Pacific: environmental change, vulnerability and human security." *Environmental Conservation* 27 (2000), 4, pp. 392–403.

Urban water is typically sourced from freshwater lenses, and if these are over pumped, saltwater contamination can occur and render the water unsuitable for human use. Because of the geology of the islands, waste disposal also affects the environment. Other forms of water supply contamination can occur (*e.g.,* by chemicals, sewerage), which in turn affects human health. The most important environmental issue for these cities in the 21st century is climate change, especially global warming. The low-lying islands, and their cities, are at risk of inundation because of sea level rise. Climate change is also believed to involve increased storm activity, accelerated coastal erosion, saltwater intrusion into reserves of fresh water, and increased landward reach of storm waves. Each of these events has the potential to dismantle city infrastructure and threaten urban livelihoods. Environmental hazards are further exacerbated by social vulnerabilities, especially limited institutional capacities for urban planning. In 2010 at the U.N. Framework Convention on Climate Change meeting in Cancun, the Deputy Prime Minister of Tuvalu classed climate change as a "life or death survival issue," threatening the very existence of this Pacific island nation. The highest point on Tuvalu's capital island, Funafuti, is less than 14 ft (4.3 m) above sea level.

The global economic context is crucial to urban economies in the Pacific. Many nations, like Fiji, have turned to tourism for economic survival, with urban consequences. Global commodities and mining, as well as the presence of wealthy expatriates, underpin the urban hierarchy of Papua New Guinea (PNG). And finally, global migration and in particular out-migration, can relieve some of the social, economic, and environmental pressures in cities. In Tonga especially, migration to Aotearoa/New Zealand, Australia, and the United States operates as an urban "safety valve," allowing Tongans to realize economic opportunities overseas rather than in overcrowded and economically limited urban areas. This safety valve has also become part of new, informal, urban economic activities.

In sum, cities of the island Pacific are places of vulnerability and opportunity. In a largely non-urban context, in which effective urban planning and coordination is non-existent at worst and problematic at best, urban living is still sought as a chance for a better quality of life. Though officially derided, life in informal settlements remains attractive.

Australia

The dominance of state capital cities remains the defining characteristic of Australia's urban system. The primary drivers of urban develop-

Figure 12.5 As Australia's second-largest city and a major industrial center, Melbourne prides itself on being more traditional than exuberant Sydney, its long-standing rival. (Photo courtesy of Australian government)

ment in the 20th century—industrialization, migration and, latterly, globalization—have only reinforced the importance of the state capitals and fueled their population growth. Between 1947 and 1971, the population of Australia's five largest cities doubled, and growth has continued since then. Historically, Sydney (capital of New South Wales) and Melbourne (capital of Victoria) have been the island continent's largest and most economically dominant cities. Australia's manufacturing growth after World War II was centered in Melbourne, which, until recently, housed the majority of Australian corporate headquarters (fig. 12.5). Other state capitals served their rural and resource-based hinterlands, with smaller and less diversified economic bases. In the immediate postwar period, Adelaide was somewhat of an exception, as the center of Australia's car industry.

Aboriginal Australians are much less likely to be urbanized than the broader Australian population. They are also more likely to live in small towns rather than large cities. Indeed, a little over 1 percent of Sydney's total population, and 1.7 percent of Perth's population are indigenous. Indigenous movement to capital cities is often temporary, and linked to kinship and friendship ties with rural areas. Aboriginal people have long dwelled on the fringes of cities, often in substandard housing. Places of residence within the city are related to the provision of public housing and also localities with strong identification for indigenous Australians. One of these places is "The Block," in Sydney's inner city Redfern, where housing and other cultural services are concentrated.

The past 25 years have seen some shifts in the distribution of economic and population

Box 12.2   The Geography of Everyday Life in Suburban Sydney

Australia is a suburban nation. Despite increasing urban consolidation and gentrification, more than 72 percent of Sydney's population lives in detached housing, and 33 percent in areas more than 15 kilometers (9 mi) from the city center. Suburban Sydney, unlike North American suburbs, is heterogeneous. Sydney's greatest concentration of migrants is found in its suburbs, and hence we see pockets of affluence and poverty neighboring each other. What is everyday life like in this differentiated world city?

Suburban Sydney residents live in houses of varying age and design. New houses are more likely to be large, averaging 893 sq ft (83 sq m) per person—27 percent of houses have four or more bedrooms, a double garage, formal and informal living areas, separate rooms for each child, perhaps a games/media room, and a backyard that may just be able to accommodate a cricket pitch. Family members—both adults and children—typically know their immediate neighborhood and participate in local sporting and recreational activities. Increasingly, this neighborhood may be gated in some way, with a large proportion of socializing done within the home. The family shops locally, sometimes at a small corner shop or on a main street, but more likely at a supermarket in a large shopping mall. Here, not only can they pick up their weekly provisions, but they can also eat a meal and see a movie.

Daily travel patterns are increasingly complex spatially and socially. One adult (more likely male) may commute to the CBD for his job in the finance or business sector or to another suburb for manufacturing employment. The woman is likely to work in this or a nearby suburb, most likely in retailing or a similar service sector job in banking, hospitality, or education. The limited availability of public transport in certain parts of suburban Sydney, and the generally poor servicing of cross-suburban travel mean that these journeys to work are most likely to be undertaken by car (71 percent). For mothers of young children the importance of the car is even more pronounced, as she drops children at school/childcare on her way to work, and takes them to social and sporting activities on the way home (fig. 12.6). For these

growth across Australia's large cities. Two factors underpinned these slight alterations in the urban system. The first was the influx of people into Australian cities through international migration. For the past twenty years, more than 100,000 people annually have migrated to Australia from around the world, most of these to the capital cities, particularly Sydney, Brisbane, and Perth. Cities that have not received substantial numbers of migrants, like Adelaide and Hobart, have declined in relative terms.

The second factor was globalization, or more specifically changing urban functions as the Australian economy became increasingly tied to, and driven by, global flows of commodities and money, and increasingly reliant on globally networked business services. Globalization has seen Sydney rise in prosperity and prominence to become Australia's only world city. The headquarters of Australian-based businesses, and the regional offices of multinationals, are now more likely to be in Sydney than in Mel-

suburbanites, the time and cost of car travel is becoming an increasing burden, though one for which no relief is in sight.

The Car as a Management Tool

Figure 12.6   New roles for women, and new problems, have emerged in Australian cities over the past three decades. (Courtesy of Robyn Dowling)

bourne. The relative growth of Brisbane and its surrounding region during the same period can be attributed to internal migration (principally from Sydney), the rise of a tourist-based economy, growing economic ties between Brisbane and the Asia-Pacific region, and Queensland government incentives for business to relocate to Australia's sunbelt.

Australia's state capitals are highly suburbanized and geographically expansive by international standards (box 12.2). Historically, the predominant housing preference is for a detached house, producing sprawling suburban conurbations (fig. 12.7) like that between Brisbane and the Gold Coast, 37 miles (60 km) away. The continued proliferation of suburban housing is currently under some threat. The high energy demands of suburban life—use of the private car, heating, cooling, and the water use demands of large houses—are increasingly questioned. Limited availability of land and the high costs of servicing the social and physical

Figure 12.7   Sydney is known as a city of suburbs and single-family homes such as this one. (Photo by Robyn Dowling)

infrastructure needs of new suburbs have led to policies of urban consolidation across the nation (box 12.3). Mixed-use residential and commercial developments on old industrial land are increasing, and in some years the construction of new apartments outstrips that of detached houses. Equally important is a cultural and economic re-evaluation of living in Australia's inner cities. Australian inner cities are vibrant, cosmopolitan spaces, with a wealth of retail, social, and recreational opportunities; and they are highly accessible by public transport.

The internal structure of Australian cities has changed over the past three decades. Based on an analysis of social and economic characteristics, metropolitan localities may be divided into seven types of places (fig. 12.8): three advantaged and four disadvantaged. In *new economy* localities are found people employed in new global industries and many educated professionals. *Gentrifying* localities are found across Australia's inner cities, and are home to those with ties to the global

economy but with a sizeable proportion of low-income residents as well. *Middle-class suburbia* houses many educated professionals, though with a low density of connections to the global economy. *Working-class battler* communities have trades people, often homeowners, while *battling family communities* have above average levels of single-parent and non-family households. In *old economy* localities, primarily suburban and especially in Adelaide, the decline of manufacturing has seen concentrations of unemployment. Finally, *peri-urban* localities on the fringe of the capitals, attract low-income people seeking cheaper housing or homes for retirement.

While state capitals have, on average, been growing, small towns in rural and regional Australia have exhibited divergent patterns. Many rural towns, traditionally operating as service centers for the surrounding farms, have experienced population declines. Decreasing farm incomes, the closure of many public and commercial services such as banks, and limited employment and education opportunities for

Box 12.3 Using GIS to Analyze Urban Consolidation

Since the late 1980s urban consolidation—increasing the density of dwellings or population (or both)—has been a central part of Sydney's official planning policy, aimed to minimize fringe development, limit car dependence, contain resource use and emissions, and provide greater housing choices. Despite the potential environmental benefits, many communities in traditionally low-density Sydney have reacted negatively to consolidation, particularly where higher densities have been seen as insensitive to neighborhoods' urban character and landscape. However, adapting consolidation policies to balance ecological needs with community concerns requires detailed spatial information and analysis of the location, intensity, spread, and characteristics of higher-density residential development. Researchers at the University of Western Sydney have used GIS to provide this analysis. For the period since 1981, they combined census information on population and housing stock, data on the dynamics of building approvals and local government information on development applications for different types of dwellings to analyze the impact of urban consolidation and to assess its impact over time on the built form of three specific local government areas (LGAs) in south and southwestern Sydney: Campbelltown, Sutherland, and Hurstville.

Their examination of the general distribution of higher density dwellings showed the concentration of consolidation in established higher-density areas such as central and south Sydney, the inner west, inner north, and eastern suburbs, and along the major transport routes (particularly rail lines). More importantly, they could show the diffusion of higher density housing forms through the middle suburbs as local authorities implement policy changes allowing dual occupancy and multiple small lot developments. But the researchers also wished to understand how each LGA's context produced different contributions to Sydney's consolidation, reflecting the LGAs' different histories and the differential effect of state and local policies. The older suburb of Hurstville in Sydney's south was shown to have been consolidated through major in-fill redevelopment and renewal of the aging built environment, especially through the development of flats and units along rail and transport routes. High-rise apartment towers increasingly replaced the more traditional 1-story bungalows or 3-story flats developments that had characterized the neighborhood. The more recent development of neighboring Sutherland was reflected in the dominance of greenfield, low-density, separate housing development until the early 1990s when development pressure brought a steep increase in higher density forms of residential development, especially in established town centers, along rail routes, and on the coastal fringe of the LGA. In contrast, Campbelltown's location on the city's southwestern fringe meant it was still dominated by greenfield low-density development with little high-density development although, a scattering of medium density development has occurred on sites set aside in local planning policies for this purpose.

Applying GIS may not directly change community concerns in Sydney about urban consolidation but it has armed planners with a fine-grained, integrated description and analysis of how market forces and planning policies of consolidation have combined to deliver dwelling stock in various patterns over the last two decades, and to compare the characteristics of urban consolidation achieved in the distinctive circumstances of each LGA.

Source: D. Holloway and R. Bunker, "Using GIS as an aid to understanding urban consolidation," *Australian Geographer,* 41 (2003), pp. 44–57

Locality Types in Australia's Large Cities

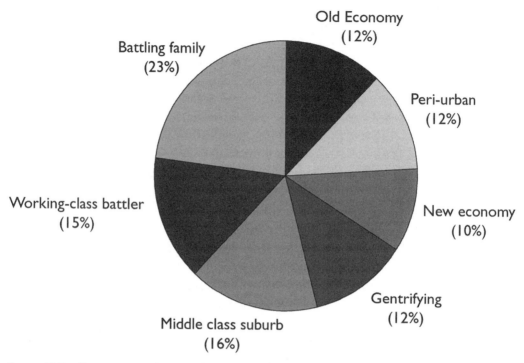

Figure 12.8   Changes over the past three decades have produced new types of urban localities in Australia. *Source:* Compiled by authors from statistics in Scott Baum, Kevin O'Connor and Robert Stimson, *Faultlines Exposed* (Melbourne: Monash University ePress, 2005)

young people have encouraged migration out of these towns and into larger regional centers or, more commonly, capital cities. A counter trend of growth in Australia's coastal towns is also evident. The 21st century boom in resource prices has meant that towns in coastal Australia have grown rapidly, instigating severe housing shortages and consequent escalations in house prices. The "sea change" phenomenon, in which city dwellers swap a hectic city lifestyle, transport congestion, and high housing costs for a slower pace of life and cheaper housing in coastal towns is also important. Initially confined to older people, principally retirees and those nearing retirement, sea changes are now undertaken by young professionals able to run

businesses outside the major cities, as well as less affluent families seeking cheaper home ownership. Towns like Byron Bay, Coffs Harbour and Port Macquarie in New South Wales, Barwon Heads in Victoria, and Denmark in West Australia are commonly identified sea-change locations. "Tree change" is a more recent but similar phenomenon in which urban dwellers move to greener locations like rural Tasmania, inland New South Wales (*e.g.,* Orange, Mudgee) or Victoria (*e.g.,* Daylesford).

Aotearoa/New Zealand

After World War II, the growth trajectories of the cities in Aotearoa/New Zealand largely paralleled those of Australia. The four largest cities of Auckland, Wellington, Christchurch, and Dunedin continued to grow, as did the primacy of Auckland (table 12.1). A number of processes underpinned this pattern. Market reforms since the 1980s have strengthened global economic, cultural, and social ties, which in turn have transformed large cities. Second, immigrants, initially from the Pacific Islands but also more recently from China and India, have flowed into the large cities, especially Auckland and Christchurch. The third factor is the internal shift in economic activity. Whilst a general process of deindustrialization in Aotearoa/New Zealand occurred in the late 20th century, employment losses in manufacturing were more severe in Wellington, Christchurch, and Dunedin; and some manufacturing relocated to Auckland. Finally, entrepreneurial urban governance processes were deployed to make cities more attractive and to stem population decline. In Wellington, for example, the waterfront was redeveloped using both public and private sector investment. The aim was for the city to become an international conference venue, and the government also located the new Te Papa National Museum there.

Aotearoa/New Zealand cities are low density, though suburban living is no longer the only residential option as high- and medium-rise apartments are becoming more common. The proportion of Maoris living in urban Aotearoa/New Zealand is now almost on par with that of the non-Maori population, because of the loss of Maori land and consequent rural-to-urban migration. Maoris face significant disadvantages in the cities, with high rates of unemployment and lower levels of home ownership and education. Increasing ethnic diversity is also an important urban characteristic.

DISTINCTIVE CITIES

Sydney: Australia's World City

With a population currently of about 4.5 million, and projected to reach 5.7 million by 2031, Sydney is the most populous and most prosperous city in Australia. The city is home to some of Australia's most widely recognized iconic landmarks: the Harbour Bridge (fig. 12.9), the Opera House (fig. 12.10), and Bondi Beach. It is an international finance market, it attracts a growing concentration of corporate headquarters, and it is Oceania's highest value-generating economy and dominant world city. Equally, it demonstrates some of the defining characteristics of contemporary Australian urban life: suburbia, urban-based prosperity arising from an advanced service economy, multiculturalism, and environmental threat.

Sydney entered the 20th century as the primate city and highest order service center in the state of New South Wales (fig. 12.11). By 1911, just 123 years after first European settlement, it had a population of 652,000 and was already a city of suburbs. Sydney's post-World War II "long boom" brought unprecedented economic and population growth and set in motion the formative settlement patterns that have shaped the contemporary city. Between 1947 and 1971, population expanded by 65 percent to reach 2.8 million; it grew to 4.1 million by 2006. The vast majority of growth has been accommodated in expansive suburban developments, including large-scale public housing estates built mainly across the city's western suburbs. Despite planned expansions of public transport networks, the rate of urban expansion and rising levels of car ownership meant that the city quickly assumed the car-oriented form of autosuburbia, connected by networks of freeways rather than

Figure 12.9   Completed in 1932, the Sydney Harbour Bridge opened up the city's North Shore. Tourists, tethered by lifelines, have been climbing the arch since 1998. (Photo by Donald Zeigler)

public transport corridors. Speculative developers' and housing consumers' preferences for low-density, detached dwellings meant that the city assumed a sprawled metropolitan form, poorly served by the existing rail network radiating from the Central Business District (CBD) (fig. 12.12). Twenty years of urban consolidation policy has contained the extent of sprawl but strong population growth (50,000 per year since the late 1990s) has meant that fringe expansion has continued. Sydney's employment, retailing, and services have been decentralizing since at least the 1970s. The development of regional centers of commercial activity, such as Ryde, North Sydney, Parramatta, Penrith, and Liverpool

has given the city an increasingly polycentric form. Indeed, the current metropolitan planning strategy labels Sydney a "city of cities."

Despite Sydney's predominantly low-rise suburban form, the city center is characterized by high-rise office towers, global tourist landscapes and, lately, residential towers tightly grouped on the edges of one of the world's most spectacular natural harbors (fig. 12.13). Since the late 1960s significant waves of international property investment—in commercial office and hotel developments—have transformed the CBD's built environment, as has the transformation of Sydney's economic base to one dominated by increasingly globally connected financial and other advanced

Figure 12.10   Now a UNESCO World Heritage site, the Sydney Opera House has become a symbol of the island continent. (Photo by Donald Zeigler)

services. Sydney has become one of the most significant financial centers in the Asia-Pacific realm, making up 40 percent of Australia's telecommunications market. Employment in the global city sectors of finance, insurance, property, and business services is concentrated in and around the city center where many of the estimated 600 multinational companies who run their Asia-Pacific operations from Sydney are clustered, along with the headquarters of approximately 200 of Australasia's top companies. The economy of the city center now generates 30 percent of the value of metropolitan Sydney's economic output and contains 28 percent of all metropolitan employment, with high concentrations in the highly paid professional and managerial occupations.

Concentrated in Sydney's city center are high-paid, advanced-services workers, as increasingly globalized connections have driven long-standing processes of gentrification, the recent resurgence of high-rise luxury residential dwellings, and the multiplication of globalized consumer spaces. Inner suburbs of 19th-century housing have been revitalized. New up-market residential locales have been built in high-density, previously used land on the edges of the CBD (fig. 12.14), and in a host of high-rise high-density towers throughout the CBD. These developments have meant that the resident population of Sydney's inner city has increased by 40 percent since 1996. The development of a range of globalized consumer spaces, catering both to global tourists and to inner-city residents, have also transformed the city center. In the 1980s the New South Wales government redeveloped Darling Harbour container terminal

Figure 12.11   In 1916, Sydney's government decided to locate the Taronga Zoo right on Sydney Harbor. As a result, many long-necked ungulates have a better view of the city's skyline than human bipeds. (Photo by Donald Zeigler)

as an international conference center, festival shopping, and entertainment precinct. In the 1990s, special legislation was passed to enable redevelopment of heritage wharves at Walsh Bay as an exclusive residential, commercial office, and restaurant precinct. The current state government proposal for a major urban renewal scheme for Redfern-Waterloo just south of the CBD is likely to be similarly contentious as growing demand for "world city"

Figure 12.12   Australia is a land of open space, a fact of geography reflected in its sprawling suburbs. (Photo by Rowland Atkinson)

Figure 12.13   Sydney's skyline, typical of a world city, dominates the capacious harbor. Can you identify Sydney Tower? (Photo by Donald Zeigler)

spaces clashes with the needs of socially disadvantaged populations whose inner-city location is now under threat. A special-purpose Redfern Waterloo Authority (RWA) is overseeing redevelopment of vast tracts of publicly owned land into residential, commercial, and consumer spaces suited to Sydney's trajectory as a dominant world city. Controversially, much of the public land to be redeveloped is currently occupied by public housing. More controversially still, the area targeted for redevelopment contains The Block, an aboriginal-owned site, renowned center of Sydney's aboriginal community, and a symbolic locus of urban aboriginal identity. The RWA has been granted exceptional powers as the sole planning authority for the area, exempt from heritage protection legislation. Political clashes between government plans and community demands seem inevitable.

Sydney's world city status is also reflected in the fact that about 40 percent of all migrants to Australia settle there, thus deepening and diversifying the long-established multicultural nature of the urban area's population. Eight out of every ten residents of Sydney were either born overseas or are the children of immigrants. The UK, China, and Aotearoa/New Zealand are the dominant source countries, though there are also substantial numbers of residents born in Vietnam, Lebanon, India, Philippines, Italy, Korea, and Greece.

Figure 12.14   Jackson's Landing is being developed on the former site of a sugar refinery in Pyrmont, an inner suburb of Sydney. (Photo by Rowland Atkinson)

Historically, particular migrant groups—especially those of non-English-speaking backgrounds—have tended to settle initially in particular Sydney suburbs: Greeks in Marrickville and Italians in Leichardt in the 1950s and 1960s, Vietnamese in Cabramatta in the 1970s and 1980s, and Lebanese in Auburn in the 1990s. However, recent research has shown that Sydney's settlement is characterized more by multi-ethnic suburbs, such as Auburn, rather than ethnic minority concentrations, and by intermixing of different ethnic minority groups both with each other and with the host society rather than by ethnic segregation. Over time, spatial and social assimilation of migrants into a predominantly multicultural city has been the dominant pathway.

Whether growing evidence of social polarization in Sydney will produce more entrenched socio-spatial segregation along lines of class and ethnicity is a concern both to Sydney's planners and citizens. In a trend common to many global cities, Sydney's median dwelling price rose by 100 percent between 1996 and 2003, before leveling out, so that housing stress (i.e., paying more than 30 percent of household income on housing costs) now affects nearly 170,000 households across the city. As the median house price has crept up, lower income groups, including recent migrants, have been increasingly confined either to rental housing or to less-accessible suburbs removed from employment opportunities and services. It remains to be seen whether Sydney's social divides, traditionally nowhere near as pronounced as in U.S. cities, are set to become increasingly stark.

Nonetheless Sydney remains renowned for its quality of life. It habitually enjoys a top-ranking position in international benchmarking exercises assessing physical and cultural lifestyle assets. However, the city's beautiful natural environment, open spaces, and national parks belie the environmental challenges generated by Sydney's car-dependent nature and population pressure, especially regarding air quality and water supply. Car ownership is ubiquitous and 71 percent of work trips are taken by private motor vehicle. Consequently, air quality suffers due to photochemical smog producing ozone at levels that, while improving, still regularly exceed the 4 hour standard for ozone concentration on 21 days a year. In addition, despite falling rates of water use per capita, Sydney's population growth is challenging the adequacy of the city's water supply (box 12.4). In 2002, Sydney's water consumption was at 106 percent of the amount that can be sustainably drawn from the drainage basin. Continuing urban development poses a significant threat to Sydney's water quality.

Box 12.4   Heat, Fire, and Flood

Nature and culture—geographical location, environmental conditions, and human practices—combine to ensure that the cities of Oceania face a daunting array of environmental hazards including bushfires, drought, and the impacts of climate change.

When the yearly Australian bushfire season crosses paths with urban development the impacts can be devastating. There have been eight major bushfire events—mainly caused by human agency—in the Sydney region since the late 1950s. They have resulted in extensive losses of property, wildlife and human life. In the catastrophic Sydney fires of 1993–1994, 800 fires resulted in 4 deaths and the loss of 206 homes. Some 800,000 hectares were burnt, including most of Sydney's historic Royal National Park. Fires reached within 10 km (6 mi) of the CBD and 25,000 people were evacuated as smoke shrouded the city and black tide-marks of ash were washed up on Sydney's famous beaches. The historic drought conditions that afflicted eastern Australia from 2001 heightened the severity of individual fire events, producing "unstoppable" fire conditions. January 18, 2003, saw Australia's capital city, Canberra, engulfed by fires that had been triggered by lightning strikes in drought-affected vegetation areas and had burned around the city for several days previously. High temperatures and strong winds took the fires out of control as they reached the urban limits. As the city was set alight, more than 500 homes were destroyed and 4 people killed; many more were afflicted by smoke inhalation and fire-related illnesses. At the margins of Australia's cities, as residents seek to live close to high-quality natural environments at the edge of the bush, the interface of fire-prone ecosystems and urban development continues to expand, meaning that the bush-fire hazard will be an ongoing feature of Australian urban life.

Many additional forms of environmental hazards arise from climate change. By 1999 Australia was the world's worst per capita emitter of climate-change-inducing greenhouse gases, with about half of those emissions attributed to urban activities including energy generated predominantly from coal. Despite research suggesting that carbon dioxide emissions could be significantly reduced through carbon pricing, application of technological advances in coal burning, and shifts to renewable energy sources, the national political

Figure 12.15   In Newcastle, NSW, this ClimateCam billboard, located in the city center, broadcasts figures on the city's electricity consumption. These are updated hourly as a way of raising awareness about the city's contribution to resource use, GHG emissions and climate change. (Photo by Kathy Mee)

will to secure these transformations is still lacking. However, individual cities have begun taking significant steps to address emissions reductions. [fig. 12.15] Meanwhile, both Australian cities and those of neighboring Pacific Island nations continue to live with the threat of the escalating hazards of climate change. For the Pacific Island nations the most devastating of these is sea level change, already disrupting tourism and local livelihoods, and pushing islanders to migrate. Australian cities are liable to become increasingly vulnerable to flooding related to intensive storms and sea level change--a particular concern to a nation in which the population is intensely concentrated in coastal cities. Simultaneously, rising temperatures will bring intensified photochemical smog, water shortages, and rising numbers of days with very high temperatures. In a peculiar circularity, this will limit the number of days in which controlled burns to reduce the fuel load of bushland on the urban margins are possible and the product of this will be intensified risk of severe bushfires.

Figure 12.16　Perth sits in splendid isolation on the southwest coast of Australia, the primate city for a vast, underpopulated region. (Photo courtesy Australian government)

Perth: Isolated Millionaire

With a population of 1.7 million, Perth may be the world's most isolated large city (fig. 12.16). Located on Australia's west coast, Perth was established in 1829 along the banks of the Swan River and laid out according to a grid pattern commonly associated with colonial planning. As the colonial capital of Western Australia until 1901 (when the states were united as a Commonwealth), Perth grew slowly for its first one hundred years. Throughout its history Perth served both a rural and mining hinterland, with much of Australia's key mineral resources located in Western Australia—gold and bauxite, for example. It is mining and other global connections that have shaped the city over the past fifty years. The mining boom of the 1960s and 1970s, coupled with immigra-

tion (primarily from the United Kingdom but also from parts of southeast Asia), instigated an acceleration of the city's economic and population growth. The location of offices of mining companies and associated services saw tall buildings emerge on the city skyline (fig. 12.17). In Australia the 1980s were a decade characterized by an entrepreneurial spirit embraced by both government and business. A consumption and leisure-based economy emerged, aided by the city's hosting of the 1987 America's Cup Challenge. For the past two decades, the Perth economy has continued to thrive on its economic base of mining and tourism, boosted by substantial immigration.

Now capital of the state of Western Australia, Perth today is far removed from its colonial beginnings. Not only does it have a modern skyscraper-dominated skyline; but

Figure 12.17   Kings Park in Perth offers a view of the skyline that serves the commercial interests of Western Australia and the Indian Ocean rim. (Photo by Stanley Brunn)

the entrepreneurial governance of the 1980s and 1990s involved substantial redevelopment of older parts of the city as tourist and leisure spaces. The redevelopment of the old Swan Brewery site in inner Perth is one example of these processes. The State Government's development corporation chose to redevelop the site which was once home to the factory making Perth's famous beer. The Old Swan Brewery complex now hosts a myriad of leisure activities including theatres, dining, and office space, as well as car parking. Across Australian cities such redevelopment plans are invariably contested. Conflict over the Swan Brewery redevelopment project is representative of indigenous struggles to claim space within urban Australia. In this particu-

lar case, Aboriginal protesters drew attention to the symbolic significance the site held for them; and they wanted the brewery buildings demolished and the land returned to parkland. Their point was made in a variety of ways, including an eleven-month period in which they camped on the site. The protests were unsuccessful, with the government authority going ahead with the redevelopment and incorporating elements of Aboriginal culture into the design. At another level the protest was successful for the ways it brought an Aboriginal presence into the urban world.

Perth, like other Australian cities, is a sprawling city. Population growth has spawned metropolitan growth, initially to the east and more recently southward toward the municipality of Mandurah. For much of the 20th century it was presumed that the private car would adequately cater to the transportation needs of this growing population. More recently, however, the need for better public transportation has been recognized. A new, profitable and well-patronized suburban railway line to east Perth was opened. Perth is also home to a wide variety of other sustainable transport initiatives. Foremost here are "TravelSmart" programs, run by employers, schools, universities, or workplaces. These programs encourage individuals to consider non-car travel, and sometimes provide incentives to do so. Like Auckland's "walking school buses," they have been successful in reducing private automobile travel in Perth, and in raising awareness of the city's precarious environmental future.

Gold Coast: Tourism Urbanization

Australian sociologist Patrick Mullin has used the term "tourism urbanization" to describe a scenario of tourism-sustained urban growth, where (1) urban development is based primarily on tourist consumption of goods and services for pleasure, and (2) urban form is shaped by the city's function as a leisure space. The Gold Coast on Australia's Queensland coast can be understood in these terms.

In Australia's Gold Coast—25 miles (40 km) south of Queensland's capital city, Brisbane—white settlement began in the 1840s with timber-getting and agricultural development. By the 187' wealthy Brisbane residents were already   'ng the area as a leisure destin'   ᵊply as the South Coast. Th   ʰil connection from   ʰe area's app   ᵃ resorts er   ᵃom of the 1ᵇ   name "Gold

Coast" and began its development as Australia's highest intensity, high-rise tourist destination. Through many cycles of boom and bust, intense real estate investment in tourist accommodation, retail, restaurants, and entertainment ventures along this 35-mile (56 km) strip of spectacular surfing beaches, transformed the Gold Coast into the most intensely developed coastal tourist strip in Australia and a key international tourist destination.

By the 1980s, the area—especially Surfers Paradise at the heart of the Gold Coast—had gained a dubious reputation as a place of relaxed social norms, brashly opulent neon-lit landscapes, and get-rich-quick real estate deals. Nonetheless, the region matured as a tourist destination. Large-scale foreign direct investment in real estate, especially from Japanese interests in the 1980s and more recent Middle Eastern investment, brought significant diversification to the array of tourist products and consumption landscapes in Surfers Paradise and its hinterland. The area developed a series of integrated tourist resorts such as the Marina Mirage and the golf-themed Sanctuary Cove; large-scale retail malls such as Pacific Fair, Conrad Jupiters casino; multiple golf courses; and its multiple theme parks, including Movieworld, Sea World, Dream World, and Wet'n'Wild Waterworld.

The Gold Coast (incorporated as a city since 1959) has had a rapidly expanding resident population, which now stands at about half a million. Its more than 13,000 accommodation rooms in hotels and serviced apartments accommodate an additional 3.5 million domestic visitors and 800,000 international visitors annually, primarily from Asian countries and Aotearoa/New Zealand. But the Gold Coast today is underlain by more than a consumption-driven tourist economy. It is also one of the most rapidly developing cities in Australia, characterized by sustained rapid population growth rates of around 2 percent annually. Its growth is largely migrant-driven as lifestyle attractions have drawn in-migrants from across Australia, many of whom have found housing in low-density canal-estates built behind the high-rise coastal strip. More recently, as the Gold Coast has expanded, more conventional forms of suburbia have developed including a major new-town development in Robina to the southwest. As this has occurred, the initial dominance of retirees amongst in-migrants—prompting one author to label the city "God's waiting room"—has subsided such that the largest in-migrant group now ranges between 20–29 years old. The city's population is expected to reach nearly 789,000 by 2031; but the Gold Coast is also blending into the extended urban region of southeast Queensland (SEQ), a conurbation which stretches 150 miles (240 km) from Noosa southward through Brisbane and the Gold Coast to Tweed in northern New South Wales. SEQ's population is approaching 3 million, representing more than two-thirds of Queensland's population. The population of SEQ is projected to reach 4.4 million by 2031.

As the Gold Coast blends into this urban region, its economy is diversifying. Tourism-related industries have tended to support lower-skilled occupations and low-paid and/or casual employment, prone to seasonal fluctuation. Now, the state-supported Pacific Innovation Corridor initiative aims to promote the region's hi-tech, biotech, computing, and multi-media industries that will integrate the region into a globalized knowledge economy and improve rail and road connections to Brisbane's larger economy. Nonetheless, Gold Coast is still one of the lowest income cities in Australia and has higher levels of socioeconomic disadvantage than other Australian cities, in part a product of its occupational structure. The tourism-dominated economy

is reflected in lower-skilled occupations, low rates of higher education, high rates of low-paid casual employment, and high rates of unemployment. As the conurbation expands, challenges emerge: managing disadvantage, enabling economic diversification, building roads and transit systems, developing sustainable communities, and balancing environmental protection against development.

Auckland: Economic Hub of Aotearoa/New Zealand

Whilst not the nation's capital, Auckland has dominated Aotearoa/New Zealand's urban system since overtaking Dunedin and Christchurch as the country's largest city in the late 19th century. Like Sydney, it developed on an aesthetically and economically advantageous harbor, and is similarly renowned for its natural beauty. Historically, it too served a rich agricultural and forested hinterland. The deregulation of the national economy in the 1980s paved the way for the transformation of Auckland. It is Aotearoa/New Zealand's largest, most prosperous and economically active city. By the 1990s, it hosted more than a third of the nation's employment in manufacturing, transport, communication, and business services. It increasingly occupies a strategic position in the national economy, through its operation as a place in which and through which the global economy operates. It is the location of multinationals, international financial transactions, global property investments, and a hub for international tourists. Global rather than local connections are also important in explaining a number of other facets of urban life in Aotearoa/New Zealand.

The 1980s saw the transformation of the Auckland residential and commercial landscapes. High-rise residential towers (like the famous Sky Tower, tallest building in the Southern Hemisphere) were built around the city's CBD, often financed in foreign currencies, designed by architects outside Aotearoa/New Zealand, and managed by global property conglomerates. High-rise residential living has become increasingly popular. The building of medium-density housing has added to the city's density. Sometimes modeled on "new urbanist" ideas imported directly from the United States, these new suburbs modify the conventional suburban way of life with smaller houses, a gridded street pattern, and sometimes a communal open space. Though not gated communities in the strictest sense, the role of these new suburbs in fostering social exclusion is an ongoing issue. In fact, the same issue often arises as inner city neighborhoods undergo gentrification (box 12.5)

Lifestyle television programs and home-focused magazines are hugely popular and foster expenditure on household items and renovation projects in Auckland and its suburbs. Suburban backyards may be getting smaller, but they still serve the important purposes of providing a place for children to play, domestic vegetable cultivation, and the fulfillment of aesthetic and economic aspirations. Some new groups of migrants do aspire to and do fulfill these suburban ideals, like residence in a detached house. Migration has also transformed suburban landscapes. Suburbs like Sandringham, with new places of worship and retail landscapes, have been the destination of many migrants from Asia.

The sustainability of a large, dynamic city like Auckland is attracting increasing scholarly and policy attention. Contradictions between reliance on the private motor vehicle and a strong environmental consciousness have seen the widespread adoption of "walking school buses." Rather than children being driven indi-vidually to school, they are dropped at locations along a designated route and walk to school with other children with parental supervision. Walking school buses now operate in many Auckland suburbs, more likely to be middle-class neighborhoods. They have been credited with removing cars from the road, reducing air pollution, reducing obesity, and enhancing community. Official urban policies of sustainability have already influenced the building of medium-density housing and housing with a small ecological footprint. A more widespread implementation of urban sustainability in Auckland has also recently been discussed.

Port Moresby and Suva: Island Capitals

Port Moresby and Suva are the largest cities, and political capitals, of their respective nations of Papua New Guinea (PNG) and Fiji (fig. 12.18). They have parallel histories, urban patterns, and contemporary influences. While their current political instabilities may be unique, their other characteristics are broadly representative of cities in the island Pacific.

Neither PNG nor Fiji has a prosperous economy. They have weak manufacturing sectors, are reliant on an agricultural enterprises beset with inefficiencies and at the mercy of low globalization, and are plagued by political instability. Hence, both Suva and Port Moresby have fragile economic bases. While population has been steadily growing in both cities, employment opportunities have not. Consequently, unemployment is high, with one estimate putting unemployment in Port Moresby at around 60 percent. These fragile economic circumstances underpin the most salient characteristics of Pacific Island cities: a large informal sector, including informal settlements, plus political problems and unrest (fig. 12.19).

Box 12.5   Gentrification and Ponsonby Road, Auckland

Whether the claim that gentrification is now a global phenomenon is valid or not, this urbanization process has certainly reshaped the inner suburbs of many of Australia's and New Zealand/Aotearoa's cities. The process has witnessed middle-class renovation and resettlement of formerly working-class housing in inner city neighborhoods in all the major metropolitan centers and regional cities. Gentrification is not merely a residential phenomenon but one involving re-fashioning local shopping streets, leisure and recreation facilities, and neighborhood services as residents' aesthetics, ethos, and consumption patterns combine to mold local streetscapes. These impacts are evident on King Street in Sydney's Newtown, Brunswick Street in Melbourne's Fitzroy's, Boundary Road in Brisbane's West End, Darby Street in Newcastle's Cooks Hill, and Ponsonby Road in Auckland's Ponsonby.

The suburb of Ponsonby is located less than a mile west of Auckland's CBD. After World War II, many of Ponsonby's more prosperous residents relocated to the expanding outer suburbs and were replaced by lower-income Pacific Island and Maori migrants. However, waves of gentrification commenced in the 1970s, as diverse groups of young, well-educated, Pakeha (white New Zealanders of European descent) were attracted to the area by its cheap property, low rents, and social and ethnic diversity. Ironically that diversity can be threatened by the very process of gentrification. In Ponsonby's case, gentrification overlapped with an Auckland-wide housing boom and property price inflation in the 1990s; the result has been significant displacement of lower-income, less-educated inhabitants, driven out by rising rents and spiraling house prices. Ponsonby's population has, proportionately, become distinctly "whiter" and higher income. Nonetheless, despite price inflation, the area has maintained a relatively young population and a significant proportion of rental housing.

Certainly, diversity is characteristic of the dramatic transformation of the consumption spaces and public culture of Ponsonby Road (http://www.ponsonbyroad.co.nz/ponsonby-road/). Gentrification has combined with changes to licensing laws to see the birth of a thriving agglomeration of over 90 cafes, restaurants and bars, interspersed with specialty stores, greengrocers, butchers, and newsagents. Alan Latham's (2003) research has shown how the plethora of cafes and bars—often flamboyantly and expensively styled and open to the street—departs from the more traditional, enclosed spaces of pubs and the culture of hard-drinking masculinity they accommodate more readily than other forms of sociality. As a result, gentrification has seen Ponsonby Road develop a range of more ambiguous spaces for consumption and sociability that are welcoming to women, gay-friendly, and less confined to traditional norms of gendered identity. In this particular site of gentrification, working class displacement and middle-class colonization have been accompanied by the development of a diverse public culture that, while definitely accessible (most easily to those with disposable income), is open to diverse expressions of identity and diverse ways of inhabiting the city.

Source: Latham A. "Urbanity, lifestyle and making sense of the new urban cultural economy," *Urban Studies*, 40 (2003), pp. 1699–1724.

Figure 12.18   Located in the CBD, Suva's old town hall now houses Greenpeace, a champion of Pacific Ocean conservation. (Photo by Richard Deal)

Informal, squatter-like settlements are common in these cities. Port Moresby has at least 84 agglomerations of substandard, poorly serviced housing, in which urban poverty is concentrated; Suva has just a little less. Basic urban infrastructure—water, sewerage, electricity, and garbage collection—is either completely lacking or minimally provided in such settlements. Problems are exacerbated by a lack of formal employment opportunities. Urban poverty is rising, exemplified by the increasing number of street children. Informal employment, particularly prostitution, has arisen to counter the lack of formal-sector employment opportunities.

Policy responses to urban poverty and marginalization in Suva and Port Moresby have been small and problematic. The under funding of basic infrastructure has contributed to the problem. There is widespread opposition to the urban poor and street prostitution. The government's response to prostitution, the prevalence of street children, and informal settlement has been largely negative. In PNG, problem settlements have been bulldozed rather than adequately resourced. More gen-erally, these cities have been sites of social and political unrest, which has had implications for the internal structures of these cities. In Port Moresby, for example, security concerns have seen European and other expatriates withdraw further into barricaded residential estates on the hillsides of the city.

TRENDS AND CHALLENGES

Many of the cities in Australia and the Pacific are cradled by fragile ecosystems and are extremely vulnerable to the multifaceted impacts of climate change. Australia's largest cities are further challenged by the fact that they area all located in areas where climate change is inducing significant declines in rainfall levels. All of the capital cities have desalination plants in operation or nearing completion, to convert seawater to drinking water. This may be one solution to water supply, but has other environmental impacts due to the voluminous energy demands of the desalination process. In addition, the geographic expansion of urbanized areas involves

Figure 12.19   Because it is an overseas territory of France, decisions in Paris affect workers in Nouméa, sometimes precipitating protest, in this case led by the labor unions. (Photo by Richard Deal)

the loss of productive land, loss of biodiversity, and increased energy use. The imperatives in all cities have thus become reduced energy consumption and emissions reduction alongside increased use of renewable energy.

Urban governance provides many challenges across the region. The challenge is the establishment of effective urban governments able to meet environmental and security challenges and fashion positive outcomes (fig. 12.20). Governance processes that contribute to social cohesion are also key. In Australia and Aotearoa/New Zealand, urban governance is now characterized by a variant of neoliberalism in which market processes and solutions underpin policy. Waterfront redevelopments in many cities are classic outcomes of neoliberal policies. The extent to which such governance is equitable remains questionable, and ways to produce more "just" cities within such a framework are still being sought. Equitable outcomes for indigenous peoples of these cities are especially important.

Finally, the provision of adequate, appropriate, and affordable housing is a pressing issue for all cities in the region. In Sydney and Melbourne particularly, where house-price escalation has been intense, affordability has now reached historic lows. Mortgage stress—where households are paying more than 30 percent of gross household income on housing—has risen, most particularly in the suburbs. The impacts of the affordability crisis include displacing younger people and lower-paid workers from high-cost urban areas, labor shortages, and growing debt burdens on households with mortgages.

SUGGESTED READINGS

Baum, Scott, Robert Stimson, and Kevin O'Connor. 2005. *Fault Lines Exposed: Advantage and Disadvantage across Australia's Settlement System.* Clayton, Victoria: Monash University ePress. Uses quantitative analysis to identify categories of places that are differentially advantaged and disadvantaged by processes of global change.

Connell, John, and J. P. Lea. 2002. *Urbanisation in the Island Pacific.* London: Routledge. Third

Figure 12.20 One of the challenges of urban governance in Australia is maintaining safe streets. Signs like this one in Sydney have been increasing rapidly as people everywhere become more security conscious. (Photo by Donald Zeigler)

edition. An overview of urbanization in 11 independent island states.

Forster, Clive. 2004. *Australian Cities: Continuity and Change.* South Melbourne: Oxford University Press. Explores the urban experience across the Pacific Islands, including the role of cities in national development and as centers of globalization.

Jacobs, Jane M. 1996. *Edge of Empire: Postcolonialism and the City.* London and New York: Routledge. An analysis of how the connections built by globalization and the postcolonial world shape and reshape the composition of cities.

Le Heron, Richard, and E. Pawson. 1996. *Changing Places: New Zealand in the Nineties.* Auckland: Longman Paul. Provides an overview of the transforming geography of New Zealand including its cities and regions in the context of globalization.

Major Cities Unit. 2010. *State of Australian Cities 2010.* Infrastructure Australia. Canberra: Australian Government. Provides a detailed empirical snapshot of demographic, economic, social, environmental and governance dynamics of Australia's major cities.

McGillick, Paul. 2005. *Sydney, Australia: The Making of a Global City.* Singapore: Periplus. A photographically illustrated history of Sydney's built environment.

McManus, Phil. 2005. *Vortex Cities to Sustainable Cities: Australia's Urban Challenge.* Sydney: University of New South Wales Press. Examines the histories and planning decisions that have contributed to the unsustainability of Australian cities.

Newton, P. (ed) 2008. *Transitions: Pathways toward Sustainable Urban Development in Australia.* Canberra: CSIRO Publishing. Examines demographic and developments trends and their implications for resource demands and potential pathways to sustainable urban development.

O'Connor, Kevin, Robert Stimson, and Maurice Daly. 2001. *Australia's Changing Economic Geography: A Society Dividing.* Melbourne: Oxford University Press. Provides an overview of the impacts of economic and social change, including globalization, on spatial patterns of economic performance across Australia.

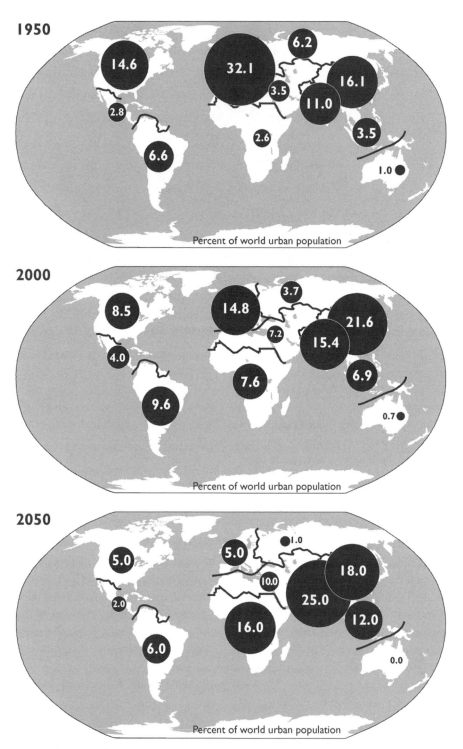

Figure 13.1   Urban Populations: 1950, 2000, and 2050. *Source:* UN, *World Urbanization Prospects: 2001 Revision*, http://www.un.org/esa/population/publications/wup2001/wup2001dh.pdf Projections for 2050 by authors

13

Cities of the Future

STANLEY D. BRUNN, RINA GHOSE, AND MARK GRAHAM

KEY URBAN FACTS

Largest in Population (in millions: 2010–2025 projections)	Tokyo (37.1), Delhi (28.6), Mumbai (25.8), São Paulo (21.6), Dhaka (20.9), Mexico City (20.7)
Urban Areas Adding Most Residents (in millions: 2010–2025 projections)	Delhi (6.4), Dhaka (6.3), Kinshasa (6.3), Mumbai (5.8), Karachi (5.6), Lagos (5.2)
Fastest Annual Growth Rates (%) (2010–2025 projections)	Ouagadougou (8.5), Lilongwe (7.1), Blantyre-Limbe (7.1), Yamoussoukro (6.9), Niamey (6.7), Kampala (6.6)
Slowest Annual Growth Rates (%) (2010–2025 projections)	Dnipropetrovsk (-0.25), Saratov (-0.20), Donetsk (-0.17), Zaporizhzhya (-0.15), Havana (-0.11), Volgograd (-0.09)
Sharpest Declines (in thousands: 2010–2025 projections)	Dnipropetrovsk (37), Havana (35), Saratov (24), Donetsk (24), St. Petersburg (18), Zaporizhzhya (17)

KEY CHAPTER THEMES

1. Two established urban trends will continue: slow and no growth and graying populations in much of the developed world and rapid growth of youthful cohorts in much of the developing world.

2. Regional and global migrations will continue, with marked Asianization and Africanization occurring in Australia, Europe, and North and South America.

3. Increased cultural homogenization will characterize cities in the Arab, Chinese, Japanese, and Indian worlds, and increased cultural diversity will characterize cities in Europe, Australia, the U.S. and Canada.

4. The cultural, social, and economic faces of globalization will result in both core and peripheral urban locations having similar built environments, while making difficult the preservation and conservation of indigenous cultures and practices.

5. Both K-knowledge and S-survival economies will be paramount in looking at the impacts of globalization where fluidity and dynamism will be norms in daily life, corporate and institutional life, and governing structures.

6. Time-space convergence attributed to improved information and communications technologies is providing increased flexibility in the location of key banking, health, and education institutions and in the empowerment of traditionally marginalized populations.

7. Flash points of potential ethnic and cultural conflict, including anti-globalization and anti-Western worldviews, and "gaps" between rich and poor countries are likely in some cities of the periphery: the Caribbean and U.S.-Mexico borderlands, the Mediterranean, and Southeast Asia.

8. Regional and global environmental problems, including global warming, pollution, solid waste management, and water quality and availability, are certain to be center stage in planning the futures of the world's cities.

9. Human security issues are destined to be major problems facing cities heavily dependent on imported food and water and skilled and semiskilled immigrants from the developing world.

10. Commitments to solve global urban problems will call for the growth and export of knowledge-economies and new cohorts of internationally trained, experienced, and traveled professionals, plus creative grassroots empowerment initiatives.

We have approached cities from historical and contemporary perspectives in the previous chapters. Now it is time to think about the future. One prediction that has been borne out in the first decade of this century is that the majority of the world's residents now live in urban regions. That milestone was reached sometime in 2007. That transition from majority-rural to majority-urban was significant on a global scale, but not so much in more developed regions where more than half of the residents have been living in cities for a half century or more. The United States, for example, reached the urban-majority mark in 1920; the developing world will reach this mark before 2020. What does the future hold for such a world and its people? Will present trends continue or will there be a sea change in the very nature of urban life? Will cities in More Developed Countries (MDCs)

Table 13.1   Projected Populations of the Largest Urban Agglomerations in 2025

Rank	City	Pop. (milllions)	Rank	City	Pop. (milllions)
1	Toyko, Japan	37.09	16	Los Angeles-Long Beach, U.S.	13.68
2	Delhi, India	28.57	17	Al-Qahirah (Cairo), Egypt	13.53
3	Mumbai (Bombay), India	25.81	18	Rio de Janeiro, Brazil	12.65
4	São Paulo, Brazil	21.65	19	Istanbul, Turkey	12.11
5	Dhaka, Bangladesh	20.94	20	Osaka-Kobe, Japan	11.37
6	Mexico City, Mexico	20.71	21	Shenzhen, China	11.15
7	New York-Newark, U.S.	20.64	22	Chongqing, China	11.07
8	Kolkata (Calcutta), India	20.11	23	Guangzhou, China	10.96
9	Shanghai, China	20.02	24	Paris, France	10.88
10	Karachi, Pakistan	18.73	25	Jakarta, Indonesia	10.85
11	Lagos, Nigeria	15.81	26	Moscow, Russia	10.66
12	Kinasha, Dem. Rep. Congo	15.04	27	Bogotà, Colombia	10.54
13	Beijing, China	15.02	28	Lima, Peru	10.53
14	Manila, Philippines	14.92	29	Lahore, Pakistan	10.31
15	Buenos Aires, Argentina	13.71	30	Chicago, U.S.	9.94

continue to stabilize in population and eventually decline, while those in Less Developed Countries (LDCs) experience continued high population growth? What will be the consequences of divergent development paths on urban economies, communication and transportation, and human organizations? These questions intrigue urban geographers as we examine the early 21st century.

URBAN POPULATIONS: WINNERS AND LOSERS

The United Nations periodically publishes reports on urban futures, with particular emphasis on numbers, housing, food, health, and environmental quality. A 2010 U.N. publication provides data on the projected populations of major world regions and countries within them. This report also provides information on 595 agglomerations that had more

than 750,000 in 2009. Data for these cities are provided for 1950, 2005, and 2025. There are also UN data for the thirty largest agglomerations in 2025 (table 13.1)

In chapter 1 is a table listing the 30 largest agglomerations in 1950, 1975, and 2000. Two distinct differences are observed in the largest agglomerations in these years and in 2025. First is the changing "mix" or numbers of agglomerations in the More Developed Countries and the Less Developed Countries. In 1950, 19 of the largest were in the developed world, especially Europe and North America; by 2000 that number was reduced to 9. By 2025, only 7 of the largest agglomerations will be in More Developed Countries. The rapid increases are in the developing world, especially countries in South, Southeast and East Asia and also in the Greater Middle East and in Sub-Saharan Africa.

The second dramatic change from 1950–2025 is in the total number of people living in the thirty largest agglomerations on the planet. In 1950 the total population for these thirty cities was 117 million, by 2000 it was 347 million, and by 2025 it is projected to be 479 million. Most of that growth, as we have seen in the previous chapters is in the developing world and especially in large cities with more than 5 million inhabitants. In 2010 there were 53 cities in the world that had more than 5 million inhabitants; by 2025 it is estimated there will be 72 with a combined population of 753 million. Almost one-half (37) of the 72 will be in South, Southeast, and East Asia (fig. 13.1).

What is also noteworthy about these largest agglomerations is the appearance of some cities on all lists, the disappearance of some cities that were prominent early, and the appearance of other cities in the most recent decades. In 1950 the largest agglomeration was New York-Newark with a population exceeding 12 million; that number will qualify Osaka-Kobe to be the 20th largest agglomeration in 2025. Glasgow was the thirtieth largest city in 1950 with only 1.76 million people; 60 cities in China will have that many people in 2025. Even cities that are currently considered major world cities are dropping in rank. Paris was ranked 4th in 1950 and is projected to be 24th in 2025; London was 3rd in 1950 and will not be among the 30 largest cities in 2025.

What clearly emerges in looking at the 2010–2025 projections for those agglomerations over 750,000 today are two distinct patterns (table 13.2). First, there will be some "clear losers" on the world scene, that is, those cities that are actually expected to lose population between 2010 and 2025. Second, a somewhat larger number of cities will grow very slowly, and third, a larger number of cities will actually experience rapid growth.

These are results one can obtain from studying these UN data:

- 15 cities will actually lose population; 9 of these are in Russia and include Saratov, St. Petersburg, Samara, Omsk, and Perm, all heavy industrial cities; 3 similar cities in Ukraine, Dnipropetrovsk, Donetsk, and Kharkiv will also lose inhabitants; the combined losses of the 15 cities are anticipated to be 308,000 inhabitants.

- 64 cities will add fewer than 100,000 inhabitants each between 2010–2025; most of these are also old industrial centers in Europe (42 cities) and East Asia (South Korea 11 and Japan 7); the combined gains of all 64 cities will be *only* 2.7 million or about what Abidjan is expected to experience from 2010–2025.

- 80 cities are expected to add between 101,000–200,000 inhabitants each from 2010–2025; 30 of these will be in the United States and Canada, 14 in South America and 10 in Central America; the combined growth of these 80 cities will be *only* 11.8 million, which is less than the combined growth of Delhi and Dhaka in this same fifteen year period.

- 71 agglomerations will add more than 1 million residents each; their combined growth will be 82.3 million new residents; the countries adding the most "new million residents" are China (14) and India (11); no European or North America city will add more than one million residents. These cities will add more than 5 million *each*: Delhi (6.4 million), Dhaka and Kinshasa (6.3 million each), Mumbai and Lagos (5.7 million each), and Karachi (5.6 million); these six cities will

Table 13.2   Projected Urban Losers and Gainers 2010–2025

REGION	Number of Cities Over 750,000 in Each Region		
	Expected Population Decrease*	Expected Medium Growth**	Expected Highest Growth ***
Sub-Saharan Africa	0	2	24
East Asia	0	22	14
South Asia	0	2	16
Southeast Asia	0	3	6
Greater Middle East	0	10	6
South America	1	13	3
Central America	1	2	1
U.S. and Canada	0	40	0
Europe	1	38	0
Russia and Ukraine	12	1	0
Australia/Pacific	0	1	0

*Number of cities expected to decline in population.
**Number of cities expected to grow, but by less than 200,000 inhabitants.
***Number of cities expected to increase by more than 1 million inhabitants.
Source: UN, *World Urbanization Prospects: 2009 Revision*, http://esa.org/unpd/wup/index.htm

add 36 million residents between 2010–2025; another way of looking at Delhi's rapid growth is that it will add 100,000 inhabitants per month, 3,500 per day and 145 new residents per hour!! Delhi will add more residents from 2010–2025 than the population of Paris in 1950.

- 52 cities will increase more than 50 percent from 2010–2025; most of these are in Sub-Saharan Africa. Ouagadougou in Burkina Faso, Lilongwe and Blantyre-Limbo in Malawi, and Niamey in Niger are projected to increase more than 100 percent. Another 7 cities will increase from 50–99 percent. At the other extreme are 101 cities that will grow less than 10 percent; most of these are in Europe, Russia, Japan, and South Korea. Modest growth (about 30 percent) is an-

ticipated for both Chinese and Indian cities and 10–15 percent for those in the U.S. and Canada.

In 1950, urban residents numbered approximately 750 million worldwide. By 2000 the figure was 2.8 billion, and by 2050 it is expected to be 6.5 billion. In 1950, Africa and Southwest Asia had 6 percent of all urban residents, while Europe had 32 percent and Asia about 30 percent. Between 1950 and 2000, Africa more than doubled its city residents, while Asia's urban population grew by 40 percent. In 2000, Europe's share of the world's urban population had declined to about 15 percent and North America's to only 8.5 percent. By 2050, more than one in four urban residents will be living in African and Middle Eastern cities, and one in two will be living in Asian

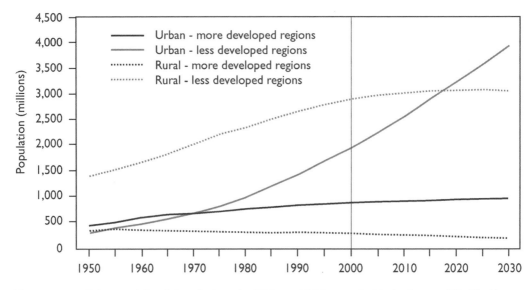

Figure 13.2   Urban and Rural Populations in MDCs and LDCs, 1950–2030. *Source:* UN, *World Urbanization Prospects: 2001 Revision*, http://www.un.org/esa/population/publications/wup2001/wup2001dh.pdf

cities. Europe and Latin America will have almost the same number of city residents. U.S. and Canadian urbanites will almost equal the number living in urban West Africa. In aggregate, there are currently more rural than urban residents in LDCs, but this pattern will be reversed by about 2020 (fig. 13.2). Cities of the developing world, and especially Asia, are accounting for an increasingly larger share of the world's urban population (box 13.1).

The world's largest cities by 2050 will be largely in the LDCs and especially Asia. Significant questions emerge about the implications of these changes for human and natural resource bases and the future of humanity. Tokyo (27 million) is expected to remain the world's largest agglomeration, but it will be followed by cities in the LDCs: Dhaka and Mumbai (each about 23 million), São Paulo (21 million), and Delhi (20 million) (fig. 13.3).

TEN HUMAN GEOGRAPHIES OF THE EARLY 21ST CENTURY

Economic, social, and political futures in any region are affected by local cultures and events as well as external regional and global actions and institutions (box 13.2). These futures are likely to surface first in large cities; they are not listed in any order of magnitude.

1. *An urbanizing world.* Urbanization will continue to concentrate a growing population on comparably less land. Urban agglomerations will grow bigger and urban institutions will increasingly dominate even rural areas. Interactions and associations are likely to increase between cities near and distant rather than between cities and rural trade areas.

2. *Urban connectedness, anomie, and placelessness.* Faster transportation and information

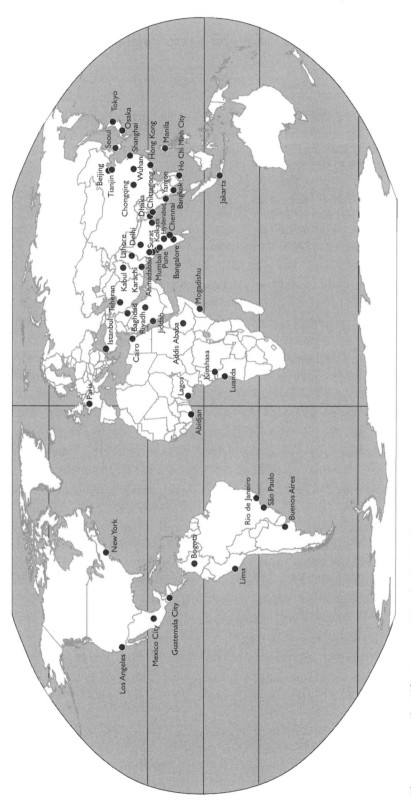

Figure 13.3 The Fifty Largest Cities in the World, 2050. *Source:* UN, *World Urbanization Prospects: 2001 Revision,* http://www.un.org/esa/ population/publications/wup2001/wup2001dh.pdf

Box 13.1   A Global Urban Village

How has the composition of the world's urban population changed during the past half century? One way to answer this question in a meaningful way is to consider, using U.N. data, the 30 largest agglomerations in 1950, 1975, 2000, and 2015 and reduce the huge numbers into a "global urban village" of only 100 people. The largest number of residents in our village in our first year would be from New York (10 residents), but then Tokyo would have more residents in the other years (9, then 8, and then 7). Both cities would be losing citizens in each year, so that by 2015 New York would have only 4 residents and Tokyo only 7.

Not unexpected from what we have seen described elsewhere in this chapter are declines in the numbers from European and U.S. cities and increases from LDC regions. Europe and the United States would account for half of all villagers in 1950, but only 13 combined by 2015. Manchester, Birmingham, Rhine-Ruhr, and Boston would have lost representation. Another 19 villagers would have come from East Asia in 1950, with 30 by 1975, but only 29 by 2015. The largest number of newcomers would come from South and Southeast Asia, growing from 6 in 1950 (all these from South Asia) to 36 from 10 different cities in these regions in 2015. The lion's share of these additions would hail from Dhaka, Mumbai, and Delhi with 5, 5, and 4 residents, respectively, in 2015.

There would be no Russian residents in this village after 1975 (4 would come from Moscow and 2 from St. Petersburg in this year) and there would be none any year from Oceania. Sub-Saharan Africa would have none in 1950, but 6 residents (from Lagos and Kinshasa) in 2015. Cairo would have 1 resident in 1950 and by 1975 it would be joined by Tehran as the only cities with a resident from the Middle East; these two cities would have 6 residents combined in 2015. Mexico City would be the sole representative of Central America; there would be 2 from this agglomeration in 1950 and 5 in each succeeding year. South American cities would always be represented, initially by Buenos Aires and Rio de Janeiro, but later by São Paulo, Lima, and Bogotá. By 2015, there would be 15 residents in our village from these 5 cities, with most from São Paulo. What is clear from the cities of origin of residents in our "global village" is that the number of MDC regions would decrease and the number of LDC regions would increase. Whereas in 1950 there would be 61 villagers from 18 different cities in North America, Europe, and Russia, but 2015 there would be only 13 citizens from just 4 cities in those regions. By contrast, the numbers and composition from LDC regions would increase, from 12 cities and 39 residents of our village population of 100 in 1950 to 26 cities and 87 residents by 2015. (Note: Japan is included in the East Asia region. If counted with the MDC regions, the numbers would change somewhat.)

Box 13.2   Seeing Cities on the Soles of Your Feet

Donald J. Zeigler

The cities of the world await your visit. Start learning urban geography the way it should be learned: on the soles of your feet. Be attentive to details, but also use the wide-angle lens your education has provided. Look for patterns and processes on the landscape and document them for the future. When geographers are traveling, they are doing research. Here's how:

Take pictures. Photograph people and landscapes—ordinary and extraordinary. You can't cover everything, but you can find a few topics of personal interest and follow them from city to city: skylines, waterfronts, signs, buskers, open space, monuments, maps on the landscape, etc. Meander off the "high streets" and slow down. What story is the city trying to tell you? Help tell that story in pictures. To do that, you will have to become a master of field notes as well.

Keep a journal. Document your observations in words. When you stop for coffee or lunch, take out your journal (or open your netbook or tablet) and record a few notes and impressions. Then, at the end of the day, synopsize what you have learned. It will take discipline but your essays will bring back the experience like nothing else you can do. And remember the first axiom of learned travel: writing it down helps you think it out.

Open your ears. Attune ears to new languages, new music, and new cacophony. Each city has an auditory signature of its own. Listen to how the local place names are pronounced. Find out what CDs are selling fast. Talk to people you meet, listen to their accents, and pay attention to how they tell their stories, not just what they have to say. Record what you can.

Educate your taste buds. Eating is a learning activity, and the cities you visit will be anxious to educate you. Learn about the regional cuisines, patronize the microbreweries, search out the farmers' markets, and become a locavore. Ask questions about what you are eating. Use words and pictures as gustatory *aides memoires*, field documents that will activate your taste buds' memory.

Acquire some mementos. But be selective. Look for post cards, buy a local newspaper, find a local language dictionary or children's book, save a few coins or bills, and buy a few stamps at the post office (the visit alone will be worth the experience). Go to the supermarket for a box of cereal: eat the cereal and save the box. Go to the tourist kiosks for a souvenir: see if you can find one that was actually made in the country you are visiting.

Figure 13.4   Wireless fidelity, commonly known as wi-fi, is a local area network that has been used since 1997 to turn cities, even small ones, into Internet hubs. Here in West Palm Beach, surfers depend on waves not wires to keep them connected. (Photo by Donald Zeigler)

and communication technologies may lead people to lose their sense of place. The result could be anomie—alienation and social instability. Cell phones, fax machines, the Internet, and wireless communications will diminish the significance of place (fig. 13.4).

3. *Meshings of the local and the global in daily life.* Scale meshings will be evident in transactions and interactions—where one works, with whom one works, and the destinations of goods and services produced, including luxury crops, telecommunications equipment, digitized health records, or components in a global product (computers or motor vehicles). While some urban residents will interact predominantly at very local levels, others are operating at international and transnational scales.

4. *Asianization and Africanization of Europeanized worlds.* An ongoing contemporary cultural global process is the impress of Asian and African diasporas on traditionally Europeanized worlds (fig. 13.5). While non-Asian cultural groups are also involved in worldwide migration, Asians are having a significant visible impact, especially the large numbers of skilled and unskilled, legal and illegal residents into gateway cities of urban Europe, North America, and Oceania (fig. 13.6). Africans are also emerging as new diaspora communities grow in Latin American, Asian, and Australian cities. These new cultures add new layers of food, music, entertainment, and intellectual diversity to city life.

5. *Increased regional and global awareness.* The diffusion of Information and Communica-

Figure 13.5   The first generation of Vietnamese immigrants to Australia came after 1975 when South Vietnam fell, beginning a chain of Vietnamese migration to a Europeanized world, in this case Brisbane. Can you identify the two flags? (Photo by Donald Zeigler)

tion Technologies (ICTs) and instant global reporting of major crises will increasingly transcend cultural, national, and political boundaries and raise the awareness of planetary concerns. International boundaries will further diminish with globalization. Examples include the free flow of goods, services, labor, and capital throughout the European Union, the importance of global and regional environmental treaties, and global pressure to assure basic human rights of women, children, elderly, disabled, and cultural minorities.

6. *Competitive K-economies.* "K" is for knowledge. It symbolizes the transition from handware to brainware and the importance of images and symbols in product consumption. These brain economies will be of increased significance in globally competitive and creative cities.

7. *Contested legal structures.* Increased volumes and densities of transborder urban networks and circulations will raise questions about the effectiveness of traditional city and state governments in daily life, including individual versus group rights and the legal structures affecting temporary and permanent residents. Additional unsettled issues will include the rights of those without property and statehood, legal status of NGOs' international employees, and the ownership of natural resources and cultural properties.

8. *Redefining norms and abnorms.* Urban cultural and political clashes (subtle and violent) are likely to continue to emerge among groups: those calling for tolerance and diversity in workplace, living, lifestyle, and social spaces versus those seeking to retain traditional norms based on religious and rural values or outdated modes of authority. Extremists may have a profound impact on future urban life.

9. *Continued technological breakthroughs and limits.* Now that we have faster transportation and nearly instant communications (at least in MDCs), will there be constraints on the adoption and dissemination of new technologies because of adverse social impacts on a culture, such as high fixed costs and government security? Will the technology "gaps" between the haves and the have-nots begin to narrow? What are the impacts of those cultures

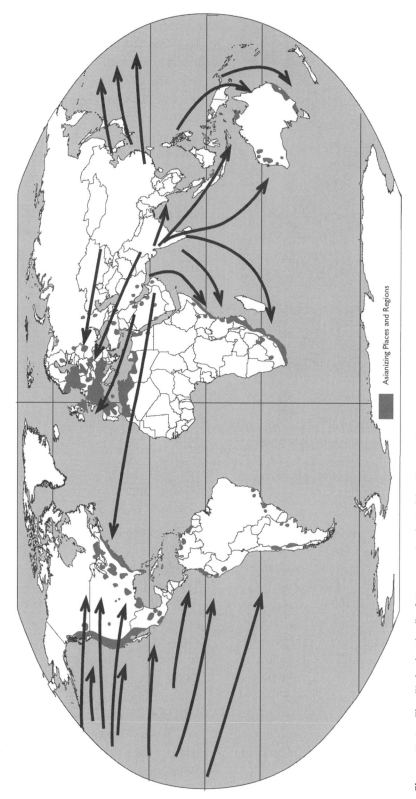

Figure 13.6 The "Asianization" of Europeanized Worlds. *Source:* Stanley Brunn

Asianizing Places and Regions

experiencing "leapfrog innovations," for example, cultures experiencing the wireless and digital worlds without having first had fixed land lines or solar energy or without the experience of fossil fuel dependency?

10. *Veneers of homogeneity amidst diversity.* The "McDonaldization of the World," or the creation of a Western globalized consumer world dominated by major Western food, music, fashion, and entertainment, will reflect a certain visible sameness in many urban landscapes. But beneath and alongside these landscapes and icons of Western and American hegemony in LDCs will be the rich historical and cultural mosaics of enduring regional cultures.

We next posit some distinguishing features of future cities and city life. Much of this thinking comes from social scientists, planners, and engineers in the MDCs, who design new built environments and technologies, plan urban infrastructures, collaborate with governments and universities, and serve as consultants with LDCs. They also learn new ways to apply applications to improve life in MDCs.

GLOBALIZATION AND URBAN LINKAGES

Urban form has always been shaped, in part, by communication and transportation systems and structures. Canals, tramlines, subways, and expressways have all significantly altered not just physical distance, but instead, the time it takes to travel between any two connected places. The time that it takes to move people, goods, and information within or between cities (*i.e.,* distance in time-space) affords places distinct advantages and disadvantages. The contemporary and future linkages between and among urban places can be illustrated using a number of different programs and initiatives that promote greater regional awareness, understanding, and activities. Many examples of linkages exist at the international and corporate levels, others exist at the scale of organizations and groups, and still others exist at the local levels. Local and global worlds are becoming increasingly fluid and fused with top-down and bottom-up initiatives linking places people work to places their consumer goods come from to places they spend leisure time and money. We illustrate these networks and linkages using several different examples, realizing, of course, that many more examples exist in cities and between cities.

Globalization and Anti-Globalization

Globalization is, and will likely continue to be, one of the most frequently repeated themes of the 21st century. The idea that cities are becoming economically, culturally, and politically more globally connected is grounds for hope, desire, fear, and despair. Globalization can lead to shared advances in science and technology, economic growth, and the exchange of philosophic, political, and artistic ideas between people and places that were once disconnected. However, increased connections can also result in economic and political exploitation and the destruction of cultural exchange (fig. 13.7).

Globalization is not something that spreads across the surface of the globe like fog rolling over a chain of hills. Rather it is evident only in specific times and places. In

Figure 13.7   McDonald's has become the symbol of globalization around the world, but anti-globalization forces envision a future without the "golden arches." (Photo by Donald Zeigler)

other words, if we take the example of any city experiencing transformations because of global connections, that city is not necessarily spatially proximate to the sources of those transformations or spatially proximate to other cities experiencing similar transformations. A useful way to conceptualize cities is through the concept of a wormhole. Geographer, Eric Sheppard, used it to imagine how connections between cities are specific and contingent while ignoring in-between places.

A widely repeated idea is that the absence of globalization is responsible for global inequities. In other words, cities that remain disconnected from global processes and flows of trade are unlikely to have high living standards. Such ideas are usually based on non-liberal economic theories that rely on "the logic of the marketplace." By allowing the global market to regulate society instead

of being regulated by society, it is argued that market forces will enrich people in poor cities by effectively governing and creating wealth for all participants. The counterpoint of this argument is the idea that ever-increasing globalization and integration of cities will only exacerbate global inequalities. Within the globalized economy, capital and jobs can now be moved rapidly from place to place, but the actual populations of cities remain rooted in their cities. The effects of globalizing processes have led some scholars to refer to "a race to the bottom," in which people have to accept increasingly lower wages and benefits in order to perform the same jobs. These concerns have fueled a massive global movement of protest and highlight the inequalities caused and intensified by globalization, including antiglobalization demonstrations at the World Trade Organization's conference in Seattle (fig. 13.8).

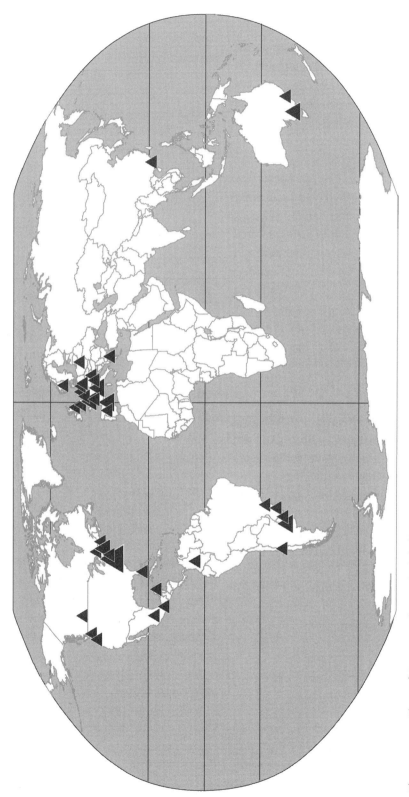

Figure 13.8 Sites of Recent Major Anti-Globalization Protests. *Source:* Mark Graham

Airline Connections

One of the most important ways that globalization can be measured is through the movement of people and cargo by air. However, the most globalized airports are not always located in cities that most people would associate with a high degree of global connectivity. Of the world's three largest cargo airports in 2010, one comes as a surprise. Memphis, Tennessee, is outranked only by Hong Kong as the world's second largest cargo-handling airport, due to FedEx's primary hub being located there. Following Memphis are Shanghai and Seoul, and, then, surprisingly, Anchorage which serves as the crossroads between North America and Asia on the "great circle route." In Europe, the leading cargo-handling airports are Frankfurt and Paris; Dubai leads in the Middle East.

When passenger flows between cities are considered, we get a more familiar picture of cities networked into the global economy. Europe and particularly North America dominate the rankings. While some highly ranked cities have high connectivity because they serve as major airline hubs (*e.g.,* Atlanta as a hub for Delta Airlines), others including London and New York are highly connected due to their central position in global economic and tourist flows. London's Heathrow ranked first in Europe; Atlanta and Chicago were at the top of the list in North America; and Beijing and Tokyo took top rank in Asia and in 2010.

Quality and Costs of Living

Many individuals, corporations, organizations, and governments are interested in quality of life measures. These parameters not only measure one's standing vìs-a-vìs other cities, but this data can also be used to promote a city's climate for investment, international conferencing and sporting events, and to attract specific groups, including artists, scientists, wealthy retirees, and tourists to these cities. The Mercer Consulting group ranked 221 cities in 2010 on the basis of Quality of Living criteria for each city; these included measures of personal health and safety, the economy and physical environment, transportation and communications, public services, and the overall political climate. Not surprisingly, the top cities were in high-income European countries, plus similar cities in Canada, Australia, and New Zealand. The top cities were Vienna, Zurich, Geneva, Vancouver, and Auckland. The highest U.S. cities were Honolulu which ranked 31st and San Francisco which ranked 32nd. The cities with the lowest rankings were Baghdad (not a surprise) followed by the African cities of Brazzaville, Bangui, and Khartoum. Some cities improved their ranking over the previous year: Vienna moved to rank 3 and Washington, D.C., to 44. Some lost their ranking; Oslo moved from 26 to 31and Madrid from 42 to 45.

In 2010 Mercer also developed an Eco-City Ranking based on water availability, water potability, waste removal, sewage, air pollution, and traffic congestion. The top five "eco-cities" were Calgary, Honolulu, Ottawa, Helsinki, and Wellington. Nordic cities ranked especially high; those in Eastern Europe ranked somewhat below their counterparts in Western Europe. U.S. and Canadian cities ranked high. In the Asia Pacific region, Adelaide, Kobe, Perth, and Auckland ranked the highest, and Dhaka the lowest. In the Middle East and North Africa, Cape Town, Muscat, and Johannesburg ranked high; Antananarivo and Baghdad were at the bottom.

TYPOLOGIES OF URBAN SYSTEMS

Future cities and city systems will have some features similar to those of today, but also some new features. Realizing that many cities may exhibit more than one form, we suggest some snapshots of future urban systems (fig. 13.9).

- *A single major urban cluster that will increase in numbers, density, and territory.* Frequently these are LDC primate cities. Examples include Addis Ababa, Nairobi, Mogadishu, Manila, Ho Chi Minh City, Tegucigalpa, and Lima.
- *Regional clusters of small and large cities within the same state that will grow and gradually coalesce.* Examples include southeast Brazil, southeast Korea, the Ganges River, the lower Nile, southwest Nigeria, southeast Australia, and many parts of the United States (fig. 13.10).
- *Transborder regional systems that cross several international boundaries.* Examples include El Paso-Ciudad Juarez, Detroit-Windsor, the Tashkent-Osh corridor, the Malay Peninsula.
- *New international gateway cities have access to new water and land frontiers and present opportunities for investors and workers.* Examples are cities in South America's "Southern Cone," Amazonia, and Northeast Asia.
- *Cross-border cities will continue to experience either symmetrical or asymmetrical growth.* Examples include Brazzaville and Kinshasa, Buenos Aires and Montevideo, San Diego and Tijuana, Seattle

and Vancouver, Vladivostok near the Russian/Chinese border, and Yanji near the Chinese/Korean border.

- *Frontier cities will emerge with development of new natural (mineral) or human resources (tourism, retirement, etc.) or hubs for new transportation and communication networks.* Examples are cities in Brazil's circumference highway, Dunedin for the Antarctic, Darwin for Southeast Asia, Kashgar and Urumqi for Central Asia, Lhasa in Tibet, Ushuaia and Punta Arenas in southern South America, and Magadan on the Sea of Okhotsk in northeast Siberia.
- *Corridor city systems include cities spaced unevenly along railroads, highways, and rivers.* These "beads on a string" experience asymmetrical growth. Examples include the Trans-Siberian railroad, the Trans-Amazonia highway, the highway between Istanbul and Ankara, cities on the revitalized Old Silk Road, the proposed Trans-African highways from Dakar to Mombasa and Cairo to Cape Town, the Brazilian "megalopolis" (Recife to Porto Alegre), and the cities on India's new Urban Quadrilateral.
- *Bypassed cities will experience economic decline, an out-migration of youth, an aging population, and dated infrastructures (bridges, highways, railroads).* Examples include many old mining and heavy-industrial centers in Eastern Europe, Siberia, the Canadian Maritimes, and Appalachia.
- *Isolated cities located at the termini of transportation and communication networks are often located in inaccessible and hostile physical environments.* Examples include the single-resource cities in Siberia and capitals in the Sahara, highland

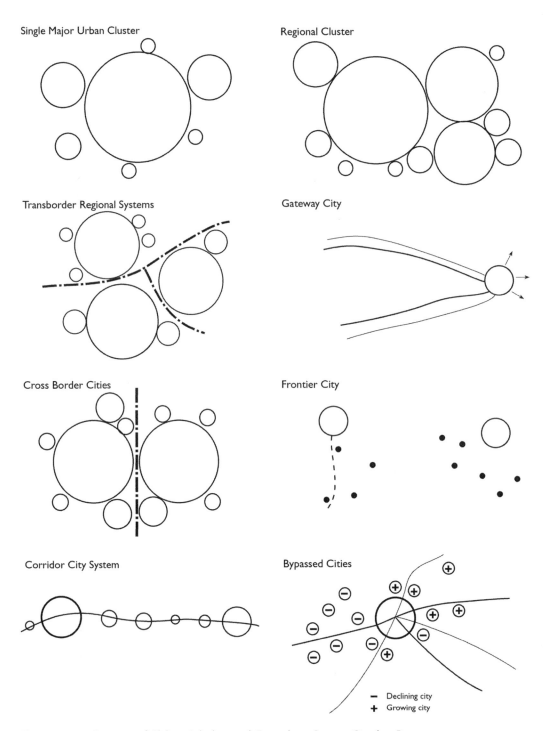

Figure 13.9   Systems of Cities: Existing and Emerging. *Source:* Stanley Brunn

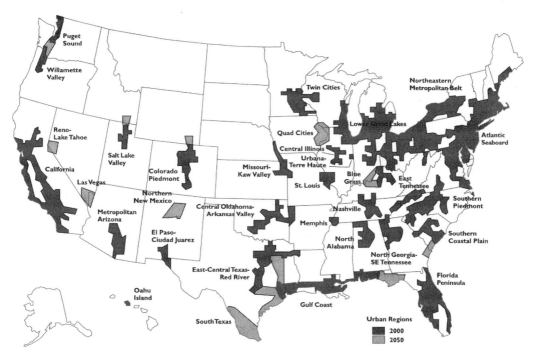

Figure 13.10   U.S. Urban Regions in 2050. *Source:* Stanley Brunn

Andes and Central America, the eastern Caribbean, and the Pacific Islands.

- *Ephemeral cities have fluctuating populations with seasonal employment.* These include specialized university, military-defense, pilgrimage, tourist, and entertainment cities.
- *Historic preservation cities celebrate their pasts and use their heritage to generate income from tourism, retirees, and investments in historic preservation.* Examples include Williamsburg (Virginia), Vienna, Timbuktou (Mali), and Fès.
- *Fun cities are centers of hedonistic activities and have economies that thrive on the pursuit of entertainment and leisure.* Examples include Orlando, FL, with Disney World, Las Vegas, Reno, Monaco, and Macau for casino gambling, Dubai in the Middle East, and cities with major

theme parks, regular sports events, or exotic and quasilegal forms of entertainment.

- *World cities anchor dense commercial and cultural networks and play leading roles in regional finance, investment, tourism, arts, and culture.*
- *Global cities—Tokyo, London, and New York—are recognized for their pinnacle ranking in international business networks.*
- *Ecumenopolis, a concept developed by Constantinos Doxiadis, the originator of the term* ekistics *(the science of human settlements), describes a worldwide urban system that links major populated areas on major transportation and communication corridors (fig. 13.11).* Such an intercontinental city is not beyond the realm of possibility.

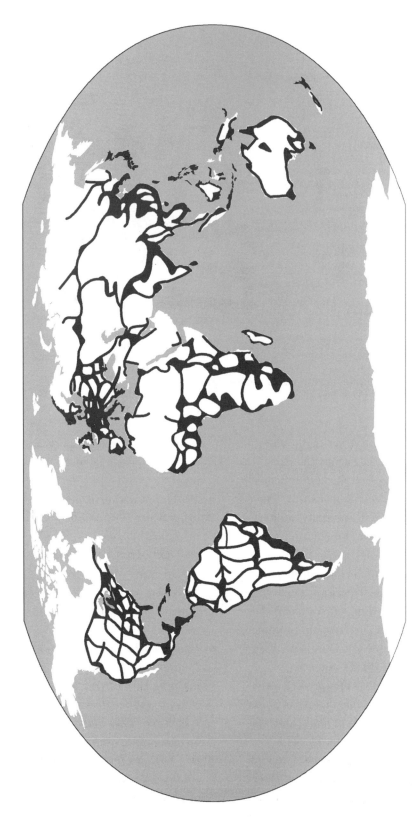

Figure 13.11 Ecumenopolis: The Global City. *Source:* Adapted from C. A. Doxiadis, "Man's Movements and His Settlements," *Ekistics* 29, no. 174 (1970), 318

SPACE-TIME TRANSFORMATIONS

The Internet and the use of cyberspace have now brought about an entirely new time-space dynamic. Communications allow for a rapid convergence of space-time between all places connected to IT networks. In theory, when discussing the movement of information, all places linked into the Internet have a time-space distance of zero from one another. All connected districts and cities are in the same cyberspace and any competitive advantages of earlier eras are supposedly diminished by a reduction in the amount of time required to conduct social exchanges. The above is not to suggest that shrinking time-space can render physical distance meaningless for cities in the future. As long as the world continues to consist of more than bits and bytes, physical distance will exert powerful effects on cities.

However, the reorganization (or shrinking) of time-space brought about by new technologies will undoubtedly cause fundamental changes to our metropolitan environments and societies as well as to individual travel. The following sections outline some of those implications.

DIGITAL TECHNOLOGY AND THE CITY

Alongside the real (or "ground truth") city, a new city is emerging in MDCs. This digital city is where digital technologies, especially geographical information systems (GIS) are used to plan, build, and manage the urban environment (box 13.3). Since the early 1970s, GIS has become routine in urban planning tasks in North America, Western Europe,

---

Box 13.3   Engineering Earth Futures

Stanley D. Brunn

The planet is replete with megaengineering projects of varying sizes, financial costs, and environmental impacts. While we generally would associate these projects with extensive transportation schemes, dams, airports, river diversion, and irrigation projects, they also can include theme parks and leisure spaces (golf courses and sports arenas), new capital cities, and towering skyscrapers. But engineering the Earth can and does include projects of a social nature as well. Are not genetically modified foods, Google Earth, GIS, the Internet, and Facebook also ways in which we choose to engineer the Earth? And what about social engineering projects such as gated communities, international "cookie cutter" suburbs, resettlement projects for "alien" newcomers and socially divisive projects designed by governments that separate groups based on skin color, religion, and ethnicity? Examples of these projects exist on all continents; and their social, environmental, and political impacts will only increase in importance in the coming decades.*

Three issues are paramount in looking at the future of megaprojects. First, is "bigger" necessarily "better"? Contemporary society, whether in the developed or developing world, seems almost addicted (and gives rewards) to projects that are huge in scale and cost. China's Three Gorges Dam, Dubai's Burj Khalifa, the Brazilian Trans Amazon highway, plus gigantic nuclear power plants, offshore oil rigs, postmodern skyscraper skylines, and outlandishly designed new capital cities, all come to mind. There is no shortage of megadreams and megaschemes to try and resolve an immediate energy or transportation problem or to appease the superego of government leaders. We think and plan "big" because we think that the size "fix" will solve the problem or appease potential users.

Second, what are the impacts of megaprojects? For each project, whether designing a dam or constructing a nonrenewable or renewable energy project or changing the course of a river or building a new city, there are social as well as environmental externalities. What are the short- and long-term consequences of relying on nuclear power or altering the course of a river or planning residential areas in environmentally sensitive wetland zones or on erosion-prone slopes? Or what about a society whose youth and wealthy are addicted to the impersonal worlds of Facebook, iPods, cell phones, and the Internet? Or a world where everything and everyone is a coded into privatized and/or governmental GIS databases? Do we really want a world *where* everything can be mapped and *where* everyone is mapped 24–7? With our fixation on "technology coming to the rescue," we often forget the environmental and human impacts of financial and engineering solutions to problems. Social, environmental, and civil engineering often seem to operate in parallel universes. Hazards to future generations may originate in the social and physical engineering projects of today.

Third, what about our future? Futures, like pasts and presents, will be engineered. The questions are *who* will be masterminding the engineering or planning and *what* do we want and expect from it? Postmodernists like to remind us that we have freedom and flexibility to plan and design the futures we want. Perhaps that is true; but, perhaps this view is also elitist, self-serving, materialistic, and unresourceful. Present and future worlds will be comprised of many living in survival conditions in rich and poor worlds whose daily lives are and will be full of uncertainty, frustration, desperation, and without a moral compass. Where these conditions exist, how can one "engineer hope"? Perhaps the human condition will be improved by placing greater focus on economic and social microengineered projects that are local, sustainable, and community based rather than highly visible and heavily financed projects by the international banking community. What is certain is that global warming, biological species decline, financial meltdowns, social underachieving, geopolitical impotence, surface conformities, and social restlessness are futures with us now. It remains for the youth and elderly, leaders and followers to decide what kind of engagement, empowerment, or engineering of the planet's resources and population we wish to pass on to the coming generations.

*There are more than one hundred megaprojects discussed in S. D. Brunn, ed. *Engineering Earth: The Impacts of Megaengineering Projects* (Dordrecht, Netherlands: Springer, 2011)

Figure 13.12   Pro-democracy urban uprisings took place not only in Syrian cities in 2011, but in large cities around the world where Syrian expatriates lived, such as here in London. (Photo by Donald Zeigler)

Japan, and Australia. These systems are also penetrating the cities of some LDCs, particularly India, China, South Africa, Senegal, Ghana, Brazil, and Mexico. GIS is the science and technology that integrates vast databases with georeferenced data (exact latitudinal/ longitudinal coordinates) to prepare maps, satellite images, and aerial photographs. These data allow researchers and planners to perform a range of statistical and spatial analyses, including modeling, visualization, and simulation to design, plan, and manage urban environments and to propose future scenarios. GIS can easily be linked to the Global Positioning System (GPS) to ensure highly accurate spatial analysis. Also, mobile and hand-held GPS units integrated into GIS programs can be taken to remote areas to conduct real-time analysis. A common use of GIS in urban planning in MDCs is citizen participation.

Another digital technology that is vital to urban economic growth and governance is wireless access to "the cloud" (fig. 13.12). Many cities are being restructured as wired cities in order to compete for a vast array of high-technology digital transfers that link businesses and households. San Francisco, Seattle, San Diego, San Jose, Los Angeles, New York, Washington, D.C., Chicago, Boston, and Miami are among the top US cities in Internet penetration. The Internet is being utilized not only for luring capital and businesses, but also for security purposes. For example, web cameras monitor London's streets and its financial district. Internet penetration in LDCs is also significant and is a deliberate strategy to attract economic investment from the MDCs. Delhi, Mumbai, and Bangalore are all wired cities in India's globalized economy.

Examples of countries making heavy investments in national and urban digital economies are Singapore, Estonia, Slovenia, the Gulf States (as part of their post-petroleum economic planning), Hong Kong, Ireland, Jamaica, and Trinidad. Digital cities invest in computer hardware and software design and support large and small companies performing various digital tasks, as well as the many intergovernmental organizations (IGOs), NGOs, and governmental institutions. These include libraries, courts, hospitals, and employment and environmental centers. Digital cities contain multiple land uses, including cafes, and venues for music, photography, education, healing, conferencing, and tourism.

In the past, cities thrived or atrophied based on their relative positionality within global flows of trade. Most large urban agglomerations still owe their prosperity to transportation facilities, including important roads, canals, and airports. While positionality will be no less important for cities of the future, prosperity will increasingly rely on an alternative structure. Cables and satellite dishes grounded in specific physical places

are the prerequisites for time-space compression. Such networks, as well as the social, economic, and political opportunities for people in cities to access such networks, are currently very unevenly distributed across the globe. Broadband and fiber optic cables are heavily concentrated in (and link) East Asia, Europe, and North America. Heavily wired cities include Seoul, London, and Boston; and they are much more likely to experience change brought about by time-space compression than cities lacking high network connectivity, such as Pyongyang or Kinshasa.

Inclusion and Exclusion

Certain cities are more included in global networks than others. However, physical connections do not ensure that a majority of a city's inhabitants will benefit from globalized networks. Large groups in a highly connected city, as measured by bandwith, can remain disconnected. Unfamiliarity with world languages that dominate the Internet, such as Chinese, Spanish, and especially English, can render network connections without meaning and value. Similarly, various cultural and political restrictions can exclude people, based on gender, ethnicity, or religion, from regular Internet use. For instance, in certain cities it is highly inappropriate for women to spend time in male dominated Internet cafes. Finally, economic barriers may be the most powerful exclusionary force. Both the means (a computer and a mobile device) and the cost of access (hourly or monthly fees) can be prohibitively expensive for many residents even in the most connected cities. Questions of inclusion and exclusion are becoming ever more important for cities, because unequal access to distant information and commu-

Figure 13.13   The camera phone was invented in the 1990s and has since come to dominate the market. Here in Athens, images of a parade are shared with far-away friends. (Photo by Donald Zeigler)

nications will likely cause other increasing urban inequalities.

Hybrid Spaces

What the above means is that the scale of space-time transformations for cities is never at a metropolitan level, but it exists instead always at an individual or group level. The city becomes a very different place for different residents. Individuals exist in a hybrid cyber/physical space, which some geographers have dubbed "DigiPlace," while others remain rooted in the physically proximate present. For example, imagine how two different people commuting to work on a San Francisco cable car could have vastly different experiences of, and with, the same city. The first person does not own a cell phone, laptop, or PDA and is restricted to direct sensory experiences arising from the proxi-

mate urban landscape: conversation between other passengers, the view from a window, advertisements, a newspaper, etc. In contrast, the second person possesses a multifunctional mobile phone (fig. 13.13). She spends the commute time immersed in both the physical- and cyber-spheres she is passing through. She is, furthermore, able to interact with true DigiPlaces in her cities, places that she can experience both physically and in cyberspace. These DigiPlaces can be restaurant reviews, localized chat rooms and forums, maps and satellite images, search engine rankings, and countless other experiences.

Hybrid spaces significantly alter the meaning and significance of many elements in urban environments. To return to the previous example, when looking for a restaurant, our first person might ask a friend, be swayed by an advertisement, or see a restaurant on a main street. Our second person might also see a physical advertisement for a restaurant, but be put off by mediocre online reviews. Prominence of a restaurant in the listings of a local search engine (Google Maps or Yahoo Maps), would more likely determine which establishment is ultimately chosen by our second person. As DigiPlaces expand and become more commonly used, we will see far reaching changes in the fabric of cities. Cities will exist as not only bricks and mortar, but also as constantly changing drapes of electronic information. The digital and the material will thus become increasingly blended and blurred in daily urban life.

Transcending the Fog of Distance

Space-time transformations afforded by new information and communication technologies alter the meaning of distance for intra- and inter-city relationships. Distance has previously always been a powerful social force. It had the power to obscure detailed knowledge about (and hinder rich communication with) distant places, thus leading geographers to talk about the "fog of distance." People are more likely to interact with those in their own neighborhood than with those in other districts or cities. Such localized interactions and spaces of knowledge have given rise to a myriad of traditions, dialects, fashions, cuisines, philosophies, and musical styles. A combination of fast, inexpensive transportation and cyberspace now allows people to cut through the fog of distance in a variety of ways. Cheap transportation had profound effects on making urban economies highly dependent on non-local products and allowing large cultural exchanges to take place through mass tourism. These developments occurred gradually over the centuries in tandem with incremental advancements in transportation

technologies. However, revolutionary changes in the ways in which distance affects urban life are now materializing with the diffusion of various Internet applications. As noted above, a number of online tools allow individuals and groups to gain local knowledge about distant regions and cities. Mapping tools such as *Mapquest* or *Google Earth* offer standardized visions of faraway places. We can proceed from looking at street level pictures of Manhattan to satellite images of Baghdad's layout in a few clicks. User-generated websites such as *Wikitravel, WikiChains, MySpace, and Face-Book* offer alternative means through which to transcend the fog of distance by allowing us to virtually experience other places through detailed, localized descriptions of people, places, and landscapes.

While technology will never fully eliminate the significance of distance, it will likely render it increasingly less opaque. As cyberspace brings about fundamentally altered space-time relationships, "distance" will take on new meanings and feed back into the ways in which urban culture is generated and reproduced.

Urban Change

Space-time transformations will undoubtedly have a profound effect on the new urban form of cities around the globe. Some areas may experience further suburbanization as office workers choose to telecommute. Cities with large concentrations of office jobs may then see workers living hundreds of kilometers or miles away as they only need to connect a day or two every week. Many jobs can be performed from anywhere and are becoming entirely unrooted from any attachment to specific cities. These developments could result in large in-migrations to cities, such as Rome,

Las Vegas, or Phuket, Thailand. The altering of space-time relationships through communications technologies is also the impetus behind large migrations of whole company divisions to cities with low labor and production costs, thus changing the employment possibilities of these places. This multiplying practice of moving certain types of jobs, such as back office work and call centers to cities with low costs suggests that a certain kind of "geographical unrootedness" is redefining urban employment. At the same time, cities may start to develop strong specializations in certain functions that can best be performed electronically. Manila might become a center for digitizing architectural drawings and Bangalore a hub for call centers.

Empowerment through GIS

Usage of Geographic Information Systems has become popular with citizen-based grassroots organizations because GIS enables them to be informed—and powerful—participants in community planning efforts. "Public Participation GIS" (PPGIS) is a critical arena for GIS researchers. It directly addresses the issues of providing equitable access to spatial data and GIS technology to marginalized groups, which have traditionally been deprived of access to GIS owing to its technical complexity and high cost. Through intermediary institutes, such as universities, governmental agencies, and NGOs, free access to data and GIS are being provided to a wider audience.

Key PPGIS applications include natural resource management and conservation efforts, community-based planning and neighborhood revitalization in urban areas, and activism organized at local, national, global and multiscalar levels. In such efforts, local, qualitative knowledge is integrated with quantifiable public data sets, and mental maps and sketch maps are integrated with official digital maps, aerial photographs, and satellite images. These practices have been used in a number of projects among indigenous societies in the non-Western world where local knowledge and cultural practices are preserved through oral histories, songs, dances, and alternative representations of the world. One Kenyan project integrates traditional Maasai songs, dances, sketch maps, and mental maps with digital video recordings, photography, and satellite imagery. The project aims to draw on the environmental knowledge and pastoralist practices of the Maasai to inform community conservation and development initiatives and ecosystem management policies.

In the context of neighborhood revitalization in the Western world, community organizers use public data sets through GIS to inform and legitimate local knowledge to obtain action and formulate strategies, monitor neighborhood conditions and predict changes, prepare for organizational tasks and funding recruitment efforts, generate new information based on their own experiential knowledge to enhance service delivery tasks, and to explore spatial relations to challenge or reshape urban policy. GIS analysis has enabled these community organizers to fight blighted housing conditions by tracking down absentee landlords through property records, fight crime issues through crime hot spots analysis, analyze land use data to track vacant and boarded up houses, track data on sanitation issues and garbage removal, combine mortgage lending data with demographic data to address discriminatory investments, and to use various indicator data to generate sophisticated multiscalar maps to measure a community's changing well-being.

PPGIS activities have been aided in the emergence of the Internet mapping sites such as Google Earth, in which high-resolution satellite images of places across the world can be accessed in minutes for free by anyone. The power of Google mapping is evident in the case of the Surui people of southern Brazil, whose Chief, Almir Surui, approached Google for high-resolution satellite imagery to monitor illegal loggers and miners on the tribe's 600,000 acre reserve. While high tech mapping has already been used to track illegal activity and record knowledge, this will be the first time an Amazon tribe will share their own vision of their territory with the rest of the world via Google Earth. In the western world, various government agencies provide easy access to public databases through Internet GIS sites. One site, COMPASS, enables citizens of Milwaukee to access, view, query, and map detailed property data, health data, crime data, community asset data, and demographic data at no cost. Such Internet GIS sites are particularly useful for resource poor organizations that have difficulties in creating and maintaining in-house GIS.

URBAN ENVIRONMENTAL PROBLEMS

A major source identifying the scale and scope of urban environmental problems and challenges is the Worldwatch Institute's *State of the World* report (table 13.3). Some salient facts drawn from the 2007 report include the following:

- Nearly 850 million people are hungry and 15 percent of the estimated 1.6 billion people who lack electricity and other modern energy services live in the world's cities.

- The U.N. Millennium Project estimates that improving the lives of 100 million slum dwellers will cost $830 billion over the next 17 years.

- According to the International Labor Organization, 184 million people on the planet do not have jobs; this figure would approximate one billion if underemployment were considered.

- The per capita footprint of high-income countries is eight times that of the low-income countries.

- An assessment of 116 cities by the World Health Organization (WHO) in 2000 estimated that only 43 percent of urban dwellers had access to piped water.

- Of the 292 large river systems in North America, Europe, and Russia, 42 percent are strongly affected by impoundments and diversions putting at risk humans and plant and animal species.

- In 1970 there were 200 million cars in the world; by 2006 the number was 800 million and that number will again double by 2030.

- The WHO estimates that each year about 1.2 billion people are killed as a result of road accidents and perhaps as many as 50 million are injured.

- The majority of the world's CO_2 emissions are traced to cities, even though they cover only 0.4 percent of the earth's surface.

Some problems that still merit attention at local, regional, and global scales are discussed below. These range from food production, sanitation and waste issues, natural disasters, and mass transit initiatives to health and diseases, slums and poverty, and global warming.

Natural Disasters and Global Warming

Natural disasters remain major problems affecting the planet's largest cities. These include flooding, hurricanes, earthquakes, and tsunamis. The Worldwatch report noted that the growing numbers of natural and other disasters are resulting in more loss of life each decade. In the past 25 years, 98 percent of the people injured by natural disasters lived in the 112 countries classified as low- or middle-income by the World Bank. These countries accounted for 90 percent of lives lost to natural disasters. Earthquakes alone affect more people in the developing world than in the developed world. The same is true for the melting of sea ice, which results from global warming. The U.S. Geological Survey estimated that the economic losses from natural disasters in the 1990s could have been reduced by $280 billion if just $40 billion had been invested in preventive measures. Less than 3 percent of LDC residents have insurance compared to about 30 percent in the rich world. The poor suffer losses to housing, crops, livestock, and housing goods.

There are many examples of large and small cities that are vulnerable to multiple natural disasters. Flooding affects Melbourne, New Orleans, Dhaka, Port-au-Prince, Mumbai, and Delhi. Earthquakes affect San Francisco, Shanghai, Tehran, Jakarta, and Tokyo. Tsunamis affect coastal cities in the western and eastern Pacific. For some residents early warning systems save lives. Following natural disasters, cities have to consider when and how to rebuild destroyed and damaged infrastructures, how to support residents returning to their homes, and the best practices for implementing enforceable security systems.

Global warming will also affect major cities of coastal North America, Europe, and Asia. Comparing a map of the world's largest and fastest growing cities with one showing shallow continental shelves reveals that many prosperous U.S. cities from Boston to Miami to Houston will be affected by a two-meter rise in sea level. Cities in northern and northwest Europe are also vulnerable, as are cities at the mouth of the Ganges, on the Malay Peninsula, in insular and mainland Southeast Asia, eastern China, and Japan.

Health Care and Diseases

This heading includes a city's sanitation system, the quality of drinking water, programs to stem air, water, and land surface pollution, as well as the spread of diseases by human contact, livestock, and food consumption. One could also include inadequate disposals of urban waste and trash, which affect the spread of diseases as well as the overall quality of urban life. Populations of all ages, in increasing numbers, are crossing international boundaries for work, pleasure, or study, thus increasing the likelihood of local epidemics reaching pandemic proportions. Hardly a year goes by without the reported outbreak of some new disease or the contamination of an urban water supply. Some cities alert their residents about high and dangerous levels of pollution. Air pollution kills an estimated 800,000 people annually; China has 16 of the 20 most polluted cities.

European, North American, Australian, and Japanese cities implement programs designed to protect the quality of the living, working, and leisure environments, whether they be indoors or outdoors. For many LDC cities, health issues are often considered of secondary importance in a country's or region's economic development goals. Polluted air from automobiles or from industries is a fact of daily life just

Table 13.3   Environmental Problems and Initiatives in Selected World Cities

CITY	Urban Farming	Natural Disasters	Mass Transit	Water	Slums and Squatters	Renewable Energy	Health and Sanitation	Grassroots Empowerment	Green City	Global Warming	Urban Violence	Waste and Garbage	Air Pollution
Accra	■			■			■	■					
Bangkok	■		■										
Beijing						■			■				
Boston	■			■	■					■			
Buenos Aires	■				■			■					
Cairo	■											■	
Copenhagen									■				
Chicago	■					■			■				
Delhi			■					■					■
Dhaka		■			■								■
Hanoi	■	■					■						
Houston		■								■			
Jakarta	■		■	■			■	■	■			■	■
Karachi			■	■			■	■			■		
Kolkata					■						■	■	■
Lagos	■				■								
Los Angeles					■			■					■
London			■					■					
Melbourne						■			■	■			
Mexico City	■			■	■			■				■	■
Mumbai				■	■		■	■	■				
Nairobi		■		■				■		■	■		
New Orleans		■						■					
New York	■							■					■

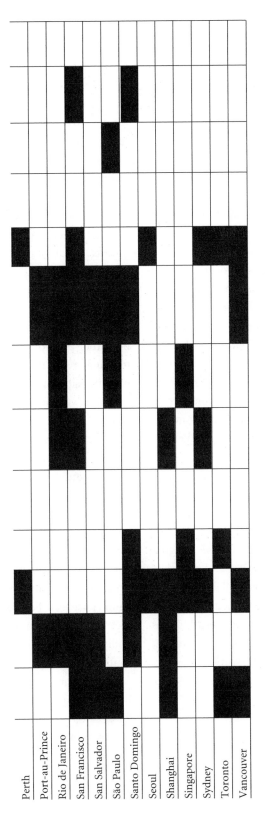

Degree of importance noted in shadings: Dark: an existing problem, White: not reported.
Source: Worldwatch Institute. *State of the World 2007: Our Urban Future.* (Washington, DC, 2007).

as much as polluted drinking water and piles of garbage that can appear anywhere. Ecological problems abound including the outbreak of diseases, insufficient health vaccinations, and sporadic public safety warnings.

Slums and Poverty

For much of the developing world, the rising number of people living in substandard housing is a fact of daily life. Many poor neighborhoods frequently lack electricity, running water, and sufficient food and energy sources. Many factors contribute to poverty, including employment in the informal sector, lack of effective urban networking to find jobs, low wages of multiple-member households, low rates of literacy, and lack of job skills. The U.N. Development Programme estimated that there are more than 800 million people on the planet who are urban farmers, most living in China and India. About 200 million provide food primarily for the market, but a great majority raises food for themselves. Growing food, the report states, is not a hobby for most people on the planet. The urban poor are also those who have to carry water to their residences (equivalent to eight suitcases a day); they experience irregular and unexpected losses of water and electricity (if available), lack readily available and safe fuel sources to heat food or their homes, use various outdoor facilities for toilets and live in constant fear of eviction, loss of income and employment, injuries from work or reckless drivers, and random violence brought on by criminal gangs or police.

Empowerment and Greening

In large and small LDC and MDC cities there are grassroots efforts to improve the qual-ity of the urban physical and human environments. Many LDC initiatives stem from the efforts of disenfranchised and previously powerless groups who see government officials and agencies as insensitive to their demands for improved living conditions. Their concerns include providing safe water sources, installing lights for safer streets, providing places where children can play (green spaces), safe transportation to places of employment (even central markets), recycling waste products and organizing neighborhood trash collection, promoting urban farming along roadsides and hillsides, organizing neighborhood music, arts, and sporting programs for youth, and developing microcredit initiatives, especially for women. In the rich world there are organizations that promote green city development by planting trees, installing bicycle lanes, placing mass transit proposals on public ballots, closing streets to vehicle use, encouraging and subsidizing renewable energy sources, and advocating effective recycling programs. Many projects are also on the agenda of Sister Cities. Examples of grassroots empowerment are legion. They include the Grameen Bank in Bangladesh which provides microcredit to poor women, the Pakistani NGO Oranji Pilot Program which works on sanitation programs in poor neighborhoods, the Villa Maria del Triunfo in Lima which supports urban farming, the National Slum Dwellers Federation in Mumbai, the Metro Cebu urban forestry initiative in the Philippines, a music program for Rio's *favela* youth in Brazil, the Slum/Shack Dwellers International which works with groups in Brazil, Kenya, Sri Lanka, Swaziland, and elsewhere. In the developed world there are many examples of local groups in Miami, Philadelphia, Seattle, Toronto, Perth, Amsterdam, and elsewhere that work with local and state governments and the private sector on various greening efforts.

Sustainable Cities

These are cities with commitments to conservation, recycling, and sustainable economies, and environmental ethics, that is, cities that are making wise use of resources for future generations to develop an ethic of recycling and reuse, as well as addressing issues of social justice, gender equality, livable wages, affordable housing, and ready access for all residents to education and health care. Integrating elements of ecology, economy, community, and social cohesion are thus the lynchpins of sustainable city development. Typical physical characteristics of sustainable cities include compact forms of residential development, mixed land use, use of mass transit, along with widespread use of pedestrian and bicycle paths, heavy use of wind and solar energy, protection of natural hydrologic systems, protection of wetlands, woodlands, natural open spaces, and habitats, use of natural fertilizers and integrated pest management, use of natural means of sewage treatment, reduction of waste, and recovery, reuse, and recycling of waste materials. These cities encourage strong citizen participation as well as coalitions with businesses.

The U.N.'s Sustainable Cities Programme (SCP) operates in scores of cities, mostly capitals. It aims at developing local capacities for environmental planning and management (EPM) as poorly managed urbanization creates serious environmental and social problems. At the municipal level, development issues generally include water resources and water supply management, environmental health risks, and solid and liquid waste management/on-site sanitation, air pollution and urban transport, drainage and flooding, industrial risks, informal sector activities, and land-use management in the context of open space/urban agriculture, tourism and coastal area resource management, and mining.

Human Security

Human security issues are emerging as important topics for those in the social and policy sciences. They are also important for those studying urban worlds as security concerns affect the daily lives of individuals and influence where they reside, work and play. Security issues are also a concern of governmental and nongovernmental agencies at the local and regional levels. Human security concerns include more than simply dealing with terrorism in its many dimensions, installing alarm systems in homes, purchasing anti-theft car insurance, or dealing with preferences for gated communities. Transnational terrorism and biological and environmental terrorism are considered high priority problems for state and multistate decision makers. Many public officials must insure that dangerous individuals, goods, and substances do not enter their country's airports and ports or cross their borders. They are ever vigilant for contaminated products, whether foods, beverages, or other materials, that might enter their country and endanger their food supplies, livestock, public water systems, and the unique natural environments that attract tourists. The security industry will remain one of the "growth industries" for much of this century, as evidenced already in airports, train stations, harbors, and border crossings.

Many residents of rich-world cities see security as dealing mainly with personal security purchases (*e.g.,* residential alarm systems, theft-proof windows and doors), policing, and law enforcement. In the developing world, security assumes some different faces: it is also a function of such daily concerns as having sufficient food, safe housing, lighted streets, and safe drinking water. Security also includes programs to prepare women and men for respectable jobs, to stop the spread of sexually transmitted diseases, to empower grassroots groups for representation in neighborhoods and city governments, to receive basic medical care, including vaccinations for communicable diseases, to protect women and children from domestic violence, to prevent the spread of violence brought on by gangs, paramilitary forces, and poorly disciplined urban police, and to reduce and eliminate the trafficking of drugs and deadly weapons. Some of these issues are of major concern to women and children, others to new or old, others to racial, religious, and ethnic minorities, and still others to political and ecological refugees who cross borders seeking a safe place to live, work, worship, and play.

SOLVING URBAN PROBLEMS

All problems have solutions. Below we suggest some, many of which will require huge financial investments, local empowerment, civic leadership, and bold, creative thinking to implement at national and global scales.

Investment in New Infrastructure

Investments to construct new transportation and telecommunications corridors linking cities not currently integrated, such as macroreginal and continental systems in the Andes and Amazon, West to East Africa, and across Central Asia into northeast Asia.

LDCs Investing in MDCs

Emerging rich, well-endowed, and powerful LDCs, including China, India, Brazil, Mexico, Kazakhstan, Saudi Arabia, Iran, and South Africa, could use revenues from natural resource extraction to assist marginalized cities in Europe and elsewhere with loans and credit for economic restructuring.

Empowerment of Women

Change will occur when and where there are strong commitments to eradicating illiteracy, HIV/AIDS, and child labor; to reducing violence in the home and the workplace; to investing in local grassroots efforts; and to supporting investments in local politics.

Living with Alternative Structures

Rather than uniform policies of rigid conformity, there could be groups and communities exploring alternative models of policing, participation, and administration. While some groups may choose very traditional and local forms of government and group participation, others may exhibit high degrees of nonconformity or totalitarianism.

Corporate Ownership of Cities

Cities could be bought and sold on international financial markets like food, energy supplies, and industrial products, and administered by transnational corporations rather than by traditional political parties or by the state.

Triage Planning Strategies

Government officials in rich and poor countries might divide cities and urban regions into three groups: those that (a) have healthy economies and societies, (b) might respond to short-term infusions of skilled labor and capital, and (c) face dismal futures and should be left to fend for themselves.

Greater Voice

Two new groups could call for increased participation and representation in urban governance: the new Latin, African, and Asian diaspora communities in MDCs, and the dispossessed, poor, and recent rural migrants in LDC cities.

Transborder Governing Systems

New forms of coordinated city government and planning could emerge in rich and poor world cities that straddle international boundaries and are already linked by thousands of daily commuters; and they could merge for mass transportation, solid waste, health care, and public safety programs.

A World Urban Cabinet

An organization of the world's largest cities could operate either separately or alongside the United Nations to address problems facing MDCs and LDCs, including transborder pollution, water and sanitation issues, and transportation and communication systems.

Environmental Awareness

To alleviate loss of human life and destruction of housing stock and other infrastructure from natural and technological disasters, there will be increased investments in disaster preparedness, post-disaster delivery systems, and effective reconstruction planning, perhaps with the assistance of PPGIS.

Transnational "Urban Response Teams"

These teams, drawing on talent from around the world, will assist communities facing major power outages, disease outbreaks, destruction of water systems, massive refugee populations, major industrial and transportation accidents, prolonged social conflicts, and collapses of governance and security.

Measuring the Ecological Footprint

Sustained efforts need to be made to accurately measure and map the impacts or "footprints" of environmental destruction on neighborhoods, communities, cities, and city systems.

LOOKING AHEAD

Numerous questions arise when contemplating cities at 2050, 2100, and beyond. Given faster and cheaper transportation, will humans always have a need for cities? While diversity has long been a distinguishing feature of the urban fabric, might new forms of identity and social separation emerge that are based on new loyalty units such as sports and lifestyles, more than cultural traditions, ethnicity, and class? Might transurban loyalties replace those that are territorially based? Could global information and communication technologies be used as vehicles to generate feelings of belonging to place-based communities rather than of placelessness? Will an urban future be a peaceful future? (box 13.4)

What will be the impacts of wireless communications on privacy and city culture? Will new behavioral norms evolve based on hypermobility? Do new technologies promote gender, class, and cross-cultural equity? Will voiceless and visual Internet worlds inhibit or resolve pressing urban problems? Will fluidity in time more than space define the urban resident of the mid-21st century? What happens to the mosaic of the world's languages if inter-

Box 13.4
Paris, France: The Peace Memorial

There's
a word for peace
in every tongue.
Bring them together.
You feel among
those who see:
the future of Earth
as conflict free.
Beyond the words:
the Eiffel Tower,
now a symbol
of Euro-power.
Beginning life
as a rallying point
for a Paris fair
that did not disappoint,
"la tour Eiffel" has now emerged:
emblem of Europe recently purged
of constant strife, now pointing the way
to a peaceful life.

D. J. Zeigler

national business is increasingly conducted in American English or if Chinese becomes the second major language of global commerce? What are the impacts of single-child households on an urban population with a growing numbers of elderly? What are the long-term impacts of increased rates of HIV/AIDS on rural and urban residents in Africa and Asia or the rapid global diffusion of unforeseen pandemics associated with greater commercial travel and new foods?

Will rural ways of life disappear? Will "wilderness" rather than rural become the antithesis of urban? Will rural areas become "future museums" and fantasy places for the urbanites, the places where die-hard traditionalists, a society's rejects, and those with nonacceptable lifestyles live? Will the production of food (long associated with supporting rural livelihoods and city food supplies) be replaced by chemical and pharmaceutical factories and signal the end of agrarian life?

How will cities be affected by major catastrophes such as September 11, 2001? Will the work environments of cities be different? Will skyscrapers reach ever upward? Will urbanites become more supportive of each other? Will neighborhoods of new immigrants become new urban landscapes of fear? Will the perception of increased domestic and international terrorism change lifestyles, modes of transportation, and discretionary spending? Will fantasy and exotic tourism be casualties of the leisure economy? What new places and types of living might emerge? Some futurists foresee huge residential ocean liners or floating cities that circle the globe or ply major tourist rivers and the coastlines of wealthy megacities. Still others suggest using the "oil platform" model to build huge permanent or temporary residences anywhere. Others envi-

sion the world's wealthy becoming peripatetic global nomads with multiple "international homes." Might terratexture (underground) spaces become more popular? These alternative settings might be used to house the wealthy, the elites, and the powerful, but also the undesirable and unwanted.

Will the world's future political agendas be dominated by cities rather than nation-states? Or will global political agendas be forged in the large urban agglomerations of the present LDCs rather than the MDCs? Will the rich MDCs and their large cities be able and willing to adjust to "A South World"? Will the world's "development divide" be increasingly urban-rural rather than North-South? What will China's urban future look like as the population becomes more urban, globally aware, and wealthy? Will China, India, and Indonesia form the "Global Growth Triangle" of the late 21st century? Does the ethnic diversity of Sub-Saharan African cities make them cauldrons for conflict or models for cooperation?

Does urban homogeneity also contribute to conflict, a feature of some agglomerations in Africa and Asia? Can cities of 10, 20, or 30 million be governed effectively? Could the future megacity be composed of fifty rival "city governments," each competing for the best talent and most financial resources? Will territory cease to be the basis for administration and be replaced by loyalties to transnational corporations, new diasporas, and regional sports? Where can one look for solutions to the problems of megacities faced with vast contrasts in income, formal and informal economies, and religious and ethnic diversity? What can MDC cities learn from LDC cities and vice-versa? How applicable are Western urban growth, economic development, trans-

World Time Regions

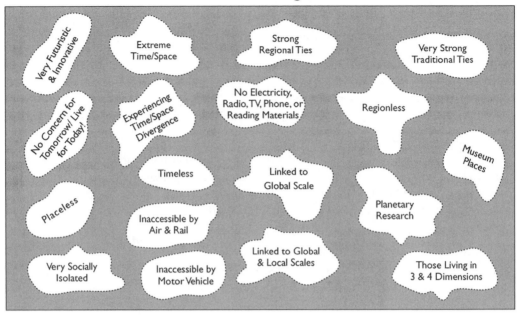

Examples of Cultural Time Regions in the 2000s

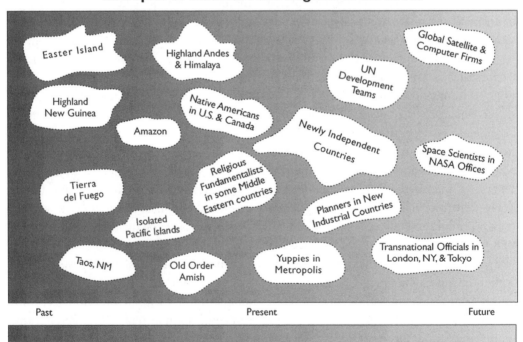

Figure 13.15   World Time Regions. *Source:* Stanley Brunn, "Human Rights and Welfare in the Electronic State," in *Information Tectonics*, ed., Mark I. Wilson and Kenneth E. Corey (New York: Wiley, 2000), 60. Reprinted with permission

portation, and community planning models to African, Latin American, and Asian cities?

How prepared are global megacities and small cities for natural and technological disasters? What top-down strategies and grassroots (bottom-up) efforts in LDCs are successful in preventing losses of life and the destruction of homes and urban infrastructures? How can GIS and PPGIS initiatives be incorporated into community and regional preparedness and responses?

Will urban and national governments, as well as non-governmental and intergovernmental lending organizations, be able and willing to consider food availability, quality health care, water and sanitation networks, and protection of the vulnerable (poor, disabled, children, women, and elderly) as being important in defining human security as thwarting domestic and international terrorism and the proliferation of dangerous weapons? What kinds of grassroots efforts by women, gangs, sports teams, artists, and religious groups, have been successful in stemming the tide of hatred, fear, and terrorism?

It is not difficult to envision future human settlements including preindustrial remnants that exist alongside postmodern cities, safe cities alongside unsafe cities, and smart (wired) cities alongside bypassed (little-wired) cities (fig. 13.15). There may even be governable and ungovernable places, innovative and laggard cities in the same region.

While pessimism about the future of cities might be justified, based on the problems detailed in previous chapters, to succumb to this view is to give up hope. Humanity survives on hope, including hope in the future of cities. It is important to remember that contemporary cities continue to resonate with the very qualities that characterized the first cities

in history. For thousands of years, humans have chosen to live in concentrated settlements of varying sizes. Cities will always be:

- Tapestries of rich cultural diversity (religious, ethnic, lifestyle, etc.);
- Places where new identities transcend traditional social-class and local and regional identities;
- Places of innovation in culture, economics, identities, and governments that may "leapfrog" some earlier forms of urban development;
- Loci of opportunities to improve the quality of life for all ages and residents; and
- Nexuses of integrated communication and information technologies that transcend existing social, political, and cultural identities and boundaries.

There never has been and never will be a single city anywhere in which all residents are happy, content, and peaceful. Utopia is unattainable. But among the new and old inhabitants there can be conscious efforts to improve the human condition and to recognize that cities can bring out the best of human qualities. International collaboration and opportunities to attain these goals offer many opportunities for current and future generations of citizens on the planet.

SUGGESTED READINGS

Brunn, Stanley D. 2011. "World Cities: Present and Future." In: Joseph P. Stoltman. Geography for the 21st Century. Thousand Oaks, CA: Sage Publications, pp. 301–314. A survey of major population trends and major problems facing the world's cities.

Castells, Manuel. 2001. *The Internet Galaxy: Reflections on the Internet, Business, and Society*. New York: Oxford University Press. One of the leading scholars on Internet culture and commerce examines how the Internet is affecting urban planning as well as society.

Craig, William J., *et al.*, eds. 2002. *Community Participation and Geographic Information Systems*. London: Taylor and Francis. Includes numerous examples of community groups using GIS for empowerment and gaining voices in public policy.

Graham, Stephen, ed. 2004. *The Cybercities Reader*. London: Routledge. A splendid collection of short and lengthy essays on the built environment, work and living, technologies and politics.

Hall, Peter G. 1988. *Cities of Tomorrow. An Intellectual History of City Planning in the Twentieth Century*. Oxford, England: Blackwell. A history of the ideology and practice of urban planning, regarded by some as the successor to Lewis Mumford's *The City in History*.

Kotkin, Joel. 2000. *The New Geography: How the Digital Revolution Is Reshaping the American Landscape*. New York: Random House. Explores how the digital revolution is changing the way we live and work, including the role of cities.

Leinbach, Thomas R., and Stanley D. Brunn, eds. 2001. *The Worlds of E-Commerce: Economic, Geographical and Social Dimensions*. New York: Wiley. Discusses developments in banking and finance, retailing, job searches, government regulations, and international development.

Ruchelman, Leonard I. 2006. *Cities in the Third Wave: The Technological Transformation of Urban America*, 2nd ed. Lanham, MD: Rowman and Littlefield. Considers the impact of new space-adjusting computer technologies on cities and emerging city types.

Worldwatch Institute. 2007. *State of the World 2007*. Washington, DC: Worldwatch Institute. The single most valuable reference on problems facing urban humankind; many valuable tables, footnotes, and references.

Zook, Matthew. 2005. *The Geography of the Internet Industry: Venture Capital, Dot-Coms, and Local Knowledge*. Malden, MA: Blackwell. Explores the history of global Internet networks and their impacts on contemporary economic and social geographies.

Zook, Matthew, and Mark Graham. 2007. "From Cyberspace to DigiPlace: Visibility in an Age of Information and Mobility." In: H. J. Miller, ed. *Societies and Cities in the Age of Instant Access*. London: Springer, 231–244. Introduces 'DigiPlace' as a way of conceptualizing movement through hybrid spaces comprised of physical and virtual elements.

SUGGESTED WEBSITES

Center for Sustainable Urban Development
www.earth.columbia.edu/csud
Projects and research papers on sustainable development around the world.

Mega-Cities Project
www.megacitiesproject.org
Identifies both challenges and plans to address problems in the world's mega-cities.

Mercer Cost of Living Survey
www.mercer.com/press-releases/quality-of-living-report-2010
Ranks 221 cities on cost of housing, transport, food, clothing, entertainment, and an eco-city ranking.

Places OnLine (Association of American Geographers)
http://www.placesonline.org/
A map-based portal to websites from around the world.

Sister Cities International
www.sister-cities.org
Includes a global sister city directory.

U.N. Sustainable Cities Programme
www.unhabitat.org/caategories.a:catid=540
Excellent source of data on the urban environment.

Urban Age: A Worldwide Investigation into the Future of Cities
www.urban-age.net

Aims to shape the thinking and practice of sustainable urban development.

Wikitravel
www.wikitravel.org/en
An evolving "free, complete, up-to-date and reliable worldwide travel guide."

Appendix: Populations of Urban Agglomerations with 750,000 Inhabitants or More in 2005, by Country: 1950, 1975, 2000, and 2025 (projected)

Country/Agglomeration	Population (thousands)			
	1950	1975	2010	2025
UNITED STATES AND CANADA				
CANADA				
Calgary	132	457	1,182	1,364
Edmonton	163	543	1,113	1,274
Montréal	1,343	2,791	3,783	4,165
Ottawa-Gatineau	282	676	1,182	1,333
Toronto	1,068	2,770	5,449	6,029
Vancouver	556	1,150	2,220	2,479
UNITED STATES OF AMERICA				
Atlanta	513	1,386	4,691	5,153
Austin	137	320	1,215	1,373
Baltimore	1,168	1,650	2,320	2,579
Boston	2,551	3,233	4,593	5,034
Bridgeport-Stamford	415	703	1,055	1,193
Buffalo	899	1,041	1,045	1,181
Charlotte	142	315	1,043	1,183
Chicago	4,999	7,160	9,204	9,936
Cincinnati	881	1,216	1,686	1,887
Cleveland	1,392	1,848	1,942	2,166
Columbus, Ohio	441	813	1,313	1,478
Dallas-Fort Worth	866	2,234	4,951	5,421
Dayton	350	637	800	909
Denver-Aurora	505	1,198	2,394	2,662
Detroit	2,769	3,885	4,200	4,608
El Paso	139	394	779	887
Hartford	425	700	942	1,067
Honolulu	250	511	812	923
Houston	709	2,030	4,605	5,051
Indianapolis	505	829	1,490	1,674
Jacksonville, Florida	246	564	1,022	1,157
Kansas City	703	1,079	1,513	1,697
Las Vegas	35	325	1,916	2,147
Los Angeles–Long Beach–Santa Ana	4,046	8,926	12,762	13,677
Louisville	476	751	979	1,108
McAllen	31	121	789	901
Memphis	409	720	1,117	1,262
Miami	622	2,590	5,750	6,275
Milwaukee	836	1,228	1,428	1,603

Country/Agglomeration	Population (thousands)			
	1950	1975	2010	2025
Minneapolis-St. Paul	996	1,748	2,693	2,984
Nashville-Davidson	261	484	911	1,034
New Orleans	664	1,021	858	1,044
New York-Newark	12,338	15,880	19,425	20,636
Oklahoma City	278	573	812	923
Orlando	75	427	1,400	1,575
Philadelphia	3,128	4,467	5,626	6,135
Phoenix-Mesa	221	1,117	3,684	4,063
Pittsburgh	1,539	1,827	1,887	2,106
Portland	516	925	1,944	2,173
Providence	703	963	1,317	1,482
Raleigh	69	179	769	879
Richmond	260	558	944	1,070
Riverside-San Bernardino	139	645	1,807	2,021
Rochester	411	604	780	888
Sacramento	216	714	1,660	1,861
Salt Lake City	230	573	997	1,129
San Antonio	454	859	1,521	1,707
San Diego	440	1,442	2,999	3,316
San Francisco-Oakland	1,855	2,590	3,541	3,900
San Jose	182	1,103	1,718	1,922
Seattle	795	1,663	3,171	3,504
St. Louis	1,407	1,865	2,259	2,511
Tampa-St. Petersburg	300	1,094	2,387	2,653
Tucson	77	368	853	970
Virginia Beach	391	996	1,534	1,720
Washington, D.C.	1,298	2,626	4,460	4,891

CENTRAL AMERICA AND THE CARIBBEAN

COSTA RICA

San José	148	440	1,461	1,923

CUBA

La Habana (Havana)	1,142	1,848	2,130	2,094

DOMINICAN REPUBLIC

Santo Domingo	180	911	2,180	2,691

EL SALVADOR

San Salvador	194	596	1,565	1,891

GUATEMALA

Ciudad de Guatemala (Guatemala City)	287	715	1,104	1,690

HAITI

Port-au-Prince	133	575	2,143	3,246

HONDURAS

Tegucigalpa	73	292	1,028	1,493

MEXICO

Aguascalientes	94	233	926	1,073
Chihuahua	87	344	840	971
Ciudad de México (Mexico City)	2,883	10,690	19,460	20,713
Ciudad Juárez	123	474	1,394	1,575
Culiacán	49	230	836	950
Guadalajara	403	1,850	4,402	4,902
Hermosillo	44	232	781	909
León de los Aldamas	123	589	1,571	1,791

	Population (thousands)			
Country/Agglomeration	*1950*	*1975*	*2010*	*2025*
Mérida	143	351	1,015	1,164
Mexicali	66	302	934	1,075
Monterrey	356	1,589	3,896	4,351
Puebla	227	858	2,315	2,620
Querétaro	49	159	1,031	1,198
Saltillo	70	217	801	928
San Luis Potosí	132	378	1,049	1,206
Tampico	136	378	761	871
Tijuana	60	355	1,664	1,915
Toluca de Lerdo	54	309	1,582	1,776
Torreón	189	556	1,199	1,367
NICARAGUA				
Managua	110	443	944	1,192
PANAMA				
Ciudad de Panamá (Panama City)	171	528	1,378	1,758
PUERTO RICO				
San Juan	451	1,069	2,743	2,763

SOUTH AMERICA

ARGENTINA				
Buenos Aires	5,098	8,745	13,074	13,708
Córdoba	429	905	1,493	1,638
Mendoza	246	537	917	1,016
Rosario	554	883	1,231	1,354
San Miguel de Tucumán	224	424	831	924
BOLIVIA				
La Paz	319	703	1,673	2,156
Santa Cruz	42	234	1,649	2,261
BRAZIL				
Aracaju	69	231	782	902
Baixada Santista	246	770	1,819	2,045
Belém	242	706	2,191	2,460
Belo Horizonte	412	1,906	5,852	6,463
Brasília	36	827	3,905	4,474
Campinas	152	773	2,818	3,146
Cuiabá	27	162	772	861
Curitiba	158	922	3,462	3,953
Florianópolis	68	221	1,049	1,233
Fortaleza	264	1,136	3,719	4,170
Goiânia	53	527	2,146	2,439
Grande São Luís	120	342	1,283	1,440
Grande Vitória	85	493	1,848	2,109
João Pessoa	117	362	1,015	1,151
Londrina	51	263	814	944
Maceió	123	342	1,192	1,353
Manaus	90	411	1,775	2,009
Natal	108	367	1,316	1,545
Norte/Nordeste Catarinense	64	278	1,069	1,230
Pôrto Alegre	488	1,727	4,092	4,469
Recife	661	1,867	3,871	4,259
Rio de Janeiro	2,950	7,557	11,950	12,650
Salvador	403	1,341	3,918	4,411
São Paulo	2,334	9,614	20,262	21,651
Teresina	54	276	900	1,004

| | Population (thousands) | | | |
Country/Agglomeration	*1950*	*1975*	*2010*	*2025*
CHILE				
Santiago	1,322	3,138	5,952	6,503
Valparaíso	328	581	873	973
COLOMBIA				
Barranquilla	294	830	1,867	2,255
Bogotá	630	3,040	8,500	10,537
Bucaramanga	110	408	1,092	1,375
Cali	231	1,047	2,401	2,938
Cartagena	107	332	962	1,223
Cúcuta	70	265	774	963
Medellín	376	1,536	3,594	4,494
ECUADOR				
Guayaquil	258	890	2,690	3,328
Quito	206	628	1,846	2,316
PARAGUAY				
Asunción	258	654	2,030	2,715
PERU				
Arequipa	128	348	789	953
Lima	1,066	3,696	8,941	10,530
URUGUAY				
Montevideo	1,212	1,403	1,635	1,657
VENEZUELA				
Barquisimeto	127	475	1,180	1,413
Caracas	694	2,342	3,090	3,605
Maracaibo	282	787	2,192	2,593
Maracay	58	342	1,057	1,266
Valencia	126	546	1,770	2,103
EUROPE				
AUSTRIA				
Wien (Vienna)	1,615	1,583	1,706	1,801
BELGIUM				
Antwerpen	759	868	965	985
Bruxelles-Brussel	1,415	1,610	1,904	1,948
BULGARIA				
Sofia	522	977	1,196	1,215
CZECH REPUBLIC				
Praha (Prague)	935	1,126	1,162	1,173
DENMARK				
København (Copenhagen)	1,216	1,172	1,186	1,238
FINLAND				
Helsinki	366	582	1,117	1,174
FRANCE				
Bordeaux	430	614	838	913
Lille	751	936	1,033	1,107
Lyon	731	1,173	1,468	1,575
Marseille-Aix-en-Provence	756	1,253	1,469	1,577
Nice-Cannes	400	698	977	1,059
Paris	6,522	8,558	10,485	10,884
Toulouse	269	511	912	1,003
GERMANY				
Berlin	3,338	3,130	3,450	3,499
Hamburg	1,608	1,721	1,786	1,825
Köln (Cologne)	598	908	1,001	1,018

Country/Agglomeration	Population (thousands)			
	1950	*1975*	*2010*	*2025*
München (Munich)	831	1,296	1,349	1,413
GREECE				
Athínai (Athens)	1,347	2,738	3,257	3,346
Thessaloniki	292	617	837	886
HUNGARY				
Budapest	1,618	2,005	1,706	1,711
IRELAND				
Dublin	626	833	1,099	1,337
ITALY				
Milano (Milan)	1,883	3,133	2,967	2,981
Napoli (Naples)	1,498	2,096	2,276	2,293
Palermo	594	783	875	896
Roma (Rome)	1,884	3,300	3,362	3,376
Torino (Turin)	1,011	1,838	1,665	1,680
NETHERLANDS				
Amsterdam	851	978	1,049	1,110
Rotterdam	764	927	1,010	1,057
NORWAY				
Oslo	468	644	888	1,019
POLAND				
Kraków (Cracow)	339	644	756	756
Warszawa (Warsaw)	768	1,444	1,712	1,722
PORTUGAL				
Lisboa (Lisbon)	1,304	2,103	2,824	3,009
Porto	730	1,008	1,355	1,473
ROMANIA				
Bucuresti (Bucharest)	652	1,702	1,934	1,963
SERBIA				
Beograd (Belgrade)	432	975	1,117	1,168
SPAIN				
Barcelona	1,809	3,679	5,083	5,477
Madrid	1,700	3,890	5,851	6,412
Valencia	506	695	814	873
SWEDEN				
Stockholm	741	1,015	1,285	1,345
SWITZERLAND				
Zürich (Zurich)	494	713	1,150	1,217
UNITED KINGDOM				
Birmingham	2,229	2,365	2,302	2,415
Glasgow	1,755	1,601	1,170	1,245
Liverpool	1,382	1,018	819	878
London	8,361	7,546	8,631	8,816
Manchester	2,422	2,370	2,253	2,364
Newcastle upon Tyne	909	838	891	954
West Yorkshire	1,692	1,618	1,547	1,637
RUSSIA				
BELARUS				
Minsk	284	1,120	1,852	1,917
RUSSIAN FEDERATION				
Chelyabinsk	573	966	1,094	1,095
Kazan	514	942	1,140	1,164
Krasnoyarsk	290	734	961	999
Moskva (Moscow)	5,356	7,623	10,550	10,663

Country/Agglomeration	Population (thousands)			
	1950	1975	2010	2025
Nizhniy Novgorod	796	1,273	1,267	1,253
Novosibirsk	719	1,250	1,397	1,398
Omsk	444	933	1,124	1,112
Perm	498	937	982	972
Rostov-na-Donu (Rostov-on-Don)	484	874	1,046	1,038
Samara	658	1,146	1,131	1,119
Sankt Peterburg (Saint Petersburg)	2,903	4,325	4,575	4,557
Saratov	473	816	822	797
Ufa	418	887	1,023	1,016
Volgograd	461	884	977	964
Voronezh	332	732	842	838
Yekaterinburg	628	1,135	1,344	1,377
UKRAINE				
Dnipropetrovsk	536	981	1,004	967
Donetsk	585	963	966	941
Kharkiv	758	1,353	1,453	1,444
Kyiv (Kiev)	815	1,926	2,805	2,915
Odesa	532	982	1,009	1,011
Zaporizhzhya	315	730	775	758

GREATER MIDDLE EAST

Country/Agglomeration	1950	1975	2010	2025
ALGERIA				
El Djazaïr (Algiers)	516	1,507	2,800	3,595
Wahran (Oran)	269	466	770	970
ARMENIA				
Yerevan	341	911	1,112	1,143
AZERBAIJAN				
Baku	897	1,429	1,972	2,291
EGYPT				
Al-Iskandariyah (Alexandria)	1,037	2,241	4,387	5,648
Al-Qahirah (Cairo)	2,494	6,450	11,001	13,531
GEORGIA				
Tbilisi	612	992	1,120	1,138
IRAN (ISLAMIC REPUBLIC OF)				
Ahvaz	85	313	1,060	1,317
Esfahan	184	767	1,742	2,161
Karaj	7	124	1,584	2,038
Kermanshah	97	274	837	1,029
Mashhad	173	685	2,652	3,277
Qom	78	227	1,042	1,299
Shiraz	128	418	1,299	1,590
Tabriz	235	662	1,483	1,814
Tehran	1,041	4,273	7,241	8,387
IRAQ				
Al-Basrah (Basra)	116	350	923	1,267
Al-Mawsil (Mosul)	145	397	1,447	2,092
Baghdad	579	2,620	5,891	8,043
Irbil (Erbil)	30	191	1,009	1,447
Sulaimaniya	37	163	836	1,249
ISRAEL				
Hefa (Haifa)	204	350	1,036	1,195
Jerusalem	121	343	782	944
Tel Aviv-Yafo (Tel Aviv-Jaffa)	418	1,206	3,272	3,823

Country/Agglomeration	Population (thousands)			
	1950	*1975*	*2010*	*2025*
JORDAN				
Amman	90	500	1,105	1,364
KAZAKHSTAN				
Almaty	354	860	1,383	1,612
KUWAIT				
Al Kuwayt (Kuwait City)	63	688	2,305	2,956
KYRGYZSTAN				
Bishkek	150	485	864	1,034
LEBANON				
Bayrut (Beirut)	322	1,500	1,937	2,135
LIBYAN ARAB JAMAHIRIYA				
Tarabulus (Tripoli)	106	580	1,108	1,364
MOROCCO				
Agadir	11	100	783	1,020
Dar-el-Beida (Casablanca)	625	1,793	3,284	4,065
Fès	165	433	1,065	1,371
Marrakech	209	367	928	1,198
Rabat	145	641	1,802	2,288
Tanger	100	225	788	1,030
SAUDI ARABIA				
Ad-Dammam	20	136	902	1,197
Al-Madinah (Medina)	51	208	1,104	1,456
Ar-Riyadh (Riyadh)	111	710	4,848	6,196
Jiddah	119	594	3,234	4,138
Makkah (Mecca)	148	383	1,484	1,924
SYRIAN ARAB REPUBLIC				
Dimashq (Damascus)	367	1,122	2,597	3,534
Halab (Aleppo)	319	879	3,087	4,244
Hamah	125	215	897	1,307
Hims (Homs)	101	312	1,328	1,881
TUNISIA				
Tunis	384	551	767	911
TURKEY				
Adana	138	471	1,361	1,635
Ankara	281	1,709	3,906	4,591
Antalya	27	128	838	1,022
Bursa	148	345	1,588	1,906
Gaziantep	104	299	1,109	1,341
Istanbul	967	3,600	10,525	12,108
Izmir	224	1,046	2,723	3,224
Konya	97	247	978	1,186
UNITED ARAB EMIRATES				
Dubayy (Dubai)	20	167	1,567	2,076
Sharjah	19	54	809	1,096
UZBEKISTAN				
Tashkent	755	1,612	2,210	2,616
YEMEN				
Sana'a'	46	141	2,342	4,296

SUB-SAHARAN AFRICA

Country/Agglomeration	1950	1975	2010	2025
ANGOLA				
Huambo	15	95	1,034	1,789
Luanda	138	665	4,772	8,077
BENIN				
Cotonou	20	240	844	1,445

Country/Agglomeration	Population (thousands)			
	1950	1975	2010	2025
BURKINA FASO				
Ouagadougou	33	150	1,908	4,332
CAMEROON				
Douala	95	433	2,125	3,131
Yaoundé	32	292	1,801	2,664
CHAD				
N'Djaména	22	231	829	1,445
CONGO				
Brazzaville	83	329	1,323	1,878
CONGO (DEMOCRATIC REPUBLIC OF)				
Kananga	24	374	878	1,583
Kinshasa	202	1,482	8,754	15,041
Kisangani	38	262	812	1,461
Lubumbashi	96	396	1,543	2,744
Mbuji-Mayi	70	327	1,488	2,658
CÔTE D'IVOIRE				
Abidjan	65	966	4,125	6,321
Yamoussoukro	1	38	885	1,797
ETHIOPIA				
Addis Ababa	392	926	2,930	4,757
GHANA				
Accra	177	738	2,342	3,497
Kumasi	99	397	1,834	2,757
GUINEA				
Conakry	31	534	1,653	2,906
KENYA				
Mombasa	94	298	1,003	1,795
Nairobi	137	677	3,523	6,246
LIBERIA				
Monrovia	15	226	827	932
MADAGASCAR				
Antananarivo	177	454	1,879	3,148
MALAWI				
Blantyre-Limbe	14	191	856	1,766
Lilongwe	2	71	865	1,784
MALI				
Bamako	89	363	1,699	2,971
MOZAMBIQUE				
Maputo	92	456	1,655	2,722
Matola	52	161	793	1,326
NIGER				
Niamey	24	198	1,048	2,105
NIGERIA				
Aba	48	258	785	1,203
Abuja	19	77	1,995	3,361
Benin City	49	233	1,302	1,992
Ibadan	450	980	2,837	4,237
Ilorin	114	323	835	1,279
Jos	31	224	802	1,229
Kaduna	35	408	1,561	2,362
Kano	123	855	3,395	5,060
Lagos	325	1,890	10,578	15,810
Maiduguri	50	300	970	1,480
Ogbomosho	132	428	1,032	1,576

Country/Agglomeration	Population (thousands)			
	1950	*1975*	*2010*	*2025*
Port Harcourt	60	358	1,104	1,681
Zaria	50	320	963	1,471
RWANDA				
Kigali	18	90	939	1,690
SENEGAL				
Dakar	201	782	2,863	4,338
SIERRA LEONE				
Freetown	92	284	901	1,420
SOMALIA				
Muqdisho (Mogadishu)	69	445	1,500	2,588
SOUTH AFRICA				
Cape Town	618	1,339	3,405	3,824
Durban	484	1,019	2,879	3,241
Ekurhuleni (East Rand)	546	997	3,202	3,614
Johannesburg	900	1,547	3,670	4,127
Port Elizabeth	192	531	1,068	1,222
Pretoria	275	624	1,429	1,637
Vereeniging	117	372	1,143	1,313
SUDAN				
Al-Khartum (Khartoum)	183	886	5,172	7,953
TANZANIA (UNITED REPUBLIC OF)				
Dar es Salaam	67	572	3,349	6,202
TOGO				
Lomé	33	257	1,667	2,763
UGANDA				
Kampala	95	398	1,598	3,189
ZAMBIA				
Lusaka	31	385	1,451	2,267
ZIMBABWE				
Harare	143	532	1,632	2,467

SOUTH ASIA

Country/Agglomeration	*1950*	*1975*	*2010*	*2025*
AFGHANISTAN				
Kabul	129	674	3,731	6,888
BANGLADESH				
Chittagong	289	1,017	4,962	7,265
Dhaka	336	2,221	14,648	20,936
Khulna	41	472	1,682	2,511
Rajshahi	39	150	878	1,328
INDIA				
Agra	369	681	1,703	2,313
Ahmadabad	855	2,050	5,717	7,567
Aligarh	139	280	863	1,189
Allahabad	327	568	1,277	1,742
Amritsar	340	520	1,297	1,771
Asansol	93	289	1,423	1,941
Aurangabad	65	218	1,198	1,641
Bangalore	746	2,111	7,218	9,507
Bareilly	207	374	868	1,192
Bhiwandi	25	93	859	1,186
Bhopal	100	491	1,843	2,497
Bhubaneswar	16	144	912	1,258
Chandigarh	40	301	1,049	1,440
Chennai (Madras)	1,491	3,609	7,547	9,909

Country/Agglomeration	Population (thousands)			
	1950	1975	2010	2025
Coimbatore	279	810	1,807	2,449
Delhi	1,369	4,426	22,157	28,568
Dhanbad	71	525	1,328	1,812
Durg-Bhilainagar	20	330	1,172	1,604
Guwahati (Gauhati)	43	252	1,053	1,445
Gwalior	237	465	1,039	1,423
Hubli-Dharwad	192	437	946	1,299
Hyderabad	1,096	2,086	6,751	8,894
Indore	302	663	2,173	2,939
Jabalpur	251	621	1,367	1,862
Jaipur	294	778	3,131	4,205
Jalandhar	166	340	917	1,262
Jammu	82	187	857	1,184
Jamshedpur	214	538	1,387	1,891
Jodhpur	177	388	1,061	1,454
Kanpur	688	1,420	3,364	4,501
Kochi (Cochin)	163	532	1,610	2,184
Kolkata (Calcutta)	4,513	7,888	15,552	20,112
Kota	64	266	884	1,216
Kozhikode (Calicut)	156	412	1,007	1,378
Lucknow	489	892	2,873	3,858
Ludhiana	151	479	1,760	2,387
Madurai	361	790	1,365	1,856
Meerut	228	432	1,494	2,035
Moradabad	160	302	845	1,166
Mumbai (Bombay)	2,857	7,082	20,041	25,810
Mysore	237	404	942	1,293
Nagpur	473	1,075	2,607	3,505
Nashik	149	330	1,588	2,165
Patna	277	642	2,321	3,137
Pune (Poona)	581	1,345	5,002	6,649
Raipur	88	255	943	1,298
Rajkot	126	356	1,357	1,855
Ranchi	103	342	1,119	1,533
Salem	197	458	932	1,281
Solapur	272	445	1,133	1,552
Srinagar	248	494	1,216	1,662
Surat	234	642	4,168	5,579
Thiruvananthapuram	182	454	1,006	1,377
Tiruchirappalli	287	522	1,010	1,383
Tiruppur	59	176	795	1,101
Vadodara	207	571	1,872	2,536
Varanasi (Benares)	349	680	1,432	1,947
Vijayawada	155	419	1,207	1,647
Visakhapatnam	105	452	1,625	2,206
NEPAL				
Kathmandu	104	180	1,037	1,915
PAKISTAN				
Faisalabad	168	907	2,849	4,200
Gujranwala	118	427	1,652	2,464
Hyderabad	232	667	1,590	2,373
Islamabad	36	107	856	1,295
Karachi	1,055	3,989	13,125	18,725
Lahore	836	2,399	7,132	10,308
Multan	186	599	1,659	2,474
Peshawar	153	347	1,422	2,128

Country/Agglomeration	Population (thousands)			
	1950	*1975*	*2010*	*2025*
Quetta	83	190	841	1,272
Rawalpindi	233	670	2,026	3,008

SOUTHEAST ASIA

CAMBODIA				
Phnum Pénh (Phnom Penh)	364	100	1,562	2,427
INDONESIA				
Bandar Lampung	28	228	799	972
Bandung	511	1,304	2,412	2,925
Bogor	113	406	1,044	1,344
Jakarta	1,452	4,813	9,210	10,850
Malang	208	457	786	959
Medan	284	876	2,131	2,586
Palembang	277	660	1,244	1,456
Pekan Baru	37	148	769	967
Semarang	371	783	1,296	1,528
Surabaya	679	1,736	2,509	2,923
Ujung Pandang	223	502	1,294	1,621
LAO PEOPLE'S DEMOCRATIC REPUBLIC				
Vientiane	121	205	831	1,501
MALAYSIA				
Johore Bharu	47	183	999	1,382
Klang	42	148	1,128	1,603
Kuala Lumpur	208	645	1,519	1,938
MYANMAR				
Mandalay	167	442	1,034	1,484
Nay Pyi Taw	—	—	1,024	1,499
Yangon	1,302	2,151	4,350	6,022
PHILIPPINES				
Cebu	178	415	860	1,162
Davao	124	488	1,519	2,080
Manila	1,544	4,999	11,628	14,916
Zamboanga	107	209	854	1,201
SINGAPORE				
Singapore	1,016	2,263	4,837	5,362
THAILAND				
Krung Thep (Bangkok)	1,360	3,842	6,976	8,470
VIETNAM				
Da Nang—CP	63	249	838	1,291
Hà Noi	465	806	2,814	4,530
Hai Phòng	194	972	1,970	2,722
Thành Pho Ho Chí Minh (Ho Chi Minh City)	1,213	2,431	6,167	8,957

EAST ASIA

CHINA				
Anshan, Liaoning	455	878	1,663	2,120
Anyang	113	261	1,130	1,417
Baoding	174	361	1,213	1,628
Baotou	104	801	1,932	2,388
Beijing	1,671	4,828	12,385	15,018
Bengbu	168	310	914	1,222
Benxi	335	559	969	1,215
Changchun	765	1,426	3,597	4,673
Changde	89	164	849	1,064
Changsha, Hunan	577	811	2,415	3,066

Country/Agglomeration	Population (thousands)			
	1950	1975	2010	2025
Changzhou, Jiangsu	149	397	2,062	2,624
Chengdu	646	1,860	4,961	6,224
Chifeng	112	212	842	1,092
Chongqing	1,567	2,402	9,401	11,065
Cixi	3	43	781	994
Dalian	716	1,294	3,306	4,132
Dandong	133	320	795	1,014
Daqing	181	442	1,546	2,112
Datong, Shanxi	201	493	1,251	1,602
Dongguan, Guangdong	92	125	5,347	6,852
Dongying	22	137	949	1,334
Foshan	112	233	4,969	6,242
Fushun, Liaoning	637	906	1,378	1,647
Fuxin	172	375	821	1,070
Fuyang	46	93	874	1,119
Fuzhou, Fujian	301	638	2,787	3,727
Guangzhou, Guangdong	1,049	1,698	8,884	10,961
Guilin	126	346	991	1,317
Guiyang	249	754	2,154	2,679
Haerbin	727	1,738	4,251	5,080
Haikou	107	239	1,586	2,065
Handan	69	245	1,249	1,764
Hangzhou	610	1,083	3,860	4,735
Hefei	145	588	2,404	3,029
Hengyang	97	272	1,099	1,488
Hohhot	122	419	1,589	2,258
Huai'an	64	179	998	1,278
Huaibei	45	144	962	1,364
Huainan	253	555	1,396	1,854
Huizhou	49	119	1,384	1,828
Huludao	6	77	795	1,120
Jiamusi	132	305	817	1,092
Jiangmen	88	143	1,103	1,448
Jiaozuo	87	223	900	1,236
Jieyang	5	46	855	1,158
Jilin	394	672	1,888	2,489
Jinan, Shandong	576	1,150	3,237	4,044
Jingzhou	77	190	1,039	1,392
Jining, Shandong	79	159	1,077	1,394
Jinjiang	8	38	858	1,303
Jinzhou	160	347	857	1,068
Jixi, Heilongjiang	111	335	1,042	1,366
Kunming	334	515	3,116	3,915
Lanzhou	336	945	2,285	2,896
Lianyungang	130	239	878	1,183
Linyi, Shandong	15	89	1,427	1,827
Liuzhou	118	370	1,352	1,788
Lufeng	16	95	889	1,276
Luoyang	145	459	1,539	1,999
Luzhou	74	167	850	1,123
Maoming	21	62	803	1,053
Mianyang, Sichuan	74	163	1,006	1,331
Mudanjiang	137	321	783	1,000
Nanchang	343	688	2,701	3,436
Nanchong	156	212	808	1,078

Country/Agglomeration	Population (thousands)			
	1950	*1975*	*2010*	*2025*
Nanjing, Jiangsu	1,037	1,589	4,519	5,845
Nanning	143	509	2,096	2,669
Nantong	243	328	1,423	1,850
Nanyang, Henan	44	154	867	1,135
Neijiang	185	250	883	1,165
Ningbo	282	401	2,217	2,959
Panjin	54	179	813	1,101
Pingdingshan, Henan	5	121	1,024	1,307
Puning	15	42	911	1,255
Putian	79	186	1,085	1,327
Qingdao	751	933	3,323	4,159
Qinhuangdao	91	203	893	1,165
Qiqihaer	313	693	1,588	2,019
Quanzhou	104	136	1,068	1,462
Rizhao	7	64	816	1,086
Shanghai	4,301	5,627	16,575	20,017
Shantou	270	380	3,502	4,222
Shaoguan	75	154	845	1,066
Shaoxing	40	103	853	1,153
Shenyang	2,148	3,291	5,166	6,457
Shenzhen	3	36	9,005	11,146
Shijiazhuang	272	825	2,487	3,235
Suzhou, Jiangsu	457	530	2,398	3,021
Taian, Shandong	4	65	1,239	1,653
Tainan	320	518	777	959
Taiyuan, Shanxi	197	829	3,154	4,043
Taizhou, Jiangsu	58	109	795	1,101
Taizhou, Zhejiang	315	612	1,338	1,671
Tangshan, Hebei	448	692	1,870	2,487
Tianjin	2,467	3,527	7,884	9,713
Ürümqi (Wulumqi)	102	715	2,398	3,231
Weifang	129	319	1,698	2,271
Wenzhou	151	853	2,659	3,650
Wuhan	1,069	2,265	7,681	9,347
Wuhu, Anhui	163	304	908	1,252
Wuxi, Jiangsu	366	682	2,682	3,405
Xiamen	193	445	2,207	3,112
Xi'an, Shaanxi	575	1,063	4,747	5,726
Xiangfan, Hubei	57	227	1,399	1,786
Xiangtan, Hunan	171	315	926	1,236
Xianyang, Shaanxi	127	225	1,019	1,334
Xining	70	371	1,261	1,761
Xinxiang	155	340	1,016	1,355
Xuzhou	341	588	2,142	3,015
Yancheng, Jiangsu	46	175	1,289	1,731
Yangzhou	170	270	1,080	1,529
Yantai	92	234	1,526	1,958
Yichang	59	244	959	1,210
Yichun, Heilongjiang	335	663	779	917
Yinchuan	59	199	911	1,312
Yingkou	120	253	848	1,148
Yiyang, Hunan	83	127	820	1,043
Yueyang	15	98	1,096	1,408
Zaozhuang	130	220	1,175	1,574
Zhangjiakou	214	386	1,043	1,384

Country/Agglomeration	Population (thousands)			
	1950	1975	2010	2025
Zhanjiang	152	314	996	1,281
Zhengzhou	196	645	2,966	3,734
Zhenjiang, Jiangsu	199	257	1,007	1,399
Zhongshan	55	188	2,211	3,114
Zhuhai	3	25	1,252	1,516
Zhuzhou	116	250	1,025	1,330
Zibo	164	434	2,456	3,192
Zigong	189	287	918	1,142
Zunyi	94	157	843	1,198
CHINA (REPUBLIC OF)				
Kaohsiung	256	986	1,611	1,971
Taichung	198	537	1,251	1,642
Taipei	503	2,023	2,633	3,102
CHINA, HONG KONG SAR				
Hong Kong	1,682	3,943	7,069	7,969
JAPAN				
Fukuoka-Kitakyushu	954	1,853	2,816	2,834
Hiroshima	503	1,774	2,081	2,088
Kyoto	1,002	1,622	1,804	1,804
Nagoya	992	2,293	3,267	3,295
Osaka-Kobe	4,147	9,844	11,337	11,368
Sapporo	754	1,751	2,687	2,721
Sendai	538	1,566	2,376	2,413
Tokyo	11,275	26,615	36,669	37,088
KOREA (DEMOCRATIC REPUBLIC OF)				
P'yongyang	516	1,348	2,833	2,941
KOREA (REPUBLIC OF)				
Bucheon	31	108	909	961
Busan	948	2,418	3,425	3,409
Daegu	355	1,297	2,458	2,481
Daejon	131	501	1,509	1,562
Goyang	39	122	961	1,026
Gwangju	174	601	1,476	1,525
Incheon	258	791	2,583	2,631
Seongnam	54	268	955	984
Seoul	1,021	6,808	9,773	9,767
Suweon	74	220	1,132	1,194
Ulsan	29	246	1,081	1,117
MONGOLIA				
Ulaanbaatar	70	356	966	1,202
AUSTRALIA AND THE PACIFIC ISLANDS				
Adelaide	429	881	1,168	1,307
Brisbane	442	928	1,970	2,245
Melbourne	1,332	2,561	3,853	4,261
Perth	311	770	1,599	1,810
Sydney	1,690	2,960	4,429	4,852
NEW ZEALAND				
Auckland	319	729	1,404	1,671

Source: Population Division of the Department of Economic and Social Affairs of the United Nations Secretariat, *World Urbanization Prospects: The 2009 Revision*, http://esa.un.org/unup, Wednesday, May 25, 2011.

Cover Photo Credits

All photographs appearing on the cover were taken by Donald J. Zeigler.

Clockwise from top left:

New York, NY, United States: The "Ground Zero" cross at the World Trade Center site, now in the 9/11 Memorial and Museum.

Marrakech, Morocco: A heavily veiled woman sells winter caps in a Moroccan souk, or market.

Hong Kong, China: The Kowloon skyline from the observation platform atop Victoria Peak on Hong Kong Island.

Amman, Jordan: *Kunafah*, a honey-sweet pastry, for sale on the streets of Amman.

Belfast, Northern Ireland: The "Belfast Wheel" next to City Hall.

Mexico City, Mexico: An Aztec, or Mexica, purification ritual taking place on Mexico City's Zócalo, or plaza.

Toronto, Ontario, Canada: Canoes and kayaks in for the season along Toronto's anthocyanin-drenched waterfront.

Sydney, New South Wales, Australia: The waterfront Opera House, the signature landmark for which Sydney is known.

Hong Kong, China: Keeping foliage green is a challenge for many subtropical cities under the onslaught of summer heat.

Amsterdam, Netherlands: A space-economizing mini-car parked along an Amsterdam canal.

Kuala Lumpur, Malaysia: Two skylines, old and new, both add color to the urban landscape of KL.

London, England, United Kingdom: Red telephone boxes have been a London trademark, but they are rapidly disappearing.

Jerusalem, Israel: One of the minarets on a mosque in the walled city of Jerusalem glows in the late afternoon sun.

Mumbai, India: In labor-rich India, much urban freight is still hauled around cities by men on foot.

Malacca, Malaysia: The "floating mosque" is a new addition to the Malacca waterfront.

Istanbul, Turkey: Fruit for sale in one of the working neighborhoods near the Golden Horn.

Geographical Index

Index to Subjects

About the Editors and Contributors

Amal K. Ali is associate professor in the Department of Geography and Geosciences at Salisbury University in Salisbury, Maryland, where she teaches a variety of courses on planning and world cities. Among her specialties are parks and open spaces, the decentralization of decision-making in the planning process, and international development.

Lisa Benton-Short is an associate professor of geography at George Washington University. She teaches courses on cities and globalization, urban planning, and urban sustainability. As an urban geographer, she has research interests in urban sustainability, environmental issues in cities, parks and public spaces, and monuments and memorials.

Alana Boland is associate professor in the Department of Geography and Program in Planning at the University of Toronto. She teaches courses on China, and environment and development. Her research examines the changing relationship between the economy and environment in Chinese cities. Her current work focuses on state regulatory initiatives aimed at improving urban environmental conditions, particularly at the community-level.

Tim Brothers is a physical geographer specializing in human-environment relations in the Caribbean. He has worked primarily in the Dominican Republic, Cuba and Haiti. His published work includes the book "Caribbean Landscapes" (Caribbean Studies Press, 2008), an interpretive presentation of the most characteristic natural and human landscapes of the Caribbean based on satellite imagery, photographs, and interpretive essays.

Stanley D. Brunn is professor of geography at the University of Kentucky. He has taught classes on world cities, future worlds, political geography, and information and communications. He has traveled in nearly sixty countries and has taught in fifteen. His recent research focuses on images of places, creative cartographies, U.S. elections, the geographies or religion, disciplinary histories, Wal-Mart, post 9–11 worlds, the global financial crises, and megaengineering projects.

Kam Wing Chan is Professor in Geography at University of Washington. His research focuses on China's urbanization, migration, employment, and the household registration system. In recent years, he has also served as a consultant for the United Nations, World Bank, Asian Development Bank, and McKinsey & Co. on a number of policy projects. His recent commentaries and interviews have appeared in the BBC, China Radio International, CBC Radio, PBS, The Seattle Times, Caixin and other media.

Ipsita Chatterjee is an assistant professor in the Department of Geography and the Environment, University of Texas at Austin. She is interested in issues relating to globalization, urban transformation, urban renewal, conflict and violence, and social movements. Her research has focused on issues of class and ethnic segregation, ghettoization, and other forms of urban exclusions.

Megan L. Dixon holds doctoral degrees in Geography and Russian Literature. She is an adjunct instructor of geography, geology and writing at The College of Idaho. Her research to date is in Russian urban geography and cultural geography, especially the impacts of Chinese migration to St. Petersburg and the shift between Soviet and post-Soviet notions of green space. Her scholarly interests also involve competing notions of place and land use in the intermountain west (central Idaho). Her ongoing interest in Russian people and landscapes began with a visit to Kalinin/Tver in 1989, just before the fall of the Berlin Wall.

Robyn Dowling is an urban cultural geographer at Macquarie University in Sydney, Australia. Her primary research interests are in cultures of everyday urban life, focusing on gender, home and suburbs. She publishes widely on issues such as home ownership, suburban gender identities, and cultures of transport. Her current research explores the contours of privatization and privatism in Sydney's residential life.

Ashok K. Dutt is professor emeritus of geography, planning, and urban studies at the University of Akron in Ohio. His research focuses on religion, language, development, crime, and medical geographies of Indian cities. He has authored, coauthored, edited or co-edited 23 books, and authored or co-

authored more than 80 journal articles and 60 book chapters.

Irma Escamilla is a doctoral candidate in geography at the National Autonomus University of Mexico (UNAM). She serves as an academic technician in the Department of Social Geography at the Institute of Geography of UNAM. Her research focuses on urban-regional geography, urban labor markets, population geography, gender and geography and historical geography.

Rina Ghose is an associate professor of geography at the University of Wisconsin-Milwaukee, where she teaches courses on GIS, South Asia, and urban geography. She has done extensive research and published widely on Public Participation GIS, GIS in local government and in non-western societies, and inner-city citizen participation in collaborative planning programs.

Brian Godfrey is Professor of Geography and Director of the Urban Studies program at Vassar College. He specializes in historical and urban geography with a particular focus on cities of the Americas, including the U.S., Latin America, Brazil, and the Amazon Basin. His scholarship has addressed issues of global cities, community change, historic preservation and memory, sustainable development, and public space. Currently Professor Godfrey is writing a book on the heritage-based redevelopment of Brazil's historic cities.

Mark Graham is a research fellow at the Oxford Internet Institute, University of Oxford. His work focuses on the economic, social, and spatial effects of technology, and the spatiality of networked communications. Mark's Ph.D. is from the University of Kentucky and he has published over twenty articles and book chapters.

Angela Gray-Subulwa is an assistant professor in the Department of Geography and Urban Planning at the University of Wisconsin–Oshkosh. Her research interests include examining the processes of forced displacement as it relates to development, gender, political, and cultural geographies. Most of her research focuses on southern Africa, with a concentration on Zambia.

Jessica K. Graybill is an assistant professor of geography at Colgate University. Her interdisciplinary pedagogy and research is in environmental geography, urban geography and ecologies of the city. Investigations in ecologies of the city in the US have been conducted in a rapidly growing region (Pacific Northwest) and in a diminishing region (upstate New York). Long-term interest in the former Soviet states, especially Russia, has carried her from Moscow to the Russian Far East pursuing understandings of human experiences of the environment, both in and outside urban locations.

Maureen Hays-Mitchell is professor of geography at Colgate University. Her teaching and scholarly interests include international development, political geography, feminist thought, and spatial justice. She has conducted grassroots fieldwork in Andean America on the urban informal economy and on micro-enterprise development. Her current scholarship involves post-conflict landscapes, truth commissions, and gendered dimensions of post-conflict reconciliation and reconstruction in several Latin American countries.

Corey Johnson is an assistant professor of geography at the University of North Carolina, Greensboro. He teaches courses on Europe, the European Union, and political and urban geography. He has lived and worked in Germany. His current research is on geopolitics of energy, borders in the European Union, and urban transformation in eastern Germany.

Nathaniel M. Lewis is a Ph.D. candidate in the Department of Geography at Queen's University in Kingston, Ontario, Canada. He researches urban geography, health, sexuality, and relationships between mental health and migration. He is also interested in the governance of urban development, particularly through business improvement districts and "creative city" strategies.

Linda McCarthy is associate professor of geography and urban studies at the University of Wisconsin–Milwaukee. She is also a certified planner. Her teaching interests include Europe, cities, and globalization. Her research focuses on economic development and planning. Recent publications have been on regional cooperation, government subsidies for automobile plants, environmental justice, and the globalization.

Pauline McGuirk is professor of human geography and director of the Centre for Urban and Regional Studies at the University of Newcastle, New South Wales, Australia. Her research focuses on the politics, development and governance of metropolitan cities, but she has also published widely on aspects of urban development, city politics, planning, urban identity and place marketing. She is currently investigating new forms of governance and residential housing in Sydney.

Garth A. Myers is director of the Kansas African Studies Center, professor of geography and African/African-American studies at the University of Kansas, author of three books and coeditor of a fourth on African cities. He has published more than three dozen articles and book chapters on African urban development, and he teaches a variety of courses on African geography.

Arnisson Andre Ortega is a PhD candidate in geography at the University of Washington. He specializes in urban geography, spatial demography and cultural geography. He spent most of his life in the Philippines. He taught at the University of the Philippines prior to going to the United States. He is currently doing research on neoliberalism, real estate booms, and cultural geographies of difference in the Philippines.

Francis Owusu is an associate professor in the Department of Community and Regional Planning at Iowa State University. He has conducted research in several African countries and has published extensively on urban livelihood strategies, development policy, and public sector reforms. He teaches geography and planning courses, including the geography of Africa. He is originally from Ghana.

George Pomeroy is professor of geography-earth science at Shippensburg University of Pennsylvania. He teaches land-use planning, environmental planning, and courses related to both South and East Asia. His research focuses on new towns planning and on building capacity for local government planning activities.

Joseph L. Scarpaci is associate professor of marketing at the Gary E. West College of Business in West Liberty, WV; Executive Director of The Center for the Study of Cuban Culture and the Economy; and Emeritus Professor of Geography at Virginia Tech. Along with Emilio Morales, he is co-author of *Advertising without Marketing: Brand Preference and Consumer Choice in Cuba* (Routledge 2012).

James Tyner is professor of geography at Kent State University, where he teaches East and Southeast Asia, political geography, and other courses on social, population, and urban geography. His current research focuses on political violence in Southeast Asia and international population movements.

Dona J. Stewart is the former director of the Middle East Institute at Georgia State University and professor of Middle East Studies and Geography. A former Fulbright scholar to Jordan, she is the author of *The Middle East Today: Political, Geographical and Cultural Perspectives.* Currently, she operates Fidelis Analytics and Research, a consulting firm that provides research and policy analysis.

Donald J. Zeigler is professor of geography at Old Dominion University—Virginia Beach. He teaches courses on the Middle East, cultural and urban geography. He has held fellowships in Morocco, Syria, and Jordan. His current research focuses on Middle Eastern cities, maps in society, and the connections between geography and history in the Mediterranean world.